Concepts, Methods and Applications of Quantum Systems in Chemistry and Physics

Progress in Theoretical Chemistry and Physics

VOLUME 31

Honorary Editors

Rudolph A. Marcus (*California Institute of Technology, Pasadena, CA, USA*)
Roy McWeeny (*Università di Pisa, Pisa, Italy*)

Editors-in-Chief

J. Maruani (*formerly Laboratoire de Chimie Physique, Paris, France*)
S. Wilson (*formerly Rutherford Appleton Laboratory, Oxfordshire, UK*)

Editorial Board

More information about this series at http://www.springer.com/series/6464

Yan A. Wang · Mark Thachuk
Roman Krems · Jean Maruani
Editors

Concepts, Methods and Applications of Quantum Systems in Chemistry and Physics

Selected Proceedings of QSCP-XXI
(Vancouver, BC, Canada, July 2016)

 Springer

Editors
Yan A. Wang
University of British Columbia
Vancouver
Canada

Mark Thachuk
University of British Columbia
Vancouver
Canada

Roman Krems
University of British Columbia
Vancouver
Canada

Jean Maruani
Laboratoire de Chimie Physique
CNRS & UPMC
Paris
France

ISSN 1567-7354   ISSN 2215-0129   (electronic)
Progress in Theoretical Chemistry and Physics
ISBN 978-3-319-74581-7   ISBN 978-3-319-74582-4   (eBook)
https://doi.org/10.1007/978-3-319-74582-4

Library of Congress Control Number: 2017964436

Printed on acid-free paper

This Springer imprint is published by the registered company Springer International Publishing AG part of Springer Nature
The registered company address is: Gewerbestrasse 11, 6330 Cham, Switzerland

PTCP Aim and Scope

Progress in Theoretical Chemistry and Physics

A series reporting advances in theoretical molecular and material sciences, including theoretical, mathematical and computational chemistry, physical chemistry and chemical physics and biophysics.

Aim and Scope

Science progresses by a symbiotic interaction between theory and experiment: Theory is used to interpret experimental results and may suggest new experiments; experiment helps to test theoretical predictions and may lead to improved theories. Theoretical chemistry (including physical chemistry and chemical physics) provides the conceptual and technical background and apparatus for the rationalization of phenomena in the chemical sciences. It is, therefore, a wide ranging subject, reflecting the diversity of molecular and related species and processes arising in chemical systems. The book series *Progress in Theoretical Chemistry and Physics* aims to report advances in methods and applications in this extended domain. It will comprise monographs as well as collections of papers on particular themes, which may arise from proceedings of symposia or invited papers on specific topics as well as from initiatives from authors or translations.

The basic theories of physics—classical mechanics and electromagnetism, relativity theory, quantum mechanics, statistical mechanics, quantum electrodynamics—support the theoretical apparatus which is used in molecular sciences. Quantum mechanics plays a particular role in theoretical chemistry, providing the basis for the valence theories, which allow to interpret the structure of molecules, and for the spectroscopic models, employed in the determination of structural information from spectral patterns. Indeed, quantum chemistry often appears synonymous with theoretical chemistry; it will, therefore, constitute a major part of this book series. However, the scope of the series will also include other areas of theoretical chemistry, such as mathematical chemistry (which involves the use of algebra and topology in the analysis of molecular structures and reactions); molecular mechanics, molecular dynamics and chemical thermodynamics, which play an important role in rationalizing the geometric and electronic structures of molecular assemblies and polymers, clusters and crystals; surface, interface, solvent and solid state effects; excited-state dynamics, reactive collisions and chemical reactions.

Recent decades have seen the emergence of a novel approach to scientific research, based on the exploitation of fast electronic digital computers. Computation provides a method of investigation which transcends the traditional division between theory and experiment. Computer-assisted simulation and design may afford a solution to complex problems which would otherwise be intractable to theoretical analysis, and may also provide a viable alternative to difficult or costly laboratory experiments. Though stemming from theoretical chemistry, computational chemistry is a field of research in its own right, which can help to test theoretical predictions and may also suggest improved theories.

The field of theoretical molecular sciences ranges from fundamental physical questions relevant to the molecular concept, through the statics and dynamics of isolated molecules, aggregates and materials, molecular properties and interactions, to the role of molecules in the biological sciences. Therefore, it involves the physical basis for geometric and electronic structure, states of aggregation, physical and chemical transformations, thermodynamic and kinetic properties, as well as unusual properties such as extreme flexibility or strong relativistic or quantum field effects, extreme conditions such as intense radiation fields or interaction with the continuum, and the specificity of biochemical reactions.

Theoretical chemistry has an applied branch (a part of molecular engineering), which involves the investigation of structure–property relationships aiming at the design, synthesis and application of molecules and materials endowed with specific functions, now in demand in such areas as molecular electronics, drug design or genetic engineering. Relevant properties include conductivity (normal, semi- and super-), magnetism (ferro- and ferri-), optoelectronic effects (involving nonlinear response), photochromism and photoreactivity, radiation and thermal resistance, molecular recognition and information processing, biological and pharmaceutical activities, as well as properties favoring self-assembling mechanisms and combination properties needed in multifunctional systems.

Progress in Theoretical Chemistry and Physics is made at different rates in these various research fields. The aim of this book series is to provide timely and in-depth coverage of selected topics and broad-ranging yet detailed analysis of contemporary theories and their applications. The series will be of primary interest to those whose research is directly concerned with the development and application of theoretical approaches in the chemical sciences. It will provide up-to-date reports on theoretical methods for the chemist, thermodynamician or spectroscopist, the atomic, molecular or cluster physicist, and the biochemist or molecular biologist who wish to employ techniques developed in theoretical, mathematical and computational chemistry in their research programs. It is also intended to provide the graduate student with a readily accessible documentation on various branches of theoretical chemistry, physical chemistry, and chemical physics.

Preface

This volume collects 20 selected papers from the scientific contributions presented at the Twenty-first International Workshop on Quantum Systems in Chemistry, Physics, and Biology (QSCP-XXI), organized by Yan Alexander Wang at the University of British Columbia in Vancouver, BC, Canada, on July 02–09, 2015. Over 160 scientists from 30 countries attended this meeting. The participants discussed the state of the art, new trends, and future evolution of methods in molecular quantum mechanics and their applications to a broad variety of problems in chemistry, physics, and biology.

The high-level attendance attained in this conference was particularly gratifying. It is the renowned interdisciplinary nature and friendly feeling of QSCP meetings that make them so successful discussion forums.

Highly ranked among the world best universities, the University of British Columbia (UBC) holds an international reputation for excellence in advanced research and teaching. Only 30 min from the vibrant heart of downtown Vancouver, the spectacular UBC campus is a 'must-see' for any visitor in the world: Snow-capped mountains can be seen meeting the ocean, and breathtaking vistas greet you around every corner. The UBC campus also boasts some of the city's best attractions and recreation facilities, including the Chan Centre for the Performing Arts, the Museum of Anthropology, the UBC Botanical Garden, and endless opportunities to explore forested trails in the adjoining 763-ha Pacific Spirit Regional Park.

Details of the Vancouver meeting, including scientific and social programs, can be found on the Web site: https://groups.chem.ubc.ca/qscp/. Altogether, there were 30 morning and afternoon sessions, where 48 plenary lectures and 36 invited talks were given, and one evening poster session, with 18 posters being displayed. We are grateful to all 166 participants for making the QSCP-XXI workshop a stimulating experience and a great success. QSCP-XXI followed the traditions established at previous workshops:

QSCP-I, organized by Roy McWeeny in 1996 at San Miniato (Pisa, Italy);
QSCP-II, by Stephen Wilson in 1997 at Oxford (England);

QSCP-III, by Alfonso Hernandez-Laguna in 1998 at Granada (Spain);
QSCP-IV, by Jean Maruani in 1999 at Marly-le-Roi (Paris, France);
QSCP-V, by Erkki Brändas in 2000 at Uppsala (Sweden);
QSCP-VI, by Alia Tadjer and Yavor Delchev in 2001 at Sofia (Bulgaria);
QSCP-VII, by Ivan Hubac in 2002 near Bratislava (Slovakia);
QSCP-VIII, by Aristides Mavridis in 2003 at Spetses (Athens, Greece);
QSCP-IX, by J.-P. Julien in 2004 at Les Houches (Grenoble, France);
QSCP-X, by Souad Lahmar in 2005 at Carthage (Tunisia);
QSCP-XI, by Oleg Vasyutinskii in 2006 at Pushkin (St Petersburg, Russia);
QSCP-XII, by Stephen Wilson in 2007 near Windsor (London, England);
QSCP-XIII, by Piotr Piecuch in 2008 at East Lansing (Michigan, USA);
QSCP-XIV, by G. Delgado-Barrio in 2009 at El Escorial (Madrid, Spain);
QSCP-XV, by Philip Hoggan in 2010 at Cambridge (England);
QSCP-XVI, by Kiyoshi Nishikawa in 2011 at Kanazawa (Japan);
QSCP-XVII, by Matti Hotokka in 2012 at Turku (Finland);
QSCP-XVIII, by M.A.C. Nascimento in 2013 at Paraty (Brazil);
QSCP-XIX, by Cherri Hsu in 2014 at Taipei (Taiwan);
QSCP-XX, by Alia Tadjer and Rossen Pavlov in 2015 at Varna (Bulgaria).

The lectures presented at QSCP-XXI were grouped into seven areas in the field of *Quantum Systems in Chemistry, Physics, and Biology*, ranging from Concepts and Methods in Quantum Chemistry through Relativistic Effects in Quantum Chemistry, Atoms and Molecules in Strong Electric and Magnetic Fields, Reactive Collisions and Chemical Reactions, Molecular Structure, Dynamics and Spectroscopy, and Molecular and Nano-materials, to Computational Chemistry, Physics, and Biology.

The width and depth of the topics discussed at QSCP-XXI are reflected in the contents of this volume of proceedings in *Progress in Theoretical Chemistry and Physics*, which includes four sections:

I. Quantum Chemistry Methodology (five papers);
II. Molecular Structure and Dynamics (ten papers);
III. Biochemistry and Biophysics (two papers);
IV. Fundamental Theory (three papers).

In addition to the scientific program, the workshop had its usual share of cultural events. There was a music concert, *Amor & Pasión*, by the Vancouver International Song Institute, in the afternoon of the opening day, and in the evening of July 5 Prof. Jean Maruani, accompanied by his pianist wife Marja Rantanen, delivered an entertaining lecture on *Science and Music*.

The award ceremony of the CMOA Prize and Medal took place during the banquet in the premises of the *UBC Golf Club*. The CMOA Prize for junior scientists was shared between the two selected nominees: Roman V. Krems (UBC, Canada) and Erin R. Johnson (Dalhousie, Canada). The prestigious CMOA Medal for senior scientists was awarded to Prof. Weitao Yang (Duke, USA).

During the banquet, we also celebrated the 80th birthdays of Prof. Ernest R. Davidson (born October 1936) and Prof. Josef Paldus (born November 1935).

Birthday cakes were presented to these two great scientists, accompanied by a *Happy Birthday* sung by guests attending the banquet.

Following a QSCP tradition, the venue of the next workshop was presented at the end: Changsha, Hunan, China, in October 2017.

We are most grateful to the members of the Local Organizing Committee: Mark Thachuk, Roman Krems, Delano Chong, Peter Chung and Jane Cua, as well as to the members of the Wang team, especially Jianxiong Yang, Yuzhe Chen, Dongmei Luo, Junqing Yang, Miguel Garcia-Chavez, Yiming Wang and Lin Bryan Zhang, for their work and dedication which made the stay and work of participants both pleasant and fruitful. Last but not least, we thank the members of the International Scientific and Honorary Committees for their invaluable expertise and advice.

We hope the readers will find as much interest in consulting these proceedings as the participants in attending the meeting.

The Editors

Contents

Part IV Fundamental Theory

Part I
Quantum Chemistry Methodology

A Portal for Quantum Chemistry Data Based on the Semantic Web

Bing Wang, Paul A. Dobosh, Stuart Chalk, Keigo Ito, Mirek Sopek and Neil S. Ostlund

Abstract Chemical Semantics, Inc. (CSI) is a new start-up devoted to bringing the Semantic Web to chemistry and biochemistry. The semantic web is referred to as Web 3.0 or alternatively the Web of Data or the Web of Meaning. It does not replace the existing World Wide Web but augments it, placing data on the web in a structured form such that the data has "meaning" and computers can understand it. CSI has created a demonstration portal for exploring this new technology, specifically at this point for data created by quantum chemistry calculations. This paper describes the basics of a semantic web portal and the fundamental technology we have used in developing it.

1 Overview of the Semantic Web

Data on the existing World Wide Web is encoded in documents containing text, table, images, etc. This data is not really recognized by computers but has to be interpreted by humans. Because computers cannot recognize the data in existing web documents, the data cannot properly be shared or even found. The semantic web [1] puts data onto the web in a form that allows computers to properly recognize it along with its meaning. Computers can then perform intelligent operations on the data, ultimately creating new data by using inference from existing data. The future of scientific data involves structuring the data via the semantic web or equivalent technologies so that computers can find data and understand data on our behalf.

Chemistry generates enormous amounts of data [2]. Because existing publication channels do not give credence to the data in the way they give credence to the text of a "scientific publication" much of this data is lost or discarded and not made available to other scientists. There is a trend towards journals requiring authors to

B. Wang · P. A. Dobosh · K. Ito · M. Sopek · N. S. Ostlund (✉)
Chemical Semantics Inc, 2135 NW 15th Ave, Gainesville, FL 32605, USA
e-mail: ostlund@chemicalsemantics.com

S. Chalk
Department of Chemistry, University of North Florida, Jacksonville, FL 32224, USA

© Springer International Publishing AG, part of Springer Nature 2018
Y. A. Wang et al. (eds.), *Concepts, Methods and Applications of Quantum Systems in Chemistry and Physics*, Progress in Theoretical Chemistry and Physics 31, https://doi.org/10.1007/978-3-319-74582-4_1

3

submit data files along with the text of a publication. However, since there is as yet no standard and little infrastructure adopted by mainstream publishers for dealing with this data, most of it remains in inaccessible files (e.g. PDF documents). An appropriate answer to this dilemma is the semantic web.

In conjunction with the data, the semantic web includes a vocabulary for describing the data. This is where semantics comes to the fore. This vocabulary for a field such as Quantum Chemistry needs to be encoded in a formal language. The Web Ontology Language (OWL) [3] is such a language and the "vocabulary" is really an ontology describing the formal language of a specific domain such as computational or quantum chemistry. Chemical Semantics has created such an ontology which is referred to as the Gainesville Core (http://purl.org/chem/gc). This first ontology for quantum chemistry will need to be modified and extended by scientists in the field as basic ontology ideas become more prevalent in chemistry. Our ontology is meant to be placed in the public domain so that the quantum chemistry community can both contribute to it and use it. Developing the Gainesville Core is an ongoing project with release numbers.

The semantic web allows scientists to publish their data in a structured way such that it can be found and used by anyone with access to the World Wide Web. Chemical Semantics, Inc. is creating client and server software that allows scientists to automate their publishing of data into a modern graph database [4], i.e. a Giant Global Graph (GGG), so that they and others can share their data and use it in a way that is not now possible with the existing World Wide Web (WWW). Adding semantics to scientific data and making that data available on the semantic web makes it possible to do science in a new collaborative way that has the potential to change science forever [1].

1.1 Publishing Quantum Chemistry Data

To make the technology of the semantic web available to scientists, one has to "publish data" in a way that is related to the standard model of journal publishing. That is, one puts the data (not journal text) into an appropriate form, sends it to a publisher (of data not journal text), waits for its publication, and then informs colleagues that they can access the data (not journal article) in a standard fashion (more and more via the web rather than via hard copy).

The appropriate form for data described above has been clearly defined by the World Wide Web Consortium (W3C) [5] as the Resource Description Framework (RDF) standard [6]. This standard is a Graph Database where RDF statements all take the "triple" form (subject, predicate, and object). For example, (water, has_boiling_point, 100) is of this triple form. That is, there is a graph arc called "has_boiling_point" which points from a "water" subject node to a "100" object node. This form surpasses the normal relational database in its applicability to the web. Any and all scientific data can be put into this form and data on the semantic web is stored in servers referred to as triple stores. These triple stores may contain billions of triples.

Chemical Semantics, Inc. operates servers that store scientific data in these triple stores. Our client software allows scientists to automate the publishing of their data to these triple stores. Initially, we are focusing on data produced by quantum chemistry packages although the basic ideas apply to any chemical data including experimental data. The data produced from these quantum chemistry packages depends somewhat upon the specific package but for illustrative purposes let's assume a "CompChem" package that produces results from ab initio wave function calculations, such as Self-Consistent-Field (Hartree-Fock SCF) calculations, post Hartree-Fock correlated calculations, etc. Examples of such packages are Gamess [7], Gaussian [8], NWChem [9], etc.

The publishing of the results of these calculations ought not to be more difficult than having a "Publish" button in the GUI or text input of the package. For initial demonstration, we have implemented such a button in HyperChem [10] as shown in Fig. 1. The specific use of HyperChem is not relevant and any CompChem [11]

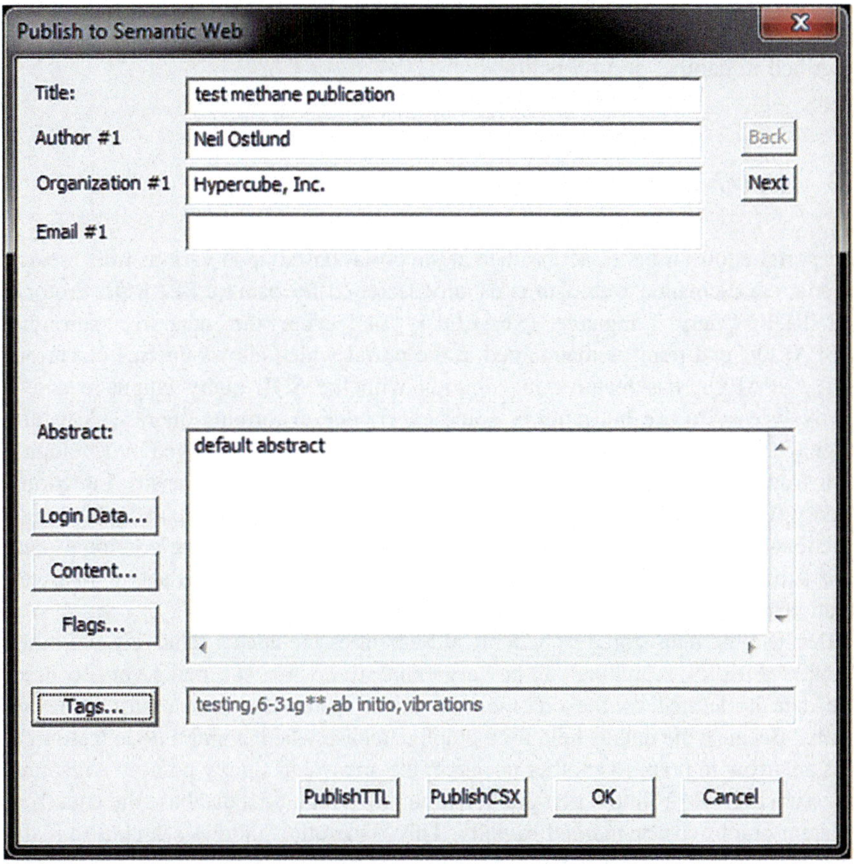

Fig. 1 Dialog box for submitting calculated quantum data to a portal

package ought perhaps to have such a button! The founders of Chemical Semantics, Inc. have a historical tie to Hypercube, Inc. and as such have used HyperChem to illustrate the basic ideas. What this button in a "CompChem" package does is create an XML file structure that includes the information about the current molecular system and any current calculation results resident in the package and sends the data to the Chemical Semantics, Inc. portal where it is published, given the Authors, Title, Abstract, Login information and other details that are part of the global setup prior to hitting the "Publish" button. Our current XML file structure is called CSX and is related in historical terms to the Chemical Markup Language (CML) file structure [12]. We use CSX because CML currently does not have certain properties that we consider not only desirable, but mandatory, such as the ability to deal with residues as independent fundamental units of biological molecules. Also, CML does not define many of the quantum chemistry concepts that we consider critical, We believe CSX encompasses CML as a subset and we could generate a CML file from CSX easily.

Our Chemical Semantics portal accepts data using a REST or SOAP [13] protocol and then publishes the data on its servers. The data is available to anyone around the world with an account at the portal. The details of the portal are described in another section below.

1.2 Searching the Data

Our portal allows users to access data at the portal based upon various rules, search criteria, etc. Semantic web data is usually searched for using a SPARQL Protocol and RDF Query Language (SPARQL) [14]. Note the recursive acronym. A SPARQL end point is maintained at the portal which allows queries of various kinds. SPARQL has features in common with the SQL query language and is relatively easy to use but requires some experience in forming queries. A natural language front end would be desirable and many groups are involved in developing such front ends. A query, for example, could ask how many Density Functional Theory (DFT) calculations have been done on a specific molecule and with which functionals and what final total energies. As opposed to querying relational database silos, the query could potentially survey the whole world's set of such calculations and return with a table of these.

Because the data stored by Chemical Semantics includes a relatively unlimited number of triples, searching can be very exhaustive while still being very focused. The data is defined by the ontology so that a search does not return irrelevant results. Because the data is held by a graph database where a graph node (resource) uses an arrow to point to another resource, the arrow can simply point to a resource in a second graph from the first graph and unlike a relational database the data from different graphs can be merged trivially. This "federation" allows searches to really use a Giant Global Graph (GGG). As part of the publish activity, data on our portal can be tagged as private, protected, or public. Any search obviously can return public data. For protected data the author can pass a key to researchers to enable them to access his/her protected data. An alternative would be to label the publication private so that only the original authors can access the data.

Data also comes with tags set by the authors that help define the search. For example an author might add a tag, "used_32bit_gpu" if he/she thought it worthwhile to distinguish calculations on a 32-bit only graphical processing unit from those on a normal CPU. Once data has been put into a semantic web form elaborate searches can be performed and the retrieved summary data could also be added as new data if so desired. Examples of SPAROL queries are shown below after the SPARQL query language is described.

2 Portal Technologies

The semantic web and our portal use a number of "new" technologies that are briefly described here. Most of these are new to chemists at this point and the following therefore provides somewhat of a primer on the semantic web so that the next section describing the actual operation of publishing and querying using the facilities of Chemical Semantics, Inc. can be better understood. There are many good reference books on the semantic web but they generally are written for computer scientists not chemists.

This tutorial will focus on describing quantum chemistry data as applied to the semantic web.

2.1 CSX—A Chemical Semantics Markup Language

The semantic web uses RDF and triple stores to hold data in a graphical database. The data of Quantum Chemistry needs to be put into this form using an ontology for Quantum Chemistry. As a practical matter it makes sense to have a way of structuring computed quantum data prior to converting it into RDF semantic web data. A fundamental issue here is that the data should always be put into a structured form from its inception so that computers can be taught this structure. We have introduced a markup language that we call a Common Standard for eXchange of quantum data (CSX).

2.1.1 Overview

Chemical Semantics, Inc. uses an Extensible Markup Language (XML) format file to capture information about quantum chemistry calculations. Specifically, a Common Standard for eXchange (CSX) file is used to transfer structured data and metadata about calculations to our web portal where it is converted to the Resource Description Framework (RDF) format appropriate to the semantic web.

While CSX has been developed to allow publication pf quantum chemistry calculations onto the semantic web, it is a useful standard in and of itself because it organizes all the important information about a calculation in a format that is readable by both humans and computers. We suggest that CSX could become a standard output format for the computational chemistry community and invite interested readers to contribute to its development.

This section thus describes the current CSX standard for describing data from quantum chemistry calculations (and possibly related computational or experimental data). The standard specifically includes information describing a publication, i.e. title, author, etc. because our portal essentially accepts "data publications" and places them onto the semantic web. CSX is still currently under development and so please be aware that the description below is dated. A CSX file has a version number and the current version described here is Version 1.0. Ongoing development is aimed at creating Version 2.0 in the near future. For example, aligning our CSX with JSON for Linked Data (JSON-LD) [15] is appropriate as this new standard for linked data is closely related to the semantic web.

2.1.2 CSX Components

A CSX file is shown in Fig. 2. The fundamental components are described below.

NameSpaces

The CSX file above includes the basic XML components including a comment that describes where the CSX file was created. The root element of the XML file is cs:chemicalSemantics. The attributes at the root include this particular version of CSX, the local file name for the CSX file, and the relevant namespaces being used.

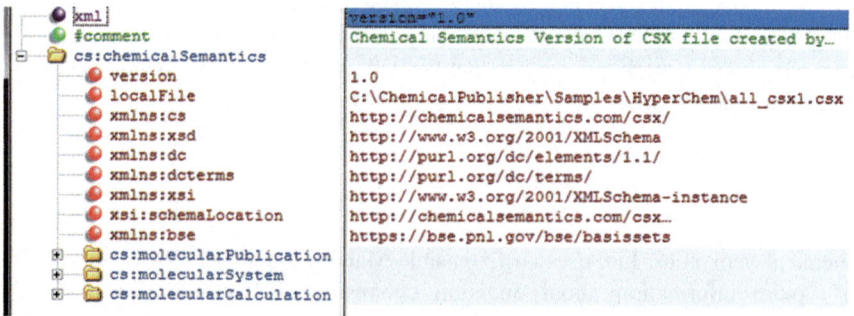

Fig. 2 A basic CSX file

The principal namespace is cs: which signifies the XML elements defined by Chemical Semantics, Inc. Other namespaces like xsd: and xsi: are associated with the XML schema (blueprint) that describes CSX. The namespaces dc: and dcterms: are part of Dublin Core, a metadata standard that is used for some of the publication parameters of CSX like title and abstract. The namespace bse: is used to describe basis sets and is subject to modification as we collaborate with Pacific NW Laboratories on semantic definitions of standard basis sets.

2.2 CSX—Sections of a File

The description given here corresponds to CSX Version 1.0. The definition of CSX is ongoing. While other sections could be added in future versions of CSX, the current standard includes three sections:

- molecularPublication
- molecularSystem
- molecularCalculation.

The Chemical Semantics portal essentially accepts "Data Publications", i.e. data from computational chemistry computations that are being published at the portal. Thus, the first section of a CSX file describes the publication itself, including copies of the input and output files used in the calculation (if available). The second section describes the molecular system (a set of molecules) that calculations were performed on. The final section describes the calculation (or calculations) that were performed on the molecular system.

2.2.1 Molecular Publication

The molecularPublication section of a CSX file is shown in Fig. 3.

A publication has a title and an abstract as indicated as defined in the Dublin Core specification. We use the Dublin Core namespace, dcterms:, to describe these. The "publisher" here is empty but would contain the name of the company/institution of the lead author.

Authors

A publication can have a number of authors described by their type. The sole author here is described as cs:corresponding, i.e. the Corresponding Author who is usually the Principal Investigator (PI) or someone with authority over the publication. The attribute "type" can also have a value of cs:submitting indicating the author submitting the publication, or could be empty indicating another author or co-author. Each author has a name described by cs:creator and an organization described by cs:organization and an email address described by cs:email.

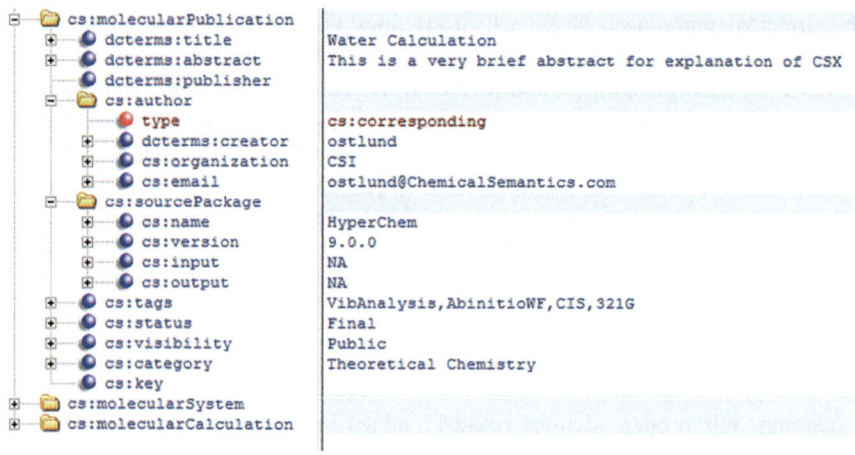

Fig. 3 The molecularPublication section of a CSX file

Source

The data being published should have an indication of its source, i.e. the software package that created the data such as Gamess, NWChem, PSI4, etc. The version of these software packages should also be indicated. In the example above, the data came from Release 9 of HyperChem.

For archival purposes, it is also possible to add the text that constitutes the input file for the calculation as well as the text of the output file. The publication data may have been extracted by parsing the output file or directly from the software package. In any event, the input and output file (if present) constitute additional archival data about the calculation that can be recovered later, if so desired.

Tags and Flags

Finally, the publication can indicate a set of arbitrary tags that can apply to the publication and provide an additional way to search for a set of publications. These tags are independent of searches based on SPARQL and are up to the authors to create. They may or may not be commonly used depending upon the preference of users.

A set of allowed values for flags is used to characterize publications. These are:

Status

- Preliminary
- Draft
- Final.

A publication can be considered preliminary (very raw data perhaps) or draft (not fully reviewed) or final. This status can be changed (edited) at any later time.

Visibility

- Private
- Protected
- Public.

Depending upon the desires of the authors, a publication can be set to private which means only the submitting author with a password can see the publication. Alternatively, a public publication can be seen by anyone with access to the portal. Finally, an intermediate visibility is available that is termed "protected". A protected publication requires a key (similar to a password) that an author can pass to a collaborator so that they can see an author's publication.

Category

Finally, it is possible to categorize a publication from one of the areas of chemistry below.

- Analytical Chemistry
- Biochemistry
- Computational Chemistry
- Inorganic Chemistry
- Materials Chemistry
- Material Science
- Molecular Chemistry
- Nanotechnology
- Neurochemistry
- Nuclear Chemistry
- Organic Chemistry
- Other
- Petro Chemistry
- Physical Chemistry
- Polymer Chemistry
- Synthetic Chemistry
- Theoretical Chemistry.

2.2.2 Molecular System

The molecularSystem section of a CSX file is shown in Fig. 4.

A molecularSystem is a collection of molecules. Each molecule is a collection of possibly residues (monomers) or perhaps just atoms. The intermediate level of

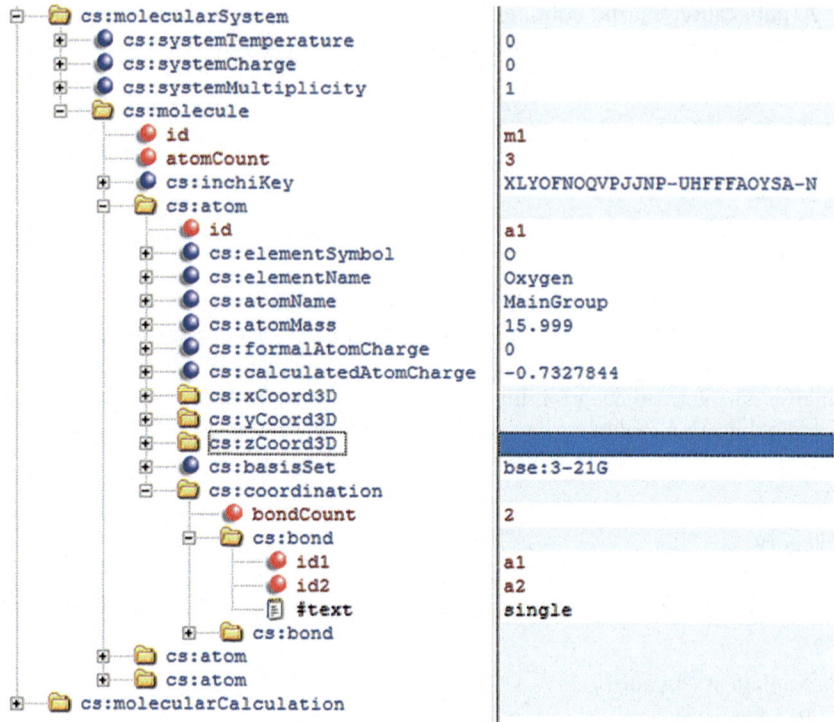

Fig. 4 The molecularSystem section of a CSX file

"group" is also possible. That is, a molecule or residue might be considered as having groups that are made up of atoms.

Normally a molecule is just a set of atoms, unless we are describing a protein or dna-like structure where the residues would be amino acids or nucleic acids. In the above example, there is only one molecule (water) made up of three atoms.

System

The molecular system currently has three properties. The traditional charge and multiplicity of a quantum calculation are two of these. In addition, we define the system temperature, as well, for statistical mechanical calculations such as molecular dynamics (MD), Monte Carlo, etc.

Molecule

Each molecule has an id which is the name or other identifier of the molecule. A default id of "m1", "m2", etc. for molecule 1, molecule 2, ... is suggested.

A molecule has an atomCount describing the number of atoms in the molecule and an InChI key [16] that is meant to be a unique identifier for each molecule in the system. The InChI string and InChI key are defined by IUPAC and may or may not exist for each molecule. The value "nil" or "" indicates that no InChI key was returned from IUPAC software. Each molecule is made up of atoms.

Atom

All atoms in a molecule must have an id attribute of the form "a1", "a2", "a3", etc. These identifiers are used to describe the resulting bonds.

An atom has a number of children including the elementSymbol, elementName, etc.

- elementSymbol—this is just normal symbol such as C, Cl, etc.
- elementName—this is the full name for the element such as Carbon, Chlorine, etc.
- atomName—for macromolecules an atom may have a pertinent name such as CA, CB for the alpha, beta carbons in a chain. For normal molecules, the default names are:

  - H
  - MainGroup
  - Metal
  - Row1TM
  - Row2TM
  - Row3TM
  - Lanthanide
  - Actinide
  - NobleGas

- atomMass—the mass in amu
- formalAtomCharge—integer charge such as +1 for the N in NH_4
- calculatedAtomCharge—as used in molecular mechanics coulomb interactions
- x/y/zCoord3D—the three Cartesian coordinates of an atom
- basisSet—the basis set is a property of each atom and may be different for different atoms
- coordination—describes the other atoms to which this atom is connected.

Bond

Bonds in CSX are a property of atoms. The XML coordination element describes these, as children of coordination. The bondCount attribute of coordination is the number of bonds that this atom participates in. Each bond is a child of the atom's coordination with attributes id1 and id2 that describe the two connected atoms. This means that each bond is described twice—as a grandchild of the atom with attribute id1 and as a grandchild of the atom with attribute id2. The content of the XML bond element describes the bond as single, double, triple, aromatic, or dative. We believe that having a bond be defined in the context of defining an atom provides better functionality that defining bonds as an isolated property as per CML.

2.2.3 Molecular Calculation

The molecularCalculation Section of a CSX file is shown in Fig. 5.

There are different types of potential calculations that need to be placed in a CSX file although many commonalities exist. In particular scf and dft calculations have much in common. The following section of a CSX file describes a simple scf calculation.

The first XML element describes the calculation as a quantum mechanical one (as opposed to a molecular mechanics). Secondly, it uses a singleReferenceState (as opposed to a multipleReferenceState such as MCSCF). Then the calculation describes a singleDeterminant as opposed to a multipleDeterminant such as CISD).

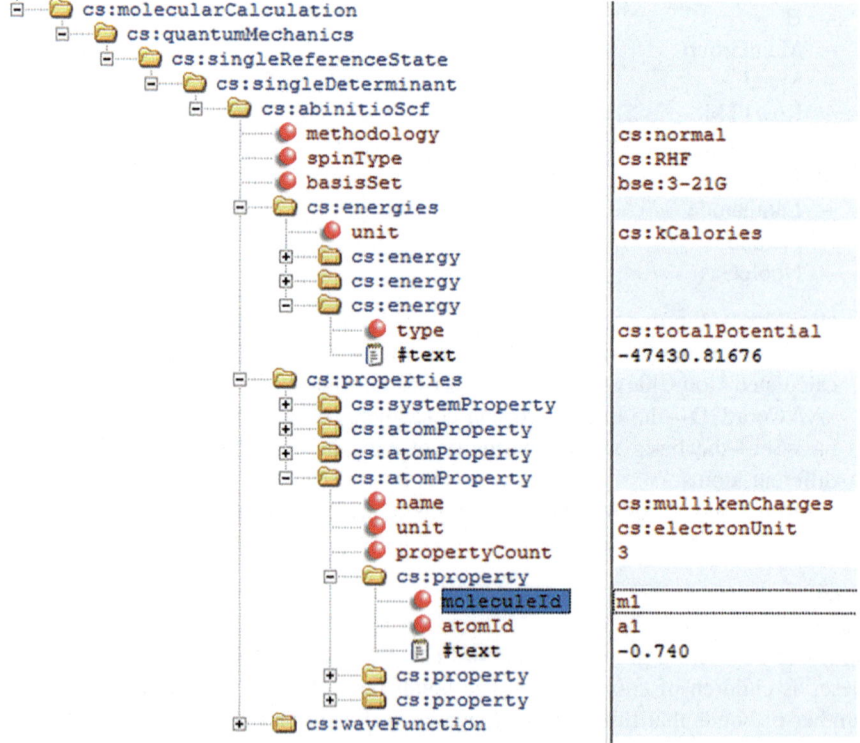

Fig. 5 The molecularCalculation section of a CSX file

Finally, the calculation is characterized as abInitioSCF. Alternatives might be semiempiricalSCF or dft.

Attributes

The attributes of such a calculation are:

- Methodology—allows for deviations from the normal type of SCF calculation.
- spinType—RHF, UHF, or ROHF
- basisSet—as defined by the PNNL Basis Set Exchange.

Child Elements of Calculation

The child elements of an SCF calculation are:

- energies—core nuclear-nuclear interaction, electronic energy, total energy, etc.
- properties—system properties and atom properties
- waveFunction—orbital energies, orbital symmetry, coefficients, etc.

The energies are all labeled by their "type" attribute. For example, cs:totalPotential is a potential energy for nuclear motion in the Born-Oppenheimer approximation and is commonly just called the total energy in quantum calculations. The advantage of using a Uniform Resource Identifier (URI) here is that it gives uniqueness to the variable being described.

A large variety of properties (possibly expectation values) could be associated with a quantum calculation. We divide these properties into system properties like dipole moment and atom properties like mullikenCharges, A systemProperty has attributes—"name" and "unit" where name, for example, is cs:dipoleMomentX denoting the X component of the total dipole moment and unit is cs:debye.

An atomProperty also has attributes "name" and "unit" in addition to the attribute propertyCount which indicates how many atoms follow with properties. The attributes of an atom property are the moleculeId and the atomId which uniquely identify the atom (atom indices or id's are unique only for the specific molecule that the atom is a member of). The value of the atomProperty is the content of the associated XML property element.

With quantum calculations there may be no connection table identifying a "molecule" since a molecule of the molecular system is defined by CSX as the collection of atoms forming a connected graph. For certain quantum calculations there may only be a molecularSystem with child atoms and no "molecule" or "coordination", "bond", etc. This is acceptable as valid CSX. It is preferred, however, to define a connection table, if possible.

waveFunction

The wave function example in Fig. 6. shows the result for the above 3–21 G calculation with its orbitals, etc. The far right side of the display is cut off and not all orbital energies, symmetries, etc. are shown.

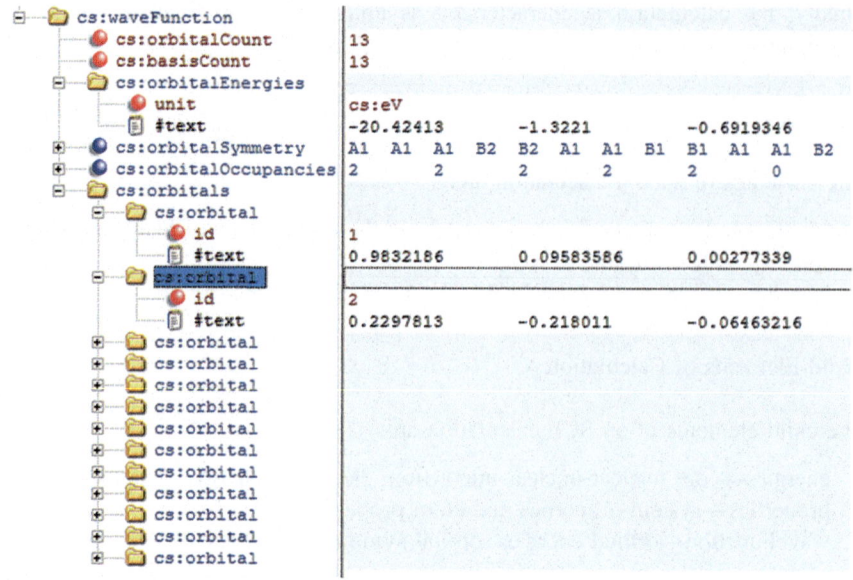

Fig. 6 The wave function section of a CSX file

The wave function has attributes orbitalCount and basisCount. These are generally the same but some calculations such as those from Gamess may differ here because of the treatment of 5 spherical or 6 Cartesian d-orbitals.

The child XML elements of the waveFunction include the orbitalEnergies, cs: orbitalSymmetry, cs:orbitalOccupancy and the orbitals themselves. Each orbital is identified by its id.

Other Calculations

If the calculation is a density functional calculation, then cs: abinitioSCF is replaced by cs:dft and two new attributes appear—cs:exchangeFunctional and cs:correlationFunctional. If the calculation is MP2, then cs:abinitioSCF is replaced by cs: secondOrderMoellerPlesset and an energy with type cs:correlation is added. Other calculations such as CCD, CCSD(T), etc. may have energies and properties but not a cs:waveFunction.

If the calculation is a molecular mechanics calculation then cs:molecularMechanics replaces cs:quantumMechanics and new attributes describing the cs:forceField and cs:parameterSet appear.

2.3 Usefulness of the CSX Description

The CSX proposed file standard is a convenient way to encapsulate computational data coming from various quantum chemistry packages, for example. It will make possible publication of the data onto the semantic web as well as portability of molecular structures and results among computational and other chemists. It is a beginning and other options may become available but we believe it is a valuable contribution.

2.4   A Quantum Chemistry Ontology

The RDF standard for a graph database and the associated serialized files, like RDF/XML or Turtle, contain the scientific data (consisting of URI's and literals). The interpretation or meaning of that data, however, requires a vocabulary for defining the data. That vocabulary is an ontology, in this case an ontology for computational chemistry. Simple ontologies use a schema for RDF (RDFS) but in general the Web Ontology Language (OWL) and OWL files are better used to describe the ontology. An OWL file is usually formatted as RDF/XML for convenience although an OWL ontology is not to be confused with the fundamental data described by RDF. It is just simply convenient to use RDF to describe an ontology. This shows the power of RDF in that the ontology can be described by the same structure as the data.

2.4.1   Simple OWL File

An OWL file begins by describing the *classes* (*and subclasses*) that are basic entities of the ontology. For example the classes might be MolecularSystem, Molecule and Atom, where the MolecularSystem is what a calculation is performed on and which is assumed to be a collection of atoms and molecules. These classes have *ObjectProperties* that relate one class to another. The ObjectProperty, hasMolecule, relates the class MolecularSystem (the *Domain*) to the class Molecule (the *Range*). The ObjectProperty, hasAtom, similarly relates the class Molecule to the class Atom. The following is a portion of such an ontology expressed in Turtle. The Gainesville Core ontology is reflected in the prefix gc:

```
@prefix rdfs: <http://www.w3.org/2000/01/rdf-schema#> .
@prefix owl: <http://www.w3.org/2002/07/owl#> .
@prefix gc: http://purl.org/gc>.

gc:hasMolecule rdf:type owl:ObjectProperty ;
          rdfs:domain gc:MolecularSystem ;
          rdfs:range gc:Molecule .

gc:hasAtom rdf:type owl:ObjectProperty ;
          rdfs:range  gc :Atom ;
          rdfs:domain gc :Molecule .
```

Using such an ontology allows the trivial inference that a MolecularSystem has atoms although that is not explicitly stated! An ontology also has DataType Properties that relate a class to a Literal rather than another class. For example, the dataType property, hasMultiplicity, might relate a class of SCF calculations to the Literal 1 or 3 to indicate singlet or triplet.

Chemical Semantics, Inc. is in the middle of defining an ontology for computational chemistry. The current Release is usable but certainly not final. The Gainesville Core web site describes the current status. A very professional version of this will not be available quickly. A view of a very preliminary current version is shown below.

One of the most important efforts of Chemical Semantics, Inc. is the development of a proper ontology for computational chemistry. The effort will require contributions from many members of this particular scientific community. Funding for such an effort is being sought.

2.5   SPARQL—Searching

The acronym SPARQL Protocol and RDF Query Language (SPARQL) is officially recursive but sometimes is referred to as Simple Protocol and RDF Query Language. It bears some similarity to SQL but is a W3C standard for querying data on the semantic web (graph databases) rather than data in relational databases.

There are a number of commercial SPARQL software packages including Virtuoso which Chemical Semantics' portal uses. The following is an example of a SPARQL query that searches for all molecules that have 5 atoms and include a Chlorine atom.

```
prefix gc: <http://purl.org/gc/>
prefix rdfs: <http://www.w3.org/2000/01/rdf-schema#>
select
  ?molecule ?moleculeLabel ?inchikey
where {
  graph ?graph {
    ?molecule rdf:type gc:Molecule ;
    rdfs:label ?moleculeLabel ;
gc:hasNumberOfAtoms "5";
gc:hasAtom ?atom;
    gc:hasInChIKey ?inchikey .

  ?atom gc:isElement "Cl" .
  }
}
order by ?molecule
```

The query begins with a definition of the namespaces gc: (Gainesville Core defined by Chemical Semantics, Inc.) and rdfs: (RDF Schema) that it will use. The quantities ? molecule, ?moleculeLabel, etc. are simply variables with arbitrary names.

The query searches for Molecules that have certain properties such as a label, number of atoms, etc. The ";" is essentially an "and" and the search ends with a "." In addition, to the required Molecule properties, a ?atom found in the molecule must also be a Chlorine according to the last requirement (Fig. 7),

```
?atom gc:isElement "Cl".
```

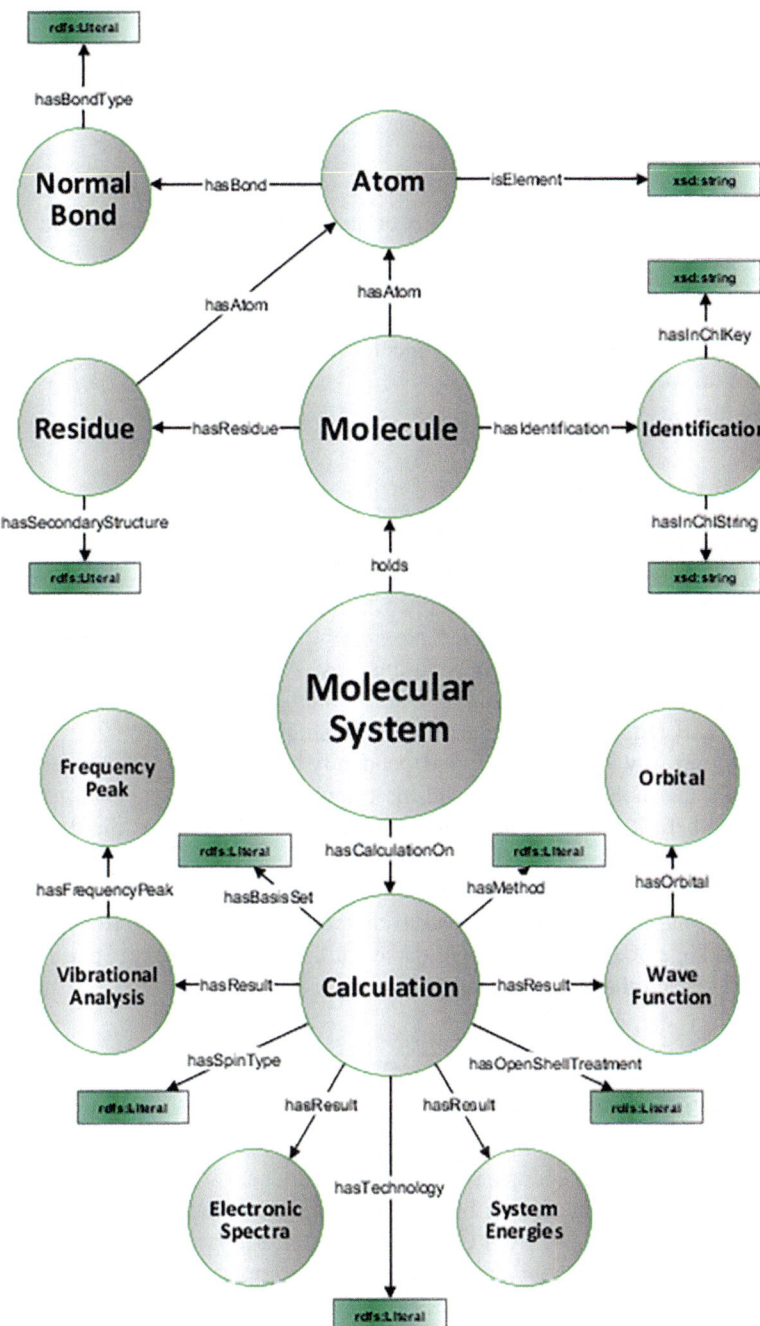

Fig. 7 Preliminary version of ontology

molecule	moleculeLabel	inchikey
http://purl.org/chem/pub/2013-08-24-hyperchem10molSys/VinylChloride%20Anion/	Molecule: VinylChloride Anion	AGSCYFRVQMCTSL-UHFFFAOYSA-N
http://purl.org/chem/pub/2013-08-24-hyperchem8molSys/MethylChloride/	Molecule: MethylChloride	NEHMKBQYUWJMIP-UHFFFAOYSA-N

Fig. 8 Result of a SPARQL query

An example of using the query at the portal of Chemical Semantics, Inc. with a small demonstration RDF graph results in a table as shown in Fig. 8.

The columns are labelled by the variables used in the query select. It returns here only the inchiKey of the two molecules but an extension of the query could return any computed properties of the two molecules that are part of our graph database on the semantic web.

3 The Web Portal

There are three ways to publish on the portal. The **first** is to publish directly from various software packages that produce computational chemistry results. The number of these will expand as time progresses. For example, versions of PSI 4, NWChem and HyperChem can publish this way. An example used to develop the basic ideas is HyperChem, Release 9 with its Publish Button. The other packages without a GUI require a simple script for publishing.

The **second** way to publish is to independently create a CSX file that defines the publication, the molecular system and the calculations and just upload that file to the portal where it will be translated to a Turtle (*.TTL) file and the data placed onto the semantic web. One way to do this is to parse an output file with software such as CSI's own ChemicalPublisher to create the CSX file as shown in Fig. 9.

Thirdly, one might directly upload to the portal the output file from a computational package and have that output parsed at the portal where it is put into CSX form initially and then translated to a TTL file.

3.1 Using a Publish Button

The first publication procedure, which uses HyperChem 9 or later, will not only publish HyperChem calculations but any third-party calculations that HyperChem has imported such as those from Gamess, Gaussian, Mopac, etc.

For example, HyperChem can parse a Gamess output file. The screen shot in Fig. 9 shows three Gamess output files containing results for a single point ab initio SCF calculation, a geometry optimization of structure and a vibrational analysis

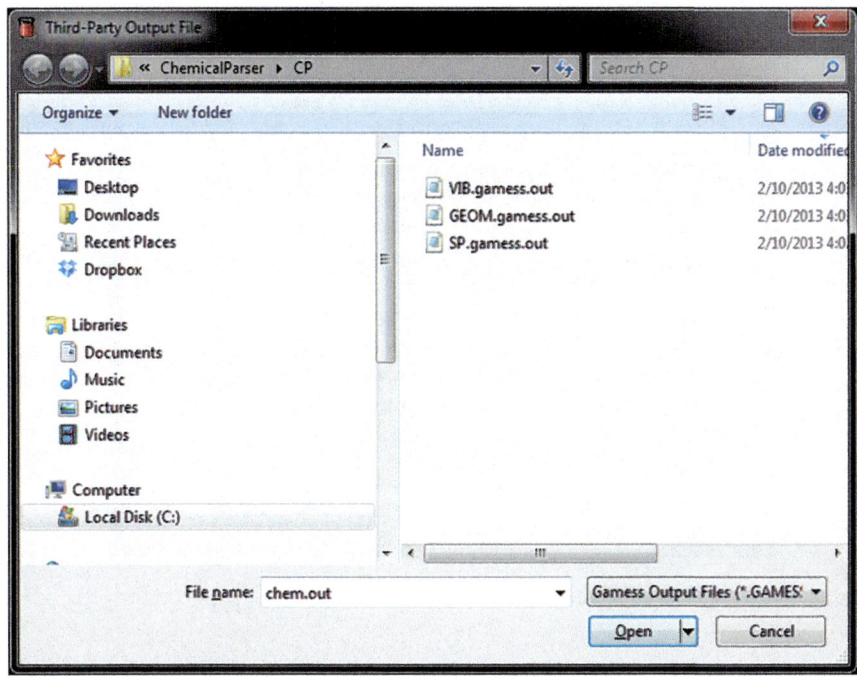

Fig. 9 Using parser software to parse an output file to create a CSX file

calculation. Once imported into HyperChem 9, these results can be published just as if they we computed by HyperChem.

While parsing output files is certainly possible, Chemical Semantics, Inc. expects to work with developers of these computational chemistry packages to help them install their own "Publish Button". The Publish Button in HyperChem is shown in Fig. 1.

In addition to the title of the publication, the authors, their organizations and e-mail, and the publication abstract, pushing the publication button (which initially creates a CSX file) adds a number of other things to the publication. The Login Data... Button allows entering data so that the Publisher Package can use a login ID and password to directly publish results. The Content Button allows choices to be made of what is published among the available results as shown in Fig. 10. The Flags button allows the author to choose to define the current state of the publication as shown in Fig. 11 or the Visibility (Private, Protected, and Public). A private publication can be seen only by the authors, a protected publication can be show to anyone that the authors send a URI to with a key, and a public publication can be seen by anyone logged into the portal.

It is also possible to add tags (essentially keywords) to any publication as shown in Fig. 12. These may help in searching. A common set of tags is available as well as custom tags set by the authors.

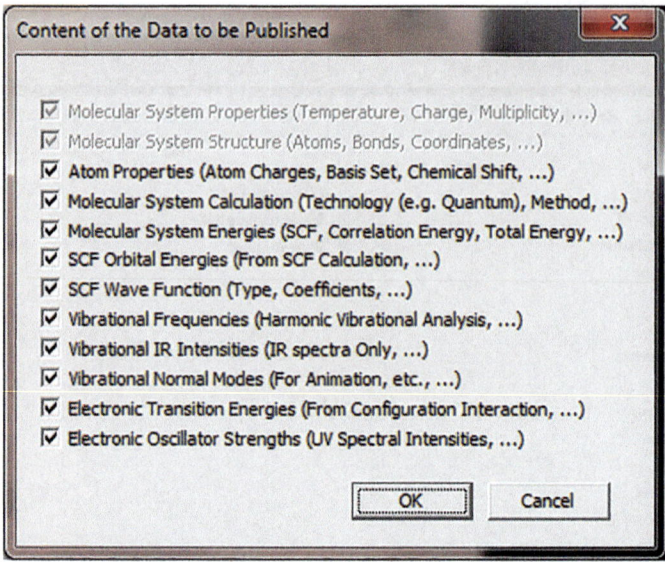

Fig. 10 Choosing content to publish

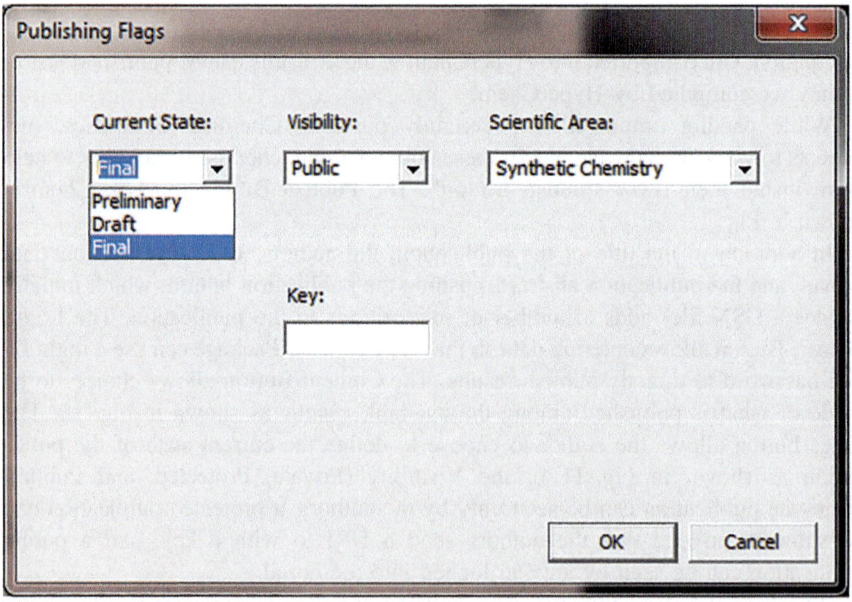

Fig. 11 Publishing flags

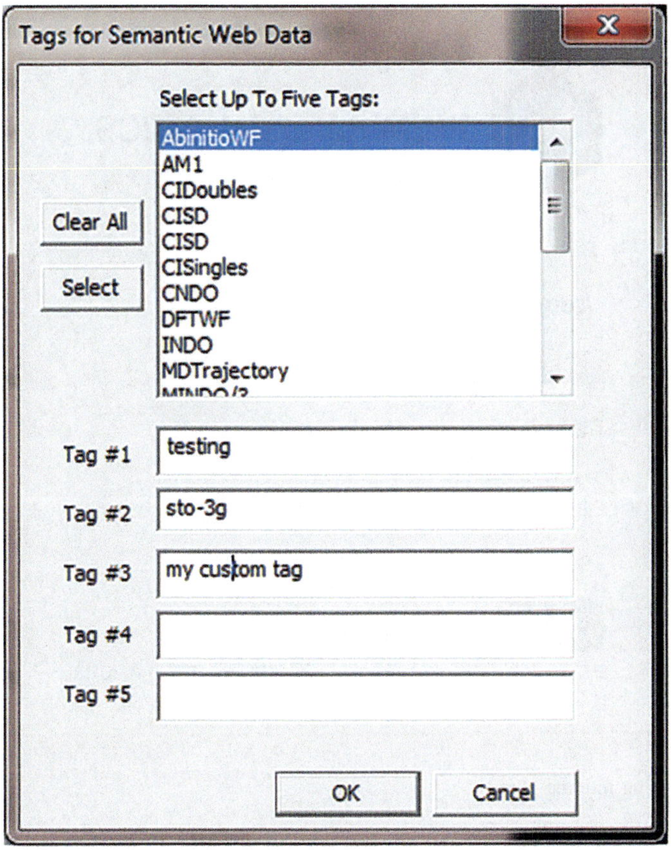

Fig. 12 Publishing tags

3.2 Logging in and Manual Upload of CSX

In addition to publishing by hitting a "Publish Button," many scientists will publish by creating a CSX file and uploading it to the portal site. As described earlier above, Chemical Semantics, Inc. has created a new CSX standard, similar to CML, for holding all the required details of a computational chemistry publication. This includes details about the authors, the title of the publication, etc. as more or less just shown in Fig. 1. The information transferred to the portal by hitting the Publish Button is that stored in a transferred CSX file. Multiple pathways can be expected to produce these CSX files as time progresses. The conversion from CSX to RDF occurs at the portal server.

One first has to log into the Portal as shown in Fig. 13. One can register if one does not yet have an ID and Password. After entering the portal one is met with a

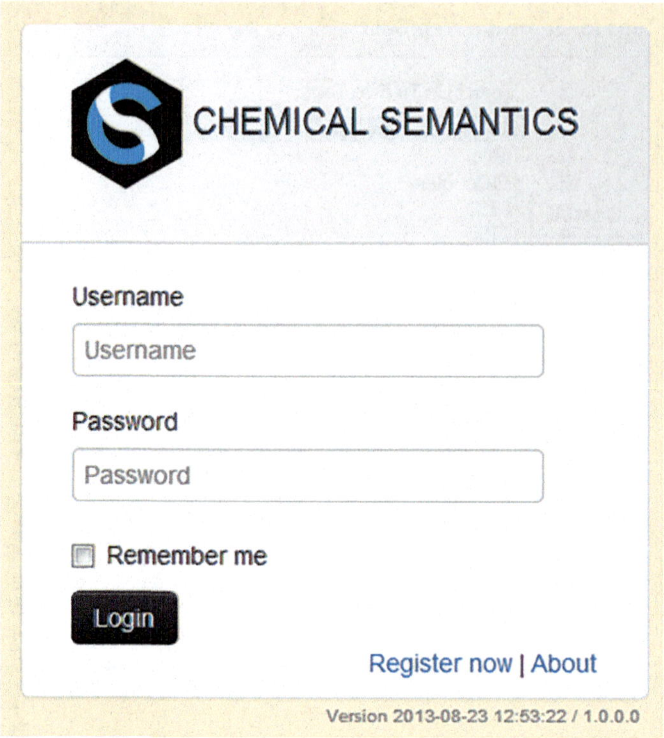

Fig. 13 Logging into the portal

list of one's own publications but one can inspect all publications as well depending upon their visibility (Private, Protected, Public).

One can peruse all publications based upon Author, Title, Category, Tag, etc. and then view any publication.

3.3   Viewing Publications

Clicking on any publication allows you to view details of a publication. Each publication requires a Unique Name to be used in generating the URI for this publication. Normally these will be assigned at the server level by the portal. An example URI is,

https://staging.chemsem.com/pub/ostlund-20161221102029/

This uniquely identifies this publication. It is dereferenceable and can be passed to friends to access the publication if so desired. The amount of data displayed with a publication depends on the type of calculation being published and is evolving with the portal.

3.4 Data Federation

One of the problems with existing databases is that the data exists in silos of isolation. The individual databases are difficult to merge and there is general difficulty in sharing data because of a lack of universal agreement on the database schema, column names, etc. A fundamental aspect of the semantic web is its ability to federate data, i.e. make data available globally. This comes about because of the global data standards that have been set, because one can merge individual ontologies easily and because two separate graph databases can be merged just by adding a single link (predicate) from one graph to another (Fig. 14).

An elementary example of this federation is available at the Chemical Semantics portal by clicking on the Data Federation Tab. If the molecule for the current publication is Methyl Chloride, then clicking on the tab brings up something like that shown in Fig. 15.

This displays the information about Methyl Chloride that exists at the ChemSpider site of the Royal Society of Chemistry (RSC) and the Chemical Entities of Biological Interest (ChEBI) site of the European Molecular Biology Lab (EMBI-EBI).

Fig. 14 List of portal publications

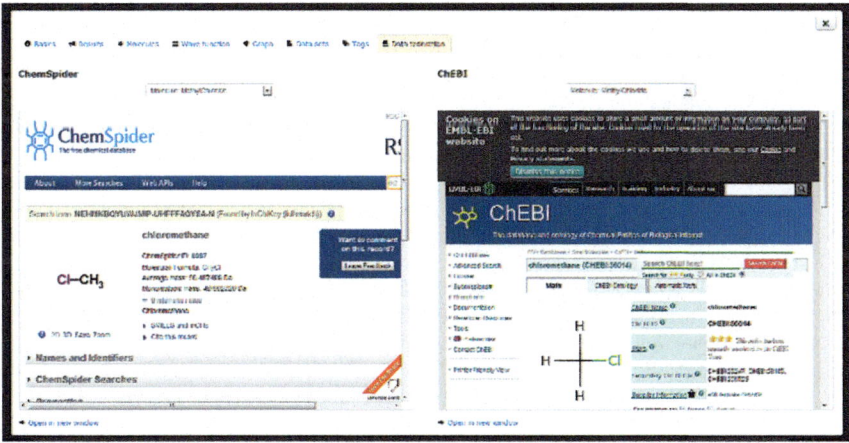

Fig. 15 Federation of a portal publication

4 Conclusion

The semantic web offers a new way to publish the data created by Quantum Chemistry calculations that matches the capabilities of the World Wide Web. As opposed to isolated silos of data that are difficult to find and/or share, the semantic web makes sharing of data a fundamental attribute. Our portal is a demonstration of this new technology and hopefully is a precursor of technology allowing scientist to finally have a vehicle for the proper sharing of scientific data leading to new and enhanced capabilities.

References

1. Berners-Lee T, Hendler J (2001) Publishing on the semantic web—the coming internet revolution will profoundly affect scientific information. Nature 410:1023–1024
2. Alvarez-Moreno M, de Graaf C, Lopez N, Maseras F, Poblet J, Bo C (2015) Managing the computational chemistry big data problem: the ioChem-BD platform. J Chem Inf Model 55:95–103
3. Web Ontology Language. https://www.w3.org/OWL/. Accessed 1 Dec 2016
4. Graph Databases. http://GraphDatabases.com. Accessed 1 Dec 2016
5. The World Wide Web Consortium. https://www.w3.org/. Accessed 1 Dec 2016
6. Resource Description Framework. https://www.w3.org/RDF/. Accessed 1 Dec 2016
7. Gordon MS, Schmidt MW (2005) Advances in electronic structure theory: gamess a decade later. In: Dykstra CE, Frenking G, Kim KS, Scuseria GE (eds) Theory and applications of computational chemistry: the first forty years. Elsevier, Amsterdam, pp 1167–1189
8. Gaussian 09, Revision A.1, Frisch MJ, Trucks GW, Schlegel HB, Scuseria GE, Robb MA, Cheeseman JR, Scalmani G, Barone V, Mennucci B, Petersson GA, Nakatsuji H, Caricato M, Li X, Hratchian HP, Izmaylov AF, Bloino J, Zheng G, Sonnenberg JL, Hada M, Ehara M, Toyota K, Fukuda R, Hasegawa J, Ishida M, Nakajima T, Honda Y, Kitao O, Nakai H, Vreven T, Montgomery JA Jr, Peralta JE, Ogliaro F, Bearpark M, Heyd JJ, Brothers E, Kudin KN, Staroverov VN, Kobayashi R, Normand J, Raghavachari K, Rendell A, Burant JC, Iyengar SS, Tomasi J, Cossi M, Rega N, Millam JM, Klene M, Knox JE, Cross JB, Bakken V, Adamo C, Jaramillo J, Gomperts R, Stratmann RE, Yazyev O, Austin AJ, Cammi R, Pomelli C, Ochterski JW, Martin RL, Morokuma K, Zakrzewski VG, Voth GA, Salvador P, Dannenberg JJ, Dapprich S, Daniels AD, Farkas Ö, Foresman JB, Ortiz JV, Cioslowski J, Fox DJ (2009) Gaussian, Inc., Wallingford, CT
9. Valiev M, Bylaska EJ, Govind N, Kowalski K, Straatsma TP, van Dam HJJ, Wang D, Nieplocha J, Apra E, Windus TL, de Jong WA (2010) NWChem: a comprehensive and scalable open-source solution for large scale molecular simulations. Comput Phys Commun 181:1477
10. HyperChem, Release 9, Hypercube, Inc. Gainesville, FL. www.hyper.com
11. The Nobel Prize in Chemistry (1998). http://www.nobelprize.org/nobel_prizes/chemistry/laureates/1998/. Accessed 1 Dec 2016
12. Murray-Rust P, Rzepa H (1999) Chemical markup, XML, and the Worldwide Web. 1. Basic principles. J Chem Inf Comput Sci 39:928–942
13. Service Station—More on REST. https://msdn.microsoft.com/en-us/magazine/dd942839.aspx. Accessed 1 Dec 2016

14. SPARQL Protocol and RDF Query Language. https://www.w3.org/TR/rdf-sparql-query/. Accessed 1 Dec 2016
15. JavaScript Object Notation for Linked Data. http://json-ld.org/. Accessed 1 Dec 2016
16. The IUPAC International Chemical Identifier. http://www.inchi-trust.org/. Accessed 1 Dec 2016

Matrix Elements for Explicitly-Correlated Atomic Wave Functions

Frank E. Harris

Abstract We refer to atomic wave functions that contain the interelectron distances as "explicitly correlated"; we consider here situations in which an explicit correlation factor r_{ij} can occur as a power multiplying an orbital functional form (a **Hylleraas** function) and/or in an exponent (producing **exponential correlation**). Hylleraas functions in which each wave-function term contains at most one linear r_{ij} factor define a method known as **Hylleraas-CI**. This paper reviews the analytical methods available for evaluating matrix elements involving exponentially-correlated and Hylleraas wave functions; attention is then focused on computation of integrals needed for the kinetic energy. In contrast to orbital-product and exponentially-correlated wave functions, no general formulas have been developed by others to relate the kinetic-energy integrals in Hylleraas-CI (or its recent extension by the Nakatsuji group) to contiguous potential-energy matrix elements. The present paper provides these missing formulas, obtaining them by using relevant properties of vector spherical harmonics. Validity of the formulas is confirmed by comparisons with kinetic-energy integrals obtained in other ways.

1 Introduction

Ever since the first days of quantum mechanics investigators have sought methods for describing the electronic structures of atoms and molecules that are more rapidly convergent than superposition-of-configurations (also called configuration-interaction) expansions of the electronic wave function in orbital products. Probably

It is a pleasure to present this work as part of a tribute to Professor Josef Paldus in celebration of his eightieth birthday.

F. E. Harris (✉)
Department of Physics, University of Utah, Salt Lake City, UT, USA
e-mail: harris@qtp.ufl.edu

F. E. Harris
Quantum Theory Project, University of Florida, Gainesville, FL, USA

© Springer International Publishing AG, part of Springer Nature 2018
Y. A. Wang et al. (eds.), *Concepts, Methods and Applications of Quantum Systems in Chemistry and Physics*, Progress in Theoretical Chemistry and Physics 31,
https://doi.org/10.1007/978-3-319-74582-4_2

29

the earliest endeavor of this type was that of Hylleraas, whose study of the He atom [1] used a wave function that included as a multiplicative factor the explicit appearance of the interelectron distance r_{12}. Although it was many years before wave functions of this type came into widespread use (probably awaiting the availability of digital computers), a few landmark studies using such wave functions (which we identify as traditional Hylleraas functions) were soon carried out, including in particular a study of the hydrogen molecule by James and Coolidge [2], published in 1936. There followed in 1960 a further study of the hydrogen molecule ground state by Kolos and Roothaan [3], in 1968 a detailed study of the lithium atom by Larsson [4], and in 1994 an essentially quantitative computation of the ground state of the hydrogen molecule by Kolos [5].

Attempts to apply Hylleraas methods to larger systems revealed that the occurrence of a wide variety of combinations of r_{ij} factors (and higher powers thereof) led to exceedingly complicated computations, and as early as 1971 it was proposed by Sims and Hagstrom [6], and independently by Woźnicki [7], to consider configurations (wave function terms) that contained at most a single, linear r_{ij} factor. Methods based on wave functions of this type, now usually referred to as **Hylleraas-CI** (Hy-CI), were over time more fully developed and applied to a variety of atomic problems. Representative work in this area is in [8–11].

An alternative to the Hy-CI development is the use of exponentially-correlated wave functions. This type of wave function was proposed in the mid-1960s by Bonham [12, 13], but at that time calculations based on it seemed impractical. However, it was practical to use exponentially-correlated Gaussian orbitals, and work in that area has been pursued by Rychlewski et al. [14]. Calculations based on exponentially-correlated Slater-type orbitals finally became practical for small atomic systems after publication of an extraordinary paper by Fromm and Hill [15].

The possibility of a practical extension to the Hy-CI method (identified by its proposers as **E-Hy-CI**) has been examined by the Wang et al. [16], who developed formulas for the "unlinked" integrals (defined in Sect. 4) that are encountered when the single r_{ij} of a Hy-CI wave function is generalized to a form of the generic type $r_{ij}^{p_{ij}} \exp(-\beta_{ij} r_{ij})$.

The present contribution reviews some aspects of electronic-structure computations by these Hylleraas-inspired methods, including a discussion of recently-discovered methods for simplifying the computation of the kinetic-energy matrix elements in both Hy-CI and E-Hy-CI.

2 Wave Functions

Before incorporation into an antisymmetrized space-spin function, the spatial parts of the exponentially-correlated three- and four-body spatial wave functions, written in terms of their internal coordinates (in which one particle, often a nucleus, is at the origin of the coordinate system), can take the form

$$\Psi(1,2) = Y_{l_1}^{m_1}(\Omega_1) Y_{l_2}^{m_2}(\Omega_2) \Phi(1,2), \tag{1}$$

$$\Psi(1,2,3) = Y_{l_1}^{m_1}(\Omega_1) Y_{l_2}^{m_2}(\Omega_2) Y_{l_3}^{m_3}(\Omega_3) \Phi(1,2,3), \tag{2}$$

with

$$\Phi(1,2) = r_1^{n_1} r_2^{n_2} r_{12}^{p} e^{-\alpha_1 r_1 - \alpha_2 r_2 - \beta r_{12}}, \tag{3}$$

$$\Phi(1,2,3) = r_1^{n_1} r_2^{n_2} r_3^{n_3} r_{12}^{p_3} r_{13}^{p_2} r_{23}^{p_1} e^{-\alpha_1 r_1 - \alpha_2 r_2 - \alpha_3 r_3 - \beta_3 r_{12} - \beta_2 r_{13} - \beta_1 r_{23}}. \tag{4}$$

Here Y_l^m are spherical harmonics, at Condon-Shortley phase [17], and Ω_i stands for the angular coordinates of Particle i. Particle i, at position $\mathbf{r}_i$, has radial coordinate r_i, $\mathbf{r}_{ij} = \mathbf{r}_i - \mathbf{r}_j$, and $r_{ij} = |\mathbf{r}_i - \mathbf{r}_j|$ is the distance between Particles i and j. Traditional Hylleraas methods use wave functions with all β equal to zero; Hylleraas-CI methods additionally require each wave-function term to have at most one p_i nonzero; that p_i, if present, has the value unity.

3   Exponential Correlation

For three-body systems exponentially-correlated wave functions do not present major problems because the three interparticle distances r_1, r_2, and r_{12} can be chosen as internal coordinates, with the position and orientation of the three-particle triangle described by specifying the center of mass (or the position of one particle) and the Euler angles of the triangle orientation. In those coordinates, the internal "volume element" is proportional to $r_1 r_2 r_{12}$, and the main computational issue is to deal with the ranges of these distances.

For four-body systems it is still possible to use all the r_i and r_{ij} as explicit coordinates, but the internal volume element (contrary to a claim in Ref. [18]) is extremely complicated, leading to integrals long thought to be intractable. Evaluation of the four-body integrals needed for exponentially-correlated electronic-structure computations was the problem solved by Fromm and Hill [15]. Fromm and Hill's solution, however, suffered from the difficulty that its implementation required laborious tracking of the branches of the dilogarithm functions it contained. Avoidance of this inconvenience and various other technical improvements were described in a paper by the present author in 1997 [19]. That 1997 paper also resolved another issue: Remiddi [20] had published completely analytical formulas for certain integrals arising in Hylleraas calculations, and they agreed numerically with the corresponding special cases of the Fromm/Hill formula. But an analytical demonstration to that effect was missing. A formula presented in [19] was shown to lead to the desired analytical result.

As indicated above, the integrals needed for Hylleraas wave functions were special cases of those needed for exponentially-correlated wave functions, so in principle all such integrals for four-body systems could now be computed analytically. However, the formulas involved were rather unweildy, and the computational process was greatly simplified by the development of recurrence formulas enabling all four-body integrals to be obtained from a single starting value. This objective was achieved, first for traditional Hylleraas integrals only, by Pachucki et al. [21], and later, for general exponential correlation, by the present author [22].

4   Hylleraas Integrals

Integrals containing various combinations of r_{ij} as factors (we call these **potential-energy integrals**) will occur in atomic Hylleraas calculations; each integral can be identified with a diagram that is formed by

(1) Introducing a vertex corresponding to each particle not at the origin of the coordinate system (usually the nucleus);

(2) For each factor r_{ij} or $1/r_{ij}$ in the integral, drawing a line in the diagram that connects vertices i and j.

Any vertex not connected to other vertices by a line corresponds to a one-particle integral that is easily evaluated. Diagrams containing closed loops (with three or more vertices) are termed **linked**, and any diagram or part of a diagram that is not contained in a closed loop is called **unlinked**. Unlinked integrals and parts can be evaluated in closed form after introducing the Laplace expansion of each $1/r_{ij}$ [23] and/or its generalization to the related quantity r_{ij} [24]. An alternative to the Laplace-type expansion for unlinked integrations is to use a coordinate system in which the unlinked r_{ij} are coordinates. That approach has been followed by Ruiz in [25–28] and in other papers. Integrations over the particles in closed loops can also be treated using Laplace-type expansions, but such linked integrations lead to infinite series that are usually evaluated numerically.

For atomic Hy-CI, the limitations in the occurrence of r_{ij} factors cause the potential-energy integrals to consist only of completely unlinked integrals involving four or fewer vertices, except for one three-vertex integral containing a linked product of the form $r_{ij}r_{ik}/r_{jk}$. This type of integral, often called a "triangle" integral, was first discussed by Szasz [29].

While the general development for exponentially correlated wave functions in principle provided closed analytic formulas for the Hylleraas triangle integrals, those formulas were often more laborious to evaluate than expansions based on Laplace-type formulas. For triangle integrals with general spherical harmonics, the most utilized current approach is probably the Levin u-transformation convergence-acceleration scheme [30] used by Sims and Hagstrom [9].

The diagrams denoting integrals arising in Hy-CI are the same as those occurring in its extension E-Hy-CI, the only difference being that each diagram line refers to a factor of type $r_{ij}^{p_{ij}} \exp(-\beta_{ij} r_{ij})$ instead of simply r_{ij}.

5 Kinetic Energy

Unexpectedly, a bottleneck in Hylleraas-CI computations has been the computation of kinetic-energy matrix elements. In completely orbital formulations, the derivatives describing the kinetic energy in quantum mechanics simply change the powers of r_i in a matrix element, making it simple to identify the kinetic energy in terms of potential-energy integrals. However, the presence of factors r_{ij}, either as powers or in exponents, generates new complications.

When no explicit angular factors are present, the kinetic-energy operator $\hat{T}$ can be reduced to the form [31]

$$\hat{T} = -\frac{1}{2} \sum_{i<j} \left(\frac{1}{m_i} + \frac{1}{m_j} \right) \left(\frac{\partial^2}{\partial r_{ij}^2} + \frac{2}{r_{ij}} \frac{\partial}{\partial r_{ij}} \right) - \sum_i \frac{1}{m_i} \sum_{\substack{j<k \\ j,k \neq i}} \cos\theta_{ijk} \frac{\partial^2}{\partial r_{ij} \partial r_{ik}}. \quad (5)$$

The indices in Eq. (5) run over all the particles (including any nuclei, whether or not they are assumed to be of infinite mass), m_i is the mass of Particle i, and $\cos\theta_{ijk}$ is the cosine of the angle between $\mathbf{r}_{ij}$ and $\mathbf{r}_{ik}$. It can be evaluated as

$$\cos\theta_{ijk} = \frac{r_{ij}^2 + r_{ik}^2 - r_{jk}^2}{2 r_{ij} r_{ik}}. \quad (6)$$

Because Eq. (5) is general, its use yields the kinetic energy even when the particles are all of finite mass, thereby removing the need for an estimate of the nonphysical quantity called "mass polarization".

For wave functions containing explicit angular factors (e.g., spherical harmonics), the kinetic-energy operator requires additional terms. This topic is discussed in [32–34].

Evaluation of the kinetic energy for exponentially-correlated wave functions was examined in 1993 by Rebane [35], who showed how the kinetic-energy matrix elements could be written in terms of suitable potential-energy contributions. Rebane's derivation was later simplified by the author's research group [36]. Unfortunately Rebane's formula involves all the exponents of the exponentially-correlated wave function and does not apply to the usual .

6 Kinetic Energy in Hylleraas-CI

In the absence of formulas relating Hylleraas kinetic-energy matrix elements to those of potential energy, expressions based on the Laplace expansion can be differentiated, leading after some complication to a set of kinetic-energy formulas. The situation is even more cumbersome if, following Ruiz, one uses the interparticle coordinates directly. In any case, relations involving kinetic-energy integrals would be useful in reducing computational effort and/or in providing additional checks for the numerical work. This section of the present paper summarizes work previously presented by the present author [37] showing how the use of concepts associated with vector spherical harmonics [38] can be applied to derive the missing kinetic-energy formulas for Hy-CI.

Our starting point is to note that after integrating the variables of a kinetic-energy matrix element that involve orbitals only, there may remain a two- or three-body integral of which the most difficult are illustrated by the following:

$$K_2 = \left\langle r_{12}\phi_a(1)\phi_b(2) \left| -\frac{1}{2}\nabla_1^2 \right| r_{12}\phi_d(1)\phi_e(2) \right\rangle, \tag{7}$$

$$K_3 = \left\langle r_{13}\phi_a(1)\phi_b(2)\phi_c(3) \left| -\frac{1}{2}\nabla_1^2 \right| r_{12}\phi_d(1)\phi_e(2)\phi_f(3) \right\rangle. \tag{8}$$

Here ϕ_a is a Slater-type orbital (STO) of the form

$$\phi_a = g_a(r)Y_{l_a}^{m_a}(\Omega), \qquad \text{with} \qquad g_a(r) = r^{n_a-1}e^{-\alpha_a r}. \tag{9}$$

For compactness in the final formulas for K_2 and K_3 we write a_l^m to denote an orbital with parameters α_a and n_a, but with the indicated quantum numbers l, m, which may differ from the values l_a, m_a of ϕ_a.

For both K_2 and K_3 we start by evaluating the application of the Laplacian for Particle 1 to $r_{12}\phi_d(1)$. Noting that $\nabla^2 Y_l^m = -l(l+1)Y_l^m/r^2$, we find

$$\nabla_1^2[r_{12}\phi_d(1)] = \left[r_{12}\nabla_1^2 g_d(1) + \left(\nabla_1^2 r_{12}\right)g_d(1) \right.$$
$$\left. - \frac{l_d(l_d+1)r_{12}g_d(1)}{r_1^2} + 2\nabla_1 g_d(1)\cdot\nabla_1 r_{12} \right]Y_{l_d}^{m_d}(1)$$
$$+ 2r_{12}\nabla_1 g_d(1)\cdot\nabla_1 Y_{l_d}^{m_d}(1) + 2g_d(1)\nabla_1 r_{12}\cdot\nabla_1 Y_{l_d}^{m_d}(1). \tag{10}$$

Equation (10) corrects a misprint in the $l_d(l_d+1)$ term of [37].

Most of the quantities in Eq. (10) are easily simplified. Defining $\mathbf{r}_{12} = \mathbf{r}_1 - \mathbf{r}_2$ and letting an overline circumflex denote a unit vector,

$$\nabla_1^2 g_d(1) = \left(\frac{n_d(n_d-1)}{r_1^2} - \frac{2\alpha_d n_d}{r_1} + \alpha_d^2 \right)g_d(1), \tag{11}$$

$$\nabla_1^2 r_{12} = \frac{2}{r_{12}}, \tag{12}$$

$$\nabla_1 g_d(1) = \left(\frac{n_d-1}{r_1} - \alpha_d \right)g_d(1)\hat{\mathbf{r}}_1, \tag{13}$$

$$\nabla_1 r_{12} = \hat{\mathbf{r}}_{12}, \tag{14}$$

$$\hat{\mathbf{r}}_1 \cdot \hat{\mathbf{r}}_{12} = \frac{r_1^2 + r_{12}^2 - r_2^2}{2 r_1 r_{12}}, \tag{15}$$

$$0 = \nabla_1 g_d(1) \cdot \nabla_1 Y_l^m(1). \tag{16}$$

With these substitutions, Eq. (10) becomes

$$\nabla_1^2[r_{12}\phi_d(1)] = \left[\frac{n_d^2 - 1 - l_d(l_d + 1)}{r_1^2} - \frac{(2n_d + 1)\alpha_d}{r_1} + \alpha_d^2 \right] r_{12}\phi_d(1)$$

$$+ \left[n_d + 1 - \alpha_d r_1 + \frac{\alpha_d r_2^2}{r_1} - \frac{(n_d - 1)r_2^2}{r_1^2} \right] \frac{\phi_d(1)}{r_{12}}$$

$$+ 2 g_d(1) \hat{\mathbf{r}}_{12} \cdot \nabla_1 Y_{l_d}^{m_d}(1). \tag{17}$$

Equation (17) corrects a sign error that was present in the corresponding equation of [37].

One further simplification can now be easily made to the final term of Eq. (17): The orthogonality of $\hat{\mathbf{r}}_1$ and $\nabla_1 Y_{l_d}^{m_d}(1)$ permit us to write

$$2 g_d(1) \hat{\mathbf{r}}_{12} \cdot \nabla_1 Y_{l_d}^{m_d}(1) = -\frac{2 r_2 g_d(1)}{r_{12}} \hat{\mathbf{r}}_2 \cdot \nabla_1 Y_{l_d}^{m_d}(1). \tag{18}$$

As explained in more detail in [37], the properties of vector spherical harmonics can now be used to complete the evaluation of Eq. (18).

Introducing the complex unit vectors

$$\hat{\mathbf{e}}_1 = -\frac{\hat{\mathbf{x}} + i\hat{\mathbf{y}}}{\sqrt{2}}, \quad \hat{\mathbf{e}}_{-1} = \frac{\hat{\mathbf{x}} - i\hat{\mathbf{y}}}{\sqrt{2}}, \quad \hat{\mathbf{e}}_0 = \hat{\mathbf{z}}, \tag{19}$$

with orthogonality relation

$$\hat{\mathbf{e}}_\mu \cdot \hat{\mathbf{e}}_{-\nu} = (-1)^\nu \delta_{\mu\nu}, \tag{20}$$

it can be shown that

$$\hat{\mathbf{r}}_2 = \sqrt{\frac{4\pi}{3}} \sum_{\mu=-1}^{1} (-1)^\mu Y_1^{-\mu}(2) \hat{\mathbf{e}}_\mu, \tag{21}$$

$$\nabla_1 Y_{l_d}^{m_d}(1) = \frac{1}{r_1} \sum_{\nu=-1}^{1} \sum_{\lambda=\pm 1} \left(\frac{l_d(l_d + 1)(2l_d + 1 - \lambda)}{2(2l_d + 1)} \right)^{1/2}$$

$$\times \left\langle \begin{matrix} l_d + \lambda & 1 & l_d \\ m_d - \nu & \nu & m_d \end{matrix} \right\rangle Y_{l_d + \lambda}^{m_d - \nu}(1) \hat{\mathbf{e}}_\nu, \tag{22}$$

and that the final term of Eq. (17) reduces to

$$2g_d(1)\hat{\mathbf{r}}_{12}\cdot\nabla_1 Y_{l_d}^{m_d}(1) = -\frac{2r_2 g_d(1)}{r_1 r_{12}}\sqrt{\frac{4\pi}{3}}\sum_{v=-1}^{1}\sum_{\lambda=\pm 1}\left(\frac{l_d(l_d+1)(2l_d+1-\lambda)}{2(2l_d+1)}\right)^{1/2}$$
$$\times\left\langle\begin{matrix}l_d+\lambda & 1 & l_d\\ m_d-v & v & m_d\end{matrix}\right\rangle Y_{l_d+\lambda}^{m_d-v}(1)Y_1^v(2).$$

(23)

In these equations the array in angle brackets is a Clebsch-Gordan coefficient as defined in the Appendix; the notation we are using for it is not standard but the author hopes it will become more widely adopted.

The spherical harmonic $Y_1^v(2)$ will in overall computations occur multiplied by the harmonic contained in $\phi(2)$; that product can be reduced to a sum of single harmonics. This reduction takes the form

$$Y_1^v(2)Y_{l_e}^{m_e}(2) = \sum_{\lambda'=\pm 1}\begin{bmatrix}l_e & 1 & l_e+\lambda'\\ m_e & v & -m_e-v\end{bmatrix} Y_{l_e+\lambda'}^{m_e+v}(2),$$

(24)

where the array in brackets is a Gaunt coefficient (also in a nonstandard notation suggested by the author). There is unfortunately no single widely-accepted definition for the Gaunt coefficients; we use here that given in the Appendix, which has the properties of being both analogous to the definition of the Wigner 3-j symbol [39] and in agreement with the Gaunt-coefficient definition chosen by Pinchon and Hoggan [40].

Combining Eqs. (17), (18), and (24), we obtain a final expression containing no differential operators for $-\frac{1}{2}\nabla_1^2 r_{12}\phi_d(1)\phi_e(2)$; it is then straightforward to insert that expression into Eqs. (7) and (8). We display here the formula for K_3; that for K_2 is included in [37]:

$$K_3 = \frac{l_d(l_d+1)-n_d^2+1}{2}\left\langle\Phi_{abc}\left|\frac{r_{12}r_{13}}{r_1^2}\right|\Phi_{def}\right\rangle - \frac{\alpha_d^2}{2}\left\langle\Phi_{abc}\left|r_{12}r_{13}\right|\Phi_{def}\right\rangle$$
$$+\frac{(2n_d+1)\alpha_d}{2}\left\langle\Phi_{abc}\left|\frac{r_{12}r_{13}}{r_1}\right|\Phi_{def}\right\rangle - \frac{n_d+1}{2}\left\langle\Phi_{abc}\left|\frac{r_{13}}{r_{12}}\right|\Phi_{def}\right\rangle$$
$$+\frac{\alpha_d}{2}\left\langle\Phi_{abc}\left|\frac{r_1 r_{13}}{r_{12}}-\frac{r_2^2 r_{13}}{r_1 r_{12}}\right|\Phi_{def}\right\rangle + \frac{n_d-1}{2}\left\langle\Phi_{abc}\left|\frac{r_2^2 r_{13}}{r_1^2 r_{12}}\right|\Phi_{def}\right\rangle$$
$$+\sqrt{\frac{4\pi}{3}}\sum_{\lambda=\pm 1}\left(\frac{l_d(l_d+1)(2l_d+1-\lambda)}{2(2\lambda+1)}\right)^{1/2}\sum_{v=-1}^{1}(-1)^{m_e+v}\left\langle\begin{matrix}l_d+\lambda & 1 & l_d\\ m_d-v & v & m_d\end{matrix}\right\rangle$$
$$\times\sum_{\lambda'=\pm 1}\begin{bmatrix}l_e & 1 & l_e+\lambda'\\ m_e & v & -m_e-v\end{bmatrix}\left\langle\Phi_{abc}\left|\frac{r_2 r_{13}}{r_1 r_{12}}\right|d_{l_d+\lambda}^{m_d-v}e_{l_e+\lambda'}^{m_e+v}f_{l_f}^{m_f}\right\rangle.$$

(25)

This equation is a corrected form of the corresponding formula in [37].

A salient feature of the formula for K_3 is that it relates the matrix element for given angular quantum numbers to those of neighboring angular indices; this behavior is indeed to be expected because of the properties of the spherical harmonics.

The formula for K_2 (not shown but given in [37]) provides an alternative to a relation identified by Kolos and Roothaan [41]. However, the technique employed by Kolos and Roothaan does not extend to cover the three-body integral represented here as K_3.

7 Kinetic Energy in Extended Hylleraas-CI

A procedure similar to that outlined in Sect. 6 can be applied to the kinetic-energy matrix elements in E-Hy-CI. We summarize here some of the results; a more complete discussion will be published elsewhere [42].

One type of integral relevant here has the form

$$
K_3^{\text{EHCI}} = \left\langle r_{13}^{p'} e^{-\beta' r_{13}} \phi_a(1)\phi_b(2)\phi_c(3) \left| -\frac{1}{2}\nabla_1^2 \right| r_{12}^p e^{-\beta r_{12}} \phi_d(1)\phi_e(2)\phi_f(3) \right\rangle. \quad (26)
$$

In a process similar to that used for Hy-CI, we start by writing

$$
\begin{aligned}
\nabla_1^2 \left[r_{12}^p e^{-\beta r_{12}} \phi_d(1) \right] =& \left[r_{12}^p e^{-\beta r_{12}} \nabla_1^2 g_d(1) + \left(\nabla_1^2 \left[r_{12}^p e^{-\beta r_{12}} \right] \right) g_d(1) \right. \\
& \left. - \frac{l_d(l_d+1) r_{12}^p e^{-\beta r_{12}} g_d(1)}{r_1^2} + 2\nabla_1 g_d(1) \cdot \nabla_1 \left[r_{12}^p e^{-\beta r_{12}} \right] \right] Y_{l_d}^{m_d}(1) \\
& + 2 r_{12}^p e^{-\beta r_{12}} \nabla_1 g_d(1) \cdot \nabla_1 Y_{l_d}^{m_d}(1) + 2 g_d(1) \nabla_1 \left[r_{12}^p e^{-\beta r_{12}} \right] \cdot \nabla_1 Y_{l_d}^{m_d}(1),
\end{aligned} \quad (27)
$$

which differs from Eq. (10) only by replacement of r_{12} everywhere it occurs by $r_{12}^p e^{-\beta r_{12}}$. Examination of Eq. (27) shows that we now need to evaluate the new quantities

$$
\nabla_1^2 \left[r_{12}^p e^{-\beta r_{12}} \right] = \left(\frac{p(p+1)}{r_{12}^2} - \frac{2\beta(p+1)}{r_{12}} + \beta^2 \right) r_{12}^p e^{-\beta r_{12}}, \quad (28)
$$

$$
\nabla_1 \left[r_{12}^p e^{-\beta r_{12}} \right] = \left(\frac{p}{r_{12}} - \beta \right) r_{12}^p e^{-\beta r_{12}} \hat{\mathbf{r}}_{12}. \quad (29)
$$

Inserting the results from Eqs. (28) and (29) into Eq. (27) and then proceeding as in Sect. 6, we reach

$$K_3^{\text{EHCI}} = \frac{l_d(l_d+1)-(n_d+p)(n_d-1)}{2}\left\langle \Phi_{abc}\left|\frac{f_{12}f_{13}}{r_1^2}\right|\Phi_{def}\right\rangle$$

$$+\frac{(2n_d+p)\alpha_d}{2}\left\langle \Phi_{abc}\left|\frac{f_{12}f_{13}}{r_1}\right|\Phi_{def}\right\rangle - \frac{\alpha_d^2+\beta^2}{2}\left\langle \Phi_{abc}\left|f_{12}f_{13}\right|\Phi_{def}\right\rangle$$

$$-\frac{p(n_d+p)}{2}\left\langle \Phi_{abc}\left|\frac{f_{12}f_{13}}{r_{12}^2}\right|\Phi_{def}\right\rangle + \frac{\alpha_d p}{2}\left\langle \Phi_{abc}\left|\frac{f_{12}f_{13}r_1}{r_{12}^2}-\frac{f_{12}f_{13}r_2^2}{r_1 r_{12}^2}\right|\Phi_{def}\right\rangle$$

$$+\frac{(n_d-1)p}{2}\left\langle \Phi_{abc}\left|\frac{f_{12}f_{13}r_2^2}{r_1^2 r_{12}^2}\right|\Phi_{def}\right\rangle + \frac{\beta(n_d+2p+1)}{2}\left\langle \Phi_{abc}\left|\frac{f_{12}f_{13}}{r_{12}}\right|\Phi_{def}\right\rangle$$

$$-\frac{\alpha_d\beta}{2}\left\langle \Phi_{abc}\left|\frac{f_{12}f_{13}r_{12}}{r_1}+\frac{f_{12}f_{13}r_1}{r_{12}}-\frac{f_{12}f_{13}r_2^2}{r_1 r_{12}}\right|\Phi_{def}\right\rangle$$

$$+\frac{\beta(n_d-1)}{2}\left\langle \Phi_{abc}\left|\frac{f_{12}f_{13}r_{12}}{r_1^2}-\frac{r_2^2 f_{12}f_{13}}{r_1^2 r_{12}}\right|\Phi_{def}\right\rangle$$

$$+\sqrt{\frac{4\pi}{3}}\sum_{\lambda=\pm1}\left(\frac{l_d(l_d+1)(2l_d+1-\lambda)}{2(2l_d+1)}\right)^{1/2}\sum_{v=-1}^{1}(-1)^{m_e+v}\begin{pmatrix}l_d+\lambda & 1 & l_d\\ m_d-v & v & m_d\end{pmatrix}$$

$$\times\sum_{\lambda'=\pm1}\begin{bmatrix}l_e & 1 & l_e+\lambda'\\ m_e & v & -m_e-v\end{bmatrix}\left[p\left\langle \Phi_{abc}\left|\frac{f_{12}f_{13}r_2}{r_1 r_{12}^2}\right|d_{l_d+\lambda}^{m_d-v}e_{l_e+\lambda'}^{m_e+v}f_{l_f}^{m_f}\right\rangle\right.$$

$$\left.-\beta\left\langle \Phi_{abc}\left|\frac{f_{12}f_{13}r_2}{r_1 r_{12}}\right|d_{l_d+\lambda}^{m_d-v}e_{l_e+\lambda'}^{m_e+v}f_{l_f}^{m_f}\right\rangle\right].$$

$$(30)$$

To keep the above formula more compact, we have defined $f_{12} = r_{12}^p e^{-\beta r_{12}}$ and $f_{13} = r_{13}^{p'}e^{-\beta' r_{13}}$.

Equation (30) confirms that the kinetic-energy matrix elements in E-Hy-CI reduce to contiguous potential-energy integrals.

8 Numerical Verification

The formulas for kinetic-energy intergrals in Hy-CI developed here and (in more detail) in [37] were confirmed by comparing integrals produced using them with similar integrals computed in other ways by Ruiz [25, 27, 28] and by Sims and Hagstrom [10]. After making some adjustments needed to achieve consistency (see [37]), complete agreement with the results of those investigators was obtained.

The errors noted in various equations of [37] arose while transcribing the formulas from computer programs and therefore did not affect the numerical verification process.

We could not find literature values of E-Hy-CI kinetic-energy integrals for comparison with the formula in Sect. 7 of the present contribution and therefore cannot present data to provide its numerical confirmation.

9 Conclusions

This paper summarizes features of a variety of types of correlated-orbital atomic calculations and identifies new formulas yielding kinetic-energy matrix elements for the Hylleeraas-CI method of Sims and Hagstron and of Woźnicki and for the extended Hylleraas-CI method proposed by the Nakatsuji group.

Acknowledgements Completion of the numerical verifications referred to in this work involved significant consultations with Drs. María Belén Ruiz and James Sims. The author is pleased and grateful to acknowledge their assistance.

Appendix. Angular-Momentum Coefficients

The spherical harmonics $Y_l^m(\theta, \phi)$, alternatively written $Y_l^m(\Omega)$, can be defined with the sign convention chosen by Condon and Shortley [17] (**Condon-Shortley phase**) by the Rodrigues formula

$$Y_l^m(\Omega) = N_{lm} \frac{(-1)^m}{2^l l!} (1 - u^2)^{m/2} \frac{d^{l+m}}{du^{l+m}} (u^2 - 1)^l e^{im\phi}, \tag{31}$$

where $u = \cos\theta$ and N_{lm} is the factor

$$N_{lm} = \sqrt{\frac{(2l + 1)(l - m)!}{4\pi(l + m)!}} \tag{32}$$

that makes the Y_l^m orthonormal. With these definitions,

$$Y_l^m(\Omega)^* = (-1)^m Y_l^{-m}(\Omega). \tag{33}$$

A product of spherical harmonics of the same argument Ω can be expanded into a sum of harmonics of that argument. The coefficients in that expansion are known as **Gaunt coefficients**. Unfortunately there is no unanimity as to the definition of the Gaunt coefficient. Choosing the definition of Pinchon and Hoggan [40], we introduce a bracket notation that we hope will become adopted:

$$\begin{bmatrix} l_1 & l_2 & l_3 \\ m_1 & m_2 & m_3 \end{bmatrix} = \int Y_{l_1}^{m_1}(\Omega) Y_{l_2}^{m_2}(\Omega) Y_{l_3}^{m_3}(\Omega) \, d\Omega. \tag{34}$$

Expansion of the spherical harmonic product $Y_{l_1}^{m_1} Y_{l_2}^{m_2}$ in the orthonormal set Y_L^M, carried out by taking scalar products with $(Y_L^M)^*$, leads after use of Eqs. (33) and (34) to

$$Y_{l_1}^{m_1}(\Omega) Y_{l_2}^{m_2}(\Omega) = \sum_{LM} (-1)^M \begin{bmatrix} l_1 & l_2 & L \\ m_1 & m_2 & -M \end{bmatrix} Y_L^M(\Omega). \tag{35}$$

Because harmonics with upper indices m_1 and m_2 form a product all of whose terms have the same value of M, Eq. (35) can be simplified by dropping the M summation, setting $M = m_1 + m_2$.

The Gaunt coefficients can be written in terms of Wigner 3-j symbols [39]. Using the standard notation for that symbol (an array of l and m values in ordinary parentheses), the Gaunt coefficients as defined here can be written

$$\begin{bmatrix} l_1 & l_2 & l_3 \\ m_1 & m_2 & m_3 \end{bmatrix} = \sqrt{\frac{(2l_1 + 1)(2l_2 + 1)(2l_3 + 1)}{4\pi}} \begin{pmatrix} l_1 & l_2 & l_3 \\ m_1 & m_2 & m_3 \end{pmatrix} \begin{pmatrix} l_1 & l_2 & l_3 \\ 0 & 0 & 0 \end{pmatrix}. \tag{36}$$

A pair of angular momenta in two independent variables can be coupled to form a quantity of definite resultant angular momentum by forming a linear combination of products of the individual angular momenta; the coefficients in that expansion are called **Clebsch-Gordan** coefficients. Coupling of the angular-momentum wave functions $\psi_{j_1}^{m_1}(1)$ and $\psi_{j_2}^{m_2}(2)$ with fixed values of j_1 and j_2 to form the combined function $\Psi_J^M(1, 2)$ is described by

$$\Psi_J^M(1, 2) = \sum_{m_1 m_2} \left\langle \begin{matrix} j_1 & j_2 & J \\ m_1 & m_2 & M \end{matrix} \right\rangle \psi_{j_1}^{m_1}(1) \psi_{j_2}^{m_2}(2), \tag{37}$$

where the array in angle brackets is our (nonstandard) notation for the Clebsch-Gordan coefficient. Here all contributing terms must satisfy $m_1 + m_2 = M$, so we can actually reduce Eq. (37) to a single sum over, say, m_2, with m_1 set to $M - m_2$.

The Clebsch-Gordan coefficients can also be written in terms of 3-j symbols:

$$\left\langle \begin{matrix} j_1 & j_2 & j_3 \\ m_1 & m_2 & m_3 \end{matrix} \right\rangle = (-1)^{j_1 - j_2 + m_3} \sqrt{2j_3 + 1} \begin{pmatrix} j_1 & j_2 & j_3 \\ m_1 & m_2 & -m_3 \end{pmatrix}. \tag{38}$$

Well-documented computer programs exist for the evaluation of the 3-j symbols, making it straightforward to evaluate expressions involving Gaunt or Clebsch-Gordan coefficients.

References

1. Hylleraas EA (1929) Z Phys 54:347
2. James HM, Coolidge AS (1936) Phys Rev 49:688

3. Kolos W, Roothaan CCJ (1960) Rev Mod Phys 32:205
4. Larsson S (1968) Phys Rev 169:49
5. Kolos W (1994) J Chem Phys 101:1330
6. Sims JS, Hagstrom SA (1971) J Chem Phys 55:4699
7. Woźnicki W (1971) Theory of electronic shell in atoms and molecules, Jucys A (ed), p 103. Mintis, Vilnius
8. Pestka G, Woźnicki W (1996) Chem Phys Lett 255:281
9. Sims JS, Hagstrom SA (2004) J Phys B 37:1519
10. Sims JS, Hagstrom SA (2007) J Phys B 40:1575
11. Sims JS, Hagstrom SA (2015) J Phys B 48:175003
12. Bonham RA (1965) J Mol Spectrosc 15:112
13. Bonham RA (1966) J Mol Spectrosc 20:197
14. Rychlewski J, Cencek W, Komasa J (1994) Chem Phys Lett 229:657
15. Fromm DM, Hill RN (1987) Phys Rev A 36:1013
16. Wang C, Mei P, Kurokawa Y, Nakashima H, Nakatsuji H (2012) Phys Rev A 85:042512
17. Condon EU, Shortley G (1951) The theory of atomic spectra, Cambridge University Press
18. Frolov AM, Smith VH (2001) J Chem Phys 115:1187
19. Harris FE (1997) Phys Rev A 55:1820
20. Remiddi E (1991) Phys Rev A 44:5492
21. Pachucki K, Puchalski M, Remiddi E (2004) Phys Rev A 70:032502
22. Harris FE (2009) Phys Rev A 79:032517
23. Arfken GB, Weber HJ, Harris FE (2013) Math Methods Phys, 7th edn. Academic Press, New York, pp 799–800
24. Jen CK (1933) Phys Rev 43:540
25. Ruiz MB (2009) J Math Chem 46:24
26. Ruiz MB (2009) J Math Chem 46:1322
27. Ruiz MB (2011) J Math Chem 49:2457
28. Ruiz MB (2016) J Math Chem 54:1083
29. Szasz L (1962) Phys Rev 126:169
30. Levin D (1973) Int J Comput Math B 3:371
31. Frost AA (1962) Theor Chim Acta (Berl) 1:36–41
32. Harris FE (2004) Fundamental world of quantum chemistry: a tribute volume to the memory of Per-Olov Löwdin: Brändas EJ, Kryachko ES (eds), vol 3. Kluwer, Dordrecht, pp 115–128
33. Harris FE (2004) In Advances in Quant Chem: Sabin JR, Brändas EJ (eds), vol 47. Academic Press, New York, pp 129–155
34. Harris FE (2005) In Advances in Quant Chem: Sabin JR, Brändas EJ (eds), vol 50. Academic Press, New York, pp 61–75
35. Rebane TK (1993) Opt Spektrosk 75:945, 557
36. Harris FE, Frolov AM, Smith VH (2003) J Chem Phys 119:8833
37. Harris FE (2016) J Chem Phys 144:204110; Erratum (2016) 145:129901
38. Morse PM, Feshbach H (1953) Methods of theoretical physics. Part II. McGraw-Hill, New York, pp 1898–1901
39. Edmonds AR (1957) Ang Moment Quant Mech. Princeton University Press, Princeton, NJ
40. Pinchon D, Hoggan PE (2007) Int J Quant Chem 107:2186
41. Kolos W, Roothaan CCJ (1960) Rev Mod Phys 32:219
42. Harris FE (to be published)

Effective Bond-Strength Indicators

Gui-Xiang Wang, Yuzhe Stan Chen, Ya-Kun Chen
and Yan Alexander Wang

Abstract To save time and computer resources, we made an attempt to design reasonable yet simple structural indicators to identify weak chemical bonds, instead of performing numerous, tedious calculations of individual bond dissociation energies (BDEs) for all bonds within a molecule. Based on the commonly available structure-property indicators for bond strength, such as bond length (R), the Mulliken interatomic electron number (MIEN), the Wiberg bond order (WBO), and BDE, we have created two new bond-strength indicators, i.e., $M = \text{MIEN}/R$ and $K = (\text{WBO} \times \text{MIEN})/R^2$, which shall be directly used to efficiently identify almost all weak bonds with BDE below 350 kJ/mol. If several bonds of the same type attain the same smallest values of M or K, values of the electron density at the bond critical points (ρ_c) alone can almost always pinpoint the weakest bond from the set of weak bonds, greatly reducing the amount of efforts in carrying out the calculations of the BDEs of the corresponding bonds.

Keywords Structural indicator · Weakest bond · Weak bond
Bond energy · Bond dissociation energy · Bond order · Explosives
Trigger bond · Trigger bond indicator · Bond critical point · Quantum theory
of atoms in molecules · Population analysis · Counterpoise correction
Basis set superposition error

G.-X. Wang (✉)
Department of Chemistry, Computation Institute for Molecules and Materials,
Nanjing University of Science and Technology, Nanjing 210094, China
e-mail: wanggx1028@163.com

G.-X. Wang · Y. S. Chen · Y.-K. Chen · Y. A. Wang (✉)
Department of Chemistry, University of British Columbia, 2036 Main Mall,
Vancouver, BC V6T 1Z1, Canada
e-mail: yawang@chem.ubc.ca

© Springer International Publishing AG, part of Springer Nature 2018
Y. A. Wang et al. (eds.), *Concepts, Methods and Applications of Quantum Systems in Chemistry and Physics*, Progress in Theoretical Chemistry and Physics 31,
https://doi.org/10.1007/978-3-319-74582-4_3

1 Introduction

Bond dissociation energy (BDE) is important to the understanding of the mechanisms of chemical reactions and crucial to the design, synthesis, and performance studies of new functional materials. In the field of high-energy density explosives, the pyrolysis of explosive molecules is considered to be the critical step in the explosion process. Especially, the BDE of the weakest (trigger) bond plays a significant role in the initial reaction of explosion. Hence, identification of the trigger bond becomes mandatory in the studies of the safety and reliability of explosives.

For a homolytic bond breaking process [1–8],

$$A - B \xrightarrow{\text{bond breaking}} A^{\bullet} + B^{\bullet},$$

the BDE can be calculated according to the formula [9–11]:

$$
\begin{aligned}
\text{BDE}_{AB} = & \left[\tilde{\text{E}}(\text{ZP})_{A(B)}^{a \cup b} + \tilde{\text{E}}(\text{ZP})_{(A)B}^{a \cup b} - \text{E}(\text{ZP})_{AB}^{a \cup b} \right] \\
& + \left[\text{E}(\text{ZP})_{A}^{a} - \tilde{\text{E}}(\text{ZP})_{A}^{a} \right] + \left[\text{E}(\text{ZP})_{B}^{b} - \tilde{\text{E}}(\text{ZP})_{B}^{b} \right].
\end{aligned} \tag{1}
$$

Here, ZP means all energies (E and $\tilde{\text{E}}$) are corrected for the zero-point vibrational effect. The capital letters A and B refer to the corresponding molecular fragments produced after breaking the bond between A and B, and the small letters a and b stand for the basis sets attached to the fragments. Before the homolysis reaction happens, the parent compound optimized within the combined basis set ($a \cup b$) has energy $\text{E}(\text{ZP})_{AB}^{a \cup b}$. After the reaction occurs, A and B radicals optimized each individually within their own basis sets, a and b, have energies $\text{E}(\text{ZP})_{A}^{a}$ and $\text{E}(\text{ZP})_{B}^{b}$, respectively. To mitigate the inconsistency of the basis sets used before and after the reaction, the counterpoise correction [10, 11] for the basis-set superposition error (BSSE) has been adopted in Eq. (1), in which four more single-point energies with molecular fragments A and B frozen in their geometries within the parent molecule are calculated: $\tilde{\text{E}}(\text{ZP})_{A(B)}^{a \cup b}$ with ghost B, $\tilde{\text{E}}(\text{ZP})_{(A)B}^{a \cup b}$ with ghost A, $\tilde{\text{E}}(\text{ZP})_{A}^{a}$, and $\tilde{\text{E}}(\text{ZP})_{B}^{b}$. For a given ghost structure, the full sets of its basis functions and numerical integration grid points are still present as before, but there are neither nuclear charges nor electrons within the ghost. Therefore, both $\tilde{\text{E}}(\text{ZP})_{A(B)}^{a \cup b}$ and $\tilde{\text{E}}$ $(\text{ZP})_{(A)B}^{a \cup b}$ are still calculated within the total combined basis set ($a \cup b$).

Equation (1) clearly shows that for every bond breaking occurrence, an accurate estimate of the BDE involves at least four single-point calculations, two geometry optimizations, and six vibrational frequency analyses for the ZP correction. Hence, calculating the BDEs of all bonds of a large molecule to identify the weakest bond is very labor-intensive and time-consuming. It is thus highly desirable to design a simpler structural indicator to replace the enormous amount of BDE calculations. In conventional chemical wisdom, bond length (R), bond order (BO), and bond energy (BE) are the three commonly used key parameters reflecting the strength of

Table 1 Average bond lengths (R_a) and bond energies (BE_a) of various common single bonds [12]

Single bond	R_a (pm)	BE_a (kJ/mol)	BE_a/R_a (kJ mol^{-1} Å^{-1})	$1/R_a$ (Å^{-1})
H–H	74.14	436	588.1	1.349
O–H	97	464	478.4	1.031
N–H	100	389	389.0	1.000
C–H	110	414	376.4	0.909
Cl–H	127.4	431	338.3	0.785
S–H	132	368	278.8	0.758
Br–H	141.4	364	257.4	0.707
C–O	143	360	251.7	0.699
C–N	147	305	207.5	0.680
C–C	154	347	225.3	0.649
I–H	160.9	297	184.6	0.622
C–Cl	178	339	190.4	0.562
Cl–Cl	199	243	122.1	0.503
Br–Br	228	193	84.6	0.439
I–I	266	151	56.8	0.376
F–H	91.7	565	616.1	1.091
N–N	145	163	112.4	0.690
N–O	136	222	163.2	0.735
O–O	145	142	97.9	0.690
F–F	143	159	111.2	0.699

a chemical bond. Empirically, there exists a good quantitative relationship between R, BO, and BE: the longer R, the less BE or the lower BO a bond has, the easier the bond breaks [12]. Thus, R or BO might be directly used to identify the trigger bond, without relying upon BE.

Along this line of reasoning, the experimental average bond lengths (R_a) and bond energies (BE_a) sampled over many different chemical species containing 20 most common single bonds are gathered in Table 1 [12]. As a whole, the data collected in Table 1 indeed validate the general observation: the longer R a bond has, the smaller BE is, i.e., the bond breaks more easily. The linear regression between BE_a/R_a and $1/R_a$ is drawn in Fig. 1 with a correlation coefficient 0.99. Despite the fact that specific bonds of the same type come with many different variations in R and BE due to local chemical environment, it is still very reassuring in discovering such a strong linear correlation between BE_a/R_a (in kJ mol^{-1} Å^{-1}) and $1/R_a$ (in Å^{-1}):

$$\left(\frac{BE_a}{R_a}\right) = 568.59 \left(\frac{1}{R_a}\right) - 150.91. \tag{2}$$

Accordingly, it might be a good idea to directly employ R to identify the weakest bond.

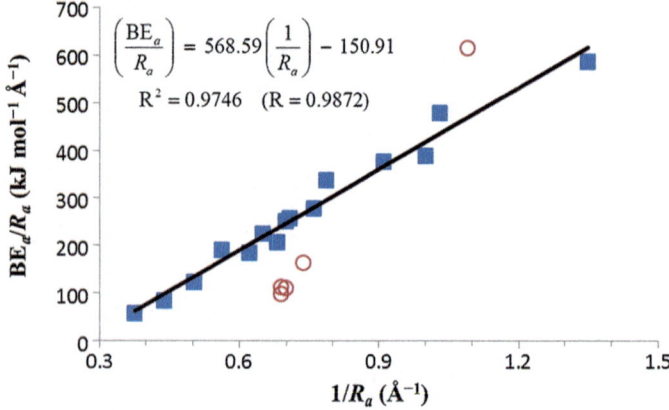

Fig. 1 Relationship between average bond energies (BE_a) and bond length (R_a). The straight line is a least-square linear fit to the data points denoted by blue squares; data points marked by red circles are excluded from the linear fit. All data are collected in Table 1

On the other hand, many previous studies [4–7] have primarily used the Mulliken interatomic electron number (MIEN), calculated via the Mulliken population analysis [13], to identify the trigger bond. Normally, only the bond with the smallest MIEN from the same type of bonds is selected for further consideration [4–7]. However, the bond with the smallest MIEN might not necessarily have the lowest BDE among the bonds within a molecule. Taking 2,4,6-trinitrotoluene (TNT) for example, the MIENs of all $C–NO_2$ bonds are among the smallest (see Table 2), indicating one of the $C–NO_2$ bonds to be the trigger bond. However, to single out which $C–NO_2$ bond to break first, calculations of the BDEs of these specific $C–NO_2$ bonds have been carried out. In Table 2, the $C_3–N_{11}$ bond has the smallest MIEN (0.1468), whereas the $C_1–N_{14}$ or $C_5–N_8$ bond has the lowest BDE, 237.75 kJ/mol. Clearly, the smallest MIEN may not correspond to the lowest BDE.

Alternatively, within the quantum theory of atoms in molecules [14, 15], Bader proposed several parameters to indicate the relative bond strength using the idea associated with bond critical points (BCPs). These BCPs are located at the inter-atomic surface between a pair of atoms, at which the electron density reaches minimum in one dimension, yet reaches maximum in the other two dimensions. As Bader originally suggested, the value of the electron density at such a bond critical point between a pair of atoms of a chemical bond, ρ_c, can be used to measure the strength of the chemical bond. On the other hand, the BE is an integral of the electron density over the associated interatomic surface between an atomic pair. Unfortunately, this BE integral has an *unknown* system-dependent dimensionless proportionality pre-factor. As a result, Bader's ρ_c indicator can only be used to compare the same kind of bonds between the same pair of atoms within very similar local chemical environment and might not be able to evaluate the strengths of different kinds of bonds even within the same molecule.

Table 2 Bond strength indicators of selected explosive molecules

Molecule	Bond	R^a	WBO[b]	MIEN[c]	W^d	M^e	K^f	ρ_c^g	BDE[h]
TNT	C_1–N_{14}	1.481	0.9138	0.1504	**0.62**	**0.10**	**0.06**	0.2621	**237.75**
	C_3–N_{11}	1.475	0.9214	**0.1468**	**0.62**	**0.10**	**0.06**	0.2653	265.01
	C_5–N_8	1.481	0.9138	0.1504	**0.62**	**0.10**	**0.06**	0.2621	**237.75**
	C_6–C_7	**1.509**	1.0375	0.3819	0.69	0.25	0.17	**0.2502**	403.65
	C_7–H_{19}	1.089	0.9062	0.3559	0.83	0.33	0.27	0.2758	360.77
	C_7–H_{20}	1.089	0.9062	0.3559	0.83	0.33	0.27	0.2758	360.77
	C_7–H_{21}	1.094	**0.8851**	0.3611	0.81	0.33	0.27	0.2699	361.26
TNM	C_1–N_{16}	1.481	**0.8880**	0.1532	**0.60**	**0.10**	**0.06**	0.2614	**232.21**
	C_3–N_{12}	1.478	0.8984	0.1458	0.61	**0.10**	**0.06**	0.2625	233.19
	C_5–N_8	1.477	0.9009	**0.1447**	0.61	**0.10**	**0.06**	0.2629	232.24
	C_2–C_{15}	**1.511**	1.0278	0.3611	0.68	0.24	0.16	0.2498	404.02
	C_4–C_{11}	**1.511**	1.0286	0.3762	0.68	0.25	0.17	**0.2495**	403.30
	C_6–C_7	**1.511**	1.0282	0.3709	0.68	0.25	0.17	0.2499	403.41
	C_7–H_{19}	1.095	0.8910	0.3614	0.81	0.33	0.27	0.2705	365.27
	C_{15}–H_{26}	1.089	0.9085	0.3518	0.83	0.32	0.27	0.2764	366.21
PNT	C_2–N_{19}	1.483	**0.8881**	0.1484	**0.60**	0.10	0.06	0.2659	208.11
	C_3–N_{16}	1.484	0.8904	0.1284	**0.60**	**0.09**	**0.05**	0.2671	200.62
	C_4–N_{13}	1.484	0.8890	0.1313	**0.60**	**0.09**	**0.05**	0.2671	**197.65**
	C_5–N_{10}	1.484	0.8912	**0.1263**	**0.60**	**0.09**	**0.05**	0.2669	200.63
	C_6–N_7	1.485	0.8856	0.1533	**0.60**	0.10	0.06	0.2650	208.40
	C_1–C_{22}	**1.510**	1.0284	0.3673	0.68	0.24	0.17	**0.2493**	419.52
	C_{22}–H_{23}	1.095	0.8878	0.3600	0.81	0.33	0.27	0.2704	365.39
	C_{22}–H_{24}	1.094	0.9039	0.3674	0.83	0.34	0.28	0.2714	365.21
	C_{22}–H_{25}	1.088	0.9033	0.3526	0.83	0.32	0.27	0.2778	365.00
TNCr	C_2–N_8	1.479	0.8908	**0.1555**	**0.60**	**0.11**	**0.06**	0.2623	**235.36**
	C_4–N_{10}	1.431	1.0545	0.1707	0.74	0.12	0.09	0.2779	294.13
	C_6–N_{12}	1.455	0.9805	0.1574	0.67	**0.11**	0.07	0.2694	250.81
	C_5–N_{11}	1.337	1.3579	0.3829	1.02	0.29	0.29	0.3413	434.30
	C_1–C_7	**1.511**	1.0321	0.3656	0.68	0.24	0.17	**0.2497**	386.34
	C_3–O_9	1.322	1.1728	0.3734	0.89	0.28	0.25	0.3223	453.39
	C_7–H_{19}	1.094	0.8892	0.3588	0.81	0.33	0.27	0.2699	367.69
	C_7–H_{20}	1.090	0.9054	0.3655	0.83	0.34	0.28	0.2742	367.31
	C_7–H_{21}	1.089	0.9044	0.3461	0.83	0.32	0.26	0.2770	367.12
	O_9–H_{22}	0.997	**0.5986**	0.2117	**0.60**	0.21	0.13	0.3106	391.96
	N_{11}–H_{23}	1.011	0.7385	0.2746	0.73	0.27	0.20	0.3264	424.61
	N_{11}–H_{24}	1.013	0.7360	0.2695	0.73	0.27	0.19	0.3245	405.53
CDNAPY	C_2–N_8	1.466	0.9397	0.1335	0.64	**0.09**	**0.06**	0.2695	**248.56**
	C_4–N_9	1.452	0.9729	**0.1237**	0.67	**0.09**	**0.06**	0.2730	278.67
	C_5–N_{10}	1.339	1.3124	0.3491	0.98	0.26	0.26	0.3415	455.55
	C_1–Cl_7	**1.738**	1.0973	0.3046	**0.63**	0.18	0.11	**0.2057**	325.22
	C_3–H_{15}	1.082	0.8624	0.3366	0.80	0.31	0.25	0.2841	470.41
	N_{10}–H_{16}	1.011	**0.7516**	0.2808	0.74	0.28	0.21	0.3255	435.63
	N_{10}–H_{17}	1.010	0.7875	0.2942	0.78	0.29	0.23	0.3284	434.71

(continued)

Table 2 (continued)

Molecule	Bond	R^a	WBO[b]	MIEN[c]	W^d	M^e	K^f	ρ_c^g	BDE[h]
PAM	C_2-N_7	**1.477**	0.9191	**0.1475**	0.62	**0.10**	**0.06**	1.1958	**217.54**
	C_4-N_9	**1.477**	0.9191	**0.1475**	0.62	**0.10**	**0.06**	1.1958	**217.54**
	C_6-N_{10}	1.474	0.9241	0.1501	0.63	**0.10**	**0.06**	1.2627	262.69
	C_3-O_8	1.340	1.0529	0.3040	0.79	0.23	0.18	1.3156	354.45
	O_8-C_{17}	1.453	**0.8424**	0.2252	**0.58**	0.16	0.09	1.0756	263.82
	$C_{17}-H_{20}$	1.090	0.9267	0.3626	0.85	0.33	0.30	0.9198	398.18
	$C_{17}-H_{21}$	1.090	0.9267	0.3626	0.85	0.33	0.28	0.9198	398.18
	$C_{17}-H_{22}$	1.090	0.9305	0.3856	0.85	0.35	0.28	**0.9158**	398.18
AMNA	N_2-N_4	1.422	0.9807	**0.1603**	0.69	**0.11**	**0.08**	0.3241	**107.87**
	N_2-N_3	1.397	1.0575	0.2451	0.76	0.18	0.13	0.3306	248.42
	C_1-N_2	**1.463**	0.9571	0.2703	**0.65**	0.18	0.12	**0.2575**	233.11
	C_1-H_7	1.088	0.9131	0.3737	0.84	0.34	0.29	0.2807	376.14
	C_1-H_8	1.099	0.9240	0.3729	0.84	0.34	0.29	0.2718	376.41
	C_1-H_9	1.090	0.9141	0.3713	0.84	0.34	0.29	0.2789	376.59
	N_3-H_{10}	1.018	**0.8192**	0.3215	0.81	0.32	0.25	0.3333	315.18
	N_3-H_{11}	1.024	0.8231	0.3109	0.80	0.30	0.24	0.3274	314.42
AMNFMC	$N_7=N_8$	1.244	1.4789	0.2938	1.19	0.24	0.28	0.4313	141.55
	N_5-N_{11}	1.449	0.9167	**0.1381**	0.63	**0.10**	**0.06**	0.3031	**124.16**
	C_1-O_3	1.361	0.9845	0.2689	0.72	0.20	0.14	0.3014	326.62
	C_1-N_5	1.414	0.9998	0.1857	0.71	0.13	0.09	0.2963	320.40
	O_3-C_4	1.443	**0.8441**	0.1903	**0.58**	0.13	0.08	**0.2393**	264.75
	C_4-N_7	**1.454**	0.9896	0.2696	0.68	0.19	0.13	0.2683	267.37
	C_4-H_{15}	1.093	0.9075	0.3719	0.83	0.34	0.28	0.2853	358.88
	N_5-C_6	1.444	0.9583	0.2199	0.66	0.15	0.10	0.2701	330.94
	C_6-F_{10}	1.378	0.8574	0.2802	0.62	0.20	0.13	0.2465	418.79
NMDACB	C_1-N_2	**1.479**	0.9456	0.2486	**0.64**	0.17	0.11	**0.2590**	196.21
	C_1-N_3	**1.479**	0.9685	0.3084	0.66	0.21	0.14	0.2704	170.88
	C_1-H_{10}	1.098	**0.9004**	0.3552	0.82	0.32	0.27	0.2774	367.45
	N_2-N_6	1.389	0.9756	**0.1878**	0.70	**0.14**	**0.10**	0.3469	**157.49**
	N_3-C_5	1.455	1.0026	0.3014	0.69	0.21	0.14	0.2704	294.30
	C_5-H_{14}	1.106	0.9212	0.3525	0.83	0.32	0.27	0.2666	378.11

[a] R is the bond length (in Å)

[b] WBO is the Wiberg bond order, defined in Refs. [21, 22]

[c] MIEN is the Mulliken interatomic electron number, defined in Ref. [13]

[d] W = WBO/R (in Å^{-1}), rounded to the second decimal place for optimal sensitivity

[e] M = MIEN/R (in Å^{-1}), rounded to the second decimal place for optimal sensitivity

[f] K = $W \times M$ (in Å^{-2}), rounded to the second decimal place for optimal sensitivity

[g] ρ_c is the electron density at the bond critical point within Bader's atoms-in-molecules analysis, defined in Refs. [14, 15]

[h] BDE (in kJ/mol) is the bond dissociation energy calculated according to Eq. (1)

To sample various single bonds within diverse chemical environment, we have chosen nine typical explosive molecules (see Fig. 2 for their structural diagrams) with varied functional groups and geometric shapes, including cyclic 1-nitro-3-methyl-1,3-diazacyclobutane (NMDACB), branched *N*-amino-*N*-methyl-nitramine (AMNA), linear and branched 1-azide methyl-*N*-nitro-*N*-fluoro methyl carbamate (AMNFMC), and aromatic TNT and its five derivatives: pentanitrotoluene (PNT), 1-methyl-3-hydroxy-6-amino-2,4,6-trinitrobenzene (TNCr), 1-methoxy-2,4,6-trinitrobenzene (PAM), 2-chloro-3,5-dinitro-6-aminopyridine (CDNAPY), and 1,3,5-trimethyl-2,4,6-trinitrobenzene (TNM).

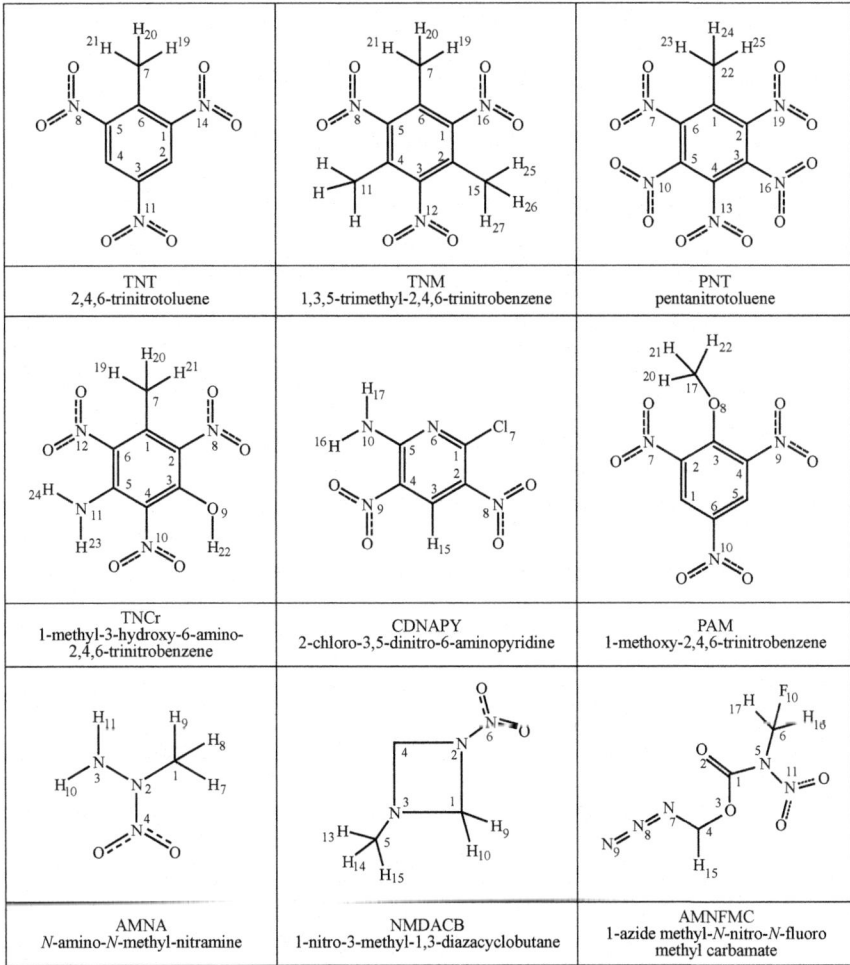

Fig. 2 Molecular structures of selected explosive compounds (some hydrogens omitted for clarity)

Hereafter, the five key bond-strength indicators (i.e., R, MIEN, BO, ρ_c, and BDE) of all single bonds of these molecules will be carefully compared to seek better relationships among them. The end goal is to propose simple, sensitive structural indicators suitable for identifying the trigger bond (or even better, all weak bonds), thus saving both computational resources and time.

2   Computational Methods

Many studies [1–8, 16, 17], have already shown that the DFT-B3LYP method [18, 19] in combination with the 6-31G* basis set [20] is able to yield accurate energetics, structures, and other molecular properties. In this paper, the same method was employed to obtain the fully optimized molecular geometries and electronic structures, including MIEN and Wiberg bond order (WBO) [21, 22], of the chosen compounds (Fig. 2) within the Gaussian09 program package [23]. To obtain the values of ρ_c, the Bader analysis was performed using AIMAll package [24].

Based on Hess's law [25], conventional counterpoise correction methods [10, 11] were employed to calculate the BDEs according to Eq. (1). Unfortunately, any existing counterpoise correction methods [10, 11] cannot address the situation when a bond within a ring is broken. Instead of trying to contemplate suitable ghost atoms and their linking strategies for ring opening scenarios, we simply supplemented extra diffuse polarization basis functions to the existing basis sets for the atoms of the broken bonds to roughly mimic the effects of ghost atoms. Taking the four-membered ring of NMDACB for example, when breaking the C_1-N_2 and C_1-N_3 bonds, the BSSE was calculated with additional aug-cc-pV5Z Diffuse ($1s$, $1p$, $1d$, $1f$, $1g$, $1h$) basis functions [26] separately placed on the C_1, N_2, and N_3 atoms. Numerical tests confirmed that the magnitude of the resulting correction to the BDEs from this basis-set local enhancement protocol was in line with the procedure of existing counterpoise correction methods [10, 11].

3   Results and Discussion

Four commonly available major bond-strength indicators (i.e., R, MIEN, WBO, and ρ_c) are listed along with BDEs in Table 2. For the selected compounds, nearly all the bonds with the longest R or the smallest WBO do not possess the lowest BDE. The bond with the lowest BDE is the $C-NO_2$ or $N-NO_2$ bond, but, according to R and WBO, the weakest bond can draw from all sorts of candidates, e.g., C–C, C–H, $C-NO_2$, C–Cl, N–H, or O–C bond. Similarly, the smallest values of ρ_c alone never correspond to the weakest bonds either. This obviously illustrates that R, WBO, or ρ_c cannot be used separately to identify the trigger bond.

Particularly, because of AMNFMC containing the azido group ($-N_7=N_8{}^+=N_9{}^-$), when the $N_7=N_8$ bond is broken to produce the N_2 gas, its BDE is the second lowest among all bonds (merely 141.55 kJ/mol). It is comforting that none of the structural parameters considered here indicates the $N_7=N_8$ bond to be the weakest. However, this also suggests that whenever exotic bonds (not listed in Table 1) are involved in a bond breaking process, their BDEs should be calculated to verify the weakest bond for certainty.

Data in Table 2 also show that MIEN alone can correctly single out the type of bonds to which the trigger bond belongs. For instance, for the selected compounds, the easiest bond to break is the $C-NO_2$ or $N-NO_2$ bond, but among which MIEN sometimes incorrectly picks up the weakest bond. For CDNAPY, the C_4-N_9 bond has the smallest MIEN (0.1237) and its BDE is 278.67 kJ/mol, but it is the C_2-N_8 bond that has the lowest BDE (248.56 kJ/mol) even with a bigger MIEN (0.1335). Therefore, much sensible structural parameters other than the existing four afore-mentioned must be proposed for trigger bond identification.

In light of the general fact that WBO and MIEN are related positively to BDE and inversely to R, $W = WBO/R$ and $M = MIEN/R$ were first conceived to be the better alternatives. Such an idea comes with no surprise, because the numerator of the $1/R_a$ term on the right-hand side of Eq. (2) can be interpreted as $WBO = 1$ for single bonds. Then, to take local chemical environment into consideration, Eq. (2) might suggest a general relationship: $BDE/R \propto WBO/R$. (Hereafter, numerical values of bond-strength indicators will be quoted without units unless otherwise noted.)

Unfortunately, the data shown in Table 2 again proclaim that the minima of W and BDE still do not align well. For example, in PAM, the O_8-C_{17} bond has the smallest W, but the C_2-N_7 or C_4-N_9 bond has the lowest BDE. In general, the results of W closely resemble those of WBO.

Fortunately, the bonds with the smallest M enclose the bond with the lowest BDE (see Table 2). For TNT, the C_1-N_{14}, C_3-N_{11}, and C_5-N_8 bonds all have the same smallest M value (0.10) and both of the C_1-N_{14} and C_5-N_8 bonds have the lowest BDE. For PNT, the C_3-N_{16}, C_4-N_{13}, and C_5-N_{10} bonds have the smallest M (0.09) and, among them, the C_4-N_{13} bond is the weakest (BDE = 197.65 kJ/mol). We then considered the product of W and M, $K = W \times M$, as a new indicator. According to M, for TNCr, the weakest bonds are the C_2-N_8 and C_6-N_{12} bonds ($M = 0.11$), whereas according to K, the weakest bond of TNCr is solely the C_2-N_8 bond ($K = 0.06$) in agreement with its lowest BDE value. Overall, K outperforms M only marginally (see Fig. 3).

Next, we observed an enlightening fact: ρ_c can almost always single out the weakest bond among the weaker bonds of the same type within a molecule iden-tified by M and K (see Table 2). The only exception might be PNT: among the three most weak bonds (C_3-N_{16}, C_4-N_{13}, and C_5-N_{10}) sorted out by M and K, ρ_c chooses the C_5-N_{10} bond (BDE = 200.63 kJ/mol) to be the weakest whereas the C_4-N_{13} bond has the lowest BDE (197.65 kJ/mol). However, this is not a huge failure because the difference in the BDEs of the three weak bonds of PNT is less than 3 kJ/mol, well within the error margin of the computational methods employed. Nonetheless, such promising results strongly advocate that M and K, as two new trigger bond indicators (TBIs), shall be first used to identify the set of most weak

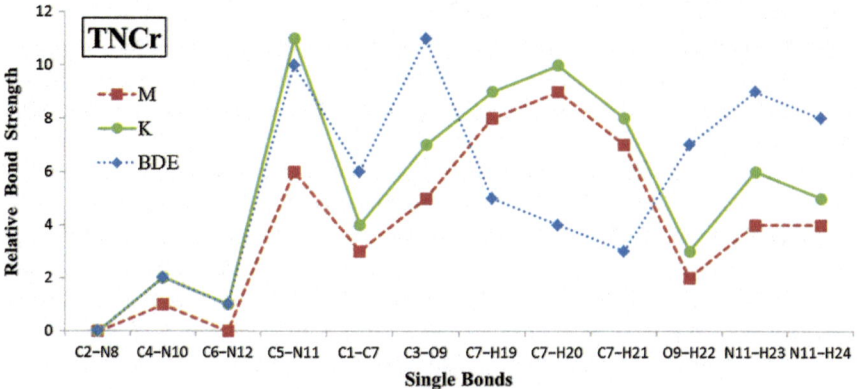

Fig. 3 Relative strengths of all single bonds within TNCr measured by bond-strength indicators: *M*, *K*, and BDE

bonds and ρ_c can subsequently narrow down the candidates whose BDEs can then be calculated to pinpoint the trigger bond among them. This can really reduce the amount of work drastically.

In general, the smaller TBI values almost perfectly map to all *weak* bonds (with BDE < 350 kJ/mol) sequentially, and both *M* and *K* do indeed capture the weakest bond once these two TBIs reach minimum (see Table 2). In spite of such a general success of these two TBIs, neither *M* nor *K* can always predict the relative order of bond strength for *strong* bonds (with BDE > 350 kJ/mol) within a molecule. Figure 4 showcases this point succinctly. Such a phenomenon is well expected because the molecular fragments after breaking a strong bond normally undergo

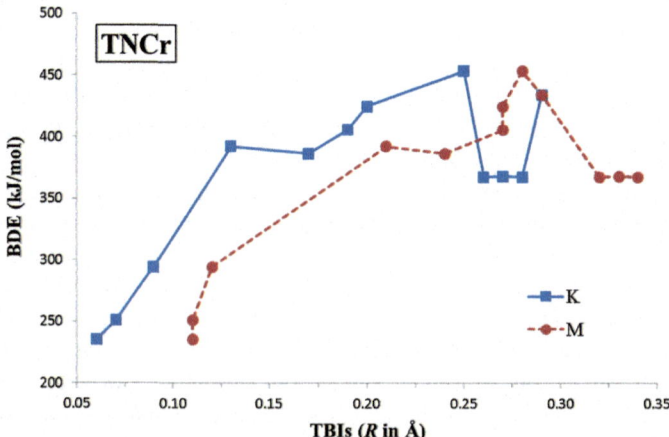

Fig. 4 Comparison of TBIs (*K* and *M*) and BDE of all single bonds within TNCr

very large structural rearrangement, which cannot be reasonably modeled by structural indicators of the parent molecule alone.

We thus have successfully designed two new structural indicators, M and K, for the identification of the trigger bond and weak bonds. If a compound has multiple weak bonds of the same type, ρ_c can help determine the weakest bond. The utilization of these two new structural indicators will become utterly efficient and effective especially for very large systems with hundreds of bonds.

4 Conclusions

In conclusion, ordinary bond-strength indicators, including R, MIEN, WBO, ρ_c, and $W = WBO/R$, are not suitable to be used alone to identify the trigger bond, whereas $M = MIEN/R$ and the better $K = W \times M$ are the indicators of choice for sorting out weak bonds (with BDE < 350 kJ/mol). When there are several bonds carrying the same smallest M or K, calculations of their ρ_c and BDEs must be subsequently performed to determine the weakest bond. Especially, these two new indicators, M and K, will become extremely helpful for analyzing bond breaking processes within large molecular systems by greatly minimizing computational efforts.

Acknowledgements We are grateful to the grant support from the Natural Sciences and Engineering Research Council (NSERC) of Canada, the National Natural Science Foundation of China (No. 21403110), and the Natural Science Foundation of Jiangsu Province (No. BK20130755). This work was mainly carried out at UBC during G.X.W.'s one-year visit to UBC from 16 August 2012 to18 August 2013.

References

1. Xu XJ, Xiao HM, Ju XH, Gong XD, Zhu WH (2006) Computational studies on polynitrohexaazaadmantanes as potential high energy density materials (HEDMs). J Phys Chem A 110:5929–5933
2. Qiu L, Xiao HM, Gong XD, Ju XH, Zhu WH (2006) Ab initio and molecular dynamics studies of crystalline TNAD (trans-1,4,5,8-Tetranitro-1,4,5,8-tetraazadecalin). J Phys Chem A 110.3797–3807
3. Qiu LM, Gong XD, Wang GX, Zheng J, Xiao HM (2009) Looking for high energy density compounds among 1,3-Bishomopentaprismane derivatives with –CN, –NC, and –ONO$_2$ groups. J Phys Chem A 113:2607–2614
4. Xu XJ, Xiao HM, Gong XD, Ju XH, Chen ZX (2005) Theoretical studies on the vibrational spectra, thermodynamic properties, detonation properties and pyrolysis mechanisms for polynitroadamantanes. J Phys Chem A 109:11268–11274
5. Wang GX, Shi CH, Gong XD, Xiao HM (2009) Theoretical investigation on structures, density, detonation properties and pyrolysis mechanism of the derivatives of HNS. J Phys Chem A 113:1318–1326

6. Wang GX, Gong XD, Yan L, Du HC, Xu XJ, Xiao HM (2010) Theoretical studies on the structures, density, detonation properties, pyrolysis mechanisms and impact sensitivity of nitro derivatives of toluenes. J Hazard Mater 177:703–710

7. Wang GX, Gong XD, Du HC, Liu Y, Xiao HM (2011) Theoretical prediction of properties of aliphatic polynitrates. J Phys Chem A 115:795–804

8. Liu Y, Gong XD, Wang LJ, Wang GX, Xiao HM (2011) Substituent effects on the properties related to detonation performance and sensitivity for 2,2',4,4',6,6'-Hexanitroazobenzene derivatives. J Phys Chem A 115:1754–1762

9. Jensen F (2007) Introduction to computational chemistry, 2nd edn. Wiley, Chichester, West Sussex, England

10. Boys SF, Bernardi F (1970) The calculation of small molecular interactions by the differences of separate total energies. Some procedures with reduced errors. Mol Phys 19:553–566

11. Simon S, Duran M, Dannenberg JJ (1996) How does basis set superposition error change the potential surfaces for hydrogen bonded dimers? J. Chem. Phys. 105:11024–11031

12. Petrucci RH, Herring FG, Madura JD, Bissonette C (2011) General chemistry, 10th edn. Pearson Publishing, Toronto, Ontario, Canada

13. Mulliken RS (1962) Criteria for the construction of good self-consistent-field molecular orbital wave functions, and the significance of LCAO-MO population analysis. J. Chem. Phys. 36:3428–3439

14. Bader RFW, Tang TH, Tal Y, Biegler-Koenig FW (1982) Properties of atoms and bonds in hydrocarbon molecules. J Am Chem Soc 104:946–952

15. Bader RFW (1994) Atoms in molecules: a quantum theory. Clarendon Press, Oxford, UK

16. Xiao HM, Xu XJ, Qiu L (2008) Theoretical design of high energy density materials. Science Press, Beijing, China

17. Wang GX, Gong XD, Liu Y, Du HC, Xu XJ, Xiao HM (2011) Looking for high energy density compounds applicable for propellant among the derivatives of DPO with $-N_3$, $-ONO_2$, and $-NNO_2$ groups. J Comput Chem 32:943–952

18. Lee C, Yang W, Parr RG (1988) Development of the Colle-Salvetti correlation-energy formula into a functional of the electron density. Phys. Rev. B 37:785–789

19. Becke AD (1992) Density-functional thermochemistry, II. The effect of the Perdew-Wang generalized-gradient correlation correction. J. Chem. Phys. 97:9173–9177

20. Hariharan PC, Pople JA (1973) Influence of polarization functions on MO hydrogenation energies. Theor. Chim. Acta 28:213–222

21. Wiberg KB (1968) Application of the Pople-Santry-Segal CNDO method to the cyclopropy-lcarbinyl and cyclobutyl cation and to bicyclobutane. Tetrahedron 24:1083–1096

22. Mayer I (1983) Change bond order and valence in the ab initio SCF theory. Chem Phys Lett 97:270–274

23. Frisch MJ, Trucks GW, Schlegel HB, Scuseria GE, Robb MA, Cheeseman JR, Barone V, Mennucci B, Cossi M, Scalmani G, Petersson GA, et al (2009) Gaussian 09, revision A. 02, Gaussian, Inc., Wallingford, CT

24. Bader RFW Atoms in molecules, McMaster University, Hamilton, Ontario, Canada. (http://www.chemistry.mcmaster.ca/bader/aim/ Accessed 3 Jan 2016)

25. Engel T, Reid P, Hehre W (2013) Physical chemistry, 3rd edn. Boston, USA, Pearson

26. Kendall RA, Dunning TH, Harrison RJ (1992) Electron affinities of the first-row atoms revisited. Systematic basis sets and wave functions. J. Chem. Phys. 96:6796–6806

Advanced Relativistic Energy Approach in Electron-Collisional Spectroscopy of Multicharged Ions in Plasmas

Alexander V. Glushkov, Vasily V. Buyadzhi, Andrey A. Svinarenko and Eugeny V. Ternovsky

Abstract We present the fundamentals of an advanced relativistic approach, based on the Gell-Mann and Low formalism, to studying spectroscopic characteristics of the multicharged ions in plasmas, in particular, computing the electron-ion collision strengths, cross-sections etc. The approach is combined with relativistic many-body perturbation theory with the Debye shielding model Hamiltonian for electron-nuclear and electron-electron systems. The optimized one-electron representation in the perturbation theory zeroth approximation is constructed by means of the correct treating the gauge dependent multielectron contribution of the lowest perturbation theory corrections to the radiation widths of atomic levels. The computation results on the oscillator strengths and energy shifts due to the plasmas environment effect, the electron-collision strengths, collisional excitation and de-excitation rates for a number of the Be- and Ne-like ions of argon, nickel and krypton embedded to different types of plasmas environment (with temperature 0.02–2 keV and electron density 10^{16}–10^{24} cm^{-3}) are presented and analyzed.

Keywords Electron-collisional processes · Multicharged ions Relativistic energy approach · Debye plasmas

1 Introduction

An accurate data about spectra, radiative decay widths and probabilities, oscillator strengths, electron-collision strengths, collisional excitation and de-excitation rates for atoms and especially ions are of a great interest for different applications, namely, astrophysical analysis, laboratory, thermonuclear plasmas diagnostics, fusion research, laser physics etc. [1–30]. It is also very important for studying energy, spectral and radiative characteristics of a laser-produced hot and dense plasmas [1, 2, 9, 10]. Above other important factors to studying electron-collisional

A. V. Glushkov (✉) · V. V. Buyadzhi · A. A. Svinarenko · E. V. Ternovsky
Odessa State Environmental University, L'vovskaya str., 15, Odessa 65016, Ukraine
e-mail: glushkovav@gmail.com

© Springer International Publishing AG, part of Springer Nature 2018   55
Y. A. Wang et al. (eds.), *Concepts, Methods and Applications of Quantum Systems in Chemistry and Physics*, Progress in Theoretical Chemistry and Physics 31,
https://doi.org/10.1007/978-3-319-74582-4_4

spectroscopy of ions one should mention the X-ray laser problem. It has stimulated a great number of papers, devoted to modelling the elementary processes in laser, collisionally pumped plasmas (see [3, 4] and Refs. therein) and construction of the first VUV and X-ray lasers with using plasmas of Li-, Ne-like ions as an active medium. Very useful data on the X-lasers problem are collected in the papers by Ivanova et al. (see [3–6] and Refs. therein). From the other side, studying spectra of ions in plasmas remains very actual in order to understand the plasmas processes themselves. In most plasmas environments the properties are determined by the electrons and the ions, and the interactions between them. The electron-ion collisions play a major role in the energy balance of plasmas. For this reason, modelers and diagnosticians require absolute cross sections for these processes. The cross sections for electron-impact excitation of ions are needed to interpret spectroscopic measurements and for simulations of plasmas using collisional-radiative models. At present time a considerable interest has been encapsulated to studying elementary atomic processes in plasmas environments (for example, see [1–30] and Refs. therein) because of the plasmas screening effect on the plasmas-embedded atomic systems. In many papers the calculations of various atomic and ionic systems embedded in the Debye plasmas have been performed [11, 12, 16, 29, 30]. Calculation of emission spectra of the plasmas ions based in the precise theoretical techniques is practical tool, which may be used instead of very expensive sophisticated experiments. Nevertheless, there are known principal theoretical problems to be solved in order to receive the correct description of master parameters of the elementary atomic processes in laser, collisionally pumped plasmas. First of all, speech is about development of the advanced quantum-mechanical models for the further accurate computing oscillator strengths, electron-collisional strengths and rate coefficients for atomic ions in plasmas, including the Debye plasmas. As usually, a correct accounting of the relativistic, exchange-correlation, a plasmas environment effects is of a great importance. To say strictly, solving of the whole problem requires a development of the quantum-electrodynamical approach as the most consistent one to problem of the Coulomb many-body system.

In this chapter we present the fundamentals of an advanced relativistic energy approach, based on the Gell-Mann and Low formalism, to studying spectroscopic characteristics of the multicharged ions in the Debye plasmas, in particular, computing the electron-ion collision strengths, cross-sections etc. The approach is combined with relativistic many-body perturbation theory (PT) with the Debye shielding model Hamiltonian for electron-nuclear and electron-electron systems. The optimized one-electron representation in the PT zeroth approximation is constructed by means of the correct treating the gauge dependent multielectron contribution of the lowest PT corrections to the radiation widths of atomic levels. It is worth to remind that the method of the relativistic many-body PT formalism is constructed on the base of the same ideas as the well-known PT approach with the model potential zeroth approximation by Ivanov-Ivanova et al. [31–44]. However there are a few fundamental differences. For example, in our case the PT zeroth approximation [51, 54] is in fact the Dirac- Debye-Hückel one. In order to calculate the radiative and collisional parameters an effective gauge-invariant version of relativistic energy approach is used [37, 45–54]. It is important to remind that a model relativistic energy approach in a case of a multielectron atom has been developed by Ivanov-Ivanova et al. [33–36]. A generalized gauge-invariant version of relativistic energy approach in a case of the multielectron atomic systems has been developed by Glushkov-Ivanov-Ivanova (see Refs. [37–39]). Earlier we have presented the fundamentals of an advanced generalized energy approach and its application to many actual problems of modern atomic, nuclear and even molecular optics and spectroscopy, including, spectroscopy of atoms in a photon vacuum and an external electromagnetic (laser) field, optics of the cooperative electron-gamma-nuclear "shake-up" processes (including processes of the NEET and NEEC: "Nuclear Excitation—Electron Transition", "Nuclear Excitation—Electron Capture"), electron-muon-beta-gamma-nuclear spectroscopy, etc. (see [55–88] and Refs. therein). Below we present and analyze the computation results on the oscillator strengths and energy shifts due to the plasmas environment effect, the electron-collision strengths, collisional excitation and de-excitation rates for a number of the Be- and Ne-like ions of argon, nickel and krypton embedded to different types of plasmas environment with the temperature 0.02–2 keV and the electron density 10^{16}–10^{24} cm^{-3}.

2   Relativistic Many-Body Perturbation Theory and Relativistic Energy Approach in Scattering Theory

2.1   Formalism of the Relativistic Perturbation Theory with Dirac-Debye Shielding Model Zeroth Approximation

Let us start our consideration from formulation relativistic many-body PT with the Debye shielding model Dirac Hamiltonian for electron-nuclear and electron-electron systems. Formally, a multielectron atomic systems (multielectron atom or multicharged ion) is described by the relativistic Dirac Hamiltonian (the atomic units are used) as follows:

$$H = \sum_i h(r_i) + \sum_{i>j} V(r_i r_j). \tag{1}$$

Here, $h(r)$ is one-particle Dirac Hamiltonian for electron in a field of a nucleus and V is potential of the inter-electron interaction.

According to Refs. [35–37] it is useful to determine the interelectron potential with accounting for the retarding effect and magnetic interaction in the lowest order on parameter α^2 (α is the fine structure constant) as follows:

$$V(r_i r_j) = exp(i\omega_{ij} r_{ij}) \cdot \frac{(1 - \alpha_i \alpha_j)}{r_{ij}}, \tag{2}$$

where ω_{ij} is the transition frequency; α_i, α_j are the Dirac matrices.

In order to take into account the plasmas environment effects already in the PT zeroth approximation we use the known Yukawa-type potential of the following form:

$$V(r_i, r_j) = (Z_a Z_b / |r_a - r_b|) exp(-\mu \cdot |r_a - r_b|) \tag{3}$$

where r_a, r_b represent respectively the spatial coordinates of particles, say, A and B and Z_a, Z_b denote their charges.

The potential (3) is (look, for example, [23–28] and Refs therein) well known, for example, in the classical Debye-Hückel theory of plasmas. The plasmas environment effect is modelled by the shielding parameter μ, which describes a shape of the long-rang potential. The parameter μ is connected with the plasmas parameters such as temperature T and the charge density n as follows:

$$\mu \sim \sqrt{e^2 n / k_B T}. \tag{4a}$$

Here e is the electron charge and k_B is the Boltzman constant. The density n is given as a sum of the electron density N_e and the ion density N_k of the k-th ion species with the nuclear charge q_k:

$$n = N_e + \sum_k q_k^2 N_k. \tag{4b}$$

It is very useful to remind the simple estimates for the shielding parameter. For example, under typical laser plasmas conditions of $T \sim 1$ keV and n $\sim 10^{23}$ cm^{-3} the parameter μ is of the order of 0.1 in atomic units. By introducing the Yukawa-type electron-nuclear attraction and electron-electron repulsion potentials, the Dirac-Debye shielding model Hamiltonian for electron-nuclear and electron-electron subsystems is given in atomic units as follows [28]:

$$H = \sum_i [\alpha c p - \beta m c^2 - Z \, exp(-\mu r_i)/r_i] + \sum_{i>j} \frac{(1 - \alpha_i \alpha_j)}{r_{ij}} exp(-\mu r_{ij}), \tag{5}$$

where c is the velocity of light and Z is a charge of the atomic ion nucleus.

The formalism of the relativistic many-body PT is further constructed in the same way as the PT formalism in Refs. [31–44]. In the PT zeroth approximation one should use a mean-field potential, which includes the Yukawa-type potential (insist of the pure Coulomb one) plus exchange Kohn-Sham potential and additionally the modified Lundqvist-Gunnarsson correlation potential (with the optimization parameter b) as in Refs. [28–30, 49, 50]. As alternative one could use an optimized model potential by Ivanova-Ivanov (for Ne-like ions) [31], which is calibrated within the special ab initio procedure within the relativistic energy approach [37, 38].

The most complicated problem of the relativistic PT computing the radiative and collisional characteristics of the multielectron atomic systems is in an accurate, precise accounting for the exchange-correlation effects (including polarization and screening effects, a continuum pressure etc.) as the effects of the PT second and higher orders. Using the standard Feynman diagram technique one should consider two kinds of diagrams (the polarization and ladder ones), which describe the polarization and screening exchange-correlation effects. The polarization diagrams take into account the quasiparticle (external electrons or vacancies) interaction through the polarizable core, and the ladder diagrams account for the immediate quasiparticle interaction. The detailed description of the polarization diagrams and the corresponding analytical expressions for matrix elements of the polarization quasiparticles interaction (through the polarizable core) potential are presented in Refs. [36, 45–50]. An effective approach to accounting for the polarization diagrams contributions is in adding the effective two- quasiparticle polarizable operator into the PT first order matrix elements. In Ref. [36] the corresponding non-relativistic polarization functional has been derived. More correct relativistic expression has been presented in the Refs. [45, 46, 49].

2.2 Generalized Relativistic Energy Approach in a Scattering Problem

In order to calculate different characteristics such as oscillator strengths and energy shifts due to the plasmas environment effect, the electron-collision strengths, collisional excitation and de-excitation rates etc. we use an advanced generalized relativistic energy approach combined with the relativistic many-body PT [28, 29, 49, 50, 67–70]. Here we briefly present the key moments of the method.

In the theory of non-relativistic atom a convenient field procedure is known for calculating the energy shifts ΔE of degenerate states. This procedure is connected with the secular matrix M diagonalization [33, 37]. In constructing M, the Gell-Mann and Low adiabatic formula for ΔE is used. The secular matrix elements are already complex in the PT second order (the first order on the inter-electron interaction). Their imaginary parts are connected with the radiation decay possibility. It is important to note that the computing the energies and radiative transition matrix elements is reduced to calculation and the further diagonalization of the complex matrix M and determination of matrix of the coefficients with eigen state vectors $B_{ie,iv}^{IK}$ [33–36, 41, 42]. To calculate all necessary matrix elements one must use the basis of the one-quasiparticle relativistic functions. In many calculations of the atomic elementary processes characteristics it has been shown that their adequate description requires using the optimized wave functions and an accurate accounting for the exchange-correlation effects. In Ref. [37] it has been proposed "ab initio" optimization principle for construction of an effective one-quasiparticle representation. The minimization of the gauge dependent multielectron contribution of the lowest QED PT corrections to the radiation widths of atomic levels, which is determined by the imaginary part of an energy shift ΔE, is used. In the fourth order of QED PT there appear diagrams, whose contribution into the $\mathrm{Im}\Delta E$ accounts for the polarization effects. This contribution describes collective effects and it is dependent upon the electromagnetic potentials gauge (the gauge non-invariant contribution ΔE_{ninv}). This value is considered to be the typical representative of the electron correlation effects, whose minimization is a reasonable criteria in the searching for the optimal one-electron basis of the PT. Let us note that this topic is of a great importance (look, for example, Refs. [49, 50, 89–98], where there are presented some alternative optimization approaches).

It is worth to remind that E. Davidson had pointed the principal disadvantages of the traditional representation based on the self-consistent field approach and suggested the optimal "natural orbitals" representation (for example, see Refs. [89, 90]). Our procedure derives an undoubted profit in the routine spectroscopic calculations as it provides the way of the refinement of the atomic characteristics calculations, based on the "first principles". The resulting expression looks as the correction due to the additional nonlocal interaction of the active quasiparticle with the closed shells. Nevertheless, its calculation is reducible to the solving of the system of the ordinary differential equations (one-D procedure) [49]. The most important refinements can be introduced by accounting for the relativistic and the density gradient corrections to the Tomas- Fermi formula (look details in Refs. [49, 50]). The minimization of the functional $\mathrm{Im}\ \Delta E_{ninv}$ leads to the integral differential equation, that is numerically solved. In result one can get the optimal one-electron representation of the PT, which is further improved within the Dirac-Kohn-Sham-Sturm approach in order to take into account for the continuum states [49, 50, 60, 61, 96–98].

As some ideas of the energy approach in application to a scattering problem have been presented in a literature (look, for example, [37, 41, 42], below we concern the most principal points. Further for definiteness, let us consider a collisional de-excitation of, say, the Ne-like ion [41]:

$$((2j_{iv})^{-1}3j_{ie}[J_iM_i],\varepsilon_{in}) \rightarrow (\Phi_o,\varepsilon_{sc}).$$

Here Φ_o is the state of the ion with the closed shells (ground state of the Ne-like ion); J_i is the total angular moment of the initial target state; indices iv, i.e. are related to the initial states of a vacancy and an electron; indices ε_{in} and ε_{sc} are the incident and scattered energies, respectively to the incident and scattered electrons. The initial state of the system "atom plus free electron" can be written as

$$|I> = a_{in}^+ \sum_{m_{iv},m_{ie}} a_{ie}^+ a_{iv}\Phi_o C_{m_{ie},m_{iv}}^{J_i,M_i} \tag{6a}$$

where $C_{m_{ie},m_{iv}}^{J_i,M_i}$ is the Clebsh-Gordan coefficient.

The final state is as follows:

$$|F> = a_{sc}^+ \Phi_o, \tag{6b}$$

where Φ_o is the state of an ion with closed electron shells (ground state of Ne-like ion), $|I>$ represents three-quasiparticle (3QP) state, and $|F>$ represents the one-quasiparticle (1QP) state.

The justification of the energy approach in the scattering problem is in details described in Refs. [38, 41–44, 49]. The scattered part of energy shift Im ΔE appears firstly in the atomic PT second order (the fourth order of the QED PT) in the form of integral over the scattered electron energy ε_{sc}:

$$\int d\varepsilon_{sc} G(\varepsilon_{iv}, \varepsilon_{ie}, \varepsilon_{in}, \varepsilon_{sc})/(\varepsilon_{sc} - \varepsilon_{iv} - \varepsilon_{ie} - \varepsilon_{in} - i0) \tag{7}$$

$$\mathrm{Im}\Delta E = \pi G(\varepsilon_{iv}, \varepsilon_{ie}, \varepsilon_{in}, \varepsilon_{sc}). \tag{8a}$$

Here G is a definite squired combination of the two-electron matrix elements (2). As usually, the value

$$\sigma = -2 \, Im\Delta E \tag{8b}$$

represents the collisional cross-section if the incident electron eigen-function is normalized by the unit flow condition and the scattered electron eigen-function is normalized by the energy δ function.

The collisional de-excitation cross section can be further defined as follows:

$$\sigma(IK \to 0) = 2\pi \sum_{j_{in}, j_{sc}} (2j_{sc} + 1)\{ \sum_{j_{ie}, j_{iv}} <0|j_{in}, j_{sc}|j_{ie}, j_{iv}, J_i > B_{ie, iv}^{IK}\}^2 \tag{9}$$

The amplitude like combination in (9) has the following form:

$$<0|j_{in}, j_{sc}|j_{ie}, j_{iv}, J_i> = \sqrt{(2j_{ie} + 1)(2j_{iv} + 1)}(-1)^{j_{ie} + 1/2} \times \sum_\lambda (-1)^{\lambda + J_i} \times$$

$$\times \{\delta_{\lambda, J_i}/(2J_i + 1)Q_\lambda(sc, ie; iv, in) + \begin{bmatrix} j_{in} \cdots j_{sc} \cdots J_i \\ j_{ie} \cdots j_{iv} \cdots \lambda \end{bmatrix} Q_\lambda(ie; in; iv, sc)\} \tag{10a}$$

$$Q_\lambda = Q_\lambda^{\text{Coul} - \text{Yuk}} + Q_\lambda^{Br}, \tag{10b}$$

where $Q_\lambda^{\text{Coul} - \text{Yuk}} + Q_\lambda^{Br}$ is the sum of the Coulomb-Yukawa and Breit matrix elements. The Coulomb part $Q_\lambda^{\text{Coul} - \text{Yuk}}$ contains the radial R_λ and angular S_λ integrals as follows:

$$Q_\lambda^{Coul-Yuk} = \{R_\lambda(1243)S_\lambda(1243) + R_\lambda(\tilde{1}243)S_\lambda(\tilde{1}243)$$
$$+ R_\lambda(1\tilde{2}43)S_\lambda(1\tilde{2}43) + R_\lambda(\tilde{1}\tilde{2}43)S_\lambda(\tilde{1}\tilde{2}43)\}. \tag{11}$$

Here the tilde designates that the large radial Dirac component f must be replaced by the small Dirac component g, and instead of l_i, $\tilde{l}_i = l_i - 1$ should be taken for $j_i < l_i$ and $\tilde{l}_i = l_i + 1$ for $j_i > l_i$.

The Breit part can be expressed as follows [39]:

$$Q_\lambda^{Br} = Q_{\lambda,\,\lambda-1}^{Br} + Q_{\lambda,\,\lambda}^{Br} + Q_{\lambda,\,\lambda+1}^{Br} \tag{12}$$

$$Q_\lambda^{Br} = \{R_\lambda(1243)S_\lambda^l(1243) + R_\lambda(\tilde{1}243)S_\lambda^l(\tilde{1}243) +$$
$$+ R_l(1\tilde{2}43)S_\lambda^l(1\tilde{2}43) + R_l(\tilde{1}\tilde{2}43)S_\lambda^l(\tilde{1}\tilde{2}43)\}. \tag{13}$$

The details of their computing can be found in Refs. [36–54]. The modified PC code 'Superatom-ISAN" (version-93) has been used in all calculations.

3 Results and Conclusions

Here we present the results of computing the radiative and collisional characteristics (energy shifts, oscillator strengths, electron-ion cross-sections and collision strengths) for the Be-, Ne-like ions of Ar, Ni and Kr ($Z = 18$–36) embedded to the plasmas environment. Let us remind (see Refs. [11, 12, 16, 28, 39]) that the Be- and Ne-like ions play an important role in the diagnostics of a wide variety of laboratory, astrophysical, thermonuclear plasmas. Firstly, we list our results on energy shifts and oscillator strengths for transitions $2s^2$-$2s_{1/2}2p_{1/2,3/2}$ in spectra of the Be-like Ni and Kr. The plasmas parameters are as follows: $n_e = 10^{22}$–10^{24}cm^{-3}, $T = 0.5$–2 keV (i.e. $\mu \sim 0.01$–0.3). In Tables 1 and 2 we list the results of

Table 1 Energy shifts ΔE (cm^{-1}) for the $2\,s^2$-$[2s_{1/2}2p_{3/2}]_1$ transition in spectra of the Be-like Ni and Kr ions for different values of the n_e (cm^{-3}) and T (in eV) (see explanations in text)

	n_e	10^{22}	10^{23}	10^{24}	10^{22}	10^{23}	10^{24}
Z	kT	Li et al.	Li et al.	Li et al.	Our data	Our data	Our data
NiXXV	500	31.3	292.8	2639.6	33.8	300.4	2655.4
	1000	23.4	221.6	2030.6	25.7	229.1	2046.1
	2000	18.0	172.0	1597.1	20.1	179.8	1612.5
	I-S	8.3	86.6	870.9			
KrXXXIII	500	21.3	197.9	2191.9	27.2	215.4	2236.4
	1000	15.5	150.5	1659.6	21.3	169.1	1705.1
	2000	11.5	113.5	1268.0	16.9	128.3	1303.8

Table 2 Oscillator strengths gf for the $2 s^2$-$[2s_{1/2}2p_{3/2}]_1$ transition in spectra of the Be-like ion of Ni for different values of the n_e (cm^{-3}) and T (in eV) (gf$_0$–the gf value for free ion)

n_e		10^{22}	10^{23}	10^{24}	10^{22}	10^{23}	10^{24}
kT	gf$_0$ Li et al.	gf Li et al.	gf Li et al.	gf Li et al.	gf$_0$:our data	gf: our data	gf: our data	gf: our data
500	0.1477	0.1477	0.1478	0.1487	0.1480	0.1480	0.1483	0.1495
1000		0.1477	0.1477	0.1482		0.1480	0.1483	0.1495
2000		0.1477	0.1477	0.1481		0.1479	0.1482	0.1493
I-S		0.1477	0.1477	0.1479				

calculation of the energy shifts ΔE (cm^{-1}) for $2s^2$-$[2s_{1/2}2p_{1/2,3/3}]_1$ transitions and oscillator strengths changes for different plasmas parameters such as the electron density n_e and temperature T.

There are also presented the available theoretical data by Li et al. and Saha-Frische: the multiconfiguration Dirac-Fock (DF) computation results and ionic sphere (I-S) model simulation data (from [11, 12, 16] and Refs. therein). The analysis shows that the presented data are in physically reasonable agreement, however, some difference can be explained by using different relativistic orbital basis and different models for accounting of the plasmas screening effect. From the physical point of view, the behavior of the energy shift is naturally explained, i.e. by increasing blue shift of the line because of the increasing the plasmas screening effect.

Further we present the results of computing the electron-collisional cross-sections and electron-collision strengths for Ne-like ion of Ar (the part of results has been presented in Refs. [28], but without the plasmas screening effect) and compare with the known theoretical data: relativistic model potential PT (RMPPT), relativistic optimized DF PT (ODFPT) [28–30, 41, 42].

In Table 3 we list the electron-collision strengths for Ne-like argon excitation from the ground state (E = 0.75 keV is the impact electron energy). The corresponding plasmas parameters (θ-pinch plasmas) are as follows: $n_e = 10^{16}$ cm^{-3}, and $T_e = 65$ eV.

It should be noted that the experimental information about the electron-collisional cross-sections for high-charged Ne-like ions is very scarce and is extracted from indirect observations. Such experimental information for a few collisional excitations of the Ne-like barium ground state has been presented in Refs. [17–19].

Let us note that the PT first order correction is calculated exactly, the high-order contributions are taken into account for effectively: polarization interaction of two above-core quasi-particles and an effect of their mutual screening (correlation effects). It is interesting to note that here the plasmas effects do not play a critical quantitative role.

Further we present the results of studying collisional characteristics for the Ne-like ions in the collisionally pumping plasmas with the parameters $T_e = 20$–40 eV and density $n_e = 10^{19-20}$ cm^{-3}. This system represents a great interest for

Table 3 The electron-collision strengths for Ne-like Ar excitation from the ground state for impact electron energy 0.75 keV (numbers in brackets denote the multiplicative powers of ten)

Transition	Level	J	[41]	[29]	Present data
01–02	2s 2p	0	1,303[−03]	1,415[−03]	1,498[−03]
3	$2p_{3/2}3s_{1/2}$	1	9,017[−03]	9,224[−03]	9,286[−03]
4	$2p_{1/2}3s_{1/2}$	0	2,587[−04]	2,724[−04]	2,783[−04]
5	$2p_{1/2}3s_{1/2}$	1	2,241[−02]	2,342[−02]	2,394[−02]
6	$2p_{3/2}3p_{3/2}$	1	3,456[−03]	3,635[−03]	3,699[−03]
7	$2p_{3/2}3p_{3/2}$	3	2,911[−03]	2,998[−03]	3,065[−03]
8	$2p_{3/2}3p_{1/2}$	2	4,795[−03]	4,922[−03]	4,988[−03]
9	$2p_{3/2}3p_{1/2}$	1	1,033[−03]	1,213[−03]	1,254[−03]
10	$2p_{3/2}3p_{3/2}$	2	6,451[−03]	6,535[−03]	6,597[−03]
11	$2p_{1/2}3p_{1/2}$	1	9,641[−04]	9,993[−04]	1,088[−03]
12	$2p_{1/2}3p_{1/2}$	0	8,794[−04]	8,927[−04]	8,992[−04]
13	$2p_{1/2}3p_{3/2}$	2	7,814[−03]	7,978[−03]	8,113[−03]
14	$2p_{1/2}3p_{3/2}$	1	8,561[−04]	8,723[−04]	9,005[−04]
15	$2p_{3/2}3p_{3/2}$	0	8,670[−02]	8,735[−02]	8,802[−02]
16	$2p_{3/2}3d_{3/2}$	0	1,136[−03]	1,244[−03]	1,296[−03]
17	$2p_{3/2}3d_{3/2}$	1	4,129[−03]	4,327[−03]	4,389[−03]
18	$2p_{3/2}3d_{5/2}$	2	5,227[−03]	5,546[−03]	5,601[−03]
19	$2p_{3/2}3d_{5/2}$	4	3,512[−03]	3,678[−03]	3,714[−03]
20	$2p_{3/2}3d_{3/2}$	3	3,994[−03]	4,133[−03]	4,185[−03]

generation of laser radiation in the short-wave spectral range [3, 4]. Besides, it is obviously more complicated case in comparison with previous one. Here an accurate account of the excited, Rydberg, autoionization and continuum states can play a critical role.

In Table 4 we present the theoretical values of the collisional excitation rates (CER) and collisional de-excitation rates (CDR) for Ne-like argon transition between the Rydberg states and from the Rydberg states to the continuum states with

Table 4 The collisional excitation (CER) and de-excitation (CDR) rates (in cm^3/s) for Ne-like argon in plasmas with parameters: $n_e = 10^{19-20}$ cm^{-3} and electron temperature $T_e = 20$ eV

Parameters	n_e, cm^{-3}	RMPPT	RMPPT	RMPPT	Present results	Present results	Present results
Transition		$1 \to 2$	$1 \to 3$	$2 \to 3$	$1 \to 2$	$1 \to 3$	$2 \to 3$
CDR ($i \to i$; k)	1.0 + 19	5.35 − 10	1.64–10	1.13 − 09	5.77 − 10	1.92 − 10	1.28 − 09
	1.0 + 20	5.51 − 10	1.60 − 10	1.12 − 09	5.94 − 10	1.78 − 10	1.25 − 09
Transition		$2 \to 1$	$3 \to 1$	$3 \to 2$	$2 \to 1$	$3 \to 1$	$3 \to 2$
CER ($i \to i$; k)	1.0 + 19	5.43 − 10	5.39 − 12	2.26 − 11	5.79 − 10	1.88 − 12	2.64 − 11
	1.0 + 20	3.70 − 10	8.32 − 12	2.30 − 11	4.85 − 10	1.13 − 11	2.78 − 11

Fig. 1 The Rydberg states zones (Ne-like ion: [Ne,i], nl); ε_0 is the boundary of the thermalized zone, neighboring to continuum; ε_3 is the ionization potential for states $nl = 3s$; $\varepsilon_i = (\varepsilon_0 + \varepsilon_{i+1})/2$, i = 1, 2

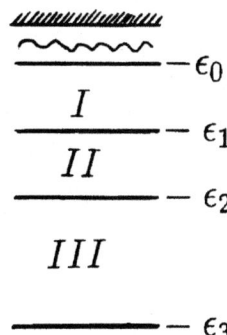

parameters: $n_e = 10^{19-20}$ cm^{-3} and electron temperature $T_e = 20$ eV (see details in Refs. [28, 42]). For comparison there are also listed the data by Ivanov et al., obtained within the RMPPT approach (without the shielding effect) [28, 41, 42].

Here we talk about the Rydberg states which converge to the corresponding lower boundary of continuum $-\varepsilon_0$ (see Fig. 1).

As it is indicated in Ref. [42], the parameter $-\varepsilon_0$ is the third parameter of the plasmas environment (together with electron density and temperature). In fact it defines the thermalized energy zone of the Rydberg and autoionization states which converge to the ionization threshold for each ion in a plasmas. Usually value ε_0 can be barely estimated from simple relation: $\varepsilon_0 = 0.1 \cdot T_e$.

In the consistent theory the final results must not be dependent on the model parameters, so the concrete value of ε_0 is usually chosen in such way that an effect of its variation in the limits $[0.01 \cdot T_e, 0.1 \cdot T_e]$ (for Ne-like ions) does not influence on the final results.

In Table 5 we present the theoretical values of the collisional excitation (CER) and de-excitation (CDR) rates (in cm^3/s) for Ne-like argon in plasmas with the parameters: $n_e = 10^{19-20}$ cm^{-3} and electron temperature $T_e = 40$ eV. Analysis of the presented data allows to conclude that the shielding effects play a definite role for the Debye plasmas. From other side, an account for the highly-lying excited states is quantitatively important for the adequate description of the collision cross-sections.

Table 5 The collisional excitation (CER) and de-excitation (CDR) rates (in cm^3/s) for Ne-like argon in plasmas with parameters: $n_e = 10^{19-20}$ cm^{-3} and electron temperature $T_e = 40$ eV (our data)

Parameters	n_e, cm^{-3}	Present results	Present results	Present results
Transition		$1 \rightarrow 2$	$1 \rightarrow 3$	$2 \rightarrow 3$
CDR (i $\rightarrow$ i;k)	1.0 + 19	3.18 − 10	8.45 − 11	6.81 − 10
	1.0 + 20	5.02 − 10	1.56 − 10	4.99 − 10
Transition		$2 \rightarrow 1$	$3 \rightarrow 1$	$3 \rightarrow 2$
CER (i $\rightarrow$ i;k)	1.0 + 19	5.33 − 10	5.63 − 10	7.11 − 11
	1.0 + 20	7.67 − 10	6.94 − 11	8.93 − 11

The calculations encourage us to believe that using energy approach combined with the relativistic many-body PT with the optimal one-electron basis is quite consistent and effective tool from the point of view of the theory correctness and results exactness. This fact was surely confirmed by other calculations of the oscillator strengths, radiative widths, hyperfine structure constants for atoms and multicharged ions (see Refs. [28–30, 49–54]).

To conclude, we have presented an effective quantum approach in electron-collisional spectroscopy of the multicharged ions in plasmas to compute the cross sections and other characteristics of the elementary collisional processes. It is based on the generalized relativistic energy approach and relativistic optimized many-body PT with the Debye shielding model Hamiltonian for electron-nuclear and electron-electron systems. The optimized one-electron representation in the PT zeroth approximation is constructed by means of the correct treating the gauge dependent multielectron contribution of the lowest PT corrections to the radiation widths of atomic levels. It is important to note that an approach is universal and, generally speaking, can be applied to quantum systems of other nature (see, for example, [57–66] and Refs. therein). Its application is especially perspective when the experimental information about corresponding properties and systems is very scarce. We have presented the illustrative results of studying spectra of some multicharged ions (Be-and Ne-like ions) in plasmas and computing the electron-ion collision strengths, cross-sections etc. The obtained data can be used in different applications, namely, astrophysical analysis, laboratory, thermonuclear plasmas diagnostics, fusion research, laser physics, quantum electronics etc.

Acknowledgements The authors are very much thankful to Prof. Jean Maruani and Prof. Alex Wang for invitation to make contributions on the QSCP-XXI workshop (Vancouver, Canada). The useful comments of the anonymous referees are very much acknowledged too.

References

1. Oks E (2010) In: Oks E, Dalimier E, Stamm R, Stehle C, Gonzalez MA (eds) Spectral line shapes in plasmas and gases. Int J Spectr 1:852581
2. Griem HR (1974) Spectral line broadening by plasmas. Academic Press, New York
3. Ivanova EP (2011) Phys Rev A 84:043829
4. Ivanova EP, Grant IP (1998) J Phys B Mol Opt Phys 31:2871; Ivanova EP, Zinoviev NA (2001) Phys Lett A 274:239
5. Khetselius OYu (2011) Quantum structure of electroweak interaction in heavy finite fermi-systems. Astroprint, Odessa
6. Glushkov AV (1991) Opt Spectrosc 70:555
7. Malinovskaya SV, Glushkov AV, Khetselius OYu, Svinarenko AA, Mischenko EV, Florko TA (2009) Int J Quant Chem 109(14):3325
8. Glushkov AV, Loboda AV, Gurnitskaya EP, Svinarenko AA (2009) Phys Scripta T135:014022
9. Glushkov AV, Khetselius OYu, Svinarenko AA (2013) Phys Scripta T153:014029
10. Svinarenko AA (2014) J Phys Conf Ser 548:012039

11. Paul S, Ho YK (2010) J Phys B At Mol Opt Phys 43:065701; Kar S, Ho YK (2011) J Phys B At Mol Opt Phys 44:015001
12. Glushkov AV (2005) Atom in electromagnetic field. KNT, Kiev
13. Yongqiang L, Jianhua W, Yong H, Jianmin Y (2008) J Phys B At Mol Opt Phys 41:145002
14. Glushkov AV, Mansarliysky VF, Khetselius OYu, Ignatenko AV, Smirnov AV, Prepelitsa GP (2017) J Phys Conf Ser 810:012034
15. Glushkov AV, Loboda AV (2007) J Appl Spectr 74:305, Springer
16. Saha B, Fritzsche S (2007) J Phys B At Mol Opt Phys 40:259
17. Marrs R, Levine M, Knapp D, Henderson J (1988) Phys Rev Lett 60:1715
18. Zhang H, Sampson D, Clark R, Mann J (1988) Atom Dat Nuc Dat Tabl 37:17
19. Reed KJ (1988) Phys Rev A 37:1791; Talukder MR (2008) Appl Phys Lasers Opt 93:576
20. Smith ACH, Bannister ME, Chung YS, Djuric N, Dunn GH, Wallbank B, Woitke O (1999) Phys Scripta T80:283
21. Buyadzhi VV (2015) Photoelectronics 24:128
22. Zeng S, Liu L, Wang JG, Janev RK (2008) J Phys B At Mol Opt Phys 41:135202
23. Okutsu H, Sako T, Yamanouchi K, Diercksen GHF (2005) J Phys B At Mol Opt Phys 38:917
24. Nakamura N, Kavanagh AP, Watanabe H, Sakaue HA, Li Y, Kato D, Curell FJ, Ohtani S (2007) J Phys Conf Ser 88:012066
25. Bannister ME, Djuric N, Woitke O, Dunn GH, Chung Y-S, Smith ACH, Wallbank B, Berrington KA (1999) Int J Mass Spectr 192:39
26. Badnell NR (2007) J Phys Conf Ser 88:012070; Griffin DC, Balance CP, Mitnik DM, Berengut JC (2008) J Phys B At Mol Opt Phys 41:215201
27. Glushkov AV, Malinovskaya SV, Prepelitsa G, Ignatenko V (2005) J Phys Conf Ser 11:199
28. Malinovskaya SV, Glushkov AV, Khetselius OYu, Loboda AV, Lopatkin YuM, Svinarenko AA, Nikola LV, Perelygina TB (2011) Int J Quant Chem 111:288
29. Glushkov AV, Khetselius OYu, Loboda AV, Ignatenko AV, Svinarenko AA, Korchevsky DA, Lovett L (2008) Spectr Line Shapes. AIP Conf Pro 1058:175
30. Buyadzhi VV, Chernyakova YuG, Antoshkina OA, Tkach TB (2017) Photoelectronics 26:94
31. Ivanov LN, Ivanova EP (1979) Atom Dat Nucl Dat Tabl 24:95
32. Driker MN, Ivanova EP, Ivanov LN, Shestakov AF (1982) J Quant Spectr Rad Transfer 28:531
33. Ivanov LN, Letokhov VS (1985) Com Mod Phys D 4:169
34. Vidolova-Angelova E, Ivanov LN, Ivanova EP, Angelov DA (1986) J Phys B At Mol Opt Phys 19:2053
35. Ivanov LN, Ivanova EP, Aglitsky EV (1988) Phys Rep 166:315
36. Ivanova EP, Ivanov LN, Glushkov AV, Kramida AE (1985) Phys Scripta 32:513
37. Glushkov AV, Ivanov LN (1992) Phys Lett A 170:33
38. Glushkov AV, Ivanov LN, Ivanova EP (1986) Autoionization phenomena in atoms. Moscow University Press, Moscow, pp 58–160
39. Ivanova EP, Glushkov AV (1986) J Quant Spectr Rad Transfer 36:127
40. Ivanova EP, Gulov AV (1991) Atom Dat Nuc Dat Tabl 49:1
41. Ivanov LN, Ivanova EP, Knight L (1993) Phys Rev A 48.4365
42. Ivanov LN, Ivanova EP, Knight L, Molchanov AG (1996) Phys Scripta 53:653
43. Glushkov AV, Ivanov LN (1992) Preprint of ISAN. AS N-1 Moscow-Troitsk
44. Glushkov AV, Ivanov LN (1993) J Phys B At Mol Opt Phys 26:L379
45. Glushkov AV (1990) Sov Phys J 33(1):1
46. Glushkov AV (1990) J Str Chem 31(4):529
47. Glushkov AV (1992) JETP Lett 55:97
48. Glushkov A V (2012) Quantum systems in chemistry and physics. In: Nishikawa K, Maruani J, Brändas E, Delgado-Barrio G, Piecuch P (eds) Progress in theoretical chemistry and physics, vol 26. Springer, Dordrecht, pp 231–252
49. Glushkov AV (2008) Relativistic quantum theory. Quantum mechanics of atomic systems. Astroprint, Odessa, p 700
50. Khestelius OYu (2008) Hyperfine structure of atomic spectra. Astroprint, Odessa, p 210

51. Glushkov AV (2013) Advances in quantum methods and applications in chemistry, physics and biology. In: Hotokka M, Maruani J, Brändas J, Delgado-Barrio G (eds) Progress in theoretical chemistry and physics, vol 27. Springer, Cham, pp 161–177

52. Glushkov AV, Khetselius OYu, Svinarenko AA, Prepelitsa GP (2010) In: Duarte FJ (ed) Coherence and ultrashort pulsed emission, InTech, Rijeka, pp 159–186

53. Svinarenko AA, Glushkov AV, Khetselius OYu, Ternovsky VB, Dubrovskaya YuV, Kuznetsova AA, Buyadzhi VV (2017) In: Orjuela JEA (ed) Rare earth element, InTech, pp 83–104

54. Glushkov AV, Khetselius OYu, Svinarenko AA, Buyadzhi VV, Ternovsky VB, Kuznetsova AA, Bashkarev PG (2017) In: Uzunov DI (ed) Recent studies in perturbation theory, InTech, pp 131–150

55. Buyadzhi VV, Glushkov AV, Lovett L (2014) Photoelectronics 23:38

56. Buyadzhi VV, Glushkov AV, Mansarliysky VF, Ignatenko AV, Svinarenko AA (2015) Sensor Electr Microsyst Techn 12(4):27

57. Glushkov AV, Khetselius OYu, Malinovskaya SV (2008) Frontiers in quantum systems in chemistry and physics. In: Wilson S, Grout PJ, Maruani J, Delgado-Barrio G, Piecuch P (eds) Progress in theoretical chemistry and physics, vol 18. Springer, Dordrecht, pp 525–541

58. Glushkov AV, Khetselius OYu, Malinovskaya SV (2008) Europ Phys J ST 160:195

59. Glushkov AV, Khetselius OYu, Malinovskaya SV (2008) Mol Phys 106:1257

60. Glushkov AV, Khetselius OY, Svinarenko AA (2012) Advances in the theory of quantum systems in chemistry and physics. In: Hoggan P, Maruani J, Brändas E, Delgado-Barrio G, Piecuch P (eds) Progress in theoretical chemistry and physics, vol 22. Springer, Dordrecht, pp 51–68

61. Glushkov AV, Khetselius OYu, Lovett L (2010) Advances in the theory of atomic and molecular systems. In: Piecuch P, Maruani J, Delgado-Barrio G, Wilson S (eds) Progress in theoretical chemistry and physics, vol 20. Springer, Dordrecht, pp 125–152

62. Glushkov AV, Khetselius YO, Loboda AV, Svinarenko AA (2008) Frontiers in quantum systems in chemistry and physics. In: Wilson S, Grout PJ, Maruani J, Delgado-Barrio G, Piecuch P (eds) Progress in theoretical chemistry and physics, vol 18. Springer, Dordrecht, pp 543–560

63. Glushkov AV, Svinarenko AA, Khetselius OYu, Buyadzhi VV, Florko TA, Shakhman AN (2015) Frontiers in quantum methods and applications in chemistry and physics. In: Nascimento M, Maruani J, Brändas E, Delgado-Barrio G (eds) Progress in theoretical chemistry and physics, vol 29. Springer, Cham, pp 197–217

64. Khetselius OYu, Zaichko PA, Smirnov AV, Buyadzhi VV, Ternovsky VB, Florko TA, Mansarliysky VF (2017) Quantum systems in physics, chemistry, and biology. In: Tadjer A, Pavlov R, Maruani J, Brändas E, Delgado-Barrio G (eds) Progress in theoretical chemistry and physics, Springer, Cham, pp 271–281

65. Glushkov AV, Buyadzhi VV, Kvasikova AS, Ignatenko AV, Kuznetsova AA, Prepelitsa GP, Ternovsky V B (2017) Quantum systems in physics, chemistry, and biology. In: Tadjer A, Pavlov R, Maruani J, Brändas E, Delgado-Barrio G (eds) Progress in theoretical chemistry and physics, vol 30. Springer, Cham, pp 169–180

66. Glushkov AV, Rusov VD, Ambrosov SV, Loboda AV (2003) In: Fazio G, Hanappe F (eds) New projects and new lines of research in nuclear physics. World Scientific, Singapore, pp 126–132

67. Glushkov O. Khetselius E Gurnitskaya, Loboda A, Florko VD, Sukharev Lovett L (2008) Frontiers in quantum systems in chemistry and physics. In: Wilson S, Grout PJ, Maruani J, Delgado-Barrio G, Piecuch P (eds) Progress in theoretical chemistry and physics, vol 18. Springer, Dordrecht, pp 507–524

68. Glushkov AV, Ambrosov SV, Loboda AV, Gurnitskaya EP, Khetselius OY (2006) Recent advances in the theory of chemical and physical systems. In: Julien P, Maruani J, Mayou D, Wilson S, Delgado-Barrio G (eds). Progress in theoretical chemistry and physics, vol 15. Springer, Dordrecht, pp 285–299

69. Glushkov AV, Ambrosov SV, Loboda AV, Chernyakova YuG, Svinarenko AA, Khetselius OYu (2004) Nucl Phys A Nucl Hadr Phys 734:21
70. Glushkov AV, Malinovskaya SV, Loboda AV, Shpinareva IM, Prepelitsa GP (2006) J Phys Conf Ser 35:420
71. Glushkov AV, Ambrosov SV, Loboda AV, Gurnitskaya EP, Prepelitsa GP (2005) Int J Quant Chem 104:562
72. Glushkov A, Malinovskaya S, Loboda A, Shpinareva I, Gurnitskaya E, Korchevsky D (2005) J Phys Conf Ser 11:188
73. Glushkov AV, Malinovskaya SV, Chernyakova YG, Svinarenko AA (2004) Int J Quant Chem 99:889
74. Glushkov AV, Ambrosov SV, Ignatenko AV, Korchevsky DA (2004) Int J Quant Chem 99:936
75. Glushkov AV, Malinovskaya SV, Sukharev DE, Khetselius OYu, Loboda AV, Lovett L (2009) Int J Quant Chem 109:1717
76. Khetselius OYu (2009) Int J Quant Chem 109:3330
77. Khetselius OYu (2009) Phys Scripta T135:014023
78. Khetselius OYu (2012) Quantum systems in chemistry and physics. In: Nishikawa K, Maruani J, Brändas E, Delgado-Barrio E, Piecuch P (eds) Progress in theoretical chemistry and physics, vol 26. Springer, Dordrecht, pp 217–229
79. Khetselius YO (2015) Frontiers in quantum methods and applications in chemistry and physics. In: Nascimento M, Maruani J, Brändas E, Delgado-Barrio E, Piecuch P (eds) Progress in theoretical chemistry and physics, vol 29. Springer, Cham, pp 55–76
80. Khetselius OYu (2008) Hyperfine structure of atomic spectra. Astroprint, Odessa
81. Khetselius OYu (2005) Hyperfine structure of radium. Photoelectronics. 14:83
82. Khetselius OYu (2012) J Phys Conf Ser 397:012012
83. Khetselius OYu, Florko TA, Svinarenko AA, Tkach TB (2013) Phys Scripta T153:014037
84. Khetselius OYu (2008) Spectral line shape. AIP Conf Proc 1058:363
85. Khetselius OYu, Glushkov AV, Gurnitskaya EP, Loboda AV, Mischenko EV, Florko TA, Sukharev DE (2008) AIP Conf Proc 1058:231
86. Khetselius OYu (2007) Photoelectronics 16:129
87. Khetselius OYu, Gurnitskaya EP (2006) Sensor Electr Microsyst Techn N3, 35–39
88. Khetselius OYu, Gurnitskaya EP (2006) Sensor Electr and Microsyst Techn Issue 2:25−29
89. Feller D, Davidson ER (1981) J Chem Phys 74:3977
90. Froelich P, Davidson ER, Brändas E (1983) Phys Rev A 28:2641
91. Rittby M, Elander N, Brändas E (1983) Int J Quant Chem 23:865
92. Yan A, Wang C, Yung Y, Ya KC, Chen GH (2011) J Chem Phys 134:241103
93. Maruani J (2016) J Chin Chem Soc 63:33
94. Pavlov R, Mihailov L, Velchev Ch, Dimitrova-Ivanovich M, Stoyanov Zh, Chamel N, Maruani J (2010) J Phys Conf Ser 253:012075
95. Dietz K, Heβ BA (1989) Phys Scripta 39:682
96. Kohn W, Sham LJ (1964) Phys Rev A 140:1133; Hohenberg P, Kohn W (1964) Phys Rev B 136:864
97. Gidopoulos N, Wilson S (2004) The fundamentals of electron density, density matrix and density functional theory in atoms, molecules and the solid state. In: Progress in theoretical chemistry and physics, vol 14. Springer, Berlin
98. Glushkov A V (2006) Relativistic and correlation effects in spectra of atomic systems. Astroprint, Odessa

Relativistic Quantum Chemistry and Spectroscopy of Exotic Atomic Systems with Accounting for Strong Interaction Effects

O. Yu. Khetselius, A. V. Glushkov, Yu. V. Dubrovskaya,
Yu. G. Chernyakova, A. V. Ignatenko, I. N. Serga
and L. A. Vitavetskaya

Abstract We present the fundamentals of a consistent relativistic theory of spectra of the exotic pionic atomic systems on the basis of the Klein-Gordon-Fock equation approach and relativistic many-body perturbation theory (electron subsystem). The key feature of the theory is simultaneous accounting for the electromagnetic and strong pion-nuclear interactions by means of using the generalized radiation and strong pion-nuclear optical potentials. The nuclear and radiative corrections are effectively taken into account. The modified Uehling-Serber approximation is used to take into account for the Lamb shift polarization part. In order to take into account the contribution of the Lamb shift self-energy part we have used the generalized non-perturbative procedure, which generalizes the Mohr procedure and radiation model potential method by Flambaum-Ginges. There are presented data of calculation of the energy and spectral parameters for pionic atoms of the ^{93}Nb, ^{173}Yb, ^{181}Ta, ^{197}Au, with accounting for the radiation (vacuum polarization), nuclear (finite size of a nucleus) and the strong pion-nuclear interaction corrections. The measured values of the Berkley, CERN and Virginia laboratories and alternative data based on other versions of the Klein-Gordon-Fock theories with taking into account for a finite size of the nucleus in the model uniformly charged sphere and the standard Uehling-Serber radiation correction are listed too.

Keywords Relativistic quantum chemistry · Spectroscopy of exotic atoms
Relativistic perturbation theory · Energy approach · Nuclear and radiative
corrections · Pionic atomic systems

O. Yu. Khetselius (✉) · A. V. Glushkov · Yu. V. Dubrovskaya · Yu. G. Chernyakova ·
A. V. Ignatenko · I. N. Serga · L. A. Vitavetskaya
Odessa State Environmental University, L'vovskaya Str., 15, 65016 Odessa, Ukraine
e-mail: okhetsel@gmail.com

© Springer International Publishing AG, part of Springer Nature 2018   71
Y. A. Wang et al. (eds.), *Concepts, Methods and Applications of Quantum Systems
in Chemistry and Physics*, Progress in Theoretical Chemistry and Physics 31,
https://doi.org/10.1007/978-3-319-74582-4_5

1 Introduction

It is well known that studying the energy, spectral, radiation parameters, including the spectral lines hyperfine structure, for heavy exotic (hadronic, kaonic, pionic) atomic systems is of a great interest for the further development as atomic and nuclear theories and quantum chemistry of strongly interacted fermionic systems (see, for example, Refs. [1–31]). Really, the exotic atoms enable to probe aspects of atomic and nuclear structure that are quantitatively different from what can be studied in the electronic ("usual") atoms. Besides, the corresponding data on the energy and spectral properties of the hadronic atomic systems can be used as a powerful tool for the study of particles and fundamental properties.

At present time one of the most sensitive tests for the chiral symmetry breaking scenario in the modern hadron's physics is provided by studying the exotic hadron-atomic systems. Nowadays the transition energies in pionic (kaonic, muonic etc.) atoms are measured with an unprecedented precision and from studying spectra of the hadronic atoms it is possible to investigate the strong interaction at low energies measuring the energy and natural width of the ground level with a precision of few meV [20–46].

The strong interaction is the reason for a shift in the energies of the low-lying levels from the purely electromagnetic values and the finite lifetime of the state corresponds to an increase in the observed level width. For a long time the similar experimental investigations have been carried out in the laboratories of Berkley, Virginia (USA), CERN (Switzerland).

The most known theoretical models to treating the hadronic (pionic, kaonic, muonic, antiprotonic etc.) atomic systems are presented in Refs. [1–46]. The most difficult aspects of the theoretical modelling are reduced to the correct description of pion-nuclear strong interaction [10–18] as the electromagnetic part of the problem can be in principle reasonably accounted for [47–60].

In the present chapter we briefly present the fundamentals of a consistent relativistic theory of spectra of the exotic pionic atomic systems (with simultaneous accounting for the electromagnetic and strong pion-nuclear interactions by means of using the generalized radiation and strong pion-nuclear optical potentials) on the basis of the Klein-Gordon-Fock. The nuclear and radiative corrections are effectively taken into account. The modified Uehling-Serber approximation is used to take into account for the Lamb shift polarization part. In order to take into account the contribution of the Lamb shift self-energy part we have used the generalized non-perturbative procedure, which generalizes the Mohr procedure and radiation model potential method by Flambaum-Ginges. There are presented data of calculation of the energy and spectral parameters for pionic atoms of the ^{93}Nb, ^{173}Yb, ^{181}Ta, ^{197}Au, with accounting for the radiation (vacuum polarization), nuclear (finite size of a nucleus) and the strong pion-nuclear interaction corrections. The measured values of the Berkley, CERN and Virginia laboratories and alternative data based on other versions of the Klein-Gordon-Fock theories with taking into account for a finite size of the nucleus in the model uniformly charged sphere and the standard Uhling-Serber radiation corrections are listed too.

2 Relativistic Theory of Exotic Quantum Systems with Accounting of the Electromagnetic and Strong Interaction Effects

2.1 The Klein-Gordon-Fock Equation. Electromagnetic Interactions and Nuclear Potential

Here we present a brief description of the key moments of our approach (more details can be found in Refs. [61–79]). The relativistic electron wave functions are determined from solution of the Klein-Gordon-Fock equation (pion is the Boson with spin 0, mass: $m_{\pi^-} = 139.57018$ МэВ, $r_{\pi_-} = 0.672 \pm 0.08$ fm) with a general potential (the latter includes an electric and polarization potentials of a nucleus plus the strong pion-nuclear interaction potential), which can be written as follows:

$$m^2 c^2 \Psi(x) = \{\frac{1}{c^2}[i\hbar\partial_t + eV_0(r)]^2 + \hbar^2\nabla^2\}\Psi(x) \tag{1}$$

where c is a speed of the light, h is the Planck constant, and $\Psi_0(x)$ is the scalar wave function of the space-temporal coordinates. Usually one considers the central potential $[V_0(r), 0]$ approximation with the stationary solution:

$$\Psi(x) = \exp(-iEt/\hbar)\varphi(x), \tag{2}$$

where $\varphi(x)$ is the solution of the stationary equation:

$$\{\frac{1}{c^2}[E + eV_0(r)]^2 + \hbar^2\nabla^2 - m^2 c^2\}\varphi(x) = 0 \tag{3}$$

Here E is the total energy of the system (sum of the mass energy mc^2 and binding energy ε_0).

In principle, the central potential V_0 is the sum of the following potentials: the electric potential of a nucleus, vacuum-polarization potential and the strong interaction potential. The nuclear potential for the spherically symmetric density $\rho(r|R)$ can be presented as follows:

$$V_{nucl}(r|R) = -(1/r)\int_0^r dr' r'^2 \rho\left(r'|R\right) + \int_r^\infty dr' r' \rho\left(r'|R\right) \tag{4}$$

Further the density can be approximated by the Gaussian function:

$$\rho(r|R) = \left(4\gamma^{3/2}/\sqrt{\pi}\right)\exp\left(-\gamma r^2\right) \tag{5}$$

$$\int_0^\infty dr r^2 \rho(r|R) = 1,$$

$$\int_0^\infty dr r^3 \rho(r|R) = R,$$

(here $g = 4/R^2$ and R is the effective nucleus radius) or by the Fermi function:

$$\rho(r) = \rho_0/\{1 + \exp[(r-c)/a)]\}, \tag{6}$$

where the parameter $a = 0.523$ fm, the parameter c is chosen by such a way that it is true the following condition for average-squared radius:

$$<r^2>^{1/2} = (0.836 \times A^{1/3} + 0.5700)\text{fm}. \tag{7}$$

Further one should use the formulas for the finite size nuclear potential and its derivatives on the nuclear radius. Here we use the known Ivanov-Ivanova et al. method of differential equations (look details in Refs. [80–83]). The effective algorithm for definition of the potential $V_{nucl}(r|R)$ is used in Refs. [65, 72] and reduced to solution of the following system of the differential equations (for the Fermi model):

$$V'nucl(r,R) = \left(1/r^2\right)\int_0^r dr' r'^2 \rho\left(r',R\right) \equiv \left(1/r^2\right)y(r,R),$$

$$y'(r,R) = r^2\rho(r,R), \tag{8}$$

$$\rho'(r) = (\rho_0/a)\exp[(r-c)/a]\{1 + \exp[(r-c)/a)]\}^2$$

with the corresponding boundary conditions. In a case of the Gaussian model the corresponding system of differential equations is as follows:

$$V'nucl(r,R) = (1/r^2)\int_0^r dr'\, r'^2\rho(r',R) \equiv (1/r^2)y(r,R)$$
$$y'(r,R) = r^2\rho(r,R) \tag{9}$$

$$\rho'(r,R) = -8\gamma^{5/2}r/\sqrt{\pi}\exp\left(-\gamma r^2\right) = -2\gamma r\rho(r,R) = -\frac{8r}{\pi r^2}\rho(r,R)$$

with the boundary conditions:

$$V_{nucl}(0, R) = -4/(\pi r),$$

$$y(0, R) = 0,$$

$$\rho(0, R) = 4\gamma^{3/2}/\sqrt{\pi} = 32/R^3 \tag{10}$$

Another, probably, more consistent approach is in using the relativistic mean-field (RMF) model, which been designed as a renormalizable meson-field theory for nuclear matter and finite nuclei [47].

2.2 Quantum Electrodynamics Effects in Pionic Atomic Systems

Consistent and accurate account of the radiation or QED effects is of a great importance and interest in spectroscopy of the pionic atomic systems. To take into account the radiation (QED) corrections, namely, the important effect of the vacuum polarization one could use the procedure, which is in details described in the Refs. [41–58, 65, 72–78]. Figure 1 illustrates Feynman diagrams, which describe a

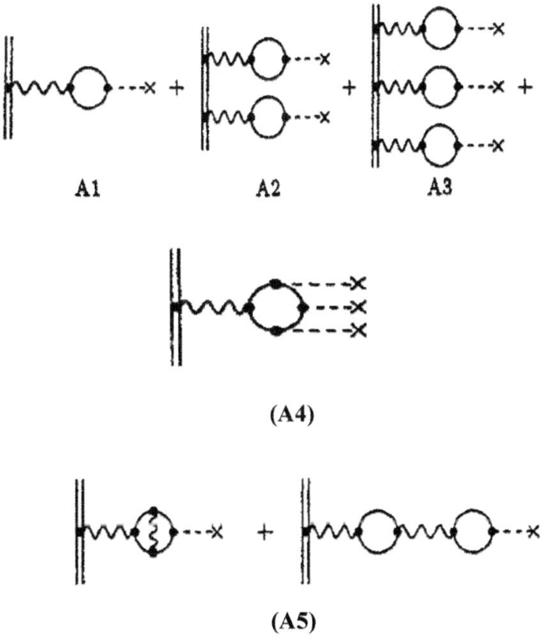

Fig. 1 Feynman diagrams, which describe a QED effect of the vacuum polarization: A1—the Uehling-Serber term; A2, A3—terms of the order $[\alpha(Z\alpha)]^n$ $(n = 2, ...)$; A4—the Källen-Sabry correction of the order $\alpha^2(\alpha Z)$; A5—the Wichmann-Kroll correction of order $\alpha(Z\alpha)^n$ $(n = 3)$

QED effect of the vacuum polarization: A1—the Uehling-Serber term; A2, A3–terms of order члены порядка $[\alpha(Z\alpha)]^n$ (n = 2, ...); A4- the Källen-Sabry correction of order $\alpha^2(\alpha Z)$; A5–the Wichmann-Kroll correction of order ка $\alpha(Z\alpha)^n$ (n = 3). An effect of the vacuum polarization is usually taken into account in the first PT theory order by means of the generalized Uehling-Serber potential with modification to take into account the high-order radiative corrections. In particular, the generalized Uehling-Serber potential can be written as follows:

$$U(r) = -\frac{2\alpha}{3\pi r} \int\limits_1^\infty dt \, \exp(-2rt/\alpha Z)(1 + 1/2t^2)\frac{\sqrt{t^2-1}}{t^2} \equiv -\frac{2\alpha}{\pi r}C(g), \quad (11)$$

where $g = r/(\alpha Z)$. More correct and consistent approach is presented in Refs. [42, 43, 52–62]. An accounting of the nuclear finite size effect modifies the potential (7) as follows:

$$U^{FS}(r) = -\frac{2\alpha^2}{3\pi} \int d^3r' \int\limits_m^\infty dt \exp(-2t|r-r'|/\alpha Z) \times \left(1 + \frac{1}{2t^2}\right)\frac{\sqrt{t^2-1}}{t^2}\frac{\rho(r')}{|r-r'|},$$

$$(12)$$

The Uehling-Serber potential, determined as a quadrature (11), may be approximated with high precision by a simple analytical function. The use of new approximation of the Uehling potential permits one to decrease the calculation errors for this term down to 0.5–1%.

A method for calculation of the self-energy part of the Lamb shift is based on an idea by Ivanov-Ivanova (see Refs. [80, 81]), which generalizes the known hydrogen-like method by Mohr and radiation model potential method by Flambaum-Ginges (look details in Refs. [41, 52, 61, 62]).

According to Ref. [9], in an atomic system the radiative shift and the relativistic part of energy are, in principle, defined by one and the same physical field. One could suppose that there exists some universal function that connects the self - energy correction and the relativistic energy. The self-energy correction for the states of a hydrogen-like ion was presented by Mohr [41] as:

$$E_{SE}(H|Z, nlj) = 0.027148\frac{Z^4}{n^3}F(H|Z, nlj) \quad (13)$$

The values of F are given at $Z = 10 - 110$, $nlj = 1s, 2s, 2p_{1/2}, 2p_{3/2}$.

These results are modified here for the states $1 s^2 nlj$ of the non-H atoms (ions). It is supposed that for any ion with nlj electron over the core of closed shells the sought value may be presented in the form [52]:

$$E_{SE}(Z, nlj) = 0.027148 \frac{\xi^4}{n^3} f(\xi, nlj) \left(cm^{-1}\right) \qquad (14)$$

The parameter $\xi = (E_R)^{1/4}$, E_R is the relativistic part of the bounding energy of the outer electron; the universal function $f(\xi, nlj)$ does not depend on the composition of the closed shells and the actual potential of the nucleus. The procedure of generalization for a case of the non-H systems with the finite nucleus consists of the following steps [9]: (1). Calculation of the values E_R and ξ for the states nlj of H-like ions with the point nucleus (in accordance with the Zommerfeld formula); (2). Construction of an approximating function $f(\xi, nlj)$ by the found reference Z and the appropriate $F(H|Z, nlj)$; (3). Calculation of E_R and ξ for the states nlj of Li-like ions with the finite nucleus; (4). Calculation of E_{SE} for the sought states by the formula (14). The energies of the states of the non-H atoms and ions are calculated twice: with a conventional constant of the fine structure $\alpha = 1/137$ and with $\tilde{\alpha} = \alpha/1000$. The results of latter calculations were considered as non-relativistic. This permitted isolation of E_R and ξ. A detailed evaluation of their accuracy may be made only after a complete calculation of $E_{SE}^n(Z, nlj)$. It may be stated that the above extrapolation method is more justified than using the widely spread expansions by the parameter αZ. The other details of the theory and computational code can be found in Refs. [61–70, 76–79].

2.3 Strong Pion-Nuclear Interactions in Pionic Atomic System

The most difficult aspect of the problem is an adequate account for the strong pion-nuclear interaction in the exotic system. Now it is well known that the most fundamental and consistent microscopic theory of the strong interactions is provided by the modern quantum chromodynamics. One should remind that here speech is about a gauge theory based on the representation of the confined coloured quarks and gluons. Naturally one could consider the regimes of relatively low and high energies (asymptotic freedom). In a case of the low energies so called coupling constant increases to the order 1 and, therefore, this perturbation methods fail to describe the interaction of strongly interacting hadrons (including pions). Naturally, to describe the strong pion-nuclear interaction (even at relatively low energies) microscopically, a different approaches can be developed (look details in Refs. [11–19, 76–79]).

More simplified and sufficiently popular approach to treating the strong interaction in the pionic atomic system is provided by the well known optical potential model (c.g. [14, 15]). On order to describe the strong π^-N interaction we have used the optical potential model n which the generalized Ericson-Ericson potential is as follows:

$$V_{\pi-N} = V_{opt}(r) = -\frac{4\pi}{2\overline{m}}\left\{q(r)\nabla\frac{\alpha(r)}{1+4/3\pi\xi\alpha(r)}\nabla\right\},$$

$$q(r) = \left(1+\frac{m_\pi}{m_N}\right)\left\{b_0\rho(r)+b_1\left[\rho_n(r)-\rho_p(r)\right]\right\}$$

$$+\left(1+\frac{m_\pi}{2m_N}\right)\left\{B_0\rho^2(r)+B_1\rho(r)\delta\rho(r)\right\}, \tag{15}$$

$$\alpha(r) = \left(1+\frac{m_\pi}{m_N}\right)^{-1}\left\{c_0\rho(r)+c_1\left[\rho_n(r)-\rho_p(r)\right]\right\}$$

$$+\left(1+\frac{m_\pi}{2m_N}\right)^{-1}\left\{C_0\rho^2(r)+C_1\rho(r)\delta\rho(r)\right\}.$$

Here $\rho_{p,n}(r)$—distribution of a density of the protons and neutrons, respectively, ξ—parameter ($\xi=0$ corresponds to case of "no correlation", $\xi=1$, if anticorrelations between nucleons); respectively isoscalar and isovector parameters b_0, c_0, B_0, b_1, c_1, C_0 B_1, C_1—are corresponding to the s-wave and p-wave (repulsive and attracting potential member) scattering length in the combined spin-isospin space with taking into account the absorption of pions (with different channels for p-p pair $B_{0(pp)}$ and p-n pair $B_{0(pn)}$), the Lorentz-Lorentz effect in the p-wave interaction and isospin and spin dependence of an amplitude π^-N scattering:

$$b_0\rho(r) \rightarrow b_0\rho(r)+b_1\left\{\rho_p(r)-\rho_n(r)\right\}, \tag{16}$$

The description of numerical values of the potential parameters will be commented below (look details in Refs. [41, 52, 61, 62]).

2.4 Complex Energy of Pionic Atomic System

Further we note that an energy of the hadronic atom can be represented as the following sum:

$$E\approx E_{KG}+E_{FS}+E_{VP}+E_N; \tag{17}$$

Here E_{KG}-is the energy of a pion in a nucleus (Z, A) with the point-like charge (dominative contribution in (17)), E_{FS} is the contribution due to the nucleus finite size effect, E_{VP} is the radiation correction due to the vacuum-polarization effect, E_N is the energy shift due to the strong interaction V_N.

It is easily to note that the strong pion-nucleus interaction contribution into energy can be directly found from the solution of the Klein-Gordon-Fock equation with the corresponding pion-nucleon potential, for example, in the optical potential approximation (8). Since the corresponding optical potential contains the complex parameters, the relevant energy eigen-values of the Klein-Gordon-Fock equation for the definite pionic state $(i = nl)$ in an atom are the complex values too, i.e. [11, 38, 76]:

$$E_i = \text{Re } E_i + i\text{Im } E_i = \text{Re } E_i - (i/2)\Gamma_i, \tag{18}$$

where the imaginary part determines a width of pionic energy level G_i. The total width of any level is determined as by the strong pion-nuclear interaction contribution Γ_i^S (pion absorption) as by the electromagnetic contribution Γ_i^{rad}. The latter is determined by a probability of the electromagnetic radiation transition (including the Auger process probability Γ_i^A) on the lower level. As an example, for the width of pion 1 s can be written:

$$\Gamma_{1s}^S = \Gamma_{2p \to 1s}^{exp} - \left(\Gamma_{2p}^{rad} + \Gamma_{2p}^S + \Gamma_{2p}^A\right) \approx \Gamma_{2p \to 1s}^{exp}, \tag{19}$$

where Γ_i^{rad} i Γ_i^A are the radiation and Auger widths respectively. Let us consider further elements of theory, associated with the implementation of the known relativistic energy formalism in our theory to calculate the electromagnetic interaction transition probabilities in spectrum of the pionic atom [80–91]. It is worth to remind that in relativistic theory of the usual many-electron systems (an energy of any excited state is a complex quantity) an shift of the total energy level is usually represented as:

$$\Delta E_i = \text{Re}\Delta E_i + i\text{Im}\Delta E_i = \text{Re}\Delta E_i - (i/2)\Gamma_i^{rad}, \tag{20}$$

where Γ_i^{rad} is a radiation width, and the corresponding radiative transition probability in the usual atomic system $P \sim \Gamma_i^{rad}$. In order to compute the latter we use the generalized relativistic energy approach.

Let us remind that an initial general energy formalism combined with an empirical model potential method in a theory of atoms and multicharged ions has been developed by Ivanov-Ivanova et al. [80–84]; further more general ab initio gauge-invariant version of relativistic energy approach has been presented by Glushkov-Ivanov [89]. The imaginary part of the energy shift of an atom is connected with the radiation decay possibility (transition probability). For the α-n radiation transition ImDE in the lowest order of the PT is determined as:

$$\text{Im } \Delta E = -\frac{1}{4\pi} \sum_{\substack{\alpha > n > f \\ [\alpha < n \leq f]}} V_{anan}^{|\omega_{an}|}, \tag{21}$$

where ω_{an} is a frequency of the α-n radiation, $(\alpha > n > f)$ for particle and $(\alpha < n < f)$ for vacancy. The matrix element V is determined as follows:

$$V_{ijkl}^{|\omega|} = \iint dr_1 dr_2 \psi_i^*(r_1)\psi_j^*(r_2) \frac{\sin|\omega|r_{12}}{r_{12}}(1 - \alpha_1\alpha_2)\psi_k^*(r_2)\psi_l^*(r_1) \qquad (22)$$

The detailed procedure for computing the matrix elements (22) is presented in Refs. [76–92]. All calculations are performed with using the numeral code Superatom (version 98).

3 Results and Conclusions

3.1 The Parameters of the Optical Potential

In Table 1 we list the values of parameters for the proton and neutron (c_p, c_n) Fermi distribution and nuclear spin I number of some nuclei.

In Table 2 there are presented the concrete values of some optical potential parameters, which have been used in different calculations [15–39]. Let us explain the used classification and abbreviations of corresponding sets of the optical potential parameters: Tauscher, $\xi = 0$—Tau1; Tauscher, $\xi = 1$—Tau2; Batty et al.—Bat.; Seki et al.—Sek; Nagels—Nag; de Laat-Konijin et al.—Laat, Serga-Shakhman—Serg-Sha, this theory—Odes [16–22, 38, 39, 76–79]. Note that the parameters of the optical potential in Table 1 (the initial set of parameters) were initially obtained by calibration of the experimental data on pion-nuclear scattering for the light and medium nuclei. Further application of the model to the heavy atoms and relatively low-lying states showed imperfections (in some cases) of these sets of the parameter values under theoretical studying heavy nuclei. This situation is called pion anomalies. For example, experimental (the low-energy scattering of pions; LAMPF) results for relatively low-lying states of pion at levels 3d, 2p, 1 s (in such heavy atoms

Table 1 Parameters for the proton and neutron (c_p, c_n) Fermi distribution and nuclear spin I number of some nuclei	Nucleus	Z	A	c_p, Φ	c_n, Φ	I^π
	^{20}Ne	10	19.9924	2.963	2.912	0
	^{24}Mg	12	23.9850	3.080	3.026	0
	^{93}Mb	41	92.906	4.986	5.127	9/2
	^{133}Cs	55	132.905	5.599	5.774	7/2
	^{165}Ho	67	164.93	6.125	6.329	7/2
	^{173}Yb	70	172.938	6.274	6.570	5/2$^+$
	^{175}Lu	71	174.941	6.274	6.570	7/2
	^{181}Ta	73	180.948	6.347	6.650	7/2$^+$
	^{197}Au	79	196.967	6.454	6.850	3/2$^+$
	^{205}Tl	81	204.974	6.587	6.881	1/2$^+$
	^{208}Pb	82	207.977	6.652	6.892	0
	^{209}Bi	83	208.98	6.688	6.870	9/2$^-$
	^{237}Np	93	237.095	7.001	7.300	5/2$^+$

Table 2 The values of some optical potential parameters, which have been used in different calculations (see text)

	Tauscher $\xi = 0$ [16]	Tauscher $\xi = 1$ [16]	Batty $\xi = 1$ [17]	Seki $\xi = 1$ [18]	Nagels $\xi = 1$ [22]	Row78 $\xi = 1$ [19]	Laat $\xi = 1$ [38, 39]	Odes $\xi = 1$ [76, 79]	
b_0	−0.0296	−0.0293	−0.017	0.003	−0.013	−0.004	0.007	0.003	m_π^{-1}
b_1	−0.077	−0.078	−0.13	−0.143	−0.092	−0.094	−0.075	−0.094	m_π^{-1}
ReB_0	0	0	−0.0475	−0.15	–	–	−0.18	−0.15	m_π^{-4}
ImB_0	0.0436	0.0428	0.0475	0.046	–	–	0.058	0.046	m_π^{-4}
c_0	0.172	0.227	0.255	0.21	0.209	0.23	0.266	0.21	m_π^{-3}
c_1	0.22	0.18	0.17	0.18	0.177	0.17	0.40	0.18	m_π^{-3}
ReC_0	–	–	–	0.11	–	–	0.07	0.11	m_π^{-6}
ImC_1	–	–	–	–	–	–	−0.34	−0.25	m_π^{-6}

as Ta, Bi et al.) have shown that the appropriate values of width due to strong interaction by a factor two and more are less than the values specified within the optical potential model with using the earliest parameterization [14, 15].

Such relatively significant deviations are typical for strong levels shifts.

A similar anomaly was detected in a case of the strong quadrupole shift of the pion energy in 3d and 4f states. It should be noted that the results of the hyperfine structure studying for heavy pionic atoms with pion at low (deep) lying states are extremely scarce.

Earlier it is indicated a possibility of improving the consistency between theory and experiment by taking into account of the strong interaction potential isovector (absorption) terms. In fact, it is a more adequate account of increasing S-wave repulsion for absorption of pion at 3d-level. In this regard, under the parameterization of the optical potential authors [33] left without changing the parameters settings that are the most reliably identified, namely: ReB_0, ImB_0, c_0, c_1, ReC_0, ImC_0.

At the same time the parameters whose values differ most strongly in different sets, in particular, b_1 (plus parameters, which are usually not included so far in the basic optical potential parameterizations, i.e. ImB_1, ImC_1), should be optimized. This is achieved by receiving preliminary calculated relationships (illustrations are given below) for the energy shifts and widths (for a number of states f of the following systems: ^{20}Ne, ^{24}Mg, ^{93}Nb, ^{133}Cs, ^{175}Lu, ^{181}Ta, ^{197}Au, ^{208}Pb) upon the b_1, ImB_1, ImC_1 parameter values; further, there were chosen such values that satisfy the smallest standard deviation of the experimental values.

In Tables 3 and 4 we list the dependences of shifts and widths for the 4f, 3d levels due to the strong pion nuclear interaction upon the parameter ImB_1 ImC_1 values for the pionic atom of ^{208}Pb (our data).

Further in Tables 5 and 6 we list the analogous dependences of shifts and widths for the 4f, 3d levels due to the strong pion nuclear interaction upon the parameter ImB_1 ImC_1 values for the pionic system ^{181}Ta.

Table 3 The dependence of shifts and widths for the 4f, 3d levels due to the strong pion nuclear interaction upon the parameter ImB$_1$ value for ^{208}Pb

ImB$_1$	ε_0^{4f} (^{208}Pb)	Γ_0^{4f} (^{208}Pb)	ε_0^{3d} (^{208}Pb)	Γ_0^{3d} (^{208}Pb)
0.00	1.596	1.11	18.2	67.7
0.02	1.603	1.09	18.9	63.6
0.04	1.625	1.07	19.7	58.7
0.06	1.633	1.05	20.5	54.8
0.08	1.641	1.03	21.6	50.5
0.10	1.652	0.96	22.8	47.2
0.12	1.723	0.91	23.8	45.3
Exp.	1.68 ± 0.04	0.98 ± 0.05	22.7 ± 2.2	47.1 ± 3.6

Table 4 The dependence of shifts and widths for the 4f, 3d levels due to the strong pion nuclear interaction upon the parameter ImC$_1$ для ядра ^{208}Pb

ImC$_1$	ε_0^{4f} (^{208}Pb)	Γ_0^{4f} (^{208}Pb)	ε_0^{3d} (^{208}Pb)	Γ_0^{3d} (^{208}Pb)
0.00	1.610	1.050	18.8	68.3
−0.05	1.622	0.949	19.4	64.4
−0.10	1.633	0.942	19.8	60.3
−0.15	1.642	0.939	20.5	56.5
−0.20	1.653	0.928	21.8	52.3
−0.25	1.665	0.918	22.6	48.2
−0.30	1.676	0.899	23.2	44.5
Exp.	1.68 ± 0.04	0.98 ± 0.05	22.7 ± 2.2	47.1 ± 3.6

Table 5 The dependence of shifts and widths for the 4f, 3d levels due to the strong pion nuclear interaction upon the parameter ImB$_1$ value for ^{208}Pb ^{181}Ta

ImB$_1$	ε_0^{4f} (^{181}Ta)	Γ_0^{4f} (^{181}Ta)	ε_0^{3d} (^{181}Ta)	Γ_0^{3d} (^{181}Ta)
0.00	0.508	0.41	13.5	34.9
0.02	0.517	0.39	14.1	30.8
0.04	0.525	0.37	14.7	27.5
0.06	0.533	0.35	15.2	24.8
0.08	0.543	0.33	15.8	22.6
0.10	0.554	0.30	16.3	20.3
0.12	0.566	0.28	16.8	18.3
Exp.	0.56 ± 0.04	0.31 ± 0.05	16.2 ± 1.3	20.1 ± 1.5

3.2 Pionic Hydrogen and the Radiation Widths of the 3d, 4f, 5g Levels in Some $\pi^- A$ Systems

In Table 7 we list the calculated (in meV) QED corrections to the energies of the 1s, 2p, 3p, 4p states for pionic hydrogen: data by Indelicato et al. and Serga et al. and

Table 6 The dependence of shifts and widths for the 4f, 3d levels due to the strong pion nuclear interaction upon the parameter ImC_1 для ядра ^{81}Ta

ImC_1	ε_0^{4f} (^{81}Ta)	Γ_0^{4f} (^{81}Ta)	ε_0^{3d} (^{81}Ta)	Γ_0^{3d} (^{81}Ta)
0.00	0.486	0.334	12.8	33.7
−0.05	0.502	0.325	13.7	30.8
−0.10	0.519	0.313	14.3	28.5
−0.15	0.534	0.302	14.9	25.7
−0.20	0.547	0.274	15.6	23.4
−0.25	0.559	0.268	16.3	20.6
−0.30	0.571	0.255	16.8	19.9
Exp.	0.56 ± 0.04	0.31 ± 0.05	16.2 ± 1.3	20.1 ± 1.5

our data. The following abbreviations are used: the Uehling-Serber vacuum polarization correction (VP-US), the Kallen-Sabry correction (VP-KS), Wichman-Kroll one (PV-WK).

Analysis of data shows that there is physically reasonable agreement between different theoretical data, namely, data by Schlesser-Indelicato et al. [31, 32, 76–79]. Naturally, the reason is obvious as the known expansion parameter aZ in the hydrogen atom is significantly less than one, and the general QED contributions into the levels energies are not large in comparison with other ones.

In Table 8 we present our calculated (the relativistic Klein-Gordon-Fock theory combined with energy approach) data on the radiation widths of the 3d, 4f, 5g levels for a number of the pionic π^-A atoms. There are also listed the analogous

Table 7 The calculated (in meV) QED corrections to the energies of the 1s, 2p, 3p, 4p states for pionic hydrogen: data by Indelicato et al., Serga et al. and our data

QED contr.	1s [31, 32]	1s [76, 77]	1s [79]
VP-US	−3240.802	−3240.799	−3240.801
VP-KS	−24.365	−24.363	−24.365
PV-WK	−4.110	−4.113	−4.112
QED contr.	2p	2p	2p
VP-US	−35.795	−35.793	−35.794
VP-KS	−0.346	−0.343	−0.345
PV-WK	−0.008	−0.010	−0.009
QED contr.	3p	3p	3p
VP-US	−11.407	−11.405	−11.406
VP-KS	−0.108	−0.105	−0.107
PV-WK	−0.002	−0.003	−0.002
QED contr.	4p	4p	4p
VP-US	−4.921	−4.918	−4.920
VP-KS	−0.046	−0.044	−0.045
PV-WK	−0.001	−0.002	−0.001

Table 8 The radiation widths of the 3d, 4f, 5g levels in some $\pi^- A$ systems

Nucleus	Γ^1_{rad} (5g) [38, 39]	Γ^2_{rad} (5g) [79]	Γ^1_{rad} (4f) [38, 39]	Γ^2_{rad} (4f) [79]	Γ^1_{rad} (3d) [38, 39]	Γ^2_{rad} (3d) [79]
^{165}Ho	–	15.2	–	56.1	–	228.8
^{173}Yb	–	17.9	–	66.8	–	275.4
^{175}Lu	–	20.7	–	77.5	–	320.3
^{181}Ta	25.7	23.5	90.9	88.6	369.9	366.1
^{203}Tl	–	37.2	–	136.8	–	557.2
^{208}Pb	41.5	39.4	146.8	143.2	587.6	583.8
^{209}Bi	43.7	41.5	156.2	153.1	617.3	613.7

data by Laat-Konijn et al. [38, 39], obtained with using the relativistic Klein-Gordon-Fock model and Hartree-Fock approximation.

3.3 Transition Energies and Energy Shift and Widths Due to the Strong Interaction in the Spectra of in Some $\pi^- A$ Systems

In Figs. 2 and 3 there are presented the parts of the X-ray spectra of the ^{165}Ho (Fig. 2), ^{181}Ta (Fig. 3) and positions of the hyperfine structure components (5 g-4f transition; experimental data from [13]).

In Tables 9a and 10 we present theoretical and experimental data (in keV) for the 4f level shift (a) and widths (b) provided by the strong pion-nuclear interaction for a number of pionic atoms. The shortened designation of the parameter sets for the strong $\pi^- N$ interaction potential: Tauscher—Tau1; Tauscher, Tau2; Batty et al.—Bat; Seki et al.—Sek; de Laat-Konijn et al.—Laat, this work—Srg-Sha [16–22, 38, 39, 76–79]. In our parameterization of the strong $p^- N$ interaction potential the most reliably defined (B_0, c_0, c_1, C_0) parameters are remained unchanged, and the parameters whose values differ greatly in different sets, in particular, b_1 ($b_1 = -0.094$) plus still not included ones ImB_1, ImC_1 have been optimized by computing dependencies of the strong shifts upon the parameters b_1, ImB_1, ImC_1 for $\pi^- - {}^{20}$Ne, ^{24}Mg, ^{93}Nb, ^{133}Cs, ^{175}Lu, ^{181}Ta, ^{197}Au, ^{208}Pb atoms. Further we have chosen the values which satisfy the smallest standard deviation of reliable experimental values.

In Table 11 the analogous theoretical and experimental data (in keV) for the 3d level shift (a) and widths for different pionic atoms are listed [16–22, 38, 39, 76–79].

In Table 12 data on the *4f-3d, 5g-4f* transition energies for pionic atoms of the ^{93}Nb, ^{173}Yb, ^{181}Ta, ^{197}Au are presented. There are also listed the measured values

Fig. 2 The fragments of the X-ray spectra of the ^{165}Ho and positions of the hyperfine structure components (5-4f transition; experimental data from [13])

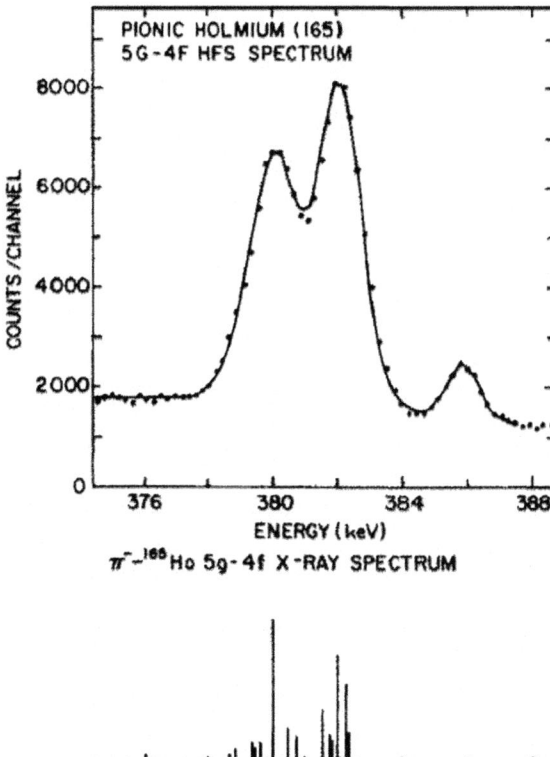

of the Berkley, CERN and Virginia laboratories and alternative data obtained on the basis of computing within alternative versions of the Klein-Gordon-Fock (KGF) theory with taking into account for a finite size of the nucleus in the model uniformly charged sphere and the standard Uehling-Serber radiation correction (see Refs. [6, 7, 13, 42, 43, 86, 89]).

The analysis of the presented data indicate on the importance of the correct accounting for the radiation (vacuum polarization) and the strong pion-nuclear interaction corrections. The contributions due to the nuclear finite size effect should be accounted in a precise theory too. More exact knowledge of the electromagnetic interaction parameters for a pionic atom will make more clear the true values for parameters of the pion-nuclear potentials. Further it allows to correct a disadvantage of widely used parameterization of the optical potential. It is especially important if one takes into account an increasing accuracy of the X-ray pionic atom spectroscopy experiments. It is interesting to note that the contributions into transition energies are about ~5 keV due to the QED effects, ~0.2 keV due to the nuclear

Fig. 3 The fragments of the X-ray spectra of the ^{181}Ta and positions of the hyperfine structure components (5g-4f transition; experimental data from [13])

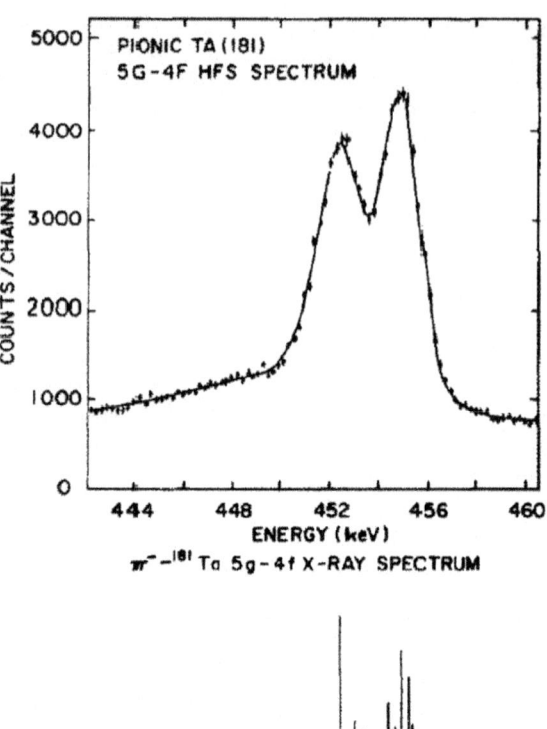

Table 9 Theoretical and experimental data for the 4f level shift (keV) provided by the strong pion-nuclear interaction for a number of pionic atoms (see text)

$\varepsilon_{4f}, \Gamma_{4f}$	Exp	H-like Func.	Tau 1 $\xi = 0$	Tau 2 $\xi = 1$	Ba $\xi = 1$	Sek $\xi = 1$	Laat $\xi = 1$	Srg-Sha $\xi = 1$	Ou r $\xi = 1$
^{165}Ho: ε	0.29 ± 0.01	0.21	0.25 0.27	0.24 0.26	0.24	0.21	0.26	0.29	0.29
^{169}Tm: ε	–	–	–	–	–	–	–	0.38	0.38
^{173}Yb: ε	–	–	–	–	–	–	–	0.44	0.44
^{175}Lu: ε	0.51 ± 0.04	0.36	0.43	0.42	0.41	0.36	0.46	0.50	0.50
^{181}Ta: ε	0.56 ± 0.04	0.47	0.57	0.54	0.53	0.47	0.60	0.55	0.55
^{197}Au: ε	1.25 ± 0.07	–	1.21	1.14	1.12	0.98	1.25	1.24	1.24
^{208}Pb: ε	1.68 ± 0.04	–	1.76	1.62	1.58	1.39	1.68	1.65	1.65
^{209}Bi: ε	1.78 ± 0.06	–	1.94	1.80	1.78	1.57	1.83	1.77	1.77

finite size effect, and ~0.07 keV due to the electron screening effect, provided by the 2[He], 4[Be], 10[Ne] electron shells [79].

To conclude, let us underline that the key factors for the physically reasonable agreement between experimental and theoretical data on the multi-electron pionic

Table 10 Theoretical and experimental data for the 4f level widths (keV) provided by the strong pion-nuclear interaction for a number of pionic atoms (see text)

$\varepsilon_{4f}, \Gamma_{4f}$	Exp	H-like Func.	Tau 1 $\xi = 0$	Tau 2 $\xi = 1$	Bat $\xi = 1$	Sek $\xi = 1$	Laat $\xi = 1$	Srg-Sha $\xi = 1$	Our $\xi = 1$
^{165}Ho: Γ	0.21 ± 0.02	0.08	0.13	0.12	0.13	0.11	0.13	0.20	0.21
^{169}Tm: Γ	–	–	–	–	–	–	–	–	0.23
^{173}Yb: Γ	–	–	–	–	–	–	–	–	0.26
^{175}Lu: Γ	0.27 ± 0.07	0.14	0.23	0.22	0.24	0.20	0.24	0.28	0.28
^{181}Ta: Γ	0.31 ± 0.05	0.16	0.31	0.30	0.31	0.27	0.31	0.30	0.31
^{197}Au: Γ	0.77 ± 0.04	–	0.73	0.68	0.69	0.58	0.67	0.75	0.77
^{208}Pb: Γ	0.98 ± 0.05	–	1.18	1.04	1.03	0.86	0.98	0.97	0.99
^{209}Bi: Γ	1.24 ± 0.09	–	1.35	1.18	1.17	0.99	1.10	1.22	1.25

Table 11 Theoretical and experimental data for the 3d level shifts and widths (keV) provided by the strong pion-nuclear interaction for a number of pionic atoms (see text)

3d	Exp.	Tau 1 $\xi = 0$	Tau 2 $\xi = 1$	Bat $\xi = 1$	Sek $\xi = 1$	Laat $\xi = 1$	Srg-Sha $\xi = 1$	Odes $\xi = 1$
^{93}Nb: ε	0.74 ± 0.02	0.66	0.67	0.73	0.66	0.75	0.75	0.73
^{169}Tm: ε	–	–	–	–	–	–	–	11.0
^{173}Yb: ε	–	–	–	–	–	–	–	12.4
^{175}Lu: ε	–	–	–	–	–	–	–	13.9
^{181}Ta: ε	16.2 ± 1.3	19.6	16.4	10.4	4.4	14.4	16.3	16.1
^{197}Au: ε	20.6 ± 1.9	27.9	22.5	13.2	5.0	20.3	21.1	20.3
^{208}Pb: ε	22.7 ± 2.2	34	25	13	3	18	22.8	22.6
^{209}Bi: ε	20 ± 3	37	27	17	5	19	23	21.1
^{93}Nb: Γ	0.40 ± 0.02	0.405	0.413	0.459	0.404	0.452	0.42	0.41
^{169}Tm: Γ	–	–	–	–	–	–	–	15.7
^{173}Yb: Γ	–	–	–	–	–	–	–	17.6
^{175}Lu: Γ	–	–	–	–	–	–	–	19.4
^{181}Ta: Γ	20,1 ± 1,5	40,5	37,5	33,4	26,2	27,6	20.3	20.2
^{197}Au: Γ	34±3.6	68	62	53	41	42	36.2	35.6
^{208}Pb: Γ	47.1 ± 3.6	88	78	65	51	51	47.2	47.0
^{209}Bi: Γ	52 ± 3.6	97	86	72	57	56	53.6	53.4

atoms are provided by a correct consideration of the nuclear, relativistic, radiative and inter-electron correlation corrections. Using different schemes for accounting of these correlations explains a difference between calculation results, obtained within the Klein-Gordon-Fock approach. To reach the further improvement of the computed data one should take into account more correctly the spatial distribution of the

Table 12 Transition energies (keV) in the spectra of some heavy pionic atoms (see text)

π–A	Trans.	Berkley E_{exp}	CERN E_{exp}	E_{KGF+EM} [6, 7]	E_{KGF-EM} [13]	E_N [6]	E_N [13]	E_N [89] a
^{93}Nb	5g-4f	–	307.79 ± 0.02	–	–	–	–	307.85
^{173}Yb	5g-4f	–	–	–	–	–	–	412.26
^{181}Ta	5g-4f	453.1 ± 0.4	453.90 ± 0.20	453.06	453.78	–	453.52 453.62	453.71
^{197}Au	5g-4f	532.5 ± 0.5	533.16 ± 0.20	528.95	–	532.87	531.88	533.08
^{93}Nb	4f-3d	–	140.3 ± 0.1	–	–	–	–	140.81
^{173}Yb	4f-3d	–	–	–	–	–	–	838.67
^{181}Ta	4f-3d	–	1008.4 ± 1.3	–	–	–	992.75	1008.80

magnetic moment inside a nucleus (the Bohr-Weisskopf effect), the nuclear-polarization corrections etc. (for example, within the Woods-Saxon model or relativistic mean filed theory). In last years this topic has been a subject of intensive interest.

References

1. PSI Experiment R-98.01. http://www.fz-juelich.de/ikp/exotic-atoms
2. Ericson T, Weise W (1988) Pions and Nuclei. Clarendon, Oxford
3. Deloff A (2003) Fundamentals in Hadronic Atom Theory. World Sci, Singapore
4. Khetselius OYu (2011) Quantum structure of electroweak interaction in heavy finite fermi-systems. Astroprint, Odessa
5. Deslattes R, Kessler E, Indelicato P, de Billy L, Lindroth E, Anton J (2003) Exotic atoms. Rev Mod Phys 75:35
6. Backenstoss G (1970) Ann Rev Nucl Sci 20:467
7. Menshikov LI, Evseev MK (2001) Phys Uspekhi 171:150
8. Scherer S (2003) In: Negele JW, Vogt EW (eds) Advances in nuclear physics, vol 27. Springer, Berlin, pp 5–50
9. Schroder H, Badertscher A, Goudsmit P, Janousch M, Leisi H, Matsinos E, Sigg D, Zhao Z, Chatellard D, Egger J, Gabathuler K, Hauser P, Simons L, El Hassani A (2001) J Phys C21:473
10. Leon M, Seki R (1974) Phys Rev Lett 32:132
11. Batty CJ, Eckhause M, Gall KP et al. (1989) Phys Rev C 40:2154
12. Chen MY, Asano Y, Cheng SC, Dugan G, Hu E, Lidofsky L, Patton W, Wu CS (1975) Nucl Phys A 254:413
13. Olaniyi B, Shor A, Cheng S, Dugan G, Wu CS (1982) Nucl Phys A 403:572
14. Erikcson M, Ericson T, Krell M (1969) Phys Rev Lett 22:1189
15. Erikcson M, Ericson T (1966) Ann Phys 36:323

16. Tauscher L (1971) Analysis of pionic atoms and the p-nucleus optical potential. In: Proceedings of the international semantic p-Meson nucleus interaction-CNRS-strasbourg, France, p 45
17. Batty C, Biagi S, Friedman E, Hoath S (1983) Phys Rev Lett 440:931; Batty C J, Friedman E, Gal A (1978) Nucl Phys A 402:411
18. Seki R, Masutani K, Jazaki K (1983) Phys Rev C 27:1817
19. Rowe G, Salamon M, Landau RH (1978) Phys Rev C 18:584
20. Anagnostopoulos D, Biri S, Boisbourdain V, Demeter M, Borchert G et al. (2003) Nucl Inst Meth B 205:9
21. Anagnostopoulos D, Gotta D, Indelicato P, Simons LM (2003) arXiv:physics. 0312090v1
22. Nagels MM, de Swart J, Nielsen H et al (1976) Nucl Phys B 109:1
23. Lauss B (2009) Nucl Phys A 827C, 401 PSI experiment R-98.01http://pihydrogen.psi.ch
24. CERN DIRAC Collaboration (2011) Search for long-lived states of $p^+ p^-$ and $p^- K$ atoms, CERN-SPSLC-2011–001 SPSLC-P-284-ADD p 22
25. Umemoto Y, Hirenzaki S, Kume K, Toki H, Tahihata I (2001) Nucl Phys A 679:549
26. Nose-Togawa N, Hirenzaki S, Kume K (1999) Nucl Phys A 646:467
27. Glushkov AV, Malinovskaya SV, Gurnitskaya EP, Khetselius OYu, Dubrovskaya YuV (2006) J Phys: Conf Ser 35:425
28. Hatsuda T, Kunihiro T (1994) Phys Rep 247:221
29. Ikeno N, Kimura R, Yamagata-Sekihara J, Nagahiro H, Jido D, Itahashi K et al. (2011) 1107.5918v1[nucl-th]
30. Kolomeitsev EE, Kaiser N, Weise W (2003) Phys Rev Lett 90:092501
31. Lyubovitskij V, Rusetsky A (2000) Phys Lett B 494:9
32. Schlesser S, Le Bigot E-Q, Indelicato P, Pachucki K (2011) Phys Rev C 84:015211
33. Sigg D, Badertscher A, Bogdan M, Goudsmit P, Leisi H, Schröder H, Zhao Z, Chatellard D, Egger J, Jeannet E, Aschenauer E, Gabathuler K, Simons L (1996) Rusi El Hassan A. Nucl Phys A 609:269
34. Gotta D, Amaro F, Anagnostopoulos D, Biri S, Covita D, Gorke H, Gruber A, Hennebach M, Hirtl A, Ishiwatari T, Indelicato P, Jensen T, Bigot E, Marton J, Nekipelov M, dos Santos J, Schlesser S, Schmid P, Simons L, Strauch H, Trassinelli M, Veloso J, Zmeskal J ed. Kania Y, Yamazaki Y (AIP) (2008) CP1037, 162
35. Gotta D, Amaro F, Anagnostopoulos D, Biri S, Covita D, Gorke H, Gruber A, Hennebach M, Hirtl A, Ishiwatari T, Indelicato P, Jensen T, Bigot E, Marton J, Nekipelov M, dos Santos J, Schlesser S, Schmid P, Simons L, Strauch H, Trassinelli M, Veloso J, Zmeskal (2008) Precision physics of simple atoms and molecules. In: Lecture notes in physics, vol 745, Springer, Berlin, Heidelberg, pp 165–186
36. Taal A, D'Achard van Enschut J, Berkhput J et al (1985) Phys Lett B 156:296
37. Taal A, David P, Hanscheid H, Koch JH, de Laat CT et al (1990) Nucl Phys A 511:573
38. de Laat CT, Taal A, Konijn J et al (1991) Nucl Phys A 523:453
39. de Laat CT, Taal A, Duinker W et al (1987) Phys Lett B 189:7
40. Khetselius OYu, Turin AV, Sukharev DE, Florko TA (2009) Sensor Electr and Microsyst Techn N1, 30–35
41. Mohr PJ (1993) Atom Dat Nucl Dat Tabl 24:453; (1983) Phys Scripta 46:44
42. Indelicato P (1996) Phys Scripta T65:57; Indelicato P, Trassinelli M (2005) arXiv: physics.0510126v1
43. Glushkov AV, Malinovskaya SV (2003) In: Fazio G, Hanappe F (eds) New projects and new lines of research in nuclear physics. World Sci, Singapore, pp 242–250
44. Santos J, Parente F, Boucard S, Indelicato P, Desclaux J (2005) Phys Rev A 71:032501
45. Mitroy J, Ivallov IA (2001) J Phys G Nucl Part Phys 27:1421
46. Strauch T (2009) High-precision measurement of strong-interaction effects in pionic deuterium, Julich
47. Serot B, Walecka J (1986) Relativistic nuclear many body problem. In: Advances in nuclear physics, Plenum Press, N-Y
48. Glushkov AV, Ivanov LN (1992) Phys Lett A 170:33

49. Glushkov AV, Ivanov LN, Ivanova EP (1986) Autoionization phenomena in atoms. Moscow University Press, Moscow, pp 58–160
50. Ivanova EP, Glushkov AV (1986) J Quant Spectr Rad Transfer 36:127
51. Glushkov AV, Khetselius OYu, Gurnitskaya EP, Loboda AV, Florko TA, Sukharev DE, Lovett L (2008) Frontiers in quantum systems in chemistry and physics. In: Wilson S, Grout PJ, Maruani J, Delgado-Barrio G, Piecuch P (eds) Progress in Theoretical Chemistry and Physics, vol 1. Springer, Dordrecht, pp 507–524
52. Flambaum VV, Ginges JSM (2005) Phys Rev A 72:052115
53. Safranova UI, Safranova MS, Johnson WR (2005) Phys Rev A 71:052506
54. Glushkov AV, Malinovskaya SV, Khetselius OYu, Loboda AV, Sukharev DE, Lovett L (2009) Int J Quant Chem 109:1717
55. Gurnitskaya EP, Khetselius OYu, Loboda AV, Vitavetskaya LA (2008) Photoelectronics 17:127
56. Glushkov AV, Ambrosov SV, Loboda AV, Gurnitskaya EP, Prepelitsa GP (2005) Int J Quant Chem 104:562
57. Johnson W, Sapirstein J, Blundell S (1993) Phys Scripta T 46:184
58. Glushkov AV (2006) Relativistic and correlation effects in spectra of atomic systems, Astroprint, Odessa
59. Glushkov AV, Khetselius OYu, Malinovskaya SV (2008) Europ Phys Journ ST 160:195
60. Buyadzhi VV, Glushkov AV, Lovett L (2014) Photoelectronics 23:38
61. Glushkov AV, Ambrosov SV, Loboda AV, Chernyakova GYu, Svinarenko AA, Khetselius OY (2004) Nucl Phys A Nucl Hadr Phys 734:21
62. Glushkov AV, Ambrosov SV, Loboda AV, Gurnitskaya EP, Khetselius OY (2006) Recent advances in the theory of chemical and physical systems. In: Julien P, Maruani J, Mayou D, Wilson S, Delgado-Barrio G (eds) Progress in theoretical chemistry and physics, vol 15. Springer, Dordrecht, pp 285–299
63. Glushkov AV (2005) AIP Conf Proceedings 796 (1): 206–210
64. Glushkov AV, Lovett L, Khetselius OYu, Gurnitskaya EP, Dubrovskaya YuV, Loboda AV (2009) Int J Mod Phys A 24:611
65. Glushkov AV, Khetselius OY and Lovett L (2009) Advances in the theory of atomic and molecular systems. Dynamics, spectroscopy, clusters and nanostructures. In: Piecuch P, Maruani J, Delgado-Barrio G, Wilson S (eds) Progress in theoretical chemistry and physics, vol 20. Springer, Dordrecht, pp 125–152
66. Khetselius OY (2012) Quantum systems in chemistry and physics. In: Nishikawa K, Maruani J, Brändas E, Delgado-Barrio G, Piecuch P (eds) Progress in theoretical chemistry and physics, vol 26. Springer, Dordrecht, pp 217–229
67. Khetselius OYu (2009) Int J Quant Chem 109:3330
68. Khetselius OYu (2009) Phys Scripta T135:014023
69. Khetselius OYu (2015) Frontiers in quantum methods and applications in chemistry and physics. In: Nascimento M, Maruani J, Brändas E, Delgado-Barrio G (eds) Progress in theoretical chemistry and physics, vol 29. Springer, Cham, pp 55–76
70. Khetselius OYu (2008) Hyperfine structure of atomic spectra. Astroprint, Odessa
71. Glushkov AV, Khetselius OYu, Svinarenko AA (2013) Phys Scripta T153:014029
72. Glushkov AV, Khetselius OY, Svinarenko AA (2012) Advances in the theory of quantum systems in chemistry and physics. In: Hoggan P, Brändas E, Maruani J, Delgado-Barrio G, Piecuch P (eds) Progress in theoretical chemistry and physics, vol 22. Springer, Dordrecht, pp 51–68
73. Khetselius O, Florko T, Svinarenko A, Tkach T (2013) Phys Scripta T153: 014037
74. Khetselius OYu (2010) AIP Conf Proc 1290:29
75. Khetselius OYu, Florko TA, Nikola LV, Svinarenko AA, Serga IN, Tkach TB, Mischenko EV (2010) Quantum Theory: reconsideration of foundations (AIP). 1232:243
76. Serga IN, Dubrovskaya YV, Kvasikova AS, Shakhman AN, Sukharev DE (2012) J Phys Conf Ser 397:012013
77. Serga IN (2013) Photoelectronics 22:71; Shakhman AN (2015) Photoelectronics 24:109

78. Sukharev DE, Khetselius OYu and Dubrovskaya YuV (2009) Sensor Electr and Microsyst Techn N3, 16–21
79. Bystryantseva AN, Khetselius OYu, Dubrovskaya YuV, Vitavetskaya LA, Berestenko AG (2016) Photoelectronics 25:56
80. Glushkov AV (2008) Relativistic quantum theory. Quantum mechanics of atomic systems, Astroprint, Odessa, 700p
81. Ivanov LN, Letokhov VS (1985) Com Mod Phys D 4:169; Ivanov LN, Ivanova EP, Aglitsky EV (1988) Phys Rep 166:315
82. Ivanova EP, Ivanov LN, Glushkov AV, Kramida AE (1985) Phys Scripta 32:513
83. Glushkov AV, Kondratenko PA, Buyadgi VV, Kvasikova AS, Sakun TN, Shakhman AN (2014) J Phys: Conf Ser 548:012025
84. Svinarenko AA, Glushkov AV, Khetselius OYu, Ternovsky VB, Dubrovskaya YuV, Kuznetsova AA, Buyadzhi VV (2017) In: Rare Earth Element, (ed) Orjuela JEA. InTech, pp 83–104
85. Glushkov AV, Khetselius OYu, Svinarenko AA, Buyadzhi VV, Ternovsky VB, Kuznetsova AA, Bashkarev PG (2017) In: Uzunov DI (ed) Recent studies in perturbation theory. InTech, pp 131–150
86. Baldwin GG, Salem JC, Goldansky VI (1981) Rev Mod Phys 53:687; Goldansky VI, Letokhov VS (1974) JETP 67:513; Ivanov LN, Letokhov VS (1975) JETP 68:1748
87. Glushkov AV, Khetselius O, Gurnitskaya E, Loboda A, Sukharev D (2009) AIP Conf Proc 1102 (1):168
88. Ivanov LN, Letokhov VS, Glushkov AV (1991) Preprint of Inst. for Spectroscopy of USSR Acad Sci (ISAN) AS-N5
89. Glushkov AV (2013) Advances in quantum methods and applications in chemistry, physics and biology. In: Hotokka M, Brändas E, Maruani J, Delgado-Barrio G (eds) Progress in theoretical chemistry and physics, vol 27. Springer, Cham, pp 161–177
90. Glushkov AV, Ambrosov SV, Loboda AV, Gurnitskaya EP, Khetselius OYu (2006) Recent advances in theoretical physics and chemistry systems. In: Julien J-P, Maruani J, Mayou D, Wilson S, Delgado-Barrio G (eds) Progress in theoretical chemistry and physics, vol 15. Springer, Dordrecht, pp 285–299
91. Glushkov AV (2012) Quantum systems in chemistry and physics. In: Nishikawa K, Maruani J, Brändas E, Delgado-Barrio G, Piecuch P (eds) Progress in theoretical chemistry and physics, vol 26. Springer, Dordrecht, pp 231–252
92. Khetselius OYu, Zaichko PA, Smirnov AV, Buyadzhi VV, Ternovsky VB, Florko TA, Mansarliysky VF (2017) Quantum systems in physics, Chemistry and biology. In: Tadjer A, Pavlov R, Maruani J, Brändas J, Delgado-Barrio G (eds) Progress in theoretical chemistry and physics, vol 30. Springer, Cham, pp 271–281

Part II
Molecular Structure and Dynamics

Difference of Chirality of the Electron Between Enantiomers of H_2X_2

Masato Senami, Ken Inada, Kota Soga, Masahiro Fukuda
and Akitomo Tachibana

Abstract The integrated chirality density of H_2X_2 molecules is studied in view-points of the internal torque for the electron spin. Since the chirality density is proportional to the zeta potential, which is the potential of the zeta force, one of the torque for the electron spin, the distribution of the chirality density affects the distribution of the internal torque in molecules. It is seen that the integrated chirality density is larger for the larger atomic number. It is found that the integrated chirality density of H_2Te_2 has the same sign as the parity-violating energy, while those of H_2O_2 and H_2S_2 are opposite to the sign of the parity-violating energy, and the dependence of the integrated chirality density of H_2Se_2 on dihedral angle is significantly different from that of the parity-violating energy.

1 Introduction

Only one form of enantiomeric pair is found in living systems. While both enantiomers of sugar and amino acids are produced in the same quantity in laboratories, only (D)-sugars and (L)-amino acids can be found in the systems. The mechanism how this bias is generated is a long-standing puzzle [1]. On the origin of this biomolecular homochirality, many hypotheses are proposed, and those are classified by some features, (1) terrestrial and extra-terrestrial, (2) biotic and abiotic, and (3) probabilistic and deterministic. We do not know the solution, and however many researchers believe that the realization of homochirality is deeply related to the parity violation of Nature, which is given by the weak interaction of the standard model of particle physics. The weak interaction is known to induce nuclear β decay. Gauge theory describes this interaction as well as electromagnetic interaction. Electromagnetic interaction is formulated as U(1) gauge theory and is mediated by the photon.

M. Senami (✉) · K. Inada · K. Soga · M. Fukuda · A. Tachibana
Department of Micro Engineering, Kyoto University, Kyoto 615-8540, Japan
e-mail: senami@me.kyoto-u.ac.jp

© Springer International Publishing AG, part of Springer Nature 2018
Y. A. Wang et al. (eds.), *Concepts, Methods and Applications of Quantum Systems in Chemistry and Physics*, Progress in Theoretical Chemistry and Physics 31,
https://doi.org/10.1007/978-3-319-74582-4_6

On the other hand, the weak interaction is formulated as SU(2) gauge theory, and W and Z bosons are mediator of this interaction. These bosons are coupled to ordinary fermions, such as electrons, only by the V-A coupling, where V and A is the vector and axial-vector currents, respectively. Hence parity is violated maximally in the weak interaction.

It is known that this parity violation makes the energy difference between enantiomers. This energy difference between them, which is called parity-violating energy shift, is very small [2]. In spite of the smallness, this energy is studied in many works by computational methods. The parity-violating energy shift is dominantly induced by virtual Z boson exchanges between nuclei and electrons. Virtual Z boson exchanges is quantum effect in relation to uncertainty principle. Since Z boson is very heavy as 90 GeV/c^2, where c is the speed of light, this particle cannot be produced energetically, and, however, uncertainty relation allows this particle to affect in a very restricted region. Single W boson exchange does not contribute to the parity-violating energy, since if occurs, it is β decay. The exchanges between electrons is known to be subdominant for particularly heavy nuclei [3]. Since nuclei are localized strongly, the existence of parity-violating energy means the existence of the nonzero chirality density around nuclei. It is surprising that low energy electrons have polarized chirality, since the electron mass, that is the interaction with Higgs field in the vacuum, vanishes chirality for free electrons.

The parity-violating energy shift is studied by many researchers as one of solutions for biomolecular homochirality. Due to this energy difference, the amount of one of enantiomers may be slightly larger than the other through an enhancement process, such as crystallization. In other viewpoints, some of the authors are interested in the total integrated chirality density of the electron in a molecule, whose existence has already been reported [4]. If nonzero integrated chirality density of the electron exists in enantiomers, its interaction rates through weak interaction are different between enantiomeric pair. This difference of the reaction rate probably generates the difference of the number density between enantiomers, which is indispensable for the problem of the biomolecular homochirality. Even though this difference is not enough, we are interested in the distribution of the chirality density in a molecule, since the chirality density is known to be proportional to the zeta potential [5]. The zeta potential is the potential for the zeta force, which is the counter force to the spin torque, defined only in quantum field theory [6, 7]. Hence chiral molecules have nonzero zeta force distribution.

In this work, we study the integrated chirality density of the electron in H_2X_2 molecules (X = O, S, Se, Te). This structure is one of the simplest chiral molecules and is chosen in many works [4, 8–10]. The total chirality of the electron of this structure is reported for H_2Te_2 in Reference [4]. In this work, the chirality is investigated also for H_2O_2, H_2S_2, and H_2Se_2 in relation to the parity-violating energy shift and the zeta potential.

2 Theory

First we briefly review the parity-violating energy. The dominant contribution to the parity-violating energy is the parity odd interaction between electrons and nuclei. This interaction is given as the coupling of vector and axial-vector currents of electrons and nucleons. For low energy nuclei, nonrelativistic approximation is good, and then the space-like component of vector current and the time-like component of axial-vector current vanishes. The internal structure of nuclei is not affected by the geometry of molecules, and hence the space-like component of axial-vector current is considered to be negligibly small. Hence, the most important contribution arises from the coupling of the time-like components of nucleon vector current and electron axial-vector current. The Hamiltonian of this interaction is given as

$$H_{PV} = \sum_n \frac{G_F}{2\sqrt{2}} Q_W^n \hat{\psi}_e^\dagger \gamma_5 \hat{\psi}_e \hat{\psi}_{N_n}^\dagger \hat{\psi}_{N_n} \tag{1}$$

where index n means the species of nuclei, $G_F = 1.166378 \times 10^{-5}$ GeV^{-2} is Fermi coupling constant [11], $\hat{\psi}_e$ and $\hat{\psi}_{N_n}$ are the field operators of electrons and nuclei, and $\gamma_5 \equiv i\gamma^0\gamma^1\gamma^2\gamma^3$. The weak charge of a nucleus Q_W^n is given as $Q_W^n = Z^n(1 - 4\sin^2\theta_W) - N^n$, where Z^n and N^n are the number of protons and neutrons in a nucleus n and θ_W is the weak-mixing angle, $\sin^2\theta_W = 0.2313$ [11]. The parity-violating energy is calculated with state vector, $|\Psi\rangle$,

$$E_{PV} = \int d^3\vec{r} \langle \Psi | H_{PV} | \Psi \rangle \tag{2}$$

and the parity-violating energy shift is defined as the energy difference between enantiomers,

$$\Delta E_{PV} = 2|E_{PV}|. \tag{3}$$

The parity-violating energy can be divided into contributions from each nucleus,

$$E_{PV} = \frac{G_F}{2\sqrt{2}} \sum_n Q_W^n \left(\int d^3\vec{r} \langle \Psi | \hat{\psi}_e^\dagger \gamma_5 \hat{\psi}_e \hat{\psi}_{N_n}^\dagger \hat{\psi}_{N_n} | \Psi \rangle \right) = \frac{G_F}{2\sqrt{2}} \sum_n Q_W^n M_{PV}^n. \tag{4}$$

This M_{PV}^n is often used for parameterizing the contribution from each nucleus. The density of nuclei is strongly localized, and hence nucleus density, $\hat{\psi}_{N_n}^\dagger \hat{\psi}_{N_n}$, can be approximated to $\hat{\psi}_{N_n}^\dagger \hat{\psi}_{N_n} = \delta^3(\vec{r} - \vec{r}_n)$ where $\vec{r}_n$ is the position of a nucleus n. As a result, the parity-violating energy is well calculated by chirality densities at the positions of nuclei, $\langle \Psi | \hat{\psi}_e^\dagger(\vec{r}_n)\gamma_5\hat{\psi}_e(\vec{r}_n) | \Psi \rangle$.

Chirality density is proportional to the zeta potential, which is the potential of the zeta force defined in Reference [5]. The definition of the zeta potential is given by

$$\hat{\phi}_5(x) = \frac{\hbar c}{2} \left[\hat{\psi}_e^\dagger(x) \gamma_5 \hat{\psi}_e(x) \right] = \frac{\hbar c}{2} \left(\hat{\psi}_{eR}^\dagger(x) \hat{\psi}_{eR}(x) - \hat{\psi}_{eL}^\dagger(x) \hat{\psi}_{eL}(x) \right), \qquad (5)$$

where $\hat{\psi}_{eL}(x) \equiv \frac{1}{2}(1 - \gamma_5)\hat{\psi}_e$ and $\hat{\psi}_{eR}(x) \equiv \frac{1}{2}(1 + \gamma_5)\hat{\psi}_e$ are fields with the left-handed and right-handed chirality, respectively. The zeta force density operator is defined with the zeta potential, $\hat{\phi}_5$, as

$$\hat{\zeta}_e^k(x) = -\partial_k \hat{\phi}_5. \qquad (6)$$

The zeta force is one of the torque in the equation of motion of the electron spin defined in quantum field theory. The equation of motion of the spin is derived from the time-derivative of the spin angular momentum. In quantum field theory, the spin angular momentum density operator is represented as

$$\hat{s}_e^k(x) = \frac{1}{2} \hbar \hat{\psi}_e^\dagger(x) \Sigma^k \hat{\psi}_e(x), \qquad (7)$$

where Σ^k is the Pauli matrix in the four-component representation. The torque density for the electron spin is derived by the time-derivative of this operator, and the time-derivative can be reduced by using Dirac equation.

$$i\hbar \gamma^\mu \hat{D}_\mu(x) \hat{\psi}_e(x) = mc\hat{\psi}_e(x), \qquad (8)$$

where m is the mass of electron. The covariant derivative $\hat{D}_\mu(x)$ is defined as $\hat{D}_\mu(x) = \partial_\mu + i\frac{Z_e e}{\hbar c}\hat{A}_\mu(x)$, where $Z_e = -1$ is the electric charge of the electron and $\hat{A}_\mu(x)$ is the gauge field operator. As a result, we obtain the equation of motion of the spin,

$$\frac{\partial \hat{s}_e^k(x)}{\partial t} = \hat{t}_e^k(x) + \hat{\zeta}_e^k(x), \qquad (9)$$

where the first term, $\hat{t}_e^k(x)$, is the spin torque density. The spin torque term is defined so that this term matches the well-known spin torque term in relativistic quantum mechanics. The Heisenberg equation of the spin angular momentum in quantum mechanics is given by $d\vec{s}_e/dt = -c\hat{\vec{\pi}} \times \vec{\alpha}$ [12]. The spin torque density operator is defined with the relativistic stress tensor density, $\hat{t}_e^{\Pi ln}(x)$, as

$$\hat{t}_e^k(x) = -\epsilon_{lnk} \hat{t}_e^{\Pi ln}(x), \qquad (10)$$

where ϵ_{lnk} is the Levi-Civita tensor. The relativistic stress tensor operator is given by [5],

$$\hat{t}_e^{\Pi ln}(x) = \frac{i\hbar c}{2} \left[\hat{\psi}_e^\dagger(x) \gamma^0 \gamma^n \hat{D}_l(x) \hat{\psi}_e(x) - \left(\hat{D}_l(x) \hat{\psi}_e(x) \right)^\dagger \gamma^0 \gamma^n \hat{\psi}_e(x) \right]. \qquad (11)$$

The stress tensor is known to classify the chemical bond [13, 14]. In the following, we call only $\hat{\tau}_e^k(x)$ the spin torque, and the sum of the terms in the right-hand side is called the torque for the spin. Our equation of motion for the spin is recently shown to be derived from the spin vorticity principle in a sophisticated way [15].

One may wonder whether this new contribution disturbs the consistency between experimental observations and a prediction by quantum mechanics. The expectation value of the zeta force is zero after the integration over the whole space, since the zeta force density operator is given as the gradient form of the zeta potential. Hence, contributions from the zeta force are considered to be negligible in past experiments. However, the contribution from the zeta force give a nonzero effect in a local region, even after the integration over a restricted local region. Hence, the zeta force is observable quantity if an experimental setup is carefully designed for this purpose. In addition, the equation of motion from quantum field theory has another advantage over that from quantum mechanics. In a time-independent stationary state of the electron spin. the spin torque and zeta force are canceled out with each other and the torque for the spin is zero at any point. Hence in quantum field theory a local picture of the spin dynamics can be correctly described. In quantum mechanics, any local spin dynamics cannot be predicted. The Heisenberg equation of the spin gives generically nonzero torque for a local region even for a spin stationary state. This is because the expectation value of quantum mechanics is defined as the integration over the whole space and hence local description is theoretically out of scope.

3   Computational Details

Our equation is defined in quantum field theory, and hence a state should also be prepared in the theory. However, a generic state based on quantum field theory is not available for our purpose, since most computation code is based on quantum mechanics. Hence in this work we use wave functions derived from ordinary electronic structure computations as a substitution. With the usage of these wave functions, computations of physical quantities, parity-violating energy, zeta potential, and so on, are performed by QEDynamics program package [16–18].

For ordinary electronic structure computations, we use DIRAC14 program package [19]. Four-component wave function by relativistic quantum mechanics can be computed by this code, which is indispensable for the study of spin. The dyall.ae2z basis set [20] is used for large components of all atoms. The small component basis set is generated by restricted kinetic balance in the code. The effect of three-component vector potential is ignored in our calculations, since it is quantitatively small for states by quantum mechanics computations [21]. The structure of H_2X_2 molecules are determined as follows. First, the geometrical optimization computations are performed by Hartree-Fock computations with Dirac-Coulomb Hamiltonian. The internuclear lengths between heavy atoms X, 1.390 Å for oxygen atoms, 2.058 Å for sulfur atoms, 2.333 Å for selenium atoms, and 2.729 Å for tellurium atoms, while the internuclear lengths between X and H atoms, 0.9439 Å for oxygen

Fig. 1 Geometry of H_2X_2 molecule and the definition of the dihedral angle

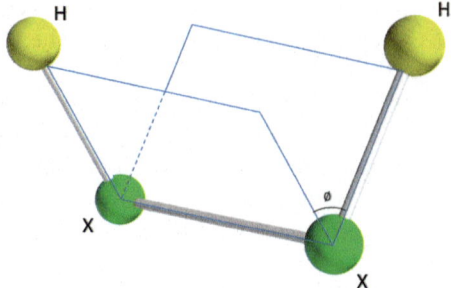

atom, 1.332 Å for sulfur atom, 1.455 Å for selenium atom, and 1.649 Å for tellurium atom. In the following, these internuclear lengths are adopted for all dihedral angles.

For wave functions derived by computations as above, parity-violating energy, M_{PV}^n, total chirality, spin torque density, zeta force, and zeta potential are computed with special attention to dihedral angle of H_2X_2. Since H_2X_2 is a chiral molecule, two choices of dihedral angle exist. Our definition of the dihedral angle is shown in Fig. 1, which is the opposite to Reference [8, 9]. Results can easily be compared by replacing the angle $360° - \phi$. Note that parity-violating energy is known to be heavily dependent on computational methods, geometry of molecules, basis sets and so on. Our basis set is smaller than previous works, and hence there are some differences between our results and previous works [4, 8–10] as discussed later.

4　Result and Discussion

For the purpose of the check of our electronic structure, our results of parity-violating energy and contributions from heavy atoms are compared to previous works. In Fig. 2, the contributions from heavy atoms to parity-violating energy, M_{PV}^X, are shown as a function of the dihedral angle for H_2X_2 (X = O, S, Se, Te) molecules. All molecules have similar pattern and it is seen that heavier X atoms give larger M_{PV}^X due to larger relativistic effects. The tendencies of these curves are consistent with previous works [4, 8–10] qualitatively. The contribution of hydrogen atoms to E_{PV} is known to be much smaller than that of heavier atoms. Our values have some deviation from previous works. This deviation is speculated to be due to the smallness of our basis set. Parity-violating energy of H_2X_2 and contributions from heavy atom X are summarized in Table 1. The dihedral angle is chosen to be 45° or −45° as the value reported in references, which have the same value with opposite sign. The abbreviation, HF, means Hartree-Fock with Dirac-Coulomb Hamiltonian, CCSD is coupled-cluster singles-and-doubles, and CISD is configuration interaction including single and double excitations. As seen from this table, larger basis sets as triple or quadruple zeta function are critically important, and results of double zeta basis sets are much smaller than that of larger ones. Larger basis set is speculated to be

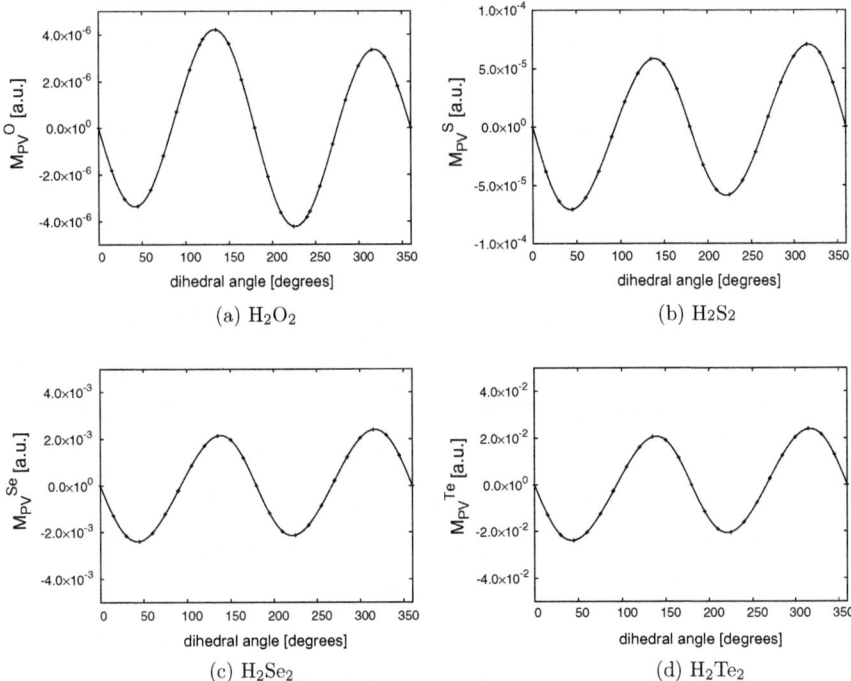

Fig. 2 Contribution from heavy atom to parity-violating energy of H_2X_2 molecules as a function of the dihedral angle

required for the description of the cusp structure near nuclei. Since our triple zeta result is well consistent with other triple zeta results, most deviation of our results from others comes from the smallness of the basis set. In addition, the effect of post Hartree-Fock computation is seen to be about 10%. Due to the limit of computational resources, the computations of the latter part of this work is restricted to the basis set, dyall.ae2z. Nevertheless, our wave functions are seen to be reasonable within our computational methods from this table.

In addition, we investigate the spin torque and the zeta force of H_2O_2. Figure 3 shows the distributions of the spin torque, the zeta force, and their sum for H_2O_2 with $\phi = 45°$. It can be seen that the sum of the spin torque and the zeta force is much smaller than the spin torque and the zeta force itself in the whole region. This result is consistent with the fact that the nonzero spin torque is in balance with the zeta force for the spin stationary state. Hence, although our computational result is derived from wave functions of quantum mechanics, it is considered that we can use these wave functions. The values of the norm of the spin torque and zeta force in the vicinity of oxygen nuclei amount to 10^{-3} [a.u.], which is much large than that in the vicinity of hydrogen nuclei. The smallness of the spin torque around hydrogen nuclei

Table 1 Parity-violating energy of H_2X_2 and contributions from heavy atom X are summarized. The dihedral angle is chosen to be $45°$ or $-45°$, which have the same value with opposite sign. HF means Hartree-Fock with Dirac-Coulomb Hamiltonian

Molecule	Method	Basis set for X	$|M_{PV}^X|$ [a.u.]	$|E_{PV}|$ [a.u.]	Reference
H_2O_2	HF	dyall.ae2z	3.3401×10^{-6}	3.8853×10^{-19}	This study
		dyall.ae3z	5.1839×10^{-6}	6.0300×10^{-19}	This study
		dyall.ae3z	–	6.051×10^{-19}	[10]
		dyall.ae4z	–	6.376×10^{-19}	[10]
		cc-pVDZ+3p [22]	5.801×10^{-6}	–	[9]
		25s25p5d [8]	6.057×10^{-6}	–	[9]
		aug-cc-pVDZ [22]	3.729×10^{-6}	–	[8]
		aug-cc-pVTZ [22]	4.239×10^{-6}	–	[8]
		aug-cc-pVQZ [22]	4.687×10^{-6}	–	[8]
		25s25p5d	6.057×10^{-6}	–	[8]
	CISD	cc-pVDZ+3p	5.410×10^{-6}	–	[9]
	CCSD	dyall.ae3z	–	5.323×10^{-19}	[10]
		dyall.ae4z	–	5.583×10^{-19}	[10]
		cc-pVDZ+3p	5.299×10^{-6}	–	[9]
H_2S_2	HF	dyall.ae2z	7.0563×10^{-5}	1.6416×10^{-17}	This study
		cc-pCVTZ [22]	–	1.825826×10^{-17}	[10]
		cc-pVDZ+3p	8.916×10^{-5}	–	[9]
		25s25p5d	9.581×10^{-5}	–	[8]
	CCSD	cc-pCVTZ [22]	–	1.82103×10^{-17}	[10]
		cc-pVDZ+3p	9.283×10^{-5}	–	[9]
H_2Se_2	HF	dyall.ae2z	2.3917×10^{-3}	1.6334×10^{-15}	This study
		25s25p5d	3.586×10^{-3}	–	[8]
	CCSD	dyall.cv3z [23]	–	2.115×10^{-15}	[10]
H_2Te_2	HF	dyall.ae2z	2.3842×10^{-2}	2.7769×10^{-14}	This study
		25s25p5d	3.149×10^{-2}	–	[8]
	CCSD	dyall.cv3z	–	3.289×10^{-14}	[10]

is consistent with our previous results [6]. The nonzero M_{PV}^O means the existence of the zeta force around oxygen nuclei, since the chirality density is proportional to the zeta potential and the zeta potential is the potential for the zeta force. Hence this result is consistent with the existence of the parity-violating energy.

In Fig. 4, the integrated chirality density as a function of the dihedral angle is shown for (a) H_2O_2, (b) H_2S_2, (c) H_2Se_2, and (d) H_2Te_2 molecules. Our result of

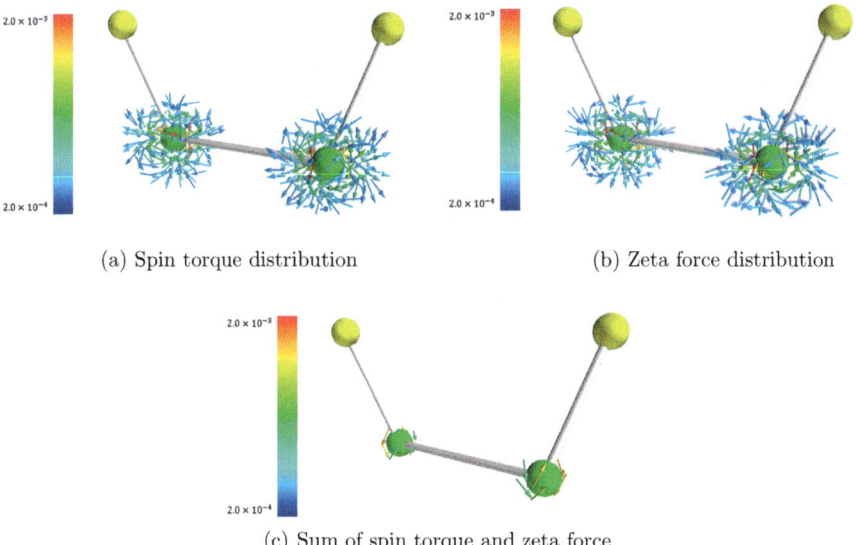

(a) Spin torque distribution   (b) Zeta force distribution

(c) Sum of spin torque and zeta force

Fig. 3 The distributions of the magnitude and the direction of **a** the spin torque distribution, **b** the zeta force, and **c** the sum of the spin torque and zeta force are shown for H_2O_2 molecule. The dihedral angle is $45°$

H_2Te_2 is consistent with the result reported in Reference [4], and it is confirmed that the electron chirality is nonzero in chiral molecules. The integrated chirality density of H_2Te_2 has almost the same dependence on the dihedral angle. However, the integrated chirality density of H_2O_2 and H_2S_2 are almost opposite to the parity-violating energy, which is determined dominantly from M_{PV}^X shown in Fig. 2. Moreover, the integrated chirality density of H_2Se_2 has different oscillation pattern from the parity-violating energy. Although we guessed that the integrated chirality density and the parity-violating energy of H_2X_2 have some correlation as in Reference [4], our integrated chirality density is not inconsistent with the parity-violating energy, since the parity-violating energy is determined dominantly only by the chirality density nearby heavy nuclei. Nevertheless, we should improve our computations with larger basis set and perform post Hartree-Fock computations in order to check our results.

In Fig. 5, the distributions of zeta potential around one Te atom of H_2Te_2 at the dihedral angle, (a) $15°$, (b) $45°$ and (c) $90°$, are shown on the xy-plane for the z coordinate on Te atoms. Our results are well consistent with those reported in Reference [4]. We have shown only the results for $\phi = 15°, 45°, 90°$, while at other dihedral angles, our results are consistent with Reference [4]. The distribution pattern of zeta potential agrees with their results well. The difference of the values arises from the factor $\hbar c/2$, which is the coefficient of the zeta potential over the chirality density. For comparison, the same figure for H_2O_2 is shown in Fig. 6 at the dihedral angle, (a) $15°$, (b) $45°$ and (c) $90°$. The results are shown on the xy-plane for the z coordinate

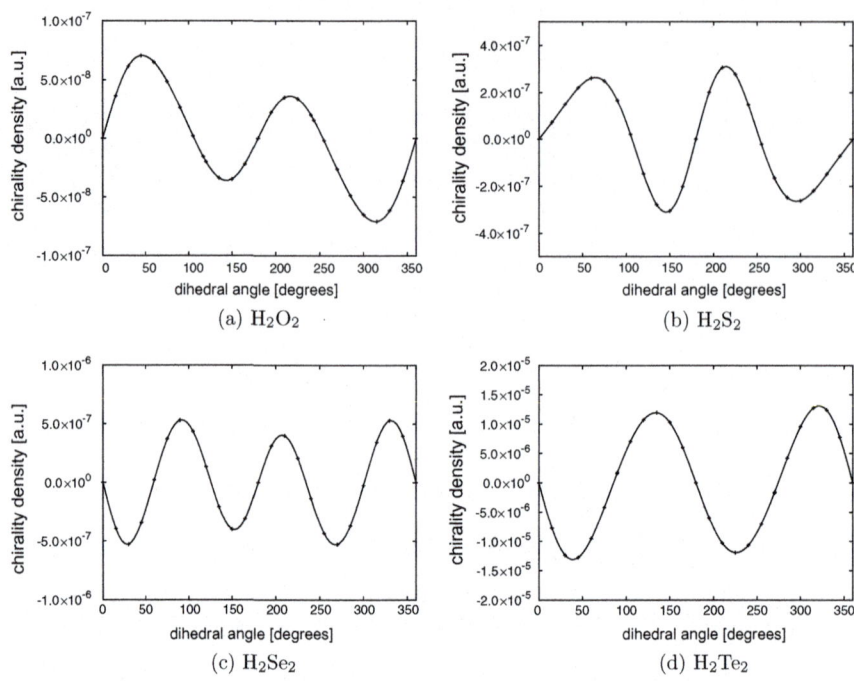

Fig. 4 The integrated chirality density as a function of the dihedral angle for **a** H_2O_2, **b** H_2S_2, **c** H_2Se_2, and **d** H_2Te_2 molecules

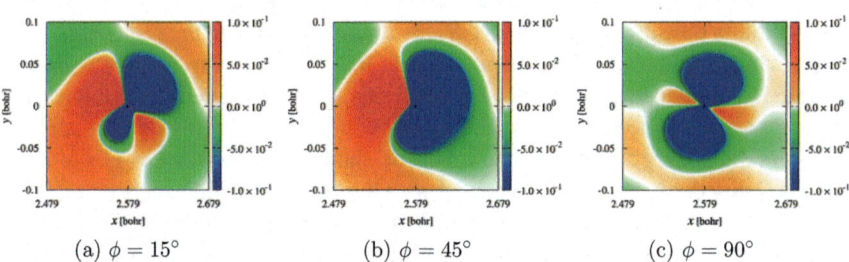

Fig. 5 The distribution of zeta potential of H_2Te_2 at the dihedral angle, **a** 15°, **b** 45° and **c** 90°. The result is shown on the xy-plane for the z coordinate on Te atoms

at one O atom. The localization of the innermost core electrons are strongly different between O and Te atoms, and hence the distribution pattern of the zeta potential is largely extended. The sign of the zeta potential at the position of a nucleus is the same for 15–45°, and opposite for 90°, this corresponds to the dependence of M_{PV}^X on the dihedral angle.

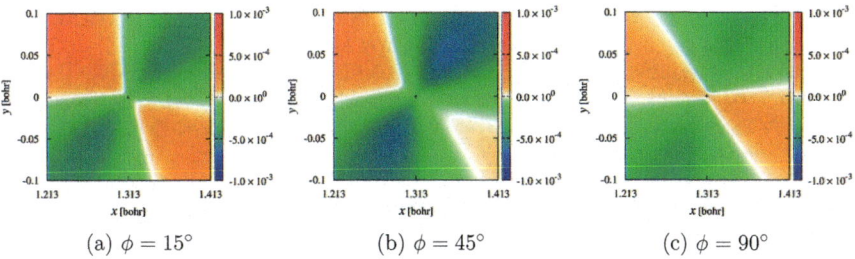

(a) $\phi = 15°$ (b) $\phi = 45°$ (c) $\phi = 90°$

Fig. 6 The distribution of zeta potential of H_2O_2 at the dihedral angle, **a** 15°, **b** 45° and **c** 90°. The result is shown on the xy-plane for the z coordinate on O atoms

5 Conclusion

In this work, we have studied H_2X_2 molecules in viewpoints of spin related local physical values. Since the chirality density is proportional to the zeta potential, which is the potential of the zeta force, one of the torque for the electron spin, the distribution of the chirality density affects the distribution of the internal torque in molecules. Our quantum states are well consistent with those in previous works within the choice of basis set, and it has been confirmed that the spin torque and the zeta force are in balance with each other. We have found that the integrated chirality density is larger for the larger atomic number as speculated from the trend of the parity violating energy. The dependence of the integrated chirality density of H_2Te_2 on dihedral angle is consistent with the previous work. We have found that the integrated chirality density of H_2Te_2 has the same sign as the parity violating energy, while those of H_2O_2 and H_2S_2 are opposite to the sign of the parity-violating energy, and moreover the dependence of the integrated chirality density of H_2Se_2 on dihedral angle is significantly different from that of the parity-violating energy.

In our future work, we should check the dependences of the integrated chirality density of H_2O_2, H_2S_2, and H_2Se_2 on dihedral angle, which are different from that of H_2Te_2. For this purpose, larger basis set and post Hartree-Fock computations are used. In addition, we investigate the relation between the distribution of the spin torque and the zeta force and the dihedral angle.

References

1. Meyerhenrich U (2008) Amino acids and the asymmetry of life. Springer, Heidelberg
2. Mason SF (1984) Nature 311:19
3. Hegstrom RA, Rein DW, Sandars PGH (1980) J Chem Phys 73:1
4. Bast R, Koers A, Gomes ASP, Iliaš M, Visscher L, Schwerdtfeger P, Saue T (2011) Phys Chem Chem Phys 13:864
5. Tachibana A (2001) J Chem Phys 115:3497; J Mol Model 11:301 (2005); J Mol Struct: (THEOCHEM), 943:138 (2010). see also Tachibana A (2017) New aspects of quantum electrodynamics. Springer

6. Fukuda M, Senami M, Tachibana A (2013) Prog Theor Chem Phys 27:131; Fukuda M, Soga K, Senami M, Tachibana A (2016) Int J Quant Chem 116:920

7. Senami M, Nishikawa J, Hara T, Tachibana A (2010) J Phys Soc Jpn 79:084302; Hara T, Senami M, Tachibana A (2012) Phys Lett A 376:1434; Fukuda M, Soga K, Senami M, Tachibana A (2013) Phys Rev A 93:012518; Fukuda M, Ichikawa K, Senami M, Tachibana A (2016) AIP Adv 6:025108

8. Laerdahrl JK, Schwerdtfeger P (1999) Phys Rev A 60:4439

9. Thyssen J, Laerdahrl JK, Schwerdtfeger P (2000) Phys Rev Lett 85:3105

10. Shee A, Visscher L, Saue T (2016) J Chem Phys 145:184107

11. Patrignani C et al (2016) Particle data group. Chin Phys C 40:100001

12. Sakurai JJ (1976) Advanced quantum mechanics. (Addison-Wesley, Reading, Mass)

13. Tachibana A (2004) Int J Quantum Chem 100:981

14. Ichikawa K, Nozaki H, Komazawa N, Tachibana A (2012) AIP Adv 2:042195

15. Tachibana A (2012) Math J Chem 50(669); In: (2013) Ghosh SK, Chattaraj PK (eds) Concepts and methods in modern theoretical chemistry. CRC, Boca Raton, FL, pp 235–251

16. Senami M, Ichikawa K, Tachibana A, QEDynamics http://www.tachibana.kues.kyoto-u.ac.jp/qed

17. Ichikawa K, Fukuda M, Tachibana A (2013) Int J Quantum Chem 113:190

18. Senami M, Miyazato T, Takada S, Ikeda Y, Tachibana A (2013) J Phys Conf Ser 454:012052; Senami M, Ogiso Y, Miyazato T, Yoshino F, Ikeda Y, Tachibana A (2013) Trans Mat Res Soc Jpn 38(4):535

19. Saue T, Visscher L, Jensen HJAa, Bast R, Bakken V, Dyall KG, Dubillard S, Ekström U, Eliav E, Enevoldsen T, Faßhauer E, Fleig T, Fossgaard O, Gomes ASP, Helgaker T, Lærdahl JK, Lee YS, Henriksson J, Iliaš M, Jacob ChR, Knecht S, Komorovský S, Kullie O, Larsen CV, Nataraj HS, Norman, P Olejniczak G, Olsen J, Park YC, Pedersen JK, Pernpointner M, di Remigio R, Ruud K, Sałek P, Schimmelpfennig B, Sikkema J, Thorvaldsen AJ, Thyssen J, van Stralen J, Villaume S, Visser O, Winther T, Yamamoto S (2014) DIRAC a relativistic ab initio electronic structure program. http://www.diracprogram.org

20. Dyall KG (2012) Theor Chem Acc 99:366 (1998); Theor Chem Acc 115:441 (2006); Theor Chem Acc 117:483 (2007); J Phys Chem A 113:12638 (2009); Theor Chem Acc 131:1217

21. Senami M, Ikeda Y, Fukushima A, Tachibana A (2010) Jpn J Appl Phys 49:115002

22. Dunning TH Jr (1989) J Chem Phys 90:1007; Kendall RA, Dunning TH Jr, Harrison RJ, J Chem Phys 96:6796 (1992); Woon DE, Dunning TH Jr, J Chem Phys 98:1358 (1993)

23. Dyall KG (2009) J Phys Chem A 113:12638; Theor Chem Acc 108:335 (2002); Chem Acc 115:441 (2006); Theor Chem Acc 117:483 (2007)

A Crystallographic Review of Alkali Borate Salts and Ab Initio Study of Borate Ions/Molecules

Cory C. Pye

Abstract The crystal structures of alkali metal borate salts are reviewed. A wide diversity of structures is noted. Structures with discrete ions containing two or more boron atoms are targeted for further study with ab initio methods (HF, B3LYP, and MP2) using modest basis sets (6-31G*, 6-31+G* and 6-311+G*). The ions identified for study are: $[B_3O_6]^{3-}$, $[B_3O_5(OH)_2]^{3-}$, $[B_3O_4(OH)_4]^{3-}$, $[B_3O_3(OH)_4]^-$, $[B_4O_5(OH)_4]^{2-}$, $[B_5O_6(OH)_4]^-$, $[B_2O_5]^{4-}$, and $[B_4O_9]^{6-}$. Some structurally related ions are examined, and an investigation of the diborates launched, including $[B_2O(OH)_6]^{2-}$, observed in the magnesium salt, and $[B_2(OH)_7]^-$, postulated as the intermediate responsible for signal exchange between borate anion and boric acid in ^{11}B NMR. The B-O bond distances and general structures are in good agreement with both crystallographic data and previous ab initio calculations.

Keywords Crystal structure · Ab initio study · Alkali borate · Polyborate

1 Introduction

In the preceding paper, the importance of the speciation of boron (III) to geochemistry, oceanography, and the nuclear industry, was discussed [1]. In addition, the crystal structures of boron oxide, boric acid, and monomeric forms of sodium and lithium borates were presented and discussed. Molecular forms exist for the orthoboric acid $[B(OH)_3]$ and metaboric acid $[B_3O_3(OH)_3]$, and certain lithium and sodium borates contain discrete $[B(OH)_4]^-$, $[BO_2(OH)]^{2-}$, and $[BO_3]^{3-}$ ions. The structure, energy, and vibrational frequencies of the crystallographically

Electronic supplementary material The online version of this chapter (https://doi.org/10.1007/978-3-319-74582-4_7) contains supplementary material, which is available to authorized users.

C. C. Pye (✉)
Department of Chemistry, Saint Mary's University, Halifax, NS B3H 3C3, Canada
e-mail: cory.pye@smu.ca

© Springer International Publishing AG, part of Springer Nature 2018
Y. A. Wang et al. (eds.), *Concepts, Methods and Applications of Quantum Systems in Chemistry and Physics*, Progress in Theoretical Chemistry and Physics 31, https://doi.org/10.1007/978-3-319-74582-4_7

characterized ortho- and metaboric acid, tetrahydroxoborate, trioxoborate, and dioxomonohydroxoborate were calculated ab initio, in addition to the postulated conjugate base of boric acid, oxodihydroxoborate $[BO(OH)_2]^-$, as well as the transition structure leading to it. The effect of additional water was also calculated. In this paper, the focus is on the polyborates.

The crystal structures of numerous sodium borates are given in Table 1. The notation (l:m:n) is shorthand for a substance of oxide formula lM$_2$O:mB$_2$O$_3$:nH$_2$O, where M is an alkali metal. Anhydrous (3:1:0) consists of discrete BO_3^{3-} ions [2]. The hydrated sodium borate (2:1:1) consists of discrete $[BO_2(OH)]^{2-}$ ions connecting sheets of edge-shared NaO polyhedral [3]. There are several hydrates of sodium metaborate (1:1:x). The octahydrate [4] consists of the tetrahedral $[B(OH)_4]^-$ anion, as does the tetrahydrate [5]. (3:3:4) consists of the $[B_3O_4(OH)_4]^{3-}$ anion with two tetrahedral and one trigonal boron atoms [6]. (3:3:2) consists of isolated $[B_3O_5(OH)_2]^{3-}$ ions with two trigonal and one tetrahedral ion [7]. The anhydrous sodium metaborate (1:1:0) contains discrete $B_3O_6^{3-}$ anions in which all boron atoms are trigonal [8]. (2:3:1) consists of the unique $[B_{12}O_{20}(OH)_4]^{8-}$ anion, which consists of 6 6-membered rings spirofused together in a cycle. There are 6 tetrahedral and 6 trigonal boron atoms. On the outer periphery of the ion, four of the trigonal boron atoms have hydroxyl groups and two have oxo groups [9]. Both (3:5:4) [10] and (3:5:2) [11] consists of layers of B$_5$O$_6$ polyhedra containing three tetrahedral and two trigonal boron atoms.

Borax (1:2:10) [12], also known as sodium tetraborate decahydrate, consist of chains of edge-shared hexaaquasodium octahedra alternating with chains of discrete $[B_4O_5(OH)_4]^{2-}$ ions hydrogen-bonded to each other [13]. The tetraborate anion consists of an oxygen-boron core with two tetrahedral and two trigonal boron atoms reminiscent of bicyclo [3.3.1] nonane, where the bridgeheads correspond to the tetrahedral boron atoms. The structure can therefore be reformulated as Na$_2$[B$_4$O$_5$(OH)$_4$] $\cdot$ 8H$_2$O. The structure was refined by neutron diffraction [14]. The determination of the hydrogen atom positions indicate that the ion has C$_2$ symmetry [15]. The oxo atoms of the anion accept hydrogen bonds either from the aquasodium chains (cross-chain) or from the hydroxyl groups (in-chain), and the hydroxyl group donate cross-chain hydrogen-bonds to the water oxygen atoms. Sodium tetraborate "pentahydrate" [16] (the mineral tincalconite, 1:2:4.667) is a reversible dehydration product of the decahydrate. The $[B_4O_5(OH)_4]^{2-}$ ions persist in this structure, however, the hydroxyls of the tetrahedral boron atoms are now directly coordinated to the sodium atoms. The structure can therefore be reformulated as Na$_2$[B$_4$O$_5$(OH)$_4$] $\cdot$ 3H$_2$O. The hydroxyls of the trigonal boron hydrogen bond to the oxo group between the two tetrahedral boron atoms. Upon refinement, incomplete occupancy of the water sites was found, corresponding to the formula Na$_2$ [B$_4$O$_5$(OH)$_4$] $\cdot$ 2.667H$_2$O [17]. The disorder was also confirmed at room temperature [18]. It was suggested that sodium tetraborate tetrahydrate (the mineral kernite, 1:2:4) was similar to borax in structure [19]; however this was shown to be incorrect, as the boron units formed an infinite chain of six-membered rings comprised of $[B_4O_6(OH)_2]_n^{2n-}$ [20, 21]. A later redetermination also included an electron population analysis [22]. The monohydrate consists of chains of

Table 1 Crystal structures of sodium borates

| Stoichiometry $Na_2B_mH_nO_{(1+3m+n)/2}$ | | Name | Synth T(K) | Space group | Unit cell (Å or deg.) | X-ray T(K) | B-O (Å) ranges | | Comments |
m	n						Tetrahedral	Trigonal	Refs.
1	0		748	$P2_1/c$	5.687,7.530,9.993 127.15		n/a	1.377–1.409	$[BO_3]^{3-}$ ions [2]
1	1		523	$Pnma$	8.627,3.512,9.863		n/a	1.351–1.439	$[BO_2(OH)]^{2-}$ ions [3]
1	8		298	$P\bar{1}$	6.126,8.1890,6.068 67.92,110.58,101.85		1.474–1.479	n/a	$[B(OH)_4]^-$ ions [4]
1	4		293	$P2_1/a$	5.886,10.566,6.146 111.60		1.463–1.483	n/a	$[B(OH)_4]^-$ ions [5]
3	4		383	Cc	12.8274,7.7276,6.9690 98.161	293	1.445–1.497	1.340–1.399	$[B_3O_4(OH)_4]^{3-}$ ions [6]
3	2	Sodium metaborate	423	$Pnma$	8.923,7.152,9.548		1.464–1.488	1.318–1.432	$[B_3O_5(OH)_2]^{3-}$ ions [7]
1	0			$R3c$	11.925,6.439 (hex) 7.212, 111.54 (rhomb)		n/a	1.280,1.433	$[B_3O_6]^{3-}$ ions [8]
2	1		523	$P2_1/c$	8.709,11.917,9.468 96.02		1.443–1.512	1.334–1.396	$[B_{12}O_{20}(OH)_4]^{8-}$ ion [9]
3	4		423	$Pbca$	8.804,18.371,10.924		1.425–1.531	1.351–1.398	Layers of B_5O_6 polyhedra [10]
3	2		523	$Pca2_1$	11.2373,6.0441,11.1336	291	1.440–1.510	1.367–1.377	Layers of B_5O_6 polyhedra [11]
1	10	Borax (sodium tetraborate decahydrate)		$C2/c$	11.89, 10.74, 12.19				[12]
				$C2/c$	11.858,10.674,12.197 106.68	298	1.46–1.54	1.32–1.40	Chains of $[B_4O_5(OH)_4]^{2-}$ ions [13]
			298	$C2/c$	11.885,10.654,12.206	297	1.464–1.500	1.363–1.374	Neutron diffraction [14]

(continued)

Table 1 (continued)

Stoichiometry $Na_tB_mH_nO_{(t+3m+n)/2}$			Name	Synth T(K)	Space group	Unit cell (Å or deg.)	X-ray T(K)	B-O (Å) ranges		Comments
t	m	n						Tetrahedral	Trigonal	Refs.
						106.623				
					C2/c	11.8843,10.6026,12.2111 106.790	145	1.4451–1.5075	1.3655–1.3784	[15]
3	6	14	Tincalconite ("pentahydrate")	353	R32	11.09,21.07		1.454–1.507	1.330–1.386	Assumed n = 15 [16]
				348	R32	11.097,21.114	108	1.442–1.495	1.361–1.374	Incomplete occupancy [17]
					R32	11.1402,21.207	291	1.445–1.493	1.364–1.370	[18]
1	2	4	Kernite (tetrahydrate)		P2_1/c	7.022,9.151,15.676, 108.83		n/a	n/a	Assumed similar to borax [19]
					P2_1/c	7.0172,9.1582,15.6774 108.861		1.48	1.37	Chain of 6-membered rings [20]
					P2_1/c	7.022,9.151,15.676, 108.8		1.428–1.521	1.353–1.384	[21]
					P2_1/c	7.0172,9.1582,15.6774 108.861		1.436–1.523	1.350–1.383	[22]
1	2	1	(Monohydrate)	423	Pbca	8.540,10.263,14.547		1.452–1.503	1.356–1.392	[23]
1	2	0		973	P1	6.5445,8.6205,10.4855 93.279,94.870,90.843	295	1.434–1.492	1.295–1.453	6- and 6,6-spiro fused rings [24]
2	5	7	Ezcurrite		P1	8.598,9.570,6.576 102.75,107.5,71.52		1.431–1.494	1.322–1.385	Chains of cond. [B5O6(OH)4]⁻ [25]
2	5	5	Nasinite	423	Pna2_1	12.015,6.518,11.173		1.440–1.496	1.345–1.399	Sheets of cond. [B5O6(OH)4]⁻ [26]
				453	Pna2_1	11.967,6.5320,11.126	153	1.4746	1.3685	[27]
				443	Pc	11.323,6.5621,12.244 91.050	298	1.444–1.534	1.351–1.434	[28]
2	5	3	Biringuccite	523	P2_1/c	11.1955,6.5607,20.7566 93.891		1.421–1.513	1.343–1.413	Condensed Spiro-B5O6 [29]
1	3	4	Ameghinite		C2/c	18.428,9.882,6.326		1.443–1.506	1.342–1.398	[B3O3(OH)4]⁻ [30]

(continued)

Table 1 (continued)

| Stoichiometry $Na_pB_mH_nO_{(l+3m+n)/2}$ | | | Name | Synth T(K) | Space group | Unit cell (Å or deg.) | X-ray T(K) | B-O (Å) ranges | | Comments |
	m	n						Tetrahedral	Trigonal	Refs.
1						104.38				
1	3	0	α-sodium triborate	998	$P2_1/c$	10.085,11.363,10.845 104.48	295	1.443–1.512	1.343–1.399	Spiro-B_5O_6 and BO_4 [31]
1			β-sodium triborate	1003–1013	$P2_1/c$	8.990,11.033,12.107 90.50	295	1.431–1.504	1.338–1.416	Spiro-B_5O_6, B_3O_5, and BO_4 [32]
1	4	0	Sodium tetraborate	Melt	$P2_1/a$	6.507,17.796,8.377 96.57		1.448–1.511	1.331–1.396	Spiro-B_5O_6 and B_3O_3 [33]
1	5	10	Sborgite		$C2/c$	11.119,16.474,13.576 112.83		1.456–1.482	1.339–1.412	$[B_5O_6(OH)_4]^-$ ions [34]
1	5	4	Sodium pentaborate	423	$P2_1/c$	8.701,8.067,12.977 106.77		1.451–1.489	1.342–1.394	$[B_5O_6(OH)_4]^-$ ions [35]

$[B_4O_6(OH)_2]^{2-}$ in which the trigonal hydroxyl of a $[B_4O_5(OH)_4]^{2-}$ ion has condensed with a tetrahedral hydroxyl of another to eliminate water [23]. The anhydrous sodium diborate (1:2:0) consists of single and spiro-fused double 6-membered rings [24]. The single rings are linked to double rings, which are linked together with oxide bridges.

Ezcurrite (2:5:7) consists of chains of $[B_5O_6(OH)_4]^-$ ions which have condensed to give two tetrahedral and three trigonal boron atoms [25]. Nasinite (2:5:5) consists of sheets of $[B_5O_6(OH)_4]^-$ ions which have condensed to give two tetrahedral and three trigonal boron atoms [26]. The structure was refined further [27] and a very similar structure identified [28]. Biringuccite (2:5:3) consists of sheets of condensed spiro-B_5O_6 units containing three trigonal and two tetrahedral boron atoms [29]. Ameghinite (1:3:4) consists of monomeric $[B_3O_3(OH)_4]^-$ ions [30]. Anhydrous α-sodium triborate (1:3:0) contains both a spiro-fused double 6-membered ring and a tetraborate core, which link to each other by oxo groups [31]. The β-modification consists of the spiro-fused double 6-membered ring, a simple 6-membered ring, and a standalone BO_4 tetrahedra [32]. Anhydrous sodium tetraborate (1:4:0) consists of layers of single and spiro-fused double 6-membered rings [33]. The single rings are linked to the double rings, which are linked together with oxide bridges. Both sborgite, $NaB_5O_8 \cdot 5H_2O$ (1:5:10) [34], and its dehydrated form of sodium pentaborate (1:5:4) [35], consist of discrete $[B_5O_6(OH)_4]^-$ ions containing one tetrahedral and four trigonal boron atoms.

To summarize these results, the following discrete polyborate ions (as a sodium salt) are known in the solid phase. The trinuclear $[B_3O_3(O)(OH)_4]^{3-}$, $[B_3O_3(O)_2(OH)_2]^{3-}$, $[B_3O_3(O)_3]^{3-}$, and $[B_3O_3(OH)_4]^-$ all contain the B_3O_3 core. The tetranuclear $[B_4O_5(OH)_4]^{2-}$ contains the B_4O_5 core. The pentanuclear $[B_5O_6(OH)_4]^-$ contains the B_5O_6 core (two B_3O_3 cores spiro fused at a boron). The dodecanuclear $[B_{12}O_{20}(OH)_4]^{8-}$ containing the $B_{12}O_{18}$ core (six B_3O_3 cores spiro fused into a macrocycle). Upon dehydration, typically chains or sheets of condensed units form (sometimes more than one unit is present). Some plausible species not observed to date in sodium salts are: $BO(OH)_2^-$, any dinuclear species; $[B_3O_3(OH)_6]^{3-}$, or $[B_3O_3(OH)_5]^{2-}$; $[B_4O_5(OH)_6]^{3-}$, or $[B_4O_5(OH)_5]^{2-}$; $[B_5O_6(OH)_6]^{3-}$, or $[B_5O_6(OH)_5]^{2-}$.

The crystal structures of numerous potassium borates are given in Table 2. The previously described discrete ions $[B_3O_6]^{3-}$, $[B_4O_5(OH)_4]^{2-}$, $[B_3O_3(OH)_4]^-$, and $[B_5O_6(OH)_4]^-$, are all represented. The only new borate anion appearing is $[B_{12}O_{16}(OH)_8]^{4-}$. For lithium borates (Table 3), only the previously-mentioned discrete ion $[B(OH)_4]^-$ is found, but two new ions, $B_2O_5^{4-}$ and $B_4O_9^{6-}$, appear. For rubidium borates (Table 4), the previously mentioned discrete ions $[B_3O_4(OH)_4]^{3-}$, $[B_3O_6]^{3-}$, $[B_4O_5(OH)_4]^{2-}$, and $[B_5O_6(OH)_4]^-$ are represented, and the new ion $[B_7O_9(OH)_5]^{2-}$ ion appears. For cesium borates (Table 5), the previously discussed $[B_3O_6]^{3-}$, $[B_4O_5(OH)_4]^{2-}$ and $[B_5O_6(OH)_4]^-$ ions are represented, but no new ions have been found. Condensation of the triborate, tetraborate, and pentaborate into chains and sheets is quite common as the water content of the borates decrease. Conversely, hydration/dissolution of the water-poor solids might be expected

Table 2 Crystal structures of Potassium Borates

Stoichiometry $K_jB_mH_nO_{(1+3m+n)/2}$			Name	Synth T(K)	Space group	Unit cell (Å or deg.)	X-ray T(K)	B–O (Å) ranges		Comments
j	m	n						Tetrahedral	Trigonal	Refs.
1	1	0	Potassium metaborate		$R3c$	7.76, 110.6		n/a	1.33–1.38	$[B_3O_6]^{3-}$ ions [36]
					$R3c$			n/a	1.331–1.398	[37]
1	2	4	Potassium tetraborate tetrahydrate	298	$P2_12_12_1$	12.899, 11.774, 6.859		1.447–1.511	1.312–1.401	$[B_4O_5(OH)_4]^{2-}$ ions [38]
1	2	0	Potassium diborate	1073		6.46, 9.56,10.38, 90.7, 102.6, 101.2				[39]
				973	$P\bar{1}$	6.484, 9.604,10.413, 89.28, 102.75,101.25		1.435–1.508	1.357–1.381	Condensed B_4O_5, B_3O_3 and BO_3 units [40]
2	5	5			$Pna2_1$	12.566,6.671,11.587		1.454–1.496	1.346–1.412	$[B_5O_8(OH)]$ unit [41]
2	5	2		443	$C2/c$	18.077,6.857,13.266 95.271	298	1.449–1.513	1.350–1.413	B_3O_3 and B_4O_5 units [42]
1	3	6		298	$C2/c$	15.540,6.821,14.273, 104.44	163	1.462–1.489	1.347–1.393	$[B_5O_5(OH)_4]^-$ ions [43]
1	3	2		428	$P4/ncc$	11.3482,15.9169		1.426–1.471	1.270–1.468	$[B_{12}O_{26}(OH)_8]^{4-}$ ions [44]
1	3	1		453	$P2_1/n$	9.036,6.6052,15.997, 91.862	298	1.451–1.482	1.337–1.398	B_3O_3 units [45]
1	3	0	Potassium triborate	1073		10.63, 10.64, 13.01, 90.7, 95.2, 114.2				[39]
				1073	$P2_1/c$	9.319, 6.648,21.094 94.38		1.455–1.492	1.332–1.406	B_3O_3 units [46]
			γ-potassium pentaborate	1073		11.59, 17.80, 12.99, 95.8				[39]
5	19	0	Pentapotassium enneakaidekaborate	1023	$C2/c$	17.888,11.479,12.973, 95.52	295	1.459–1.494	1.336–1.404	B_5O_6, B_3O_3, BO_3, and BO_4 groups [47]
1	5	8	Potassium pentaborate hydrate		$Aea2$	11.08,11.11,8.97		1.53	1.28–1.42	[48]
					$Aea2$	11.065, 11.171, 9.054				[49]

(continued)

Table 2 (continued)

Stoichiometry $K_lB_mH_nO_{(l+3m+n)/2}$			Name	Synth T(K)	Space group	Unit cell (Å or deg.)	X-ray T(K)	B-O (Å) ranges		Comments
l	m	n						Tetrahedral	Trigonal	Refs.
1	5	4			Aea2	11.062, 11.175, 9.041		1.474–1.481	1.345–1.380	[B5O6(OH)4]⁻ ions [50]
				443	$P2_1/c$	9.4824,7.5180,11.4154, 97.277	298	1.457–1.479	1.342–1.399	B5O6 units [42]
1	5	3		473	P1̄	7.5612,9.2236,11.7298, 99.038,106.595,91.314	100	1.468–1.483	1.340–1.396	Condensed B5O6 units [51]
1	5	2			$P2_1/c$	7.6690,9.0445,12.2304 119.132	100	1.464–1.482	1.339–1.391	Condensed B5O6 units [52]
1	5	0	α-potassium pentaborate	1033	Pbca	8.52,8.53,21.74				[53]
				1033	Pbca	8.383,8.418,21.540		1.469–1.477	1.319–1.405	Condensed B5O6 units [54]
			β-potassium pentaborate	973	Pbca	7.42, 11.69,14.72		1.35–1.57	1.07–1.51	[55]
					Pbca	7.418,11.702,14.745		1.40–1.49	1.33–1.41	Condensed B5O6 units [56]

Table 3 Crystal structures of Lithium Borates

Stoichiometry $Li_2B_mH_nO_{(t+3m+n)/2}$		Name	Synth T(K)	Space group	Unit cell (Å or deg.)	X-ray T(K)	B-O (Å) ranges		Comments
m	n						Tetrahedral	Trigonal	Refs.
1	0	β	905	I2/a	10.2269,4.6988,8.7862, 93.562	153	n/a	1.345–1.444	$B_2O_5^{4-}$ ions [57]
		α	905	Pca2₁	10.1497,4.7365,17.5880	153	n/a	1.340–1.454	$B_2O_5^{4-}$ ions [57]
2	0		875	P2₁/n	3.31913,23.361,9.1582, 92.650	293	n/a	n/r, 1.316–1.436 cif	Planar $B_4O_9^{6-}$ ions [58]
1	16			P3	6.5534,6.1740		1.472–1.473	n/a	$[Li(H_2O)_4]^+$ and $[B(OH)_4]^-$ ions [59]
				P3	6.555,6.177				[60]
1	4			Pbca	9.16,7.95,8.54			n/a	$[B(OH)_4]^-$ ions [61]
				Pbca	7.9362,8.5220,9.1762	120	1.4644–1.4989		[62]
1	0	Lithium metaborate	Melt	P2₁/c	5.838,4.348,6.449, 115.12		n/a	1.330–1.400	BO₃ chains [60]
		γ-LiBO₂	1223, 15 kbar	I42d	4.1961,6.5112		1.483	n/a	BO₄ units [63]
2	3			Pnma	9.7984,8.2759,9.6138				powder XRD, spiro-linked B₃O₅ chains [64]
2	0	Dilithium tetraborate	1173	I4₁cd	9.47,10.26		1.43–1.47	1.37–1.42	Condensed B₄O₅ units [65]
				I4₁cd	9.479,10.280		1.448–1.507	1.349–1.386	[66]
				I4₁cd	9.477,10.286		1.453–1.505	1.347–1.375	[67]
				I4₁cd	9.479,10.290	298	1.453–1.506	1.355–1.374	[68]
7	0	"2:5"	1113	P1	6.487,7.840,8.510, 92.11,104.85,99.47		1.443–1.502	1.347–1.392	B₃O₅ and BO₃ units [69]

(continued)

Table 3 (continued)

| Stoichiometry $Li_lB_mH_nO_{(l+3m+n)/2}$ | | | Name | Synth T(K) | Space group | Unit cell (Å or deg.) | X-ray T(K) | B-O (Å) ranges | | Comments |
l	m	n						Tetrahedral	Trigonal	Refs.
1	3	0	Lithium triborate	melt	$Pna2_1$	8.447,7.3789,5.1408	298	1.4610–1.4872	1.3481–1.3967	B_3O_3 unit [70], [71]
1	3	0		1107	$Pna2_1$	8.444,7.378,5.1416	293	1.4614–1.4847		[72]
3	11	0	"$Li_2B_8O_{13}$"	873	$P2_1/c$	17.7607,7.7737,9.6731 100.906	293	n/r	n/r	B_3O_3 and B_5O_6 units [58]
3	11	0		1273	$P2_1/a$	9.766, 7.849, 17.899		1.464–1.496	1.345–1.412	[73]
1	5	2		573, 10 MPa	$P2_1/a$	13.576,9.077,5.543, 91.47	293	1.445–1.492	1.346–1.382	B_5O_6 unit [74]

Table 4 Crystal structures of Rubidium Borates

Stoichiometry Rb$_b$B$_m$H$_n$O$_{(t+3m+n)}$			Name	Synth T(K)	Space group	Unit cell (Å or deg.)	X-ray T(K)	B-O (Å) ranges		Comments
	m	n						Tetrahedral	Trigonal	Refs.
1	3	8			Pna2$_1$	7.90,13.95,9.28		1.44-1.50	1.37-1.45	[B$_3$O$_4$(OH)$_4$]$^{3-}$ ions [75]
1	1	0			R3c	13.21, 7.78				[37]
			Rubidium metaborate	1173	R3c	13.1572,7.7434	293		1.315-1.407	[B$_3$O$_6$]$^{3-}$ ions [76]
1	2	5.6		298	Pbcn	11.276,13.097,16.751	298	1.416-1.524	1.352-1.391	[B$_4$O$_5$(OH)$_4$]$^{2-}$ ions [77]
1	2	0		980	P1	9.860,10.653,6.649, 103.4,101.4.89.1		1.41-1.50	1.33-1.40	B$_4$O$_5$, B$_3$O$_3$ and BO$_3$ units [78]
3	7	0		913	P1	6.603,6.632,30.085, 91.183,91.781,119.293		1.42-1.53	1.31-1.41	B$_5$O$_6$, B$_3$O$_3$ and BO$_4$ units [79]
1	3	0	α-rubidium triborate	873	P2$_1$2$_1$2$_1$	8.209,10.092,5.382	293	1.44-1.58	1.31-1.40	powder XRD, B$_3$O$_3$ units [80]
			β-rubidium triborate	973	P2$_1$2$_1$2$_1$	8.438,8.719,6.240		1.45-1.49	1.34-1.40	B$_3$O$_3$ units [81]
2	7	5		443	P2$_1$/n	8.047,11.785,13.952, 102.974	298	1.420-1.500	1.333-1.395	[B$_7$O$_9$(OH)$_5$]$^{2-}$ ions [82]
1	5	8		298	Aea2	11.302,10.962,9.335	298	1.473-1.474	1.344-1.375	[B$_5$O$_6$(OH)$_4$]$^-$ ions [83]
1	5	0	β-rubidium pentaborate	973	Pbca	7.50, 11.84, 14.80				B$_5$O$_6$ units [55]
				1003	Pbca	7.553, 11.857, 14.813	298	1.464-1.481	1.339-1.391	[84]

Table 5 Crystal structures of Cesium Borates

Stoichiometry $Cs_lB_mH_nO_{(l+3m+n)/2}$			Name	Synth T(K)	Space group	Unit cell (Å or deg.)	X-ray T(K)	B-O (Å) ranges		Comments	Refs.
l	m	n						Tetrahedral	Trigonal		
1	1	0	Cesium metaborate		R3c	13.68, 8.36					[37]
					R3c	13.637, 8.365			1.301–1.418	[B$_3$O$_6$]$^{3-}$ ions [85]	
1	2	5		298	P2$_1$/c	8.424,11.378,13.160, 92.06	298	1.439–1.516	1.361–1.379	[B$_4$O$_5$(OH)$_4$]$^{2-}$ ions [86]	
3	7	0		993	C2$_1$/c	59.911, 11.520.34.525, 101.052	293	1.42–1.52	1.32–1.42	3-layers, 9xsupercell [87]	
1	3	0	Cesium triborate	1073	P2$_1$2$_1$2$_1$	6.18,8.48,9.17		1.45–1.50	1.34–1.44	B$_3$O$_3$ unit [88]	
				1003	P2$_1$2$_1$2$_1$	6.213,8.521,9.170	295	1.458–1.485	1.336–1.399	[89]	
3	13	0			C2/c	23.064,13.367, 24.464, 90.281	298	1.39–1.56	1.33–1.43	B$_3$O$_3$ and B$_5$O$_6$ units [90]	
1	5	8		298	P2$_1$/c	11.584,7.174,13.959 94.61		1.46–1.49	1.33–1.40	[B$_5$O$_6$(OH)$_4$]$^-$ ions [91]	
1	5	0	α-cesium pentaborate	823	P2$_1$/n	7.117,9.634,10.391 101.160	298	1.463–1.495	1.334–1.409	B$_5$O$_6$ units [84]	
			α-HT	923	P2$_1$/c	7.122,9.640,11.411 116.64		1.466–1.497	1.346–1.387	B$_5$O$_6$ units [92]	
			γ-cesium pentaborate	733	Pbca	8.697, 8.431,21.410	298	1.46–1.48	1.30–1.45	B$_5$O$_6$ units [84]	
1	9	0	Cesium enneaborate	853	P222$_1$	8.768,15.790		1.43–1.46	1.23–1.44	B$_3$O$_3$ units [93]	

initially to result in partial hydrolysis into discrete structural motifs already present in the solid state, followed by equilibration in solution.

Ab initio calculations have been carried out on some of the discrete ions seen in the crystal structures. Gupta and Tossell fixed the symmetry of $B_2O(OH)_4$ to C_{2v} (B-O fixed at 1.353 Å (HF/STO-3G) and 1.375 Å (4-31G), and found that the B-O-B linkage was bent [94]. They also found B-O distances of 1.430 Å and 1.279 Å for $B_3O_6{}^{3-}$ at HF/STO-3G, which compares well with the crystal structure. Zhang et al. [95] completely optimized $B_2O(OH)_4$ with symmetry C_{2v} (HF/STO-3G, 6-31G*), C_s (HF/6-31G) and C_2 (HF/STO-3G, 3-21G*, 4-31G, 6-31G, 6-31G*). They also calculated the $B_2O(OH)_6^{2-}$ (C_{2v}), $B_2O(OH)_5^-$ (C_s), $B_3O_3(OH)_4^-$ (C_{2v}) and B_3O_3 $(OH)_5^{2-}$ (C_s) ions at HF/STO-3G. Oi calculated $B_2O(OH)_4$ (C_2), $B_2O(OH)_5^-$ (C_1), and $B_2O(OH)_6^{2-}$ (C_2) at HF/6-31G* [96]. In addition, Oi also calculated $B_3O_3(OH)_4^-$ (C_2), $B_3O_3(OH)_5^{2-}$ (C_1), $B_4O_5(OH)_4^{2-}$ (C_2), and $B_5O_6(OH)_4^-$ (S_4) at HF/6-31G* [97]. A combined Raman and DFT (B3LYP/aug-cc-pVDZ) investigation of $B_2O(OH)_4$, $B_2O(OH)_5^-$, $B_2O(OH)_6^{2-}$, $B_3O_3(OH)_4^-$, $B_3O_3(OH)_5^{2-}$, $B_3O_3(OH)_6^{3-}$, $B_4O_5(OH)_4^{2-}$, and $B_5O_6(OH)_4^-$ was presented by Zhou et al [98]. In addition to these ions, two heptamers $B_7O_9(OH)_5^{2-}$ were calculated by Beckett et al. at B3LYP/6-311++G(d, p) [99].

2 Methods

Calculations were performed using Gaussian 03 [100]. The MP2 calculations use the frozen core approximation. The geometries were optimized using a stepping stone approach, in which geometries at the levels HF/6-31G*, HF/6-31+G*, HF/6-311+G*, B3LYP/6-31G*, B3LYP/6-31+G*, B3LYP/6-311+G*, MP2/6-31G*, MP2/6-31+G* and MP2/6-311+G* were sequentially optimized, with the geometry and molecular orbitals reused for the subsequent level. Default optimization specifications were normally used. After each level, where possible, a frequency calculation was performed at the same level and the resulting Hessian was used in the following optimization. Z-matrix coordinates constrained to the appropriate symmetry were used as required to speed up the optimizations. Because frequency calculations are done at each level, any problems with the Z-matrix coordinates would manifest themselves by giving imaginary frequencies corresponding to modes orthogonal to the spanned Z-matrix space. The Hessian was evaluated at the first geometry (Opt = CalcFC) for the first level in a series in order to aid geometry convergence.

3   Results and Discussion

In our previous work, the structure, energy, and vibrational spectra of orthoboric acid, metaboric acid, and tetrahydroxoborate was thoroughly discussed. We now focus on the polyborate ions. First we discuss those observed in the crystal structure of alkali metal borates, followed by structurally related ions. The more highly-charged ions might be observed in low-water, high ionic-strength environments. We also discuss other possibilities.

3.1   Triborate Species

3.1.1 $[B_3O_3(OH)_4]^-$

The structure of the crystallographically-observed triborate ion, $[B_3O_3(OH)_4]^-$, is given in Fig. 1. The ion, present in the sodium borate mineral ameghinite, could be formed by the addition of hydroxide ion to metaboric acid. Firstly, four different C_{2v} structures were optimized. None of these was an energy minimum. All possess an imaginary A_2 mode, suggesting desymmetrization to C_2. Structures 3 and 4 possess also an imaginary B_2 mode, and for some MP2 levels, an imaginary B_1 mode, suggesting desymmetrization to C_s. Most of the C_2 forms are minima at some levels. None of the three C_s forms are minima, desymmetrizing instead to C_1 #5–7, respectively. C_2 #1 was not a minimum at HF/6-31+G* and MP2/6-31+G*, coalescing into C_1 #6. C_2 #2 was only a minimum ay B3LYP/6-31G* and MP2/6-31G*, morphing into C_2 #4 at HF/6-31G* and HF/6-311+G* levels, or desymmetrizing to C_1 #7 at the other levels. C_2 #3 was not a minimum at MP2/6-31+G*, becoming C_1 #4. C_2 #4 was not stable at the MP2 levels, becoming C_1 #2 instead. The order of stability was C_1 #6 (0.0 kJ/mol) < C_2 #3 (−0.4 to 1.4 kJ/mol) < C_1 #5 (2.8–3.6 kJ/mol) < C_2 #1 (4.7–9.8 kJ/mol) < C_1 #7 (12–14 kJ/mol) < C_2 #4 (11–23 kJ/mol) < C_2 #2 (17–24 kJ/mol).

3.1.2 $[B_3O_3(OH)_5]^{2-}$

The structure of the triborate ion $[B_3O_3(OH)_5]^{2-}$ is given in Fig. 1. Neither of the two C_s forms is stable and both desymmetrize to the corresponding C_1 forms. Of these, structure #2 is 2.1–3.4 kJ/mol more stable than structure #1.

3.1.3 $[B_3O_3(OH)_6]^{3-}$

The structure of the triborate ion $[B_3O_3(OH)_6]^{3-}$ is given in Fig. 1. First, two D_{3h} structures were tried. Neither were minima, and the number of imaginary

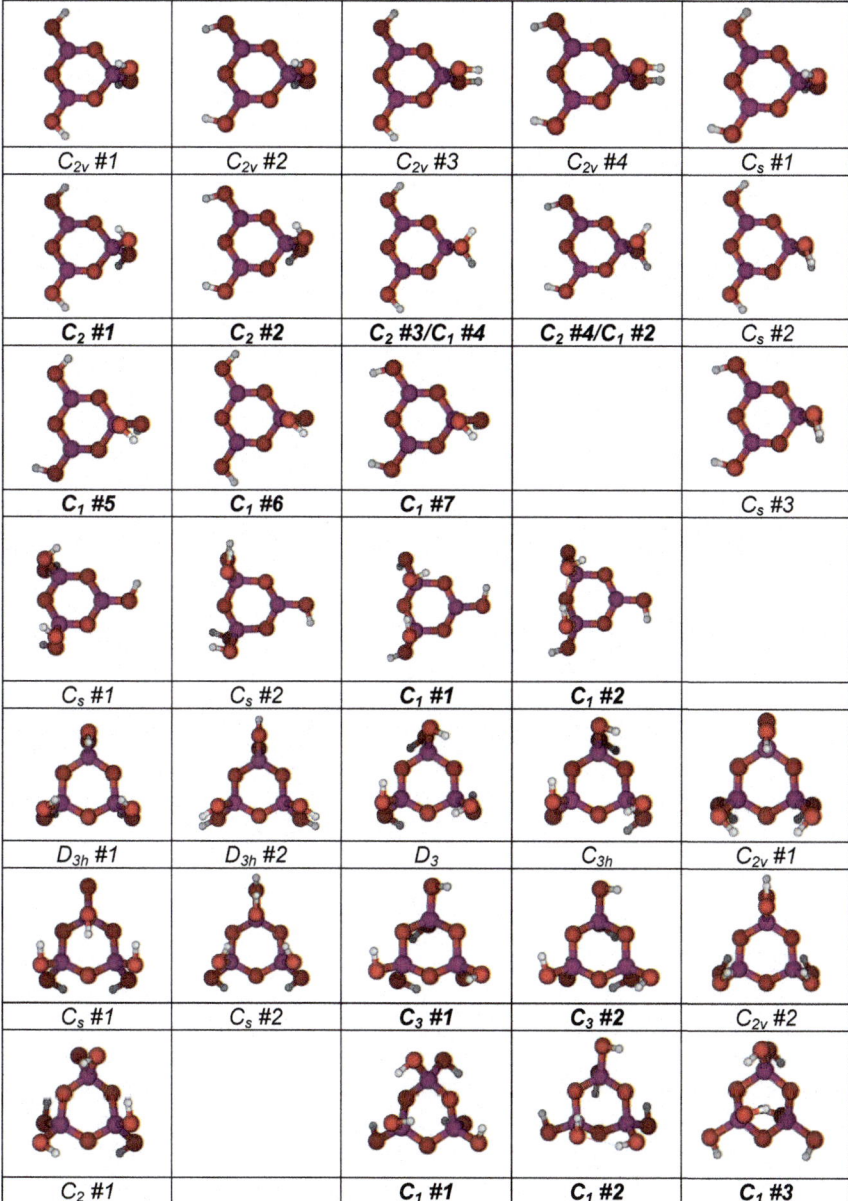

Fig. 1 Structure of Triborate Ion, $[B_3O_3(OH)_{3+n}]^{n-}$, n = 1 − 3. A bold symmetry label indicates a minimum energy structure

frequencies suggested many possible structures of lower symmetry. The two possible D_3 structures (via A_1") coalesced into one, but it was not a minimum. The two possible C_{3h} structures (via A_2') coalesced into one, and it too is not a minimum. While desymmetrization along the E" and E' modes should lead to a C_2 and C_s structure, respectively, it was thought that it would be advantageous to first desymmetrize to the corresponding C_{2v} structures (which might ascend in symmetry back to the corresponding D_{3h} structures). Neither of these was a minimum, as expected, and C_{2v} #1 did indeed ascend to D_{3h} #1 at the HF levels. The D_3 and C_{3h} forms desymmetrize into the stable C_3 #1 and #2 forms, respectively, along the A_2 and A" modes, respectively. The two possible C_{2v} structures could desymmetrize into either two C_2 structures, or into one of four possible C_s structures. The two C_2 structures coalesce into a structure that is only stable at MP2/6-31G*. C_s #3 (from C_{2v} #1) and C_s #4 (from C_{2v} #2) ascend in symmetry to either C_{3h} or C_{2v} #1. Neither of the C_s structures are stable. The C_2 structure desymmetrizes into C_1 #1 at MP2/6-31+G* and MP2/6-311+G*, and ascends in symmetry to C_3 #1 at the HF and B3LYP levels. The C_s structures desymmetrize to the stable C_1 #2 and 3 at most levels. The C_3 #1 structure is the most stable.

3.1.4 $[B_3O_4(OH)_4]^{3-}$

The structure of the triborate ion $[B_3O_4(OH)_4]^{3-}$ is given in Fig. 2. The C_{2v} structure has at least three imaginary frequencies of irreducible representation A_2, B_1 and B_2. These suggest desymmetrization to a C_2 or to two different C_s structures, respectively. None of these are energy minima (except C_s #1 at B3LYP/6-31G*) and all desymmetrize to C_1 structures #1–3, respectively. C_1 #1 is the most stable. This ion has been characterized crystallographically as the sodium and rubidium salt (see Tables 1 and 4).

3.1.5 $[B_3O_5(OH)_2]^{3-}$

The structure of the triborate ion $[B_3O_5(OH)_2]^{3-}$ is given in Fig. 2. The two C_{2v} structures are unstable and all contain an imaginary A_2 mode, leading to two C_2 structures that coalesce into the most stable minimum. The higher energy C_{2v} #2 structure also contains a B_2 mode. The lower energy C_{2v} #1 structure also contains either a B_2 mode (B3LYP/6-31G*) or B_1 mode (MP2/6-31+G* and MP2/6-311 +G*). The C_s #1 structure, derived from C_{2v} #2, is only stable at B3LYP/6-31G*. C_s #2, derived from C_{2v} #1, coalesces into C_s #1. C_s #3, also derived from C_{2v} #1, is not stable. The C_1 #1 structure, derived from C_s #1, coalesces into C_2, except at the stable B3LYP/6-31+G* and B3LYP/6-311+G* levels, whereas the C_1 #2 structure, derived from C_s #3, coalesces into C_2 at both levels. This ion has been characterized as the sodium salt (see Table 1).

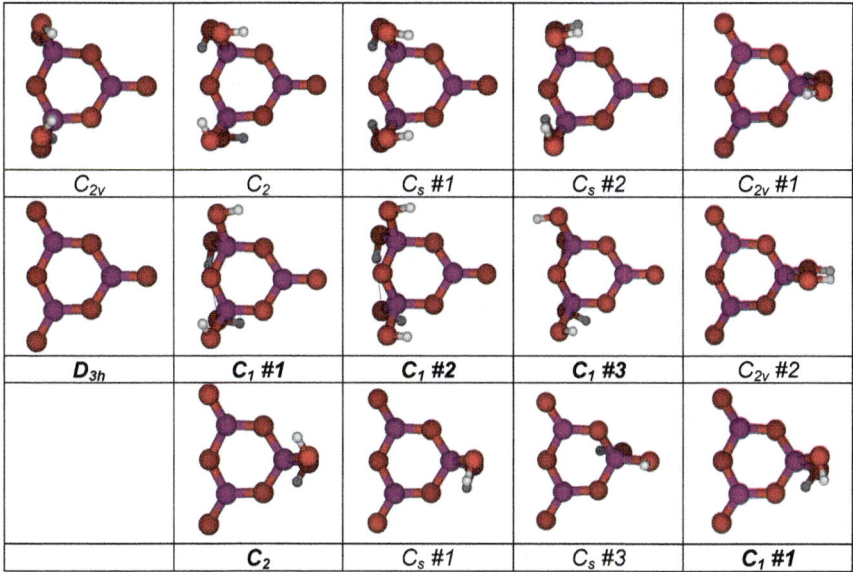

Fig. 2 Structure of triborate ion, $[B_3O_{3+n}(OH)_{6-2n}]^{3-}$, n = 1 – 3

3.1.6 $[B_3O_6]^{3-}$

The only structure of the triborate ion $[B_3O_6]^{3-}$, of D_{3h} symmetry, is given in Fig. 2. This ion is characterized crystallographically in sodium, potassium, rubidium, and cesium metaborate (see Tables 1, 2, 3, 4 and 5). There does not appear to be any systematic trend in the range of the B-O distances upon varying the alkali metal.

3.2 Tetraborate Species

3.2.1 $[B_4O_5(OH)_4]^{2-}$

The structure of the tetraborate ion, $[B_4O_5(OH)_4]^{2-}$, is given in Fig. 3. Initially, four structures of C_{2v} symmetry were tried. None of these were energy minima. All of these had both an A_2 and a B_2 imaginary mode, and, at some levels, some have a B_1 imaginary mode. The four structures of C_2 symmetry derived by desymmetrizing along the A_2 mode were all minima, but only three unique structures were found, as some coalesced. The four structures of C_s symmetry derived by desymmetrizing along the B_2 mode (#1–#4) coalesced into two different structures, but only C_s #2 was a minima, at all levels except B3LYP/6-31G* and MP2/6-31G*. The four structures of C_s symmetry derived by desymmetrizing along the B_1 mode (#5–#8) did not give minimum energy structures, and C_s #8 ascended in symmetry to C_{2v}

#3. C_s #1 desymmetrized to the stable C_1 #1. The C_2 #2, C_2 #3, and C_1 #1 are quite close in energy (5 kJ/mol). This ion has been characterized as a salt of all the alkali metal ions (except lithium), and the sodium salt is commonly known as borax.

An alternative structure might exist where two of the hydroxyls connect to the non-bridgehead boron atoms. Initially, four structures of C_{2v} symmetry (#5–#8) were tried. One of these (C_{2v} #6) was an energy minima at all levels except B3LYP/6-31G*. For this structure at this level, and for the other three structures, there exists A_2 and B_1 imaginary modes. This suggests desymmetrization to C_2 #5–#8 and C_s #9–#12, respectively. The C_2 #6 (B3LYP/6-31G* only) and #7 structures are stable. All other attempts at structures of C_2 symmetry revert to either C_2 #7 or C_{2v} #6. The C_s #10 (B3LYP/6-31G*) and C_s #11 is stable. All other attempts at structures of C_s symmetry revert to either C_s #11 or C_{2v} #6.

3.2.2 $[B_4O_9]^{6-}$

Another tetraborate ion, observed in the lithium salt (Table 3), is shown in Fig. 3. Initially we considered structures of C_{2h} and C_{2v} symmetry, of which there are two each. None of these is an energy minimum. For C_{2h} #1, there are imaginary modes of irreducible representation A_u (desymmetrization to C_2), B_g (desymmetrization to C_i), and, at MP2/6-311+G*, B_u (desymmetrization to C_s #2). For C_{2h} #2, there are A_u imaginary modes, and at MP2/6-31+G* and MP2/6-311+G*, B_u imaginary modes (desymmetrization to C_s #3). For C_{2v} #1, there are imaginary modes of irreducible representation A_2 (desymmetrization to C_2) and B_2 (desymmetrization to C_s), whereas for C_{2v} #2, there is only an A_2 imaginary mode, giving rise to the stable C_2 #4. Desymmetrization of C_{2h} #1 to a C_i structure results in ascent in symmetry to C_{2h} #2. Desymmetrization of C_{2v} #1 to a C_s structure usually results in ascent in symmetry to C_{2h} #2, except for B3LYP/6-311+G* and MP2/6-311+G*, where the C_s #1 structure remains. None of the three C_s structures are energy minima. The C_s #1 structure desymmetrizes to the stable C_1 #1, whereas attempts to desymmetrize C_s #2 and C_s #3 result in ascent in symmetry to C_2 #4.

3.3 Pentaborate Species

3.3.1 $[B_5O_6(OH)_4]^-$

The pentaborate ion, $[B_5O_6(OH)_4]^-$, observed as a sodium, potassium, rubidium, and cesium salt (Tables 1, 2, 4, and 5), is given in Fig. 4. Initially, two structures of D_{2d} symmetry were tried, and these were indeed minima at most levels. In addition, other structures formed from rotating hydroxyls by 180 degrees were also minima at most if not all levels. These are labelled C_{2v} #1, C_2 #1, C_s #1, and C_s #2. The only possible levels where these structures are not minima are MP2/6-31+G* and MP2/6-311+G*. Structure D_{2d} #1 has an imaginary E mode at MP2/6-31+G*, whereas

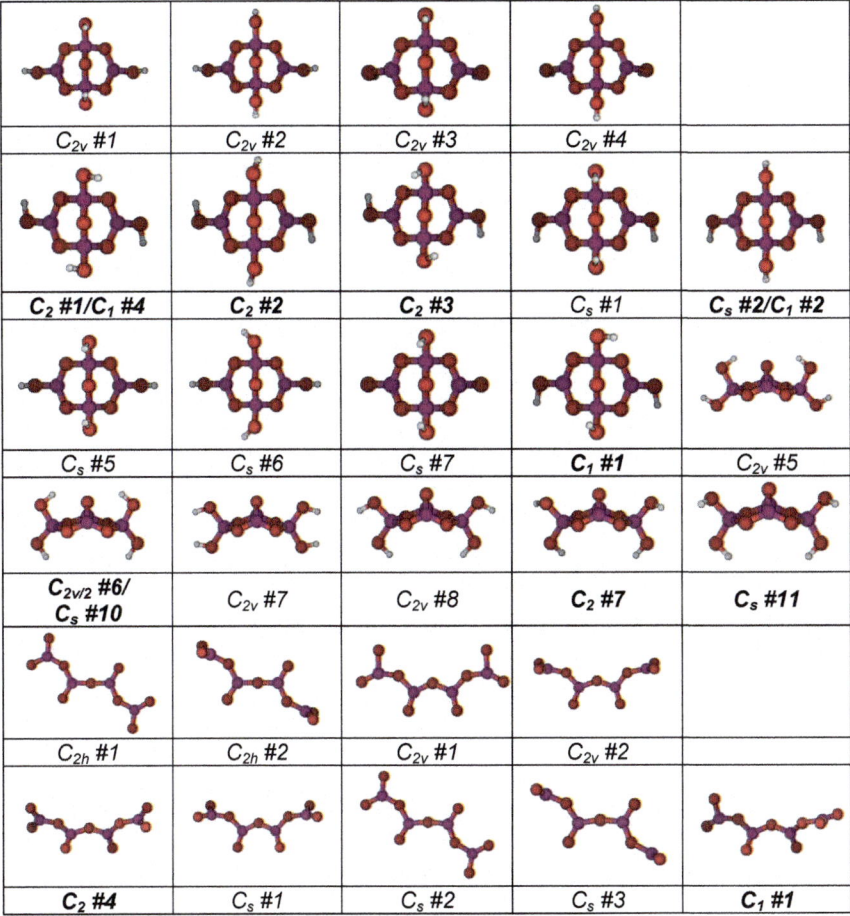

Fig. 3 Structure of tetraborate ions, $[B_4O_5(OH)_4]^{2-}$ and $[B_4O_9]^{6-}$

structure D_{2d} #2 has an imaginary E mode at both MP2/6-31+G* and MP2/6-311 +G*. These structures could either desymmetrize to C_2 #2/3 (stable), or C_s #3/4 (unstable), both of which have slightly puckered ring(s). Desymmetrization of these C_s #3 and #4 structures via C_1 #3/4 results in ascent in symmetry to C_2 #2/3. C_2 #1 is stable at all levels. C_{2v} #1 has an imaginary B_1 mode at both MP2/6-31+G* and MP2/6-311+G*, and an imaginary B_2 mode at MP2/6-31+G*. These lead to C_s #5 and 6 respectively. C_s #5 is stable at MP2/6-311+G*. Both C_s structures are unstable at MP2/6-31+G*, leading to stable C_1 #5 and 6 structures. C_s #1 and #2 are unstable at MP2/6-31+G*, converting into C_1 #1 and #2. The energy ordering is D_{2d} #1 < C_s #1 < C_2 #1 < C_{2v} #1 < C_s #2 < D_{2d} #2.

3.3.2 $[B_5O_6(OH)_6]^{3-}$

The pentaborate ion, $[B_5O_6(OH)_6]^{3-}$, is given in Fig. 4. We initially started with two C_{2v} structures and two C_2 structures. Neither C_{2v} structure is stable. Both have imaginary A_2, B_1, and B_2 modes at most levels. Both C_2 structures are stable at most levels, with the exception of B3LYP/6-311+G* (C_2 #1) and B3LYP/6-31+G* (C_2 #2), which desymmetrize to stable C_1 #1 and #2, respectively. The C_2 #3 and #4 structures, derived from C_{2v} #1 and #2, are unstable at all levels and desymmetrize to the stable C_1 #3 and #4. At B3LYP/6-31G* and MP2/6-31G*, one of the central B-O bonds in C_1 #3 breaks. Attempts to locate the C_s #1 and #2 structures also result in breaking of one of the central B-O bonds. The C_s #3 and #4 structures are not stable,

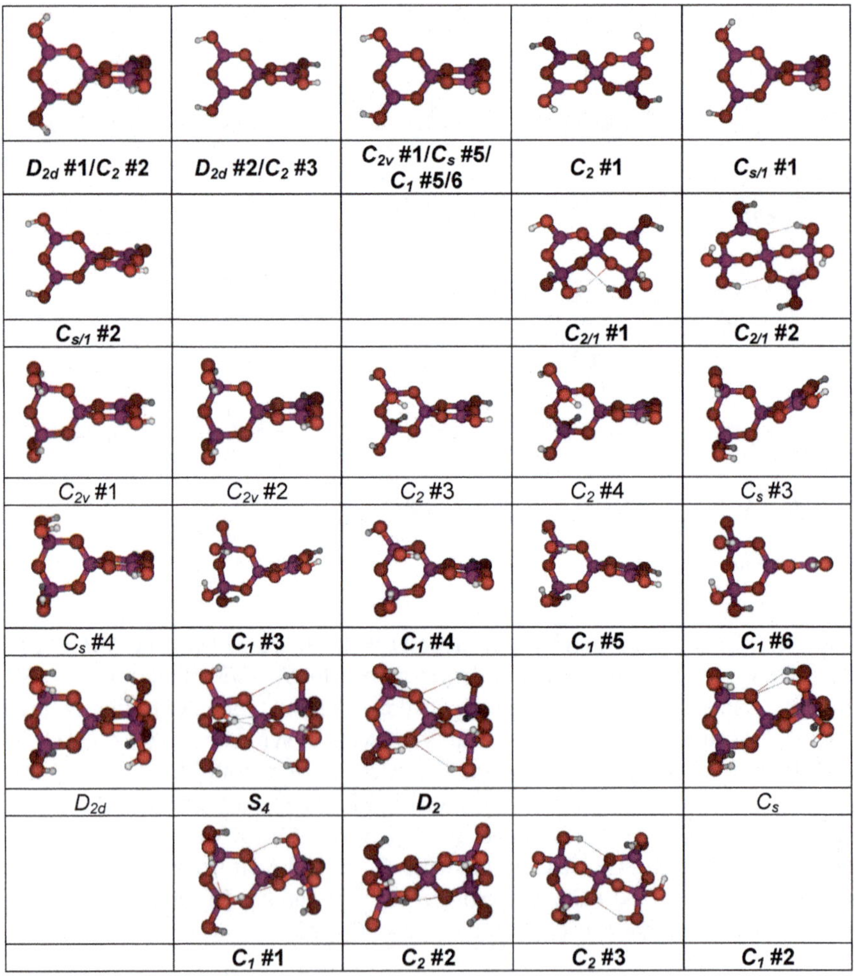

D_{2d} #1/C_2 #2	D_{2d} #2/C_2 #3	C_{2v} #1/C_s #5/ C_1 #5/6	C_2 #1	$C_{s/1}$ #1
$C_{s/1}$ #2			$C_{2/1}$ #1	$C_{2/1}$ #2
C_{2v} #1	C_{2v} #2	C_2 #3	C_2 #4	C_s #3
C_s #4	C_1 #3	C_1 #4	C_1 #5	C_1 #6
D_{2d}	S_4	D_2		C_s
	C_1 #1	C_2 #2	C_2 #3	C_1 #2

Fig. 4 Structure of pentaborate ions, $[B_5O_6(OH)_{4+2n}]^{-1-2n}$, $n = 0 - 2$

and desymmetrize to C_1 #5 and #6, respectively. Structure C_1 #5 fragments at all levels except HF/6-31+G*, but structure C_1 #6 is either stable (B3LYP/6-31+G*, MP2/6-31 +G*, and MP2/6-311+G*) or coalesces into C_1 #4.

3.3.3 $[B_5O_6(OH)_8]^{5-}$

The pentaborate ion, $[B_5O_6(OH)_8]^{5-}$, is given in Fig. 4. We initially started with a D_{2d} structure, but this was not stable and possessed imaginary B_1 (all levels), A_2 (HF and MP2), and E (MP2/6-31G* and MP2/6-311+G*) modes, to give D_2, S_4, and C_s structures, respectively. Both the D_2 and S_4 structures are stable at most levels examined. The D_2 is unstable at HF/6-31+G*, reverting via C_2 #1 to S_4, and at MP2/6-311+G* giving the stable C_2 #2 or C_2 #3. The S_4 structure is unstable at MP2/6-31G* and MP2/6-311+G*, desymmetrizing to the stable C_1 #1. The C_s structure is unstable at both levels and desymmetrizes to the stable C_1 #2.

3.4 Diborate Species

3.4.1 $[B_2O_5]^{4-}$

The diborate ion, $[B_2O_5]^{4-}$, crystallographically observed as the lithium salt (Table 3), is given in Fig. 5. Initially, structures of D_{2d} and D_{2h} symmetry were

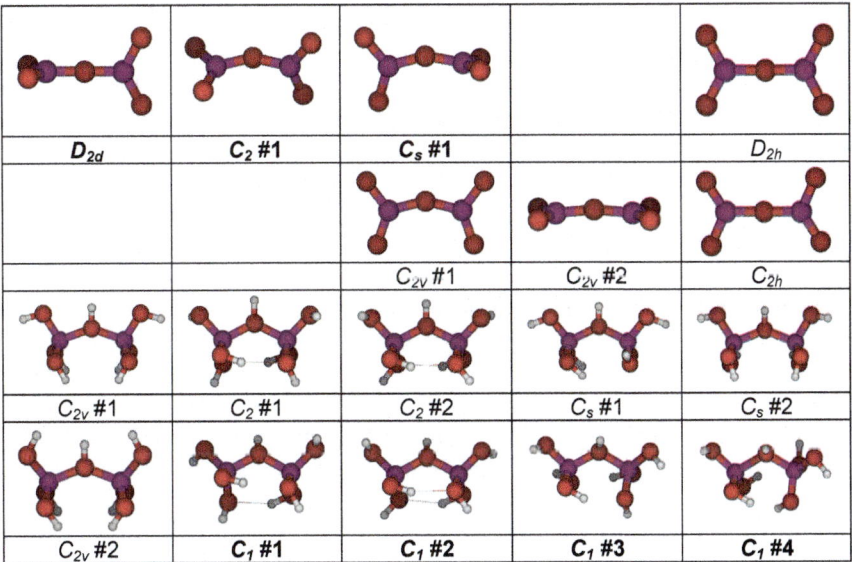

Fig. 5 Structure of diborate ions, $[B_2O_5]^{4-}$ and $[B_2(OH)_7]^-$

tried. The D_{2d} structure was stable at B3LYP/6-31G* and MP2/6-31G*, but had an imaginary E mode at the other levels, suggesting desymmetrization to either a C_2 #1 or C_s #1. The C_2 #1 is stable at all levels except HF/6-31+G* and HF/6-31+G*, for which the C_s #1 is stable. The D_{2h} structure is unstable at all levels. Desymmetrization along an imaginary A_u mode at all levels results in ascension in symmetry to the D_{2d} structure. Desymmetrization along the B_{2u}, B_{3u}, and B_{3g} modes give unstable C_{2v} #1 (all levels except B3LYP/6-31G*), C_{2v} #2 (MP2/6-31+G* and MP2/6-311+G*), and C_{2h} (B3LYP/6-31+G* and B3LYP/6-311+G*) structures. Desymmetrization of these structures leads to structures already observed.

3.4.2 [B$_2$(OH)$_7$]$^-$

In the polyborate structures discussed thus far, any oxygen that is bound to two boron atoms is not bound to a hydrogen. However, the diborate species [B$_2$(OH)$_7$]$^-$ with a bridging hydroxide was postulated as the intermediate to explain the boric acid–borate interchange in aqueous solution as studied by ^{11}B NMR [101]. Such an intermediate may lie on the reaction path to condensation to form the other polyborates. It is expected that the hydrogen of the bridging oxygen would be quite acidic and easily lost.

The diborate ion, [B$_2$(OH)$_7$]$^-$, is given in Fig. 5. Initially, two C_{2v} structures were tried. These had numerous imaginary A_2, B_1, and B_2 frequencies, which suggested desymmetrization to two C_2 and four C_s structures. Neither C_2 structure was stable, possessing an imaginary B mode. The C_s #1 and C_s #2 structures, derived from C_{2v} #1, were not stable, containing A'' imaginary frequencies. The C_s #3 and C_s #4 structures, derived from C_{2v} #2, either dissociated into a hydrogen bonded [B(OH)$_4$]$^-$...B(OH)$_3$ hydrogen bonded complex (C_s #3) or coalesced into C_s #2. The C_2 structures desymmetrized into the corresponding stable C_1 #1 and #2 structures. The C_s structures desymmetrized into the corresponding stable C_1 #3 and #4 structures. The four stable structures are quite close in energy, and the gas-phase association energy of boric acid and borate is in the range −47.9 to 108.5 kJ/mol.

3.4.3 [B$_2$(OH)$_6$]^0

We were curious to see whether a dimer of boric acid could exist possessing a single hydroxyl bridge. The bridge necessarily converts one of the boron atoms into tetrahedral, and under the constraint of C_s symmetry, results in eight possible structures. None of these are stable at the HF/6-31G* level, resulting in fragmentation of the B-O bond to give hydrogen bonded structures. In addition, structure C_s #1 is unstable at all levels investigated. These results suggest that such a structure does not exist.

Next, a dimer of boric acid containing two bridging hydroxyls was investigated. The optimized structures are shown in Fig. 6. Initially two structures of D_{2h} symmetry were considered. Neither of these structures were stable, possessing

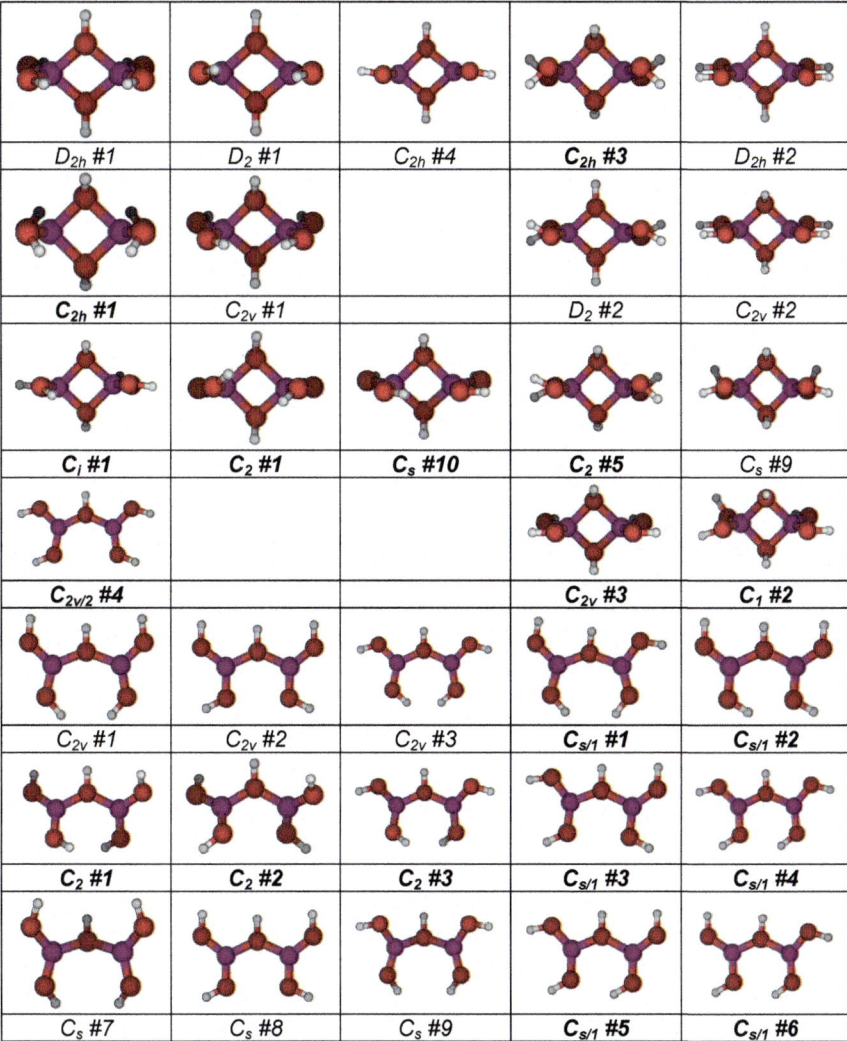

D_{2h} #1	D_2 #1	C_{2h} #4	C_{2h} #3	D_{2h} #2
C_{2h} #1	C_{2v} #1		D_2 #2	C_{2v} #2
C_i #1	C_2 #1	C_s #10	C_2 #5	C_s #9
$C_{2v/2}$ #4			C_{2v} #3	C_1 #2
C_{2v} #1	C_{2v} #2	C_{2v} #3	$C_{s/1}$ #1	$C_{s/1}$ #2
C_2 #1	C_2 #2	C_2 #3	$C_{s/1}$ #3	$C_{s/1}$ #4
C_s #7	C_s #8	C_s #9	$C_{s/1}$ #5	$C_{s/1}$ #6

Fig. 6 Structure of diboric acid, $[B_2(OH)_6]^0$, and $[B_2(OH)_5]^+$

imaginary B_{3g} and B_{2u} modes at all levels, suggesting desymmetrization to the C_{2h} #1 and #3, and C_{2v} #1 and #2 structures. The D_{2h} #2 structure possessed an A_u mode at all levels (as did D_{2h} #1 at MP2/6-311+G*), suggesting desymmetrization to D_2 #2 (D_2 #1). The D_{2h} structures also contained a B_{2g} imaginary mode at some levels suggesting desymmetrization (D_{2h} #1: B3LYP/6-311+G* to C_{2h} #2; D_{2h} #2: HF/6-311+G* to C_{2h} #4).

To our surprise, the C_{2h} #1 structure was stable at all B3LYP levels and at MP2/ 6-31+G*, but possessed a B_g imaginary mode at the other levels, suggesting desymmetrization to C_i #1. The C_{2h} #3 structure was stable at all levels! The C_{2h} #2 structure (B3LYP/6-311+G*) breaks apart into two boric acid molecules. The C_{2v} #1 structure has an A_2 imaginary mode at all levels leading to C_2 #1. The D_2 #1 structure (MP2/6-311+G*) has an imaginary B_3 and B_2 mode, leading to C_2 #2 and C_2 #3. The C_{2v} #2 structure has an A_2 and B_1 imaginary mode at all levels leading to C_2 #4 and C_s #9. The D_2 #2 structure has an imaginary B_3 and B_2 modes at all levels, leading to C_2 #5 and C_2 #6. The C_{2h} #4 structure (HF/6-311+G*) is unstable, possessing both a B_g and A_u mode, suggesting desymmetrization to C_i #2 and C_2 #7.

The C_i #1 structure is stable at all levels except HF/6-31G* and HF/6-311+G*, where it dissociated. The C_2 #1 structure is only stable at HF/6-31+G*. It dissociates at HF/6-31G*, HF/6-311+G*, and B3LYP/6-311+G*. At the other levels, it ascends in symmetry via C_1 #1 to the new C_s #10 structure, which exists at all levels. The C_2 #2 structure ascends in symmetry to C_{2h} #1. The C_2 #3 structure coalesces to C_2 #1. The C_2 #4 structure ascends in symmetry to the new C_{2v} #3. The unstable C_s #9 structure only exists at MP2/6-31G*, ascending in symmetry to C_{2v} #3 otherwise. It desymmetrizes to the stable C_1 #2 at MP2/6-31G*. The C_2 #5 structure is stable at all levels, but the C_2 #6 structure ascends in symmetry to C_{2v} #3. The C_i #2 and C_2 #7 ascend in symmetry to C_{2h} #3 and C_{2v} #3, respectively.

To summarize these results, there are at least four stable structures at all levels. Their energy ordering is as follows: C_{2v} #3 (0.0 kJ/mol) < C_{2h} #3 (5.7–9.8 kJ/mol) < C_s #10 (8.1–13.1 kJ/mol) < C_2 #5 (16.4–21.6 kJ/mol). The endothermic gas-phase dimerization energy of boric acid lies in the range 35.8–103.8 kJ/mol.

3.4.4 $[B_2(OH)_5]^+$

Another structure of interest is the cationic $[B_2(OH)_5]^+$ (Fig. 6). Initially we tried four structures of C_{2v} symmetry. None of these was a minimum except for C_{2v} #4 at all levels except MP2/6-311+G*. All non-minima had imaginary A_2 modes, and in some cases, B_1 modes. Deymmetrization along the A_2 modes led to the stable C_2 #1–#4 structures. Deymmetrization along the B_1 modes led to the unstable C_s #7–#9 structures. In addition, there are six other C_s structures obtainable by flipping the hydrogen atoms. The C_s #1 and #3 structures are unstable at all levels. The C_s #2 structure is only stable at HF/6-31G*, HF/6-31+G*, and B3LYP/6-31+G*. The C_s #4, #5, and #6 structures are stable at all levels except MP2/6-311+G*. These C_s structures desymmetrize into the corresponding C_1 #1–#9 structures. The C_1 #1–#6 structures are stable, but the C_1 #7–#9 structures convert to other structures already obtained. Of these structures, C_s #4 is the most stable, followed by C_{2v} #4 (8.0–10.2 kJ/mol). The structure (with hydroxide) is thermodynamically unstable relative to two boric acid molecules (880–1020 kJ/mol).

3.4.5 $[B_2O(OH)_6]^{2-}$

The diborate ion $[B_2O(OH)_6]^{2-}$ has been observed as the magnesium salt in the mineral pinnoite. Stadler found that the space group of pinnoite was $P4_2$ or $P4_2/m$, with a = 7.617(2) and c = 8.190(2), and proposed that the ion was $[B_2O(OH)_6]^{2-}$ [102]. The space group was confirmed as $P4_2$ (a = 7.62(1), c = 8.19(1) by Paton and MacDonald and the structure of the ion confirmed [103]. The structure was further refined by Krogh-Moe [104]. Its structure is shown in Fig. 7. We initially tried two C_{2v} structures. Both were unstable and possessed A_2 and B_2 imaginary modes. In addition C_{2v} #1 possessed an imaginary B_1 mode. The two C_2 structures thus derived were stable (with C_2 #2 coalescing into C_2 #1 at B3LYP/6-31G* and MP2/6-31G*). None of the three C_s structures were stable, and desymmetrized to the corresponding C_1 structures. C_1 #3 coalesced into C_1 #1 at B3LYP/6-31G* and MP2/6-31G*. Other possibilities include two C_s structures (#4 and #5), related to the C_{2v} structures by rotation of one of the hydroxyls. Neither of these are stable, and desymmetrize to the corresponding C_1 structures.

3.4.6 $[B_2O(OH)_5]^-$

The diborate ion $[B_2O(OH)_5]^-$ was initially considered to contain a only a single oxo bridge (Fig. 8). Eight such structures of C_s symmetry were considered, with the BO_3 unit in the plane of symmetry. None of these was stable and led to the corresponding C_1 structures. In some cases, these coalesced. It could also potentially exist as a (μ-O)(μ-OH) doubly bridged dimer of C_{2v} symmetry. This structure is unstable, and has imaginary A_2, B_1, and B_2 modes leading to C_2, C_s #9, and C_s #10 structures. In C_s #10, one of the B-O(H) bonds has broken. All of these structures also have imaginary modes, leading potentially to C_1 #9–#11,

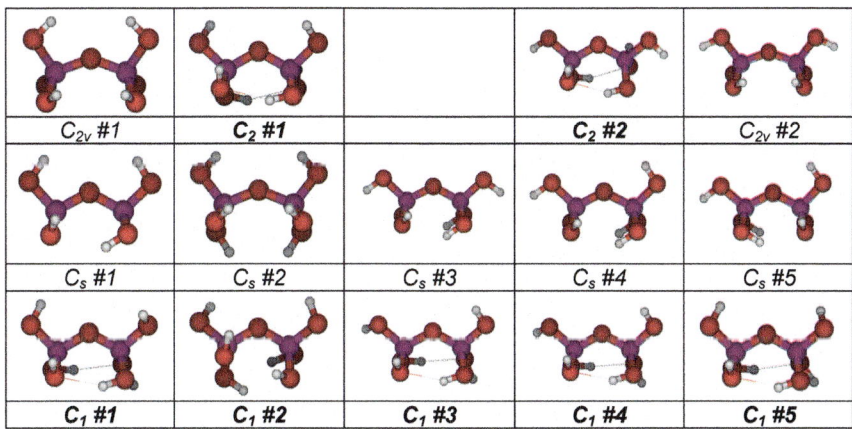

C_{2v} #1	C_2 #1		C_2 #2	C_{2v} #2
C_s #1	C_s #2	C_s #3	C_s #4	C_s #5
C_1 #1	C_1 #2	C_1 #3	C_1 #4	C_1 #5

Fig. 7 Structure of diborate ion, $[B_2O(OH)_6]^{2-}$

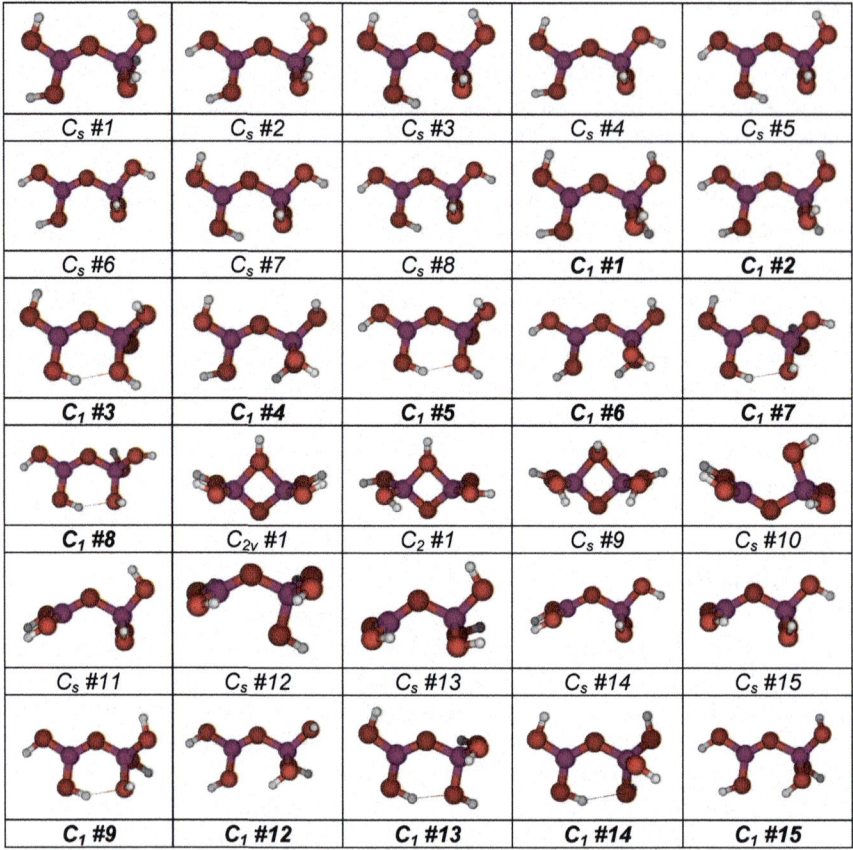

C_s #1	C_s #2	C_s #3	C_s #4	C_s #5
C_s #6	C_s #7	C_s #8	C_1 #1	C_1 #2
C_1 #3	C_1 #4	C_1 #5	C_1 #6	C_1 #7
C_1 #8	C_{2v} #1	C_2 #1	C_s #9	C_s #10
C_s #11	C_s #12	C_s #13	C_s #14	C_s #15
C_1 #9	C_1 #12	C_1 #13	C_1 #14	C_1 #15

Fig. 8 Structure of diborate ion, $[B_2O(OH)_5]^-$

respectively. Only C_1 #9 is stable, with C_1 #10 and #11 coalescing into C_1 #6 and #10, respectively. The bond breaking in C_s #10 suggests that we try three additional C_s structures (#11–#13) in which the in plane OH hydroxyls have rotated 180 degrees. To our surprise, C_s #11 and #13 resulted in inversion about the bridging oxygen. These structures, in turn, suggested trying two other C_s structures (#14 and #15), keeping the inverted oxygen but placing the hydrogen in its original position. None of these C_s structures were stable. Desymmetrization of C_s #11–15 to C_1 #12–16 gives stable structures at some levels, or coalescence to others.

3.4.7 $[B_2O(OH)_4]^0$

A dehydrated form of boric acid dimer with a bridging oxo group was considered (Fig. 9). The high-symmetry structures D_{2h} #1 and #2 has B_{2u}, A_u, and B_{3u}

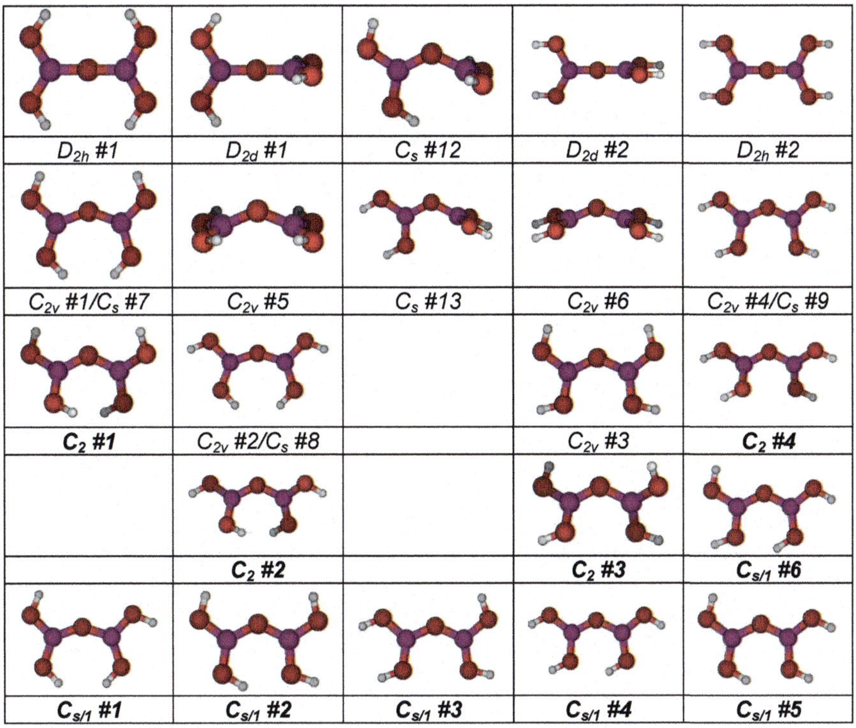

D_{2h} #1	D_{2d} #1	C_s #12	D_{2d} #2	D_{2h} #2
C_{2v} #1/C_s #7	C_{2v} #5	C_s #13	C_{2v} #6	C_{2v} #4/C_s #9
C_2 #1	C_{2v} #2/C_s #8		C_{2v} #3	C_2 #4
	C_2 #2		C_2 #3	$C_{s/1}$ #6
$C_{s/1}$ #1	$C_{s/1}$ #2	$C_{s/1}$ #3	$C_{s/1}$ #4	$C_{s/1}$ #5

Fig. 9 Structure of diboric acid, $[B_2O(OH)_4]^0$

imaginary frequencies, suggesting desymmetrization to C_{2v} #1/4, D_2 #1/2, and C_{2v} #5/6, respectively. The D_{2d} #1 and #2 had an imaginary E mode, suggesting desymmetrization to C_2 #7/8 or C_s #12/13. Two additional C_{2v} structures were also tried (#2, #3). The D_2 structures ascended in symmetry to the corresponding D_{2d} structures. None of the C_{2v} structures #1-#4 was stable. All had an imaginary A_2 mode, which led to the stable C_2 structures #1–#4. In some cases at some levels, there was also an imaginary B_1 mode leading to a C_s structure (#7–#9) which were unstable and desymmetrized, via the corresponding C_1 structure, to one of the observed C_2 structures. The C_{2v} #5 and #6 structures had imaginary B_1 and A_2 modes. The putative C_2 #5 and #6 derived along the A_2 mode coalesced into C_2 #1 and #4, whereas the putative C_s #10 and #11 derived along the B_1 mode ascended in symmetry or coalesced to C_{2v} #1/C_s #7 and C_{2v} #4. Six additional C_s structures were optimized (#1–#6). Most (#2, #4, #5, and #6) were stable at all levels except MP2/6-311+G*. Unstable C_s structures desymmetrized to the corresponding C_1 structures (#1–#6). The C_s #12/13 structures derived from the D_{2d} #1 and #2 structures were not stable. When these were desymmetrized (C_1 #10, #11), they ascended in symmetry to C_2 #1 and C_2 #4, respectively. The C_2 #7/8 structures coalesced into C_2 #1/4 as well.

3.4.8 [B$_2$O$_2$(OH)$_4$]$^{2-}$

Structures containing two oxo bridges between the two boron atoms are now considered. The first such structure, [B$_2$O$_2$(OH)$_4$]$^{2-}$, can be considered to be the dimer of the hypothetical oxodihydroxoborate ion, [BO(OH)$_2$]$^-$ (Fig. 10). First, the high symmetry D_{2h} #1 and #2 structures were considered. These had imaginary A_u, B_{3g}, B_{2g}, and (#2) B_{1u} modes, leading to the potential D_2 #1/2, C_{2h} #1/3, C_{2h} #2/4, and C_{2v} structures.

The D_2 #1 structure is stable at all levels, whereas D_2 #2 is only stable at B3LYP/6-31G* and MP2/6-31G*. It has imaginary B_2 and B_1 modes at the other levels, giving stable C_2 #3 and #4 structures. The C_{2v} structure, derived from D_{2h} #2, ascends in symmetry to the D_{2h} #1 structure. The C_{2h} #1 structure is stable at all levels except B3LYP/6-311+G* and the HF levels, for which it has an A_u mode to desymmetrize to the putative C_2 #1 structure, which ascends in symmetry to D_2 #1. The C_{2h} #2 structure, which exists at all levels except HF/6-31+G* and HF/6-311 +G*, has an A_u and B_g mode at all levels, suggesting desymmetrization to C_2 #2 and C_i #1, respectively. These ascend in symmetry to D_2 #1 and C_{2h} #1, respectively. The C_{2h} #3 structure, which exists for calculations with diffuse basis sets, has imaginary B_g and B_u modes, desymmetrizing to C_i #2 and C_s #1, respectively, which are stable at all levels, with the exception of C_s #1 at HF/6-31+G*, which ascends in symmetry via C_1 #1 to C_2 #3. The C_{2h} #4 structure coalesced with either the C_{2h} #2 or D_{2h} #1 structure.

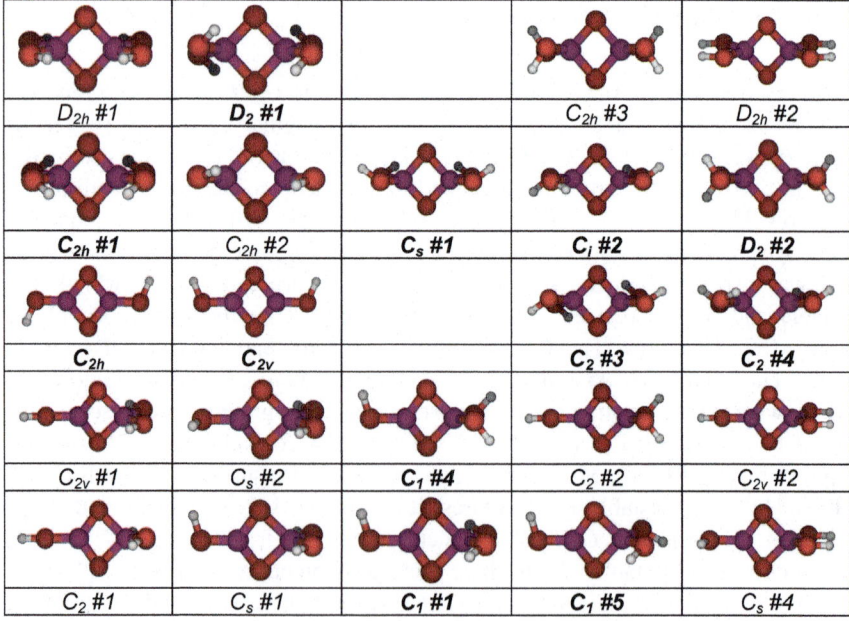

D_{2h} #1	D_2 #1		C_{2h} #3	D_{2h} #2
C_{2h} #1	C_{2h} #2	C_s #1	C_i #2	D_2 #2
C_{2h}	C_{2v}		C_2 #3	C_2 #4
C_{2v} #1	C_s #2	C_1 #4	C_2 #2	C_{2v} #2
C_2 #1	C_s #1	C_1 #1	C_1 #5	C_s #4

Fig. 10 Structure of diborate ions, [B$_2$O$_2$(OH)$_{4-n}$]$^{-2+n}$, n = 0 – 2

3.4.9 $[B_2O_2(OH)_4]^-$

The diborate ion, $[B_2O_2(OH)_3]^-$, is shown in Fig. 10. Initially, two C_{2v} structures were tried. Both had imaginary B_1, B_2, and A_2 modes. These lead to C_s #1/3, C_s #2/4, and C_2 #1/2, respectively. The structure C_s #3 coalesced into C_s #1. None of these structures were stable. The structure C_s #1 desymmetrized to C_1 #1. The structure C_s #2, desymmetrized via C_1 #2 to coalesce with C_1 #1. The structure C_2 #1 also desymmetrized via C_1 #3 to coalesce with C_1 #1. The structure C_2 #2 desymmetrized to C_1 #4. The structure C_s #4 desymmetrized via C_1 #5 to coalesce to C_1 #4 (except at MP2/6-311+G*).

3.4.10 $[B_2O_2(OH)_4]^0$

The diboric acid, $[B_2O_2(OH)_2]^0$, which is an isomer of metaboric acid, is also given in Fig. 10. The two forms examined, C_{2h} and C_{2v}, were both minima, with C_{2h} slightly lower in energy.

3.5 Structural Comparisons

A comparison of the calculated ab initio with the crystallographically observed boron-oxygen bond lengths is given in Table 6. For ease of reporting, we take high-symmetry structures in some cases and average the crystal structures according to the approximate symmetry mentioned. Inspection of Table 6 reveals that in most instances, calculations with the 6-31+G* and 6-311+G* basis sets are within 0.01 Å of each other, the exception being $[B_4O_5(OH)_4]^{2-}$, where the conformation of the hydroxyl on the tetrahedral boron is significantly different. For the singly and doubly charged anions, the 6-31G* results are, with one exception, within 0.01 Å of the 6-31+G* results, but the more highly charged anions show more deviation. The B3LYP and MP2 calculations are usually within 0.01 Å of each other, except for the highly charged $[B_4O_9]^{6-}$ anion. The Hartree-Fock calculations tend to be slightly shorter. There seems to be mostly good agreement between the crystal structures and the correlated calculations with a diffuse basis set, with the experiments usually being slightly shorter. The agreement between experiment and theory here is rather remarkable in the sense that the calculations correspond to a gas-phase ion (no medium effects were included in the calculation), whereas the experiment corresponds to the solid-state, surrounded by counterions.

We may compare our results to previous ab initio calculations (see Introduction). Although Zhang et al. [95] completely optimized $B_2O(OH)_4$ with symmetry C_{2v} (our #3), C_s, and C_2 (our #3), our lowest energy structure corresponds to a planar C_s structure with an internal hydrogen bond. However, in most cases, the previous calculations, if they exist, are comparable to ours.

Table 6 Theoretical and literature experimental B-O distances (Å) in crystallographically observed Polyborate Ions

Species	X-ray	HF			B3LYP			MP2		
		6-31G*	6-31+G*	6-311+G*	6-31G*	6-31+G*	6-311+G*	6-31G*	6-31+G*	6-311+G*
$B_2O_5^{4-}$ $\boldsymbol{D_{2d}}$ [57]	1.340	1.376	1.386	1.384	1.39	1.399	1.39	1.398	1.415	1.407
	1.454	1.485	1.438	1.442	21.526	1.440	51.445	1.528	1.443	1.445
$B_2O(OH)_6^{2-}$ C_{2v} #1 [103]	1.42,1.46,	1.431	1.431	1.429	1.444	1.445	1.443	1.447	1.449	1.443
	1.54,1.56	1.448	1.449	1.448	1.461	1.464	1.462	1.466	1.469	1.463
		1.569	1.565	1.565	1.579	1.576	1.578	1.574	1.568	1.563
$B_3O_6^{3-}$ $\boldsymbol{D_{3h}}$ [8]	1.280	1.305	1.314	1.311	1.320	1.332	1.328	1.326	1.342	1.335
	1.433	1.421	1.417	1.416	1.440	1.433	1.433	1.445	1.440	1.435
$B_3O_5(OH)_2^{3-}$ $\boldsymbol{C_2}$ #1 [7]	1.326	1.307	1.318	1.315	1.320	1.335	1.331	1.327	1.345	1.339
	1.39	1.417	1.410	1.409	1.439	1.427	1.426	1.443	1.432	1.427
	1.42	1.422	1.419	1.418	1.440	1.435	1.434	1.446	1.442	1.437
	1.47	1.446	1.447	1.446	1.461	1.461	1.460	1.467	1.468	1.463
	1.488	1.528	1.523	1.522	1.544	1.541	1.541	1.542	1.539	1.532
$B_3O_4(OH)_4^{3-}$ C_{2v} [6]	1.340	1.309	1.323	1.319	1.322	1.340	1.336	1.329	1.351	1.344
	1.397	1.418	1.411	1.410	1.438	1.428	1.427	1.442	1.433	1.428
	1.445	1.441	1.438	1.438	1.455	1.450	1.449	1.459	1.455	1.452
	1.487	1.447	1.453	1.453	1.458	1.468	1.468	1.465	1.477	1.470
	1.493	1.528	1.523	1.523	1.546	1.541	1.541	1.543	1.540	1.533
$B_3O_3(OH)_4^-$ $\boldsymbol{C_2}$ #3 [30]	1.344	1.327	1.329	1.326	1.341	1.343	1.341	1.326	1.350	1.344
	1.362	1.373	1.373	1.371	1.388	1.388	1.386	1.393	1.394	1.389
	1.396	1.381	1.380	1.377	1.396	1.396	1.393	1.400	1.400	1.394
	1.452	1.452	1.454	1.452	1.464	1.468	1.466	1.466	1.471	1.464
	1.503	1.484	1.482	1.483	1.498	1.496	1.497	1.500	1.499	1.495

(continued)

Table 6 (continued)

Species	X-ray	HF			B3LYP			MP2		
		6-31G*	6-31+G*	6-311+G*	6-31G*	6-31+G*	6-311+G*	6-31G*	6-31+G*	6-311+G*
$B_4O_9^{6-}$ C_{2h} #1 [58]	1.321	1.323	1.331	1.328	1.339	1.345	1.339	1.344	1.354	1.346
	1.344	1.359	1.373	1.370	1.373	1.374	1.368	1.378	1.394	1.385
	1.349	1.366	1.387	1.385	1.377	1.394	1.387	1.384	1.429	1.424
	1.390	1.444	1.442	1.443	1.467	1.448	1.452	1.472	1.459	1.461
	1.394	1.478	1.447	1.451	1.513	1.453	1.452	1.517	1.476	1.474
	1.434	1.548	1.481	1.484	1.607	1.473	1.469	1.608	1.485	1.485
$B_4O_5(OH)_4^{2-}$ C_2 **#1** [15]	1.3655	1.334	1.350	1.335	1.351	1.350	1.347	1.354	1.357	1.353
	1.3737	1.348	1.415	1.344	1.362	1.366	1.364	1.370	1.373	1.364
	1.3784	1.418	1.416	1.415	1.435	1.434	1.431	1.438	1.438	1.431
	1.4451	1.436	1.438	1.419	1.452	1.454	1.452	1.454	1.459	1.432
	1.4657	1.463	1.461	1.456	1.476	1.476	1.473	1.477	1.478	1.468
	1.4902	1.488	1.485	1.515	1.509	1.501	1.499	1.505	1.503	1.528
	1.5075	1.515	1.514	1.518	1.520	1.527	1.534	1.529	1.530	1.533
$B_5O_6(OH)_4^-$ D_{2d} #1 [34]	1.356	1.332	1.333	1.331	1.347	1.348	1.346	1.352	1.355	1.350
	1.367	1.371	1.371	1.370	1.386	1.386	1.384	1.391	1.393	1.387
	1.383	1.375	1.374	1.372	1.389	1.390	1.387	1.393	1.394	1.388
	1.468	1.464	1.464	1.464	1.476	1.477	1.478	1.478	1.479	1.476

4   Conclusions

Boron, when surrounded by oxygen, can be either trigonal planar or tetrahedral. An examination of existing crystal structures of the alkali metal salts reveal a wide range of possible ways in which these may condense together to form B_2O, cyclo-B_3O_3, bicyclo-B_4O_5, and spirobicyclo-B_5O_6 units. In some cases, discrete ions are formed. The structures of these ions have been calculated using Hartree-Fock, density functional, and Moller Plesset theory with modest basis sets and the predicted gas-phase geometries are shown to agree quite well with the crystal structures. In addition, several other structurally-related ions and neutral molecules have been shown to be minima on the potential energy surface and, while as yet unobserved, could potentially exist.

Acknowledgements CCP acknowledges ACENet for computational facilities.

References

1. Pye CC (2018) An ab initio study of boric acid, borate, and their interconversion. Prog Theor Chem Phys 31:143–177
2. Konig H, Hoppe R (1977) Zur Kenntnis von Na_3BO_3. Z Anorg Allg Chem 434:225–232 (in German)
3. Menchetti S, Sabelli C (1982) Structure of hydrated sodium borate $Na_2[BO_2(OH)]$. Acta Crystallogr B 38:1282–1284
4. Block S, Perloff A (1963) The direct determination of the crystal structure of NaB $(OH)_4.2H_2O$. Acta Crystallogr 16:1233–1238
5. Csetenyi LJ, Glasser FP, Howie RA (1993) Structure of sodium tetrahydroxyborate. Acta Crystallogr C 49:1039–1041
6. Andrieux J, Goutaudier C, Laversenne L, Jeanneau E, Miele P (2010) Synthesis, characterization, and crystal structure of a new trisodium triborate, $Na_3[B_3O_4(OH)_4]$. Inorg Chem 49:4830–4835
7. Corazza E, Menchetti S, Sabelli C (1975) The crystal structure of $Na_3[B_3O_5(OH)_2]$. Acta Crystallogr B 31:1993–1997
8. Marezio M, Plettinger HA, Zachariasen WH (1963) The bond lengths in the sodium metaborate structure. Acta Crystallogr 16:594–595
9. Menchetti S, Sabelli C (1979) A new borate polyanion in the structure of $Na_8[B_{12}O_{20}(OH)_4]$. Acta Crystallogr B 35:2488–2493
10. Menchetti S, Sabelli C (1977) The crystal structure of synthetic sodium pentaborate monohydrate. Acta Crystallogr B 33:3730–3733
11. Menchetti S, Sabelli C, Trosti-Ferroni R (1982) The structure of sodium borate $Na_3[B_5O_9]$. H_2O. Acta Crystallogr B 38:2987–2991
12. Font Tullot JM (1947) El borax como estructura cristalina en cadenas. Estud Geol 7:13–20 (in Spanish)
13. Norimoto N (1956) The crystal structure of borax. Mineral J 2:1–18
14. Levy HA, Lisensky GC (1978) Crystal structures of sodium sulfate decahydrate (Glauber's Salt) and sodium tetraborate decahydrate (Borax). Redetermination by neutron diffraction. Acta Crystallogr B 34:3502–3510
15. Gainsford GJ, Kemmitt T, Higham C (2008) Redetermination of the borax structure from laboratory X-ray data at 145 K. Acta Crystallogr E 64:i24–i25

16. Giacovazzo C, Menchetti S, Scordari F (1973) The crystal structure of tincalconite. Am Mineral 58:523–530

17. Powell DR, Gaines DF, Zerella PJ, Smith RA (1991) Refinement of the structure of tincalconite. Acta Crystallogr C 47:2279–2282

18. Luck RL, Wang G (2002) On the nature of tincalconite. Am Mineral 87:350–354

19. Ross V, Edwards JO (1959) On the crystal structure of kernite, $Na_2B_4O_7.4H_2O$. Acta Crystallogr 12:258

20. Giese RF Jr (1966) Crystal structure of kernite, $Na_2B_4O_6(OH)_2 \bullet 3H_2O$. Science 154:1453–1454

21. Cialdi G, Corazza E, Sabelli C (1967) La struttura cristallina della kernite, $Na_2B_4O_6(OH)_2$ $3H_2O$. Rend. Lincei Sc Fis Mat Nat Ser 8, 42:236–251 (in Italian)

22. Cooper WF, Larsen FK, Coppens P, Giese RF (1973) Electron population analysis of accurate diffraction data. V. Structure and one-center charge refinement of the light-atom mineral kernite, $Na_2B_4O_6(OH)_2 \bullet 3H_2O$. Am Mineral 58:21–31

23. Menchetti S, Sabelli C (1978) The crystal structure of $Na_2[B_4O_6(OH)_2]$. Acta Crystallogr B 34:1080–1084

24. Krogh-Moe J (1974) The crystal structure of sodium diborate, $Na_2O.2B_2O_3$. Acta Crystallogr B 30:578–582

25. Cannillo E, Dal Negro A, Ungaretti L (1973) The crystal structure of ezcurrite. Am Mineral 58:110–115

26. Corazza E, Menchetti S, Sabelli C (1975) The crystal structure of nasinite, $Na_2[B_5O_8(OH)].2H_2O$. Acta Crystallogr B 31:2405–2410

27. Wang Y, Pan S, Tian X, Zhou Z, Liu G, Wang J, Jia D (2009) Synthesis, structure, and properties of the noncentrosymmetric hydrated borate $Na_2B_5O_8(OH) \bullet 2H_2O$. Inorg Chem 48:7800–7804

28. Liu Z-H, Li L-Q, Wang M-Z (2006) Synthesis, crystal structure and thermal behavior of $Na_4[B_{10}O_{16}(OH)_2] \bullet 4H_2O$. J Alloys Compd 407:334–339

29. Corazza E, Menchetti S, Sabelli C (1974) The crystal structure of biringuccite, $Na_4[B_{10}O_{16}(OH)_2]$ $\bullet 2H_2O$. Am Mineral 59:1005–1015

30. Dal Negro A, Martin Pozas JM, Ungaretti L (1975) The crystal structure of ameghinite. Am Mineral 60:879–883

31. Krogh-Moe J (1974) The crystal structure of α sodium triborate, α-$Na_2O.3B_2O_3$. Acta Crystallogr B 30:747–752

32. Krogh-Moe J (1972) The crystal structure of a sodium triborate modification, β-$Na_2O.3B_2O_3$. Acta Crystallogr B 28:1571–1576

33. Hyman A, Perloff A, Mauer F, Block S (1967) The crystal structure of sodium tetraborate. Acta Crystallogr 22:815–821

34. Merlino S, Sartori F (1972) The crystal structure of sborgite, $NaB_5O_6(OH)_4.3H_2O$. Acta Crystallogr B 28:3559–3567

35. Menchetti S, Sabelli C (1978) The crystal structure of $NaB_5O_6(OH)_4$. Acta Crystallogr B 34:45–49

36. Zachariasen WH (1937) The crystal structure of potassium metaborate, $K_3(B_3O_6)$. J Chem Phys 5:919–922

37. Schneider W, Carpenter GB (1970) Bond lengths and thermal parameters of potassium metaborate, $K_3B_3O_6$. Acta Crystallogr B 26:1189–1191

38. Marezio M, Plettinger HA, Zachariasen WH (1963) The crystal structure of potassium tetraborate tetrahydrate. Acta Crystallogr 16:975–980

39. Krogh-Moe J (1961) Unit-cell data for some anhydrous potassium borates. Acta Crystallogr 14:68

40. Krogh-Moe J (1972) The crystal structure of potassium diborate, $K_2O.2B_2O_3$. Acta Crystallogr B 28:3089–3093

41. Marezio M (1969) The crystal structure of $K_2[B_5O_8(OH)].2H_2O$. Acta Crystallogr B 25:1787–1795

42. Zhang H-X, Zhang J, Zheng S-T, Yang G-Y (2005) Two new potassium borates, $K_4B_{10}O_{15}(OH)_4$ with stepped chain and $KB_5O_7(OH)_2 \bullet H_2O$ with double helical chain. Cryst Growth Des 5:157–161

43. Salentine CG (1987) Synthesis, characterization, and crystal structure of a new potassium borate, $KB_3O_5 \bullet 3H_2O$. Inorg Chem 26:128–132

44. Wang G-M, Sun Y-Q, Zheng S-T, Yang G-Y (2006) Synthesis and crystal structure of a novel potassium borate with an unprecedented $[B_{12}O_{16}(OH)_8]^{4-}$ anion. Z Anorg Allg Chem 632:1586–1590

45. Li H-J, Liu Z-H, Sun L-M (2007) Synthesis, crystal structure, vibrational spectroscopy and thermal behavior of the first alkali metal hydrated hexaborate: $K_2[B_6O_9(OH)_2]$. Chin J Chem 25:1131–1134

46. Bubnova RS, Fundamenskii VS, Filatov SK, Polyakova IG (2004) Crystal structure and thermal behavior of KB_3O_5. Doklady Phys Chem 398:249–253

47. Krogh-Moe J (1974) The crystal structure of pentapotassium enneakaidekaborate, $5K_2O.19B_2O_3$. Acta Crystallogr B 30:1827–1832

48. Zachariasen WH (1937) The crystal structure of potassium acid dihydronium pentaborate $KH_2(H_3O)_2B_5O_{10}$, (Potassium Pentaborate Tetrahydrate). Z Kristallogr 98:266–274

49. Cook WR Jr, Jaffe H (1957) The crystallographic, elastic, and piezoelectric properties of ammonium pentaborate and potassium pentaborate. Acta Crystallogr 10:705–707

50. Zachariasen WH, Plettinger HA (1963) Refinement of the structure of potassium pentaborate tetrahydrate. Acta Crystallogr 16:376–379

51. Gao Y-H (2011) Potassium decaborate monohydrate. Acta Crystallogr E 67:i57

52. Wu Q (2011) Potassium pentaborate. Acta Crystallogr E 67:i67

53. Krogh-Moe J (1959) On the structural relationship of vitreous potassium pentaborate to the crystalline modifications. Arkiv Kemi 14:567–572

54. Krogh-Moe J (1972) The crystal structure of the high-temperature modification of potassium pentaborate. Acta Crystallogr B 28:168–172

55. Krogh-Moe J (1959) The crystal structures of potassium pentaborate, $K_2O.5B_2O_3$, and the isomorphous rubidium compound. Arkiv Kemi 14:439–449

56. Krogh-Moe J (1965) Least-squares refinement of the crystal structure of potassium pentaborate. Acta Crystallogr 18:1088–1089

57. He M, Okudera H, Simon A, Kohler J, Jin S, Chen X (2013) Structure of $Li_4B_2O_5$: high-temperature monoclinic and low-temperature orthorhombic forms. J Solid State Chem 197:466–470

58. Rousse G, Baptiste B, Lelong G (2014) Crystal structures of $Li_6B_4O_9$ and $Li_3B_{11}O_{18}$ and application of the dimensional reduction formalism to lithium borates. Inorg Chem 53:6034–6041

59. Touboul M, Betourne E, Nowogrocki G (1995) Crystal structure and dehydration process of $Li(H_2O)_4B(OH)_4 \bullet 2H_2O$. J Solid State Chem 115:549–553

60. Zachariasen WH (1964) The crystal structure of lithium metaborate. Acta Crystallogr 17:749–751

61. Hohne E (1964) Die Kristallstruktur des $LiB(OH)_4$. Z Chem 4:431–432 (in German)

62. Fronczek FR, Aubry DA, Stanley GG (2001) Refinement of lithium tetrahydroxoborate with low-temperature CCD data. Acta Crystallogr E 57:i62–i63

63. Marezio M, Remeika JP (1966) Polymorphism of $LiMO_2$ compounds and high-pressure single-crystal synthesis of $LiBO_2$. J Chem Phys 44:3348–3353

64. Louer D, Louer M, Touboul M (1992) Crystal structure determination of lithium diborate hydrate, $LiB_2O_3(OH).H_2O$, from X-ray powder diffraction data collected with a curved position-sensitive detector. J Appl Crystallogr 25:617–623

65. Krogh-Moe J (1962) The crystal structure of lithium diborate, $Li_2O.2B_2O_3$. Acta Crystallogr 15:190–193

66. Krogh-Moe J (1968) Refinement of the crystal structure of lithium diborate, $Li_2O.2B_2O_3$. Acta Crystallogr B 24:179–181

67. Natarajan M, Faggiani R, Brown ID (1979) Dilithium tetraborate, $Li_2B_4O_7$. Cryst Struct Commun 8:367–370
68. Radaev SF, Muradyan LA, Malakhova LF, Burak YV, Simonov VI (1989) Atomic Structure and electron density of lithium tetraborate $Li_2B_4O_7$. Sov Phys Crystallogr 34:842–845
69. Jiang A, Lei S, Huang Q, Chen T, Ke D (1990) Structure of lithium heptaborate, $Li_3B_7O_{12}$. Acta Crystallogr C 46:1999–2001
70. Radaev SF, Genkina EA, Lomonov VA, Maksimov BA, Pisarevskii YV, Chelokov MN, Simonov VI (1991) Distribution of deformation electron density in lithium triborate LiB_3O_5. Sov Phys Crystallogr 36:803–807
71. Radaev SF, Maximov BA, Simonov VI, Andreev BV, D'Yakov VA (1992) Deformation density in lithium triborate, LiB_3O_5. Acta Crystallogr B 48:154–160
72. Le Henaff C, Hansen NK, Protas J, Marnier G (1997) Electron density distribution in LiB_3O_5 at 293 K. Acta Crystallogr B 53:870–879
73. Sennova N, Albert B, Bubnova R, Krzhizhanovskaya M, Filatov S (2014) Anhydrous lithium borate, $Li_3B_{11}O_{18}$, crystal structure, phase transition and thermal expansion. Z Kristallogr 229(7):497–504
74. Cardenas A, Solans J, Byrappa K, Shekar KVK (1993) Structure of lithium catena-Poly [3,4-dihydroxopentaborate-1:5-mu-oxo]. Acta Crystallogr C 49:645–647
75. Zviedre I, Ievins A (1974) Crystalline structure of rubidium monoborate $RbBO_2\bullet4/3\ H_2O$. Latvijas PSR Zinatnu Akademijas Vestis Kimijas Seria 4:395–400 (in Russian)
76. Schmid S, Schnick W (2004) Rubidium metaborate, $Rb_3B_3O_6$. Acta Crystallogr C 60:i69–i70
77. Touboul M, Penin N, Nowogrocki G (2000) Crystal structure and thermal behavior of $Rb_2[B_4O_5(OH)_4]\bullet3.6H_2O$. J Solid State Chem 149:197–202
78. Krzhizhanovskaya MG, Bubnova RS, Bannova II, Filatov SK (1997) Crystal structure of $Rb_2B_4O_7$. Crystallogr Rep 42:226–231
79. Bubnova RS, Krivovichev SV, Shakhverdova IP, Filatov SK, Burns PC, Krzhizhanovskaya MG, Polyakova IG (2002) Synthesis, crystal structure and thermal behavior of $Rb_3B_7O_{12}$, a new compound. Solid State Sci 4:985–992
80. Krzhizhanovskaya MG, Kabalov YK, Bubnova RS, Sokolova EV, Filatov SK (2000) Crystal structure of the low-temperature modification of α-RbB_3O_5. Crystallogr Rep 45:572–577
81. Krzhizhanovskaya MG, Bubnova RS, Fundamentskii VS, Bannova II, Polyakova IG, Filatov SK (1998) Crystal structure and thermal expansion of high-temperature β-RbB_3O_5 modification. Crystallogr Rep 43:21–25
82. Liu Z-H, Li L-Q, Zhang W-J (2006) Two new borates containing the first examples of large isolated polyborate anions: chain $[B_7O_9(OH)_5]^{2-}$ and ring $[B_{14}O_{20}(OH)_6]^{4-}$. Inorg Chem 45:1430–1432
83. Behm H (1984) Rubidium pentaborate tetrahydrate, $Rb[B_5O_6(OH)_4].2H_2O$. Acta Crystallogr C 40:217–220
84. Penin N, Seguin L, Touboul M, Nowogrocki G (2001) Crystal structures of three MB_5O_8 (M = Cs, Rb) borates (α-CsB_5O_8, γ-CsB_5O_8, and β-RbB_5O_8). J Solid State Chem 161:205–213
85. Schlager M, Hoppe R (1994) Darstellung und Kristallstruktur von $CsBO_2$. Z Anorg Allg Chem 620:1867–1871 (in German)
86. Touboul M, Penin N, Nowogrocki G (1999) Crystal structure and thermal behavior of $Cs_2[B_4O_5(OH)_4]\ 3H_2O$. J Solid State Chem 143:260–265
87. Nowogrocki G, Penin N, Touboul M (2003) Crystal structure of $Cs_3B_7O_{12}$ containing a new large polyanion with 63 boron atoms. Solid State Sci 5:795–803
88. Krogh-Moe J (1960) The crystal structure of cesium triborate, $Cs_2O.3B_2O_3$. Acta Crystallogr 13:889–892
89. Krogh-Moe J (1974) Refinement of the crystal structure of caesium triborate, $Cs_2O.3B_2O_3$. Acta Crystallogr B 30:1178–1180
90. Penin N, Seguin L, Touboul M, Nowogrocki G (2002) A new cesium borate $Cs_3B_{13}O_{21}$. Solid State Sci 4:67–76

91. Behm H (1984) Structure determination on a twinned crystal of cesium pentaborate tetrahydrate, $Cs[B_5O_6(OH)_4] \cdot 2H_2O$. Acta Crystallogr C 40:1114–1116
92. Bubnova RS, Fundamentsky VS, Anderson JE, Filatov SK (2002) New layered polyanion in α-CsB_5O_8 high-temperature modification. Solid State Sci 4:87–91
93. Krogh-Moe J, Ihara M (1967) The crystal structure of caesium enneaborate, $Cs_2O \cdot 9B_2O_3$. Acta Crystallogr 23:427–430
94. Gupta A, Tossell JA (1983) Quantum mechanical studies of distortions and polymerization of borate polyhedra. Am Mineral 68:989–995
95. Zhang ZG, Boisen MB Jr, Finger LW, Gibbs GV (1985) Molecular mimicry of the geometry and charge density distribution of polyanions in borate minerals. Am Mineral 70:1238–1247
96. Oi T (2000) Calculations of reduced partition function ratios of monomeric and dimeric boric acids and borates by the ab initio molecular orbital theory. J Nucl Sci Tech 37(2): 166–172
97. Oi T (2000) Ab initio molecular orbital calculations of reduced partition function ratios of polyboric acids and polyborate anions. Z Naturforsch A 55:623–628
98. Zhou Y, Fang C, Fang Y, Zhu F (2011) Polyborates in aqueous borate solution: a Raman and DFT theory investigation. Spectrochim Acta A 83:82–87
99. Beckett MA, Davies RA, Thomas CD (2014) Computational studies on gas phase polyborate anions. Comput Theor Chem 1044:74–79
100. Frisch MJ et al (2004) Gaussian 03, revision D.02. Gaussian Inc, Wallingford, CT
101. Ishihara K, Nagasawa A, Umemoto K, Ito H, Saito K (1994) Kinetic study of boric acid-borate interchange in aqueous solution by ^{11}B NMR spectroscopy. Inorg Chem 33:3811–3816
102. Stadler HP (1947) The cell dimensions and space-group of pinnoite. Mineral Mag 28:26–28
103. Paton F, MacDonald SGG (1957) The crystal structure of pinnoite. Acta Crystallogr 10:653–656
104. Krogh-Moe J (1967) A note on the structure of pinnoite. Acta Crystallogr 23:500–501

An Ab Initio Study of Boric Acid, Borate, and their Interconversion

Cory C. Pye

Abstract The chemistry of boric acid and monomeric borates is reviewed. Following a discussion of the crystal structures and nuclear magnetic resonance studies, ab initio results are presented of molecular ortho- and metaboric acid, (tetrahydroxo)borate, and the hydrates of orthoboric acid and borate. The structures and vibrational frequencies are compared with experiment. Attempts to study their interconversion lead us to a discussion of oxodihydroxoborate (the conjugate base of boric acid), and of the hydroxide-boric acid complex. It is hypothesized that the conversion of boric acid into borate proceeds via the oxodihydroxoborate intermediate. Finally, the calculated structures of hydroxodioxo- and trioxoborate are compared with experiment.

Keywords Boric acid · Borate · Ab initio

1 Introduction

The nature of boron(III) in aqueous solution has been of longstanding interest to chemists as numerous hydroxooxoborates can exist as anions in metal ion salts [1]. The isotope ^{10}B has important applications in nuclear science because of its large thermal neutron absorption cross section, and the separation of ^{10}B (natural abundance 19.58%) from the remaining ^{11}B (natural abundance 80.42%) is an important technological problem requiring knowledge of the reduced partition function ratio for isotope exchange [2–4]. Many borate minerals exist in nature and the pure compounds can sometimes be prepared in the laboratory. The boron

Electronic supplementary material The online version of this chapter (https://doi.org/10.1007/978-3-319-74582-4_8) contains supplementary material, which is available to authorized users.

C. C. Pye (✉)
Department of Chemistry, Saint Mary's University, Halifax, NS B3H 3C3, Canada
e-mail: cory.pye@smu.ca

© Springer International Publishing AG, part of Springer Nature 2018
Y. A. Wang et al. (eds.), *Concepts, Methods and Applications of Quantum Systems in Chemistry and Physics*, Progress in Theoretical Chemistry and Physics 31,
https://doi.org/10.1007/978-3-319-74582-4_8

isotope ratio is also an important consideration in geochemistry because, for marine carbonates, it is used as a paleo-pH recorder for ancient seawater [5–9]. It is also known that sound absorption in the ocean (~1 kHz) involves chemical equilibria with relaxation rates that correspond to the boric acid-borate equilibria [10]. Important information on the structure of these ions can be obtained by solubility and pH measurements on solutions, X-ray diffraction study of crystals, and nuclear magnetic resonance.

The solubilities of the simpler water-boron oxide system was investigated by Kracek et al. [11]. They identified the solid compounds ice, H_3BO_3, three modifications of HBO_2, and crystalline B_2O_3. The solubility curves of boric acid and several sodium borates were reported by Blasdale and Slansky soon thereafter [12]. The solid phases investigated were (ortho)boric acid, H_3BO_3; three hydrated forms of sodium tetraborate, $Na_2B_4O_7 \cdot nH_2O$, with n = 4 (the mineral kernite), 5, 10; sodium pentaborate pentahydrate, $NaB_5O_8 \cdot 5H_2O$; and two hydrated forms of sodium metaborate, $NaBO_2 \cdot nH_2O$, with n = 2, 4. An alternate way to denote a series of stoichiometries, common in geochemistry, is with respect to the ratio of oxides lNa_2O: mB_2O_3: nH_2O, or shortened as (l:m:n).

The crystal structures of boron oxide and its hydrates ortho- and metaboric acid are given in Table 1. Trigonal boron oxide, originally thought to contain tetrahedral boron [13], actually consists of a 3D-network of corner-linked BO_3 units [14–16]. The high-pressure orthorhombic form does consist of fused 6-membered rings of BO_4 tetrahedra [17]. Orthoboric acid was originally assumed to have the hydrogen atoms halfway between the oxygens [18]. It was also shown that there could be disorder in the layering by electron diffraction [19]. Refinement shows that both polytypes (AB [20] or ABC [21] stacking) of orthoboric acid (0:1:3) consist of stacks of approximately planar layers of hydrogen-bonded $B(OH)_3$ molecules with approximate C_{3h} symmetry. Metaboric acid (0:1:1) exists in at least three forms, two of which were discovered by Tazaki [22]. The orthorhombic α-form consists of sheets [23] of $B_3O_3(OH)_3$ with approximately C_s symmetry, held together by hydrogen bonds [24]. The monoclinic β-form was shown to contain BO_4 tetrahedra and planar B_2O_5 groups [25], and further refinement showed that it consists of endless zigzag chains of $[B_3O_4(OH)(OH_2)]$ [26, 27]. In essence, the $H_3B_3O_6$ molecules have condensed together and the water molecule produced has bonded to one of the two boron condensation sites. The cubic γ-form contains only tetrahedral boron atoms [27, 28].

The crystal structures of some sodium and lithium borates are given in Table 2. Anhydrous sodium borate (3:1:0) consists of discrete BO_3^{3-} ions [29]. The hydrated sodium borate (2:1:1) consists of discrete $[BO_2(OH)]^{2-}$ ions connecting sheets of edge-shared NaO polyhedra [30]. There are several hydrates of sodium metaborate (1:1:n), but only two of them consist of the tetrahedral $[B(OH)_4]^-$ anion, the octahydrate [31] and the tetrahydrate [32]. There are also two monomeric lithium salts (n = 16,8) [33–36]. The crystal structure of polyborates will be discussed in a separate paper.

Table 1 Crystal structures of boron oxide and boric acid

| Stoichiometry $B_mH_nO_{(3m+n)/2}$ | | Name | Synth T(K) | Space group | Unit cell (Å or deg.) | X-ray T(K) | B-O (Å) ranges | | Comments |
m	n						Tetrahedral	Trigonal	References
1	0	Boron oxide-I	513	$C3_1$	4.334, 8.334		1.31–2.145	n/a	Powder, BO_4 tetrahedra [13]
				$P3_1$	4.33, 8.39	293	n/a	1.38	Powder XRD and ED, BO_3 [14]
			573	$P3_1$	4.336,8.340		n/a	1.336–1.404	Chain of BO_3 [15]
				$P3_121$	4.3358.8.3397		n/a	1.356–1.375	[16]
		Boron oxide-II	1373 65 kb	$Ccm2_1$	4.613, 7.803,4.129		1.373–1.512	n/a	BO_4 tetrahedra [17]
1	3	Orthoboric acid (sassolite)		$P\bar{1}$	7.04,7.04,6.56 92.50, 101.16,120.0		n/a	1.36	Assumed H in middle of Os [18]
								1.36	Electron diff., small xtal [19]
				$P\bar{1}$	7.039, 7.053,6.578, 92.58,101.17, 119.83	298	n/a	1.353–1.365	AB stacked layers [20]
		Orthoboric acid-T	298	$P3_2$	7.0453,9.5608	297	n/a	1.349–1.377	ABC stacked layers [21]
1	1	Metaboric acid (orthorhombic α)	423	$Pnma$ or Pa	8.015,9.679, 6.244		n/a	1.37	[22, 23]
			443	$Pbnm$	8.019, 9.703,6.13	143	n/a	1.347–1.391	$B_3O_3(OH)_3$ layers [24]
		Metaboric acid (monoclinic β)	433	$P2_1/c$	6.76,8.80,7.15 92.67				[22]
				$P2_1/a$	7.132,8.852, 6.772 93.253		1.42–1.51	1.29–1.41	[25]
				$P2_1/a$	7.122,8.842, 6.771 93.26		1.433–1.553	1.345–1.386	$B_3O_4(OH)(OH_2)$ chains [26]

(continued)

Table 1 (continued)

Stoichiometry $B_mH_nO_{(3m+n)/2}$		Name	Synth T(K)	Space group	Unit cell (Å or deg.)	X-ray T(K)	B-O (Å) ranges		Comments
m	n						Tetrahedral	Trigonal	References
			423	$P2_1/c$	6.758,8.844, 7.075 93.5	183	1.3284– 1.5612	1.3485– 1.3812	[27]
		Metaboric acid (cubic γ)		$P\bar{4}3n$	8.886		1.436–1.505	n/a	[28]
			453	$P\bar{4}3n$	8.8811	183	1.4428– 1.5094	n/a	[27]

Table 2 Crystal structures of monomeric sodium and lithium borates

Stoichiometry $M_lB_mH_nO_{(l+3m+n)/2}$				Synth T(K)	Space group	Unit cell (Å or deg.)	X-ray T(K)	B-O (Å) ranges		Comments
M	l	m	n					Tetrahedral	Trigonal	Ref.
Na	3	1	0	748	$P2_1/c$	5.687,7.530,9.993 127.15		n/a	1.377–1.409	$[BO_3]^{3-}$ ions [29]
Na	2	1	1	523	$Pnma$	8.627,3.512,9.863		n/a	1.351–1.439	$[BO_2(OH)]^{2-}$ ions [30]
Na	1	1	8	298	$P\bar{1}$	6.126,8.1890,6.068 67.92,110.58,101.85		1.474–1.479	n/a	$[B(OH)_4]^-$ ions [31]
Na	1	1	4	293	$P2_1/a$	5.886,10.566,6.146 111.60		1.463–1.483	n/a	$[B(OH)_4]^-$ ions [32]
Li	1	1	16		$P3$	6.5534,6.1740		1.472–1.473		$[Li(H_2O)_4]^+$ and $[B(OH)_4]^-$ ions [33]
					$C3$	6.555,6.177				[34]
Li	1	1	4		$Pbca$	9.16,7.95,8.54				$[B(OH)_4]^-$ ions [35]
					$Pbca$	7.9362,8.5220,9.1762	120	1.4644–1.4989		[36]

Early work on the nuclear magnetic resonance (NMR) spectra of boron compounds (^{11}B-12.83 MHz) was done and showed a range of chemical shifts and some 1-bond ^{11}B-^{1}H and ^{11}B-^{2}H coupling constants, using $BF_3 \cdot Et_2O$ as a reference [37]. Of interest in this study are the ^{11}B chemical shifts of $NaBO_2$(aq) (-1.3 ± 0.5 ppm, due to $[B(OH)_4]^-$), NaB_5O_8(aq) (-1.3 and -14.4 ± 1.0 ppm), $K_2B_4O_7$ (-7.5 ± 1.0 ppm), $Na_2B_4O_7$ (-8.0 ± 0.5 ppm), $(NH_4)_2B_4O_7$ (-10.3 ± 0.5 ppm), KB_5O_8(aq) (-13.0 ± 0.5 ppm), and $B(OH)_3$(aq) (-18.8 ± 1.0 ppm). These were interpreted as being due to a dynamic equilibrium between $B(OH)_4^-$ and $B(OH)_3$. A later study by Momii and Nachtrieb (sat. $B(OH)_3$ (aq) reference, ^{11}B-14 MHz) reexamined these results and gave 0.090–0.900 M $NaBO_2$(aq) at 17.4 ± 0.5 ppm and 0.090–0.900 M KBO_2 at 15.5 ± 0.5 ppm [38]. These were interpreted as due to $[B(OH)_4]^-$. For sodium pentaborate solutions, the peak at 15.0 ppm was assigned to $[B_5O_6(OH)_4]^-$, and the peak at 1.1 ppm assigned to a rapid equilibrium between $B(OH)_3$, $B(OH)_4^-$ and $B_3O_3(OH)_4^-$. For the tetraborates, a single peak is observed whose chemical shift increases with concentration from 8 to 11 ppm, and this was assigned to a rapid equilibrium between $B(OH)_3$, $B(OH)_4^-$ and at least two other ions. How and coworkers showed that the chemical shift of a 50 g/L solution at 33 °C varied from -2 to -20 ppm between pH 12–2 respectively (^{11}B, 12.83 MHz, $BF_3 \cdot Me_2O$ ref.) [39]. Smith and Wiersema (^{11}B-80 MHz) noted that one NMR peak in all borate solutions was linearly related to the sodium to boron ratio and could this be interpreted as the peak of rapidly exchanging $B(OH)_3$ and $[B(OH)_4]^-$ [40]. For tetraborate solutions, three peaks could be observed, with the 5.0 ppm peak assigned to $[B_3O_3(OH)_4]^-$. Pentaborate solutions also showed three peaks, with the 5 ppm peak assigned to $[B_3O_3(OH)_4]^-$, and the peak at 18 ppm assigned to $[B_5O_6(OH)_4]^-$. Covington and Newman examined the ^{11}B spectra (28.87 MHz, rel. to infinite dilution $[B(OH)_4]^-$) of sodium and potassium borate in water and in ~0.1 mol/L added $[OH^-]$ in an effort to determine the pK_b of borate [41]. Henderson et al., in their study of the complexation of borate with diols, showed the ^{11}B NMR (12.83 MHz) of borax, boric acid, and sodium metaborate from pH 2–12, along with the line width at half height [42]. Janda and Heller examined the ^{11}B spectra (60 MHz) of sodium, potassium, and ammonium polyborates as a function of concentration and pH (0.5–13.8) and either one or two lines were observed [43]. Epperlein et al. examined the ^{10}B spectra (1.807T, 8.267 MHz) of some boron species and found $B(OH)_3$ at 0 ppm (reference), $B_5O_6(OH)_4^-$ at 17 ppm, $B_4O_5(OH)_4^{2-}$ between 70–85 ppm, and $B(OH)_4^-$ at around 140 ppm [44]. Salentine confirmed earlier results (^{11}B, 127 and 160 MHz, external reference $BF_3 \cdot Et_2O$) on the pentaborate (18, 13, 1 ppm) and tetraborate (12, 8, 1 ppm) [45]. It was proposed that the resonance at 13 ppm, due to triborate ion, and at 1 ppm, due to pentaborate ion, are due to the tetrahedral boron atoms, and the trigonal boron atoms are not observed because of quadrupolar relaxation.

We have reviewed the crystallography of boric acid and monomeric borates, and the boron NMR of boric-acid/borate containing solutions. Our remaining goals are to compare the ab initio energy, structure, and vibrational spectra to experiment (where known) of orthoboric acid, metaboric acid, and tetrahydroxoborate, and to determine the effect of hydration where appropriate. We also examine the mechanism of interconversion of orthoboric acid and tetrahydroxoborate, which leads to a discussion of the crystallography unobserved oxodihydroxoborate anion. We then discuss the related crystallographically observed trioxoborate and dioxohydroxoborate anions.

2 Methods

Calculations were performed using Gaussian 03 [46]. The MP2 calculations utilize the frozen core approximation, which is valid in nearly all cases except where excessive core/valence mixing occurs (denoted c/v). The geometries were optimized using a stepping stone approach, in which the geometries at the levels HF/6-31G*, HF/6-31+G*, HF/6-311+G*, B3LYP/6-31G*, B3LYP/6-31+G*, B3LYP/6-311+G*, MP2/6-31G*, MP2/6-31+G*and MP2/6-311+G* were sequentially optimized, with the geometry and molecular orbital reused for the subsequent level. Default optimization specifications were normally used. After each level, where possible, a frequency calculation was performed at the same level and the resulting Hessian was used in the following optimization. Z-matrix coordinates constrained to the appropriate symmetry were used as required to speed up the optimizations. Because frequency calculations are done at each level, any problems with the Z-matrix coordinates would manifest themselves by giving imaginary frequencies corresponding to modes orthogonal to the spanned Z-matrix space. The Hessian was evaluated at the first geometry (Opt = CalcFC) for the first level in a series in order to aid geometry convergence. To facilitate comparison with results from gas-phase, solution, and solid phase measurements, no solvent corrections were applied except via the supermolecule approach (explicit water molecules).

3 Results and Discussion

3.1 Orthoboric Acid, H_3BO_3

Six forms of (ortho)boric acid were investigated (Fig. 1). Two of these (C_{3h} and C_s #2) were minima, with the C_{3h} structure being lower in energy by 22–26 kJ/mol (Table 3). The C_{3v} structure, a third-order saddle point, was much higher in energy (139–152 kJ/mol). The other three structures were transition states linking the minima. Both the C_s #1 and C_s #3 linked the C_s #2 structure to itself, whereas the C_1 structure linked the C_{3h} and C_s #2 structures. The C_1 structure was 38–43 kJ/mol higher than the C_{3h} structure, whereas the C_s #1 and C_s #3 structures were 34–38 and 20–22 kJ/mol, respectively, higher than the C_s #2.

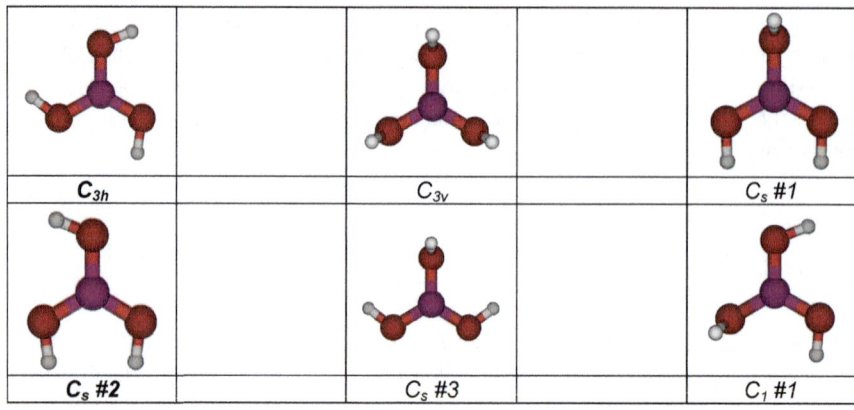

Fig. 1 Structure of boric acid, [B(OH)$_3$]. A bold symmetry label indicates a minimum energy structure

Table 3 Relative Energies of B(OH)$_3$ (kJ/mol)

	C_{3h}	C_s #2	C_s #1	C_s #3	C_1 #1	C_{3v}
HF/6-31G*	0.0	24.5	60.3	46.2	41.3	146.4
HF/6-31+G*	0.0	24.6	60.3	46.1	41.3	142.1
HF/6-311+G*	0.0	26.3	62.7	47.8	42.5	145.7
B3LYP/6-31G*	0.0	22.1	56.8	42.7	38.8	145.2
B3LYP/6-31+G*	0.0	22.8	57.6	43.4	39.4	139.6
B3LYP/6-311+G*	0.0	24.3	59.4	44.7	40.2	141.4
MP2/6-31G*	0.0	23.8	61.1	45.8	41.6	155.7
MP2/6-31+G*	0.0	24.3	61.5	46.0	41.8	149.4
MP2/6-311+G*	0.0	25.6	63.5	47.5	42.9	152.6

Of these structures, the C_{3h} structure is observed in the crystal structure of both polymorphs of boric acid (Table 1). Our calculated B-O bond lengths range from 1.3553 to 1.3780 Å, with HF < B3LYP < MP2 (Table 4). These are in good agreement with the experimentally determined (X-ray) average bond length of 1.36 Å and with previous literature values (Tables 1 and 4).

The vibrational spectra (unscaled) of the lowest-energy form of boric acid, as well as some experimental vibrational frequencies from the literature, is given in Table 5. Undistorted boric acid, of C_{3h} symmetry, has 15 modes of internal vibration and spans the vibrational representation

$$\Gamma_{vib} = 3A^{'}(R, p) + 2A^{''}(IR; R, dp) + 4E^{'}(IR) + E^{''}(R, dp).$$

There is little difference in vibrational frequency between the A′ and E′ modes of the OH stretch and HOB deformation. The computations are in reasonable

Table 4 Geometrical Parameters of the C_{3h} form of B(OH)$_3$. n/r = not reported

Level	B-O (Å)	O-H (Å)	B-O-H angle (deg.)
HF/6-31G*	1.3581	0.9466	112.56
HF/6-31+G*	1.3584	0.9471	113.69
HF/6-311+G*	1.3553	0.9401	113.99
B3LYP/6-31G*	1.3721	0.9675	110.90
B3LYP/6-31+G*	1.3729	0.9683	112.46
B3LYP/6-311+G*	1.3691	0.9630	113.01
MP2/6-31G*	1.3762	0.9700	110.38
MP2/6-31+G*	1.3780	0.9722	111.77
MP2/6-311+G*	1.3709	0.9614	112.25
Literature			
HF/STO-3G [2, 47]	1.389	n/r	114 (fixed)
HF/STO-3G [48]	1.364	0.98	110
HF/4-31G [48]	1.364	0.95	121
HF/3-21G* [49]	1.377	0.962	n/r
HF/6-31G [49]	1.370	0.947	n/r
HF/6-31G* [49]	1.358	0.947	n/r
HF/6-31G* [2, 49]	1.358	0.947	112.6
MP2/6-31G** [50]	1.357	0.942	113.0
B3LYP/6-311++G** [51]	1.380	0.971	112.6
B3LYP/6-311++G** [52]	1.370	0.962	112.8
MP2/6-311++G** [52]	1.373	0.961	110.7
B3LYP/aug-cc-pVQZ [53]	1.369	0.960	113.1
MP2/aug-cc-pVTZ [53]	1.374	0.962	111.5
MP2/aug-cc-pVQZ [53]	1.370	0.959	111.8
QCISD/6-311++G** [53]	1.371	0.959	111.1
B3LYP/aug-cc-pVDZ [54]	1.376	n/r	n/r

agreement with the IR spectra measured in an argon matrix, and with most modes of solid and aqueous boric acid. However, the in-plane BO$_3$ deformation, HOB deformation, and OH stretching frequencies differ, which would be expected, because in solid boric acid, the molecules are held together by a network of hydrogen bonds. It might be expected that aqueous solutions may exhibit similar behavior to the solid. The vibrational spectra of the higher-energy C_s #2 conformation is also given in Table 6. The E modes correlate with 2A modes of the same reflection symmetry. The main differences in the vibrational frequencies of the C_s #2 conformer are that the BOH torsion is much lower, the in plane BO$_3$ deformation is somewhat higher in frequency, the out of plane BO$_3$ deformation is slightly lower, the symmetric BO stretch is slightly higher, one component of the HOB deformation and asymmetric B-O stretch is somewhat lower, and the OH frequencies slightly higher.

Table 5 Theoretical and literature experimental vibrational frequencies (cm^{-1}) of the C_{3h} form of boric acid, B(OH)$_3$

	BO$_3$ def	BOH tors	BOH tors	BO$_3$ oop def	BO str	HOB def	HOB def	BO str	OH str	OH str
	E'	A"	E"	A"	A'	E'	A'	E'	E'	A'
Theory, C_{3h}										
HF/6-31G*	458	476	567	736	927	1110	1116	1561	4135	4138
HF/6-31+G*	452	484	572	741	924	1086	1092	1540	4129	4132
HF/6-311+G*	454	491	577	747	919	1098	1112	1533	4188	4191
B3LYP/6-31G*	428	428	535	665	877	1038	1045	1478	3800	3800
B3LYP/6-31+G*	420	443	543	669	871	1010	1015	1448	3796	3796
B3LYP/6-311++G*	420	446	544	676	869	1018	1029	1441	3829	3830
MP2/6-31G*	425	445	551	675	880	1052	1056	1494	3823	3823
MP2/6-31+G*	419	450	551	677	872	1021	1024	1458	3793	3794
MP2/6-311+G*	426	447	548	680	873	1031	1038	1459	3872	3873
Theory (Literature)										
MP2/6-31G** [50]	428	445	547	682	886	1044	1045	1509	3949	3948
B3LYP/6-311++G** [51]	427	473		669		1015		1442	3870	
Experiments										
Raman, sat.aq. [55]					875					
Raman, powder [56]					880				3172	3256
Raman, solution [57]			506		875	1130?	1060	1420		
Raman, powder			503			1155	1065			
Raman, sol. 70 °C [58]					882				3195	3290
Raman, powder		515			872		986			
Raman, powder [59]					883	1169	1032	1391	3170	3248

(continued)

Table 5 (continued)

	BO_3 def	BOH tors	BOH tors	BO_3 oop def	BO str	HOB def	HOB def	BO str	OH str	OH str
	E'	A''	E''	A''	A'	E'	A'	E'	E'	A'
IR, mull [60]	540				885	1195		1450		3270
IR, mull and xtal. [61]				648	882	1197		1450		3200
IR, KBr pellets [62]	544			639	882?	1183		1428	3150	
IR, $^{10}B(OH)_3$	545			668	882?	1195		1490		
IR, $B(OD)_3$	540			654				1428	2380	
Raman, sol			500		880			1430		3200
R, pH = 6.1–10 [63]					875					
R, pH = 6.38–10.60 [64]					872					
R, powder [65]			500		881	1168		1375		
R, pH = 0.0–9.3			492–498		877					
N_2 matrix [66]	448.9	513.8		675.0		1009.9		1426.2	3668.5	
$^{10}B(OH)_3$	450.5			700.9				1478.0		
$B(^{18}OH)_3$	430.7			668.1		999.5		1409.7	3657.6	
$^{10}B(^{18}OH)_3$	431.9			694.5				1461.2		
$B(OD)_3$	404.7	376.0						1452.7		
$^{10}B(OD)_3$	405.2					825.5		1456	2704.6	
IR+R (g), 150 °C [67]			(520) ± 5							
$H_3^{10}BO_3$					866	1017	1020	1429	3706	3705
$D_3^{11}BO_3$					866	1017	1020	1477	3706	3705
$D_3^{10}BO_3$					887	810	706	1405	2726	
Ar matrix [68]	432.1	436.0		666.4 692.2		992.4		1414.9 1471.9	3688.6	
R,NaB_5O_8,1.39 M [54]					875					

Table 6 Theoretical vibrational frequencies (cm^{-1}) of the C_s form of boric acid, B(OH)$_3$

	BO$_3$ def	BOH tors	BOH tors	BO$_3$ oop def	BO str	HOB def	HOB def	BO str	OH str	OH str
	A'	A''	A''	A''	A'	A'	A'	A'	A'	A'
HF/6-31G*	471	306	508	729	930	1006	1131	1517	4141	4168
	488		549			1116		1576	4156	
HF/6-31+G*	468	324	514	733	928	979	1099	1495	4136	4166
	483		558			1093		1554	4154	
HF/6-311+G*	471	313	506	735	923	992	1111	1488	4195	4221
	486		560			1107		1543	4211	
B3LYP/6-31G*	441	260	476	661	877	947	1058	1430	3808	3830
	458		527			1042		1494	3814	
B3LYP/6-31+G*	436	295	482	665	873	912	1022	1400	3806	3833
	450		539			1012		1463	3816	
B3LYP/6-311+G*	438	274	466	668	871	926	1031	1392	3838	3862
	452		536			1019		1454	3847	
MP2/6-31G*	439	274	491	670	879	963	1074	1446	3825	3853
	457		542			1057		1510	3842	
MP2/6-31+G*	435	289	487	672	872	928	1036	1410	3797	3830
	451		546			1024		1473	3819	
MP2/6-311+G*	443	262	467	673	874	942	1047	1411	3875	3904
	459		540			1037		1472	3894	

3.2 Metaboric Acid, B$_3$O$_3$(OH)$_3$

The structures of α-metaboric acid are analogous in symmetry to orthoboric acid (Fig. 2). The barriers are quite similar to that of orthoboric acid (Table 7), but the C_s #2 form is much more stable (only 3–4 kJ/mol above the minimum energy C_{3h} structure). If metaboric acid is ever observed in the gas phase, it should exist as a mixture of an appreciable amount of both conformers (assuming it does not decompose to form the monomer HBO$_2$). It might even be possible to observe the microwave spectrum of the C_s conformation of the molecule. It is the C_s form that is observed in the crystal structure (Table 1), so one strategy for gas-phase observation might be laser ablation.

3.3 Hydrated Orthoboric Acid, H$_3$BO$_3$ · nH$_2$O

Four C_s forms of hydrated boric acid with a water molecule directly bound to the boron atom were attempted. Alternatively, this may be viewed as protonated monoborate anion. In all cases, the water molecule dissociated. In the first two cases, a solvated boric acid was obtained, neither of which was an energy minimum (based on C_s #1 and C_s #3 naked boric acid, see Fig. 3). In the final two cases,

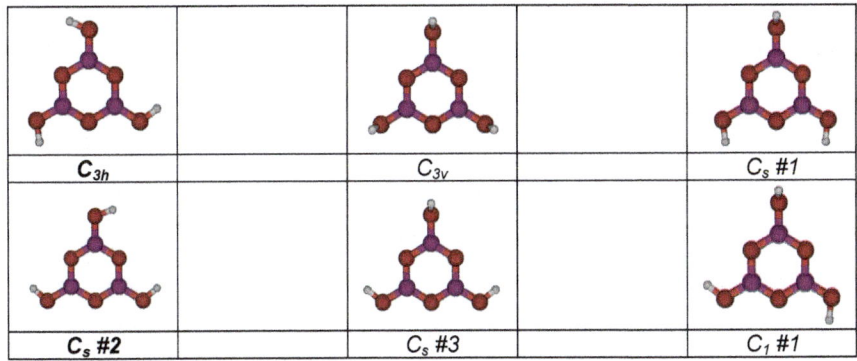

Fig. 2 Structure of metaboric acid, $[B_3O_3(OH)_3]$

Table 7 Relative Energies of $B_3O_3(OH)_3$ (kJ/mol). c/v = excessive core-valence mixing

	C_{3h}	C_s #2	C_s #1	C_s #3	C_1 #1	C_{3v}
HF/6-31G*	0.0	4.0	40.0	38.7	41.4	123.9
HF/6-31+G*	0.0	3.9	39.0	37.7	40.3	120.8
HF/6-311+G*	0.0	4.0	39.7	38.3	41.0	122.9
B3LYP/6-31G*	0.0	3.7	40.9	39.5	42.1	126.7
B3LYP/6-31+G*	0.0	3.4	39.8	38.4	40.8	122.8
B3LYP/6-311+G*	0.0	3.6	40.0	38.6	41.1	122.3
MP2/6-31G*	0.0	4.0	43.4	41.9	44.7	134.4
MP2/6-31+G*	0.0	3.8	42.3	40.8	43.4	c/v
MP2/6-311+G*	0.0	3.9	43.2	41.7	44.4	

z-matrix errors occurred as the water molecule rotated. It can be said therefore that direct protonation of $B(OH)_4^-$ results in water elimination.

Four forms of hydrated boric acid with a water molecule hydrogen-bonded to the boric acid were attempted (Fig. 3). To the naked C_{3h} structure, a water molecule may only hydrogen bond in a donor-acceptor (DA) fashion (C_1 #2), whereas to the naked C_s #2 structure, a water molecule may hydrogen bond in either a double donor (DD, C_s #1 or C_1 #3), donor acceptor (DA, C_1 #1), or double acceptor (AA, C_s #2). All were energy minima, with the exception of C_s #1 at the B3LYP/6-31G* and all MP2 levels. These desymmetrized to C_1 #3, where the water is removed from the BO_3 plane by varying amounts (significantly at the B3LYP and MP2/6-31G* levels).

For the dihydrate, a C_s and five C_1 forms were tried. The C_1 #4 and C_1 #5 forms are derived from the naked C_{3h} structure and differ in the orientation of the free hydrogen of the water molecules, either up,up (uu) or up,down (ud). The other forms are derived from the naked C_s #2 structure: namely, the $C_{s/1}$ #1 (DD,AA), the C_1 #2 (DD,DA), and the C_1 #3 (DA,AA). The C_s form was stable at the HF,

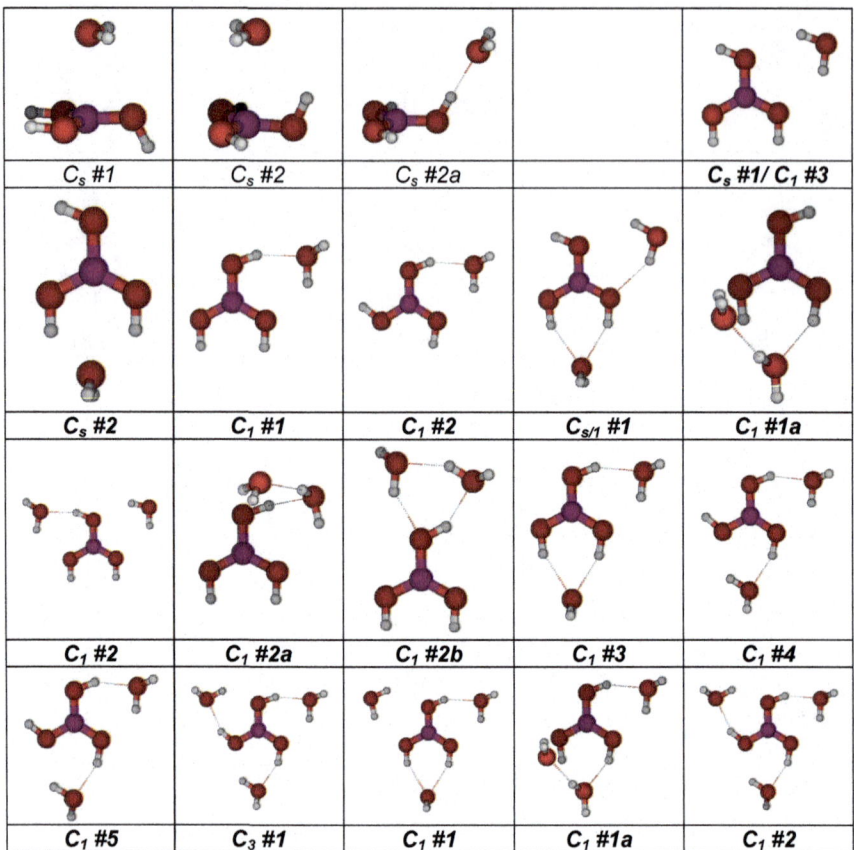

Fig. 3 Structure of hydrated boric acid, [B(OH)$_3$] · nH$_2$O

B3LYP/6-31+G*, B3LYP/6-311+G*, and MP2/6-31+G* levels but reverted to a C_1 #1 form at the other levels. For MP2/6-311+G*, the double-donor water moved slightly out of the plane, whereas for the other levels it accepted a hydrogen bond from the other water molecule (C_1 #1a). The C_1 #2 form is stable at the HF, B3LYP/6-31+G*, B3LYP/6-311+G*, MP2/6-31+G*, and MP2/6-311+G* levels. At the B3LYP/6-31G* and MP2/6-31G* levels, the DD water accepts a hydrogen bond from the free DA water to give a DDwA, DAw C_1 #2a structure (w indicates hydrogen bonding between water molecules) which flatten out at all levels except B3LYP/6-31G* and MP2/6-31G*(C_1 #2b). The C_1 #3, 4, and 5 forms were stable at all levels.

For the trihydrate, a C_3 and two C_1 forms were tried. The C_3 #1 (uuu) and C_1 #2 (uud) forms are derived from the naked C_{3h} structure and are stable at all levels, whereas the C_1 #1 is derived from the naked C_s #2 structure. The C_1 #1 structure is stable at the HF, B3LYP/6-31+G*, B3LYP/6-311+G*, MP2/6-31+G*, and

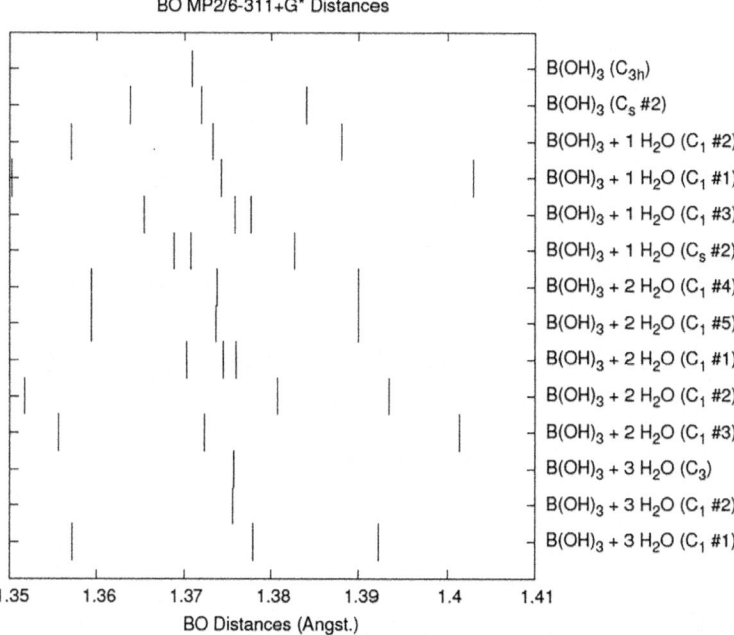

Fig. 4 Variation of B-O distance of B(OH)$_3$ as a function of hydration and structure

MP2/6-311+G* levels, but reverts to C_1 #1a by forming a hydrogen bond between the DD and AA water molecules. This structure also exists at the other levels.

The B-O distances as a function of hydration structure are plotted in Fig. 4. The average B-O distance slightly increases (<0.005 Å) upon hydration, which demonstrates how weakly boric acid is hydrated, but the variation in less-symmetric structures is up to ±0.025 Å.

The vibrational frequencies (MP2/6-311+G*) of hydrated boric acid are plotted in Figs. 5 and 6. In the OH stretching region (3500–4000 cm^{-1}, Fig. 6), it is clear that if boric acid, with a OH stretch in the range 3860–3910 cm^{-1}, donates a hydrogen bond to a DA water, then an OH stretching frequency drops to between 3600–3700 cm^{-1}, whereas if it donates to a AA water, an OH stretching frequency drops to only 3780–3830 cm^{-1}. If boric acid accepts a hydrogen bond, then the OH stretching frequency corresponding to the oxygen accepting the hydrogen bond essentially remains unchanged. In the lower-frequency region (Fig. 5), the BOH bending frequencies increase from 940–1050 to 1150–1250 cm^{-1}, and the BOH torsion increases from 260–550 to 800–830 cm^{-1}, upon accepting a hydrogen bond. The B-O symmetric (~870 cm^{-1}) and asymmetric stretching (~1410–1470 cm^{-1}) frequencies increase slightly upon hydration. The BO$_3$ in plane deformation increases from ~425 cm^{-1} to nearly 500 cm^{-1} upon hydration, whereas the out of plane deformation at ~680 cm^{-1} is hardly affected.

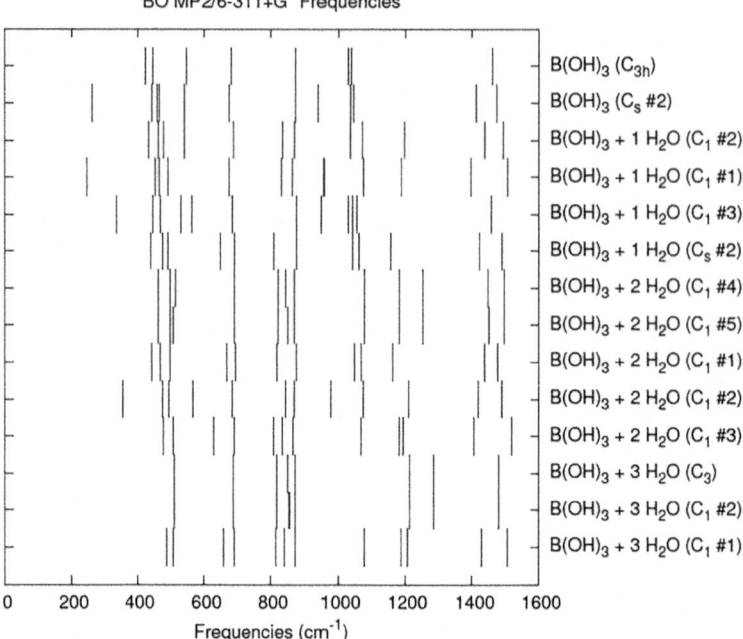

Fig. 5 Variation of vibrational frequencies of B(OH)$_3$ (0–1600 cm^{-1}) as a function of hydration and structure

Our approach to hydration of boric acid is to solvate the boric acid molecule as completely as possible with a small number of water molecules in order to efficiently model the vibrational spectra. Of course, water molecules might prefer to hydrogen-bond to other water molecules instead of to boric acid. This work is therefore somewhat complementary to that of Tachikawa [52], who studied similar clusters with up to five water molecules and found several in which water molecules were hydrogen bonded to each other.

3.4   Borate, B(OH)$_4^-$

Five forms of monoborate were investigated (Fig. 7). Two of these (D_{2d} #2 or S_4 #3, and S_4 #1) were minima, with the S_4 #1 structure being lower in energy by 7–10 kJ/mol at the Hartree-Fock levels (Table 8). The D_{2d} #2 is only a minimum at the Hartree-Fock levels. We confirm the presence of a second shallow minimum (S_4 #3) at the correlated levels, as first found by Stefani et al. [53] The S_4 #2 structure is a transition state that connects the D_{2d} #2/S_4 #3 and S_4 #1 structures. It is 2–3 kJ/mol higher in energy than D_{2d} #2. A D_2 structure, derived from D_{2d} #1, ascended in symmetry to give the D_{2d} #2 structure at all levels.

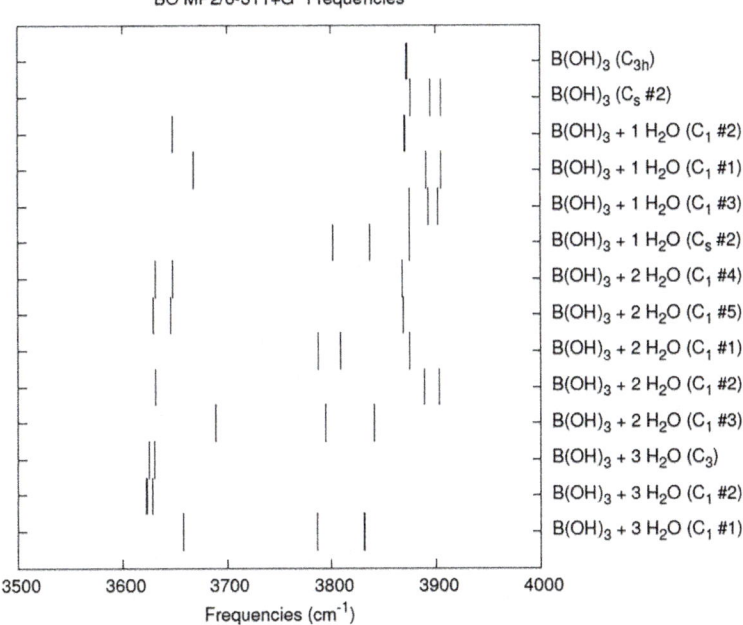

Fig. 6 Variation of vibrational frequencies of B(OH)$_3$ (3500–4000 cm^{-1}) as a function of hydration and structure

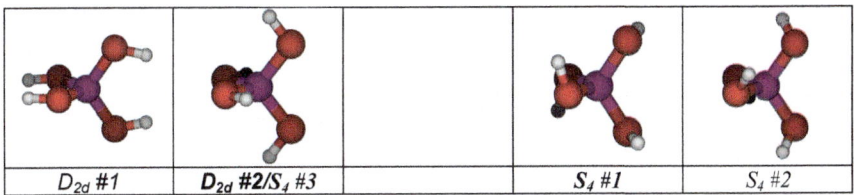

| D_{2d} #1 | D_{2d} #2/S_4 #3 | | S_4 #1 | S_4 #2 |

Fig. 7 Structure of borate, B(OH)$_4^-$

The tetrahedral borate structure is observed in the crystal structure of sodium 1:1:8 and 1:1:4, and lithium 1:1:16 and 1:1:4 (Table 2). Our calculated B-O bond lengths range from 1.4702–1.4904 Å, with HF < B3LYP < MP2 (Table 9). These are slightly longer than the experimentally determined (X-ray) bond lengths of 1.463–1.483 Å and compare favorably with previous calculations.

The vibrational spectra (unscaled) of borate, as well as some experimental vibrational frequencies from the literature, is given in Table 10. Undistorted borate, of S_4 symmetry, has 21 modes of internal vibration and spans the vibrational representation

Table 8 Relative energies of $B(OH)_4^-$ (kJ/mol)

	D_{2d} #1	D_{2d} #2	S_4 #1	S_4 #2	S_4 #3
HF/6-31G*	67.6	7.4	0.0	9.7	n/a
HF/6-31+G*	68.9	8.7	0.0	10.5	n/a
HF/6-311+G*	74.6	9.6	0.0	12.4	n/a
B3LYP/6-31G*	63.3	7.4	0.0	5.6	5.6
B3LYP/6-31+G*	66.0	8.2	0.0	8.0	7.8
B3LYP/6-311+G*	70.9	9.0	0.0	9.0	8.7
MP2/6-31G*	71.3	7.0	0.0	8.1	6.7
MP2/6-31+G*	72.3	8.1	0.0	9.1	8.0
MP2/6-311+G*	77.8	8.8	0.0	11.1	8.7

Table 9 Geometrical parameters of the S_4 #1 form of $B(OH)_4^-$

Level	B-O (Å)	O-H (Å)	B-O-H angle (deg.)
HF/6-31G*	1.4722	0.9454	105.63
HF/6-31+G*	1.4717	0.9451	107.63
HF/6-311+G*	1.4702	0.9390	107.81
B3LYP/6-31G*	1.4876	0.9674	103.05
B3LYP/6-31+G*	1.4873	0.9673	106.20
B3LYP/6-311+G*	1.4854	0.9626	106.52
MP2/6-31G*	1.4897	0.9689	102.67
MP2/6-31+G*	1.4904	0.9710	105.63
MP2/6-311+G*	1.4829	0.9610	106.00
Literature			
HF/STO-3G [69] (D_{2d})	1.48		114 (fixed)
HF/STO-3G [47] (D_{2d})	1.492		114 (fixed)
HF/6-31G* [51] (D_{2d})	1.474	0.946	104
HF/6-31G* [70]	1.472	0.945	105.6
HF/6-31G* [2]	1.472	0.954	105.4
B3LYP/aug-cc-pVDZ [54]	1.486	n/r	n/r
B3LYP/6-311++G**	1.489	0.960	n/r
MP2/6-311++G**	1.486	0.960	n/r

$$\Gamma_{vib} = 5A(R,p) + 6B(IR;R,dp) + 5E(IR;R,dp).$$

The BOH deformations appear from 1000 to 1300 cm^{-1}, the B-O stretching motions from 720 to 1100 cm^{-1}, the BO$_4$ deformations from 230 to 570 cm^{-1}, and the BOH torsions from 180 to 440 cm^{-1}.

Table 10 Theoretical and literature experimental vibrational frequencies (cm^{-1}) of the S_4 #1 form of B(OH)$_4{}^-$

	BOH tors	BO$_4$ def	BOH tors	BO$_4$ def	BOH tors	BO$_4$ def	BO$_4$ def	BO str	BO str	BO str	BOH def	BOH def	BOH def
	A	B	E	A	B	E	B	A	E	B	A	B	E
Theory													
HF/6-31G*	202	273	317	420	431	514	568	775	940	1086	1110	1169	1323
HF/6-31+G*	204	260	307	420	429	511	561	774	924	1056	1086	1141	1285
HF/6-311+G*	212	272	313	429	438	512	568	770	920	1049	1098	1151	1285
B3LYP/6-31G*	190	264	331	345	416	476	521	728	866	1018	1026	1070	1252
B3LYP/6-31+G*	181	235	307	366	407	471	513	724	851	973	1000	1048	1193
B3LYP/6-311+G*	189	245	311	380	408	473	518	721	846	964	1015	1060	1195
MP2/6-31G*	203	272	343	368	423	482	535	732	892	1041	1042	1094	1272
MP2/6-31+G*	203	249	326	372	414	473	520	725	869	990	1012	1060	1208
MP2/6-311+G*	212	265	323	406	412	477	534	729	871	993	1016	1067	1206
Theory (Literature)													
HF/6-31G* [70]	202	273	316	420	431	514	569	775	940	1086	1110	1169	1323
Experiments													
NaBO$_2$, 5N [71]								749					
NaBO$_2$ (aq) [57]				400–500				747		950		1132	1276
NaBO$_2$ · 4H$_2$O(s)								741					
H$_3$BO$_3$+NaOH								744–7		949		1150–60	1219–20
KB(OH)$_4$(aq) (T_d) [72]				379	379	533	533	754	947	947			
NaB(OH)$_4$(aq) T_d [73]				374	374	516	516	744	940	940			
sat NaB(OH)$_4$(aq) pH = 5–10 [63]								745					

(continued)

Table 10 (continued)

	BOH tors	BO₄ def	BOH tors	BO₄ def	BOH tors	BO₄ def	BO₄ def	BO str	BO str	BO str	BOH def	BOH def	BOH def
	A	B	E	A	B	E	B	A	E	B	A	B	E
NaB(OH)₄(aq), 1.5M, pH = 6–12 [64]				375	375	522	522	745	940	940			
NaB(OH)₄(s) [65]				405	405	520	520	738	952	952		1078	
Na₂B₄O₇(aq), 0.01–0.8 [65]								744					
sat Cs₂B₄O₇ · 5H₂O or CsB₅O₈ · 4H₂O [74]						480,484	511	742,745					
NaBO₂(aq), 4–5 M [54]								740					
KBO₂ · nH₂O(aq), n = 5, 10, 19 [75]								740					

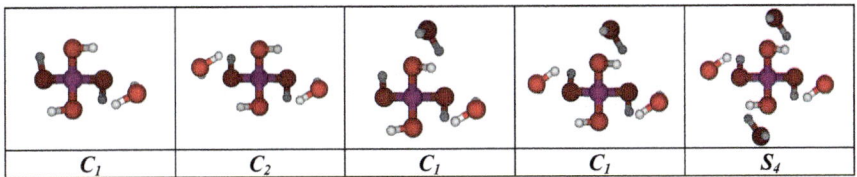

Fig. 8 Structure of hydrated borate, $B(OH)_4^- \cdot nH_2O$

3.5 Hydrated Borate, $B(OH)_4^- \cdot nH_2O$

The monoborate anion may be hydrated. The monohydrate has C_1 symmetry; the dihydrate, either C_1 or C_2; the trihydrate, C_1; and the tetrahydrate, S_4 (see Fig. 8). Upon hydration, the average B-O distance slightly decreases (0.002 Å) but the deviation from the mean can be as much as 0.025 Å (Fig. 9).

The vibrational frequencies (MP2/6-311+G*) of hydrated monoborate are plotted in Fig. 10. The OH stretching frequencies (not shown) decrease by about 50 cm^{-1} upon going from the anhydrate to the tetrahydrate. The restricted translations of the water molecules appear below 250 cm^{-1}, whereas the restricted rotations appear at approximately $300\text{–}440 \text{ cm}^{-1}$ (rock), $580\text{–}730 \text{ cm}^{-1}$ (twist), and

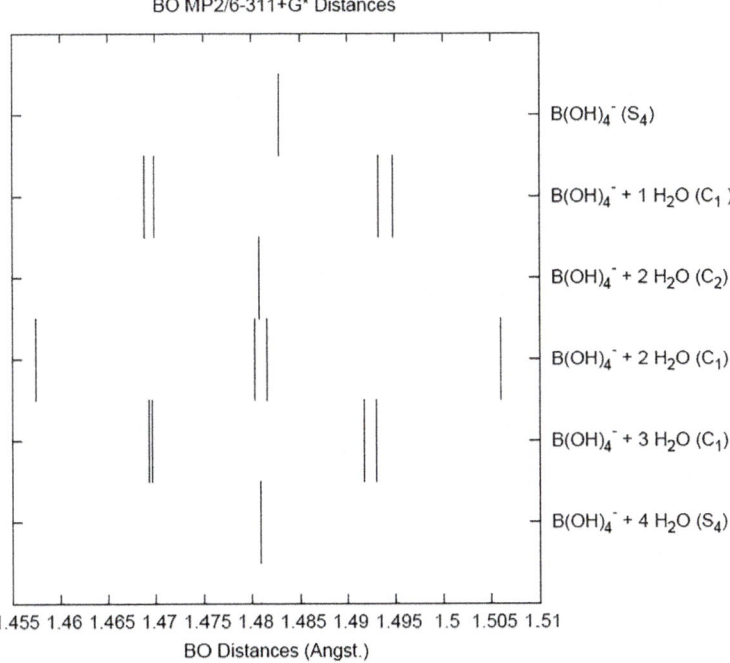

Fig. 9 Variation of B-O distance of $B(OH)_4^-$ as a function of hydration and structure

MP2/6-311+G* Frequencies

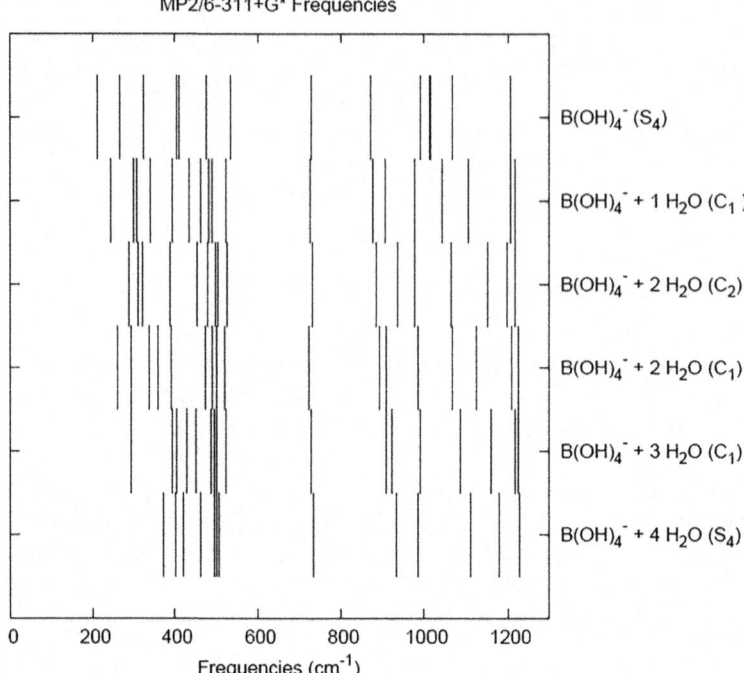

Fig. 10 Variation of vibrational frequencies of $B(OH)_4{}^-$ as a function of hydration and structure

770–820 cm^{-1} (wag). The BOH deformation frequencies are in the range from 1000–1220 cm^{-1} and tend to increase with hydration as the mode becomes stiffer upon forming a hydrogen bond to water. The B-O totally symmetric stretch only increases by a few wavenumbers, whereas the asymmetric stretch increases much more. The deformation modes tend to increase a bit, but the BOH torsional modes increase a lot because of the restrictions imposed by hydrogen bonding. There is significant mixing both between these two types of modes and with the water rocking modes.

3.6   The Reaction of Borate, $B(OH)_4{}^-$ with Hydronium, H_3O^+

Attempts to hydrate boric acid by attaching water directly to the boron (see above, equivalent to direct protonation of monoborate) resulted in the dissociation of the water molecule. It was thought that the more realistic hydronium ion might interact with the borate ion by forming an ion pair before losing a water molecule. A scan of the O...H distance from 2.0 to 0.9 Å in 0.1 Å steps was carried out in an attempt to carefully protonate the borate at HF/6-31G*; however, a different proton from the hydronium protonated a different oxygen, resulting in fragmentation of the borate molecule to form an interacting boric acid and a water dimer. Attempts to optimize an ion pair structure at the nine levels investigated invariably led to boric acid interacting with a water dimer. A typical pattern observed is that of protonation, followed by detachment, and in some cases rotation of the OH group from a C_s #2 boric acid to form C_{3h} boric acid. If such an ion pair exists, more water molecules would be needed to stabilize it. There can be no enthalpic barrier in the absence of additional water molecules.

3.7   The Dissociation of Borate, $B(OH)_4^-$

The dissociation of one of the B-O bonds of borate to form boric acid and hydroxide is a simple possibility for their interconversion. Scans were done at each level, with the zero of energy set to the optimized borate structure and are shown in Fig. 11. Typically the scans suggest that as the B-O distance increases, two of the three remaining borate hydroxyls rotate to form stabilizing hydrogen bonds with the departing hydroxide, which then swings into the plane of a C_s #2 boric acid as

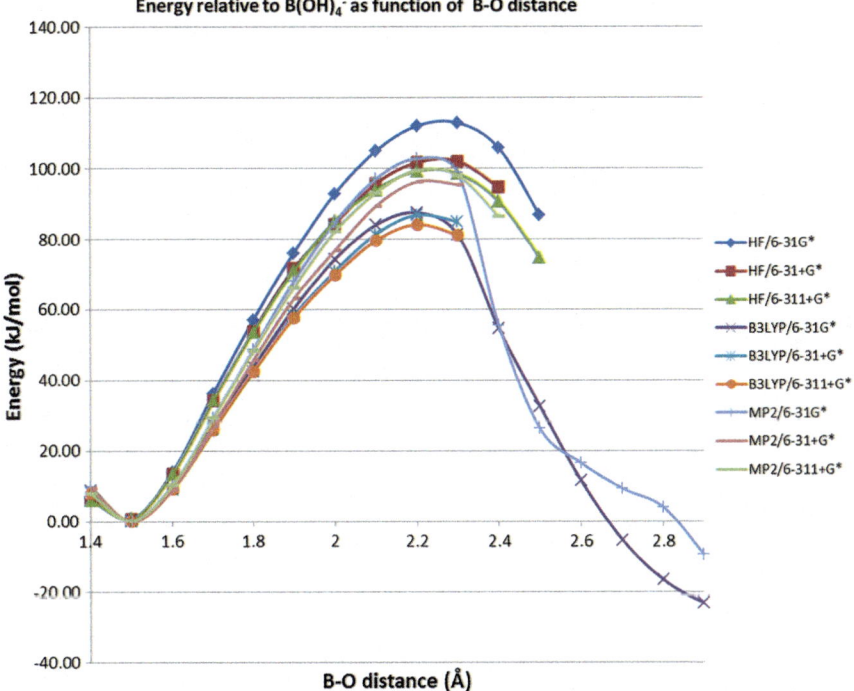

Fig. 11 Scans of the B-O distance of borate, $B(OH)_4^-$

the transition state is passed. However, for the B3LYP/6-31G* and MP2/6-31G* scans, which proceeded farther than the others, the hydroxide abstracted a proton from boric acid to give $BO(OH)_2^- + H_2O$. This suggested the need to explore the relative energies of hydrated $BO(OH)_2^-$ and of the boric acid-hydroxide complex.

3.8   Oxodihydroxoborate and Its Hydrates, $BO(OH)_2^- \cdot nH_2O$

The oxodihydroxoborate anion, $BO(OH)_2^-$, can potentially exist as one of three planar conformers, C_{2v} #1 and #2, and C_s. The C_{2v} #2 form is unstable and reverts to the C_2 form (Fig. 12). The order of stability is C_{2v} #1 (most stable) $< C_s$ #1 $< C_2$ (Table 11). Only the C_s from can be derived from the parent C_{3h} boric acid, but all can be derived from C_s #2 boric acid. To the best of our knowledge, the oxodi-hydroxoborate structure has not been observed crystallographically, but the related dioxomonohydroxoborate and trioxoborate have been observed as the sodium salts (Table 2). The B-O distances (1.28–1.31 Å and 1.42–1.45 Å, Table 12) are much shorter than those of tetraborate and bracket those of the more negatively charged deprotonated versions. These results confirm the findings of Stefani et al. [53]. This ion has been observed in the gas-phase [76].

If the oxodihydroxoborate anion can exist as a transient species in the formation of tetrahydroxoborate upon basification of boric acid, it might be possible to

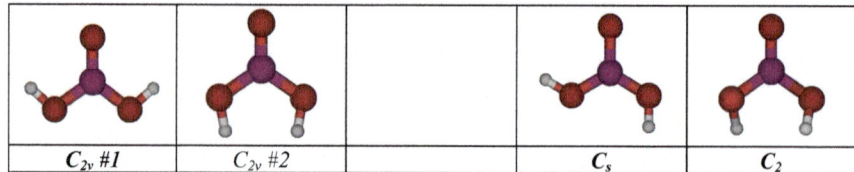

| C_{2v} #1 | C_{2v} #2 | | C_s | C_2 |

Fig. 12 Structure of oxodihydroxoborate, $BO(OH)_2^-$

Table 11 Relative energies of $BO(OH)_2^-$ (kJ/mol)

	C_{2v} #1	C_{2v} #2	C_s	C_2
HF/6-31G*	0.0	49.9	9.8	49.2
HF/6-31+G*	0.0	51.5	9.8	51.4
HF/6-311+G*	0.0	52.4	8.9	52.1
B3LYP/6-31G*	0.0	44.1	8.8	42.6
B3LYP/6-31+G*	0.0	46.6	8.6	46.6
B3LYP/6-311+G*	0.0	47.3	7.7	46.7
MP2/6-31G*	0.0	49.0	9.5	46.8
MP2/6-31+G*	0.0	50.5	9.0	50.2
MP2/6-311+G*	0.0	51.3	8.4	50.1

Table 12 Structure and vibrational frequencies (cm^{-1}) of the C_{2v} #1 form of BO(OH)$_2^-$

	B-O (Å)	B-OH (Å)	BO$_3$ def A$_1$	BO$_3$ def B$_2$	BOH tors A$_2$	BOH tors B$_1$	BO$_3$ oop def B$_1$	BOH str A$_1$	BOH def B$_2$	BOH def A$_1$	BOH str B$_2$	BO str A$_1$
HF/6-31G*	1.2791	1.4266	454	510	537	545	788	851	1075	1195	1316	1711
HF/6-31+G*	1.2866	1.4217	452	504	550	558	793	853	1072	1172	1295	1648
HF/6-311+G*	1.2831	1.4199	455	505	542	557	792	847	1077	1183	1287	1639
B3LYP/6-31G*	1.2949	1.4455	412	465	493	517	703	791	972	1106	1235	1605
B3LYP/6-31+G*	1.3041	1.4398	412	459	512	532	708	792	976	1082	1204	1532
B3LYP/6-311—G*	1.2992	1.4379	415	460	504	530	709	787	975	1093	1200	1527
MP2/6-31G*	1.3014	1.4499	415	466	508	531	712	796	989	1117	1250	1617
MP2/6-31+G*	1.3135	1.4450	412	457	511	542	710	794	982	1087	1214	1522
MP2/6-311+G*	1.3062	1.4386	420	463	485	532	708	794	989	1094	1207	1525

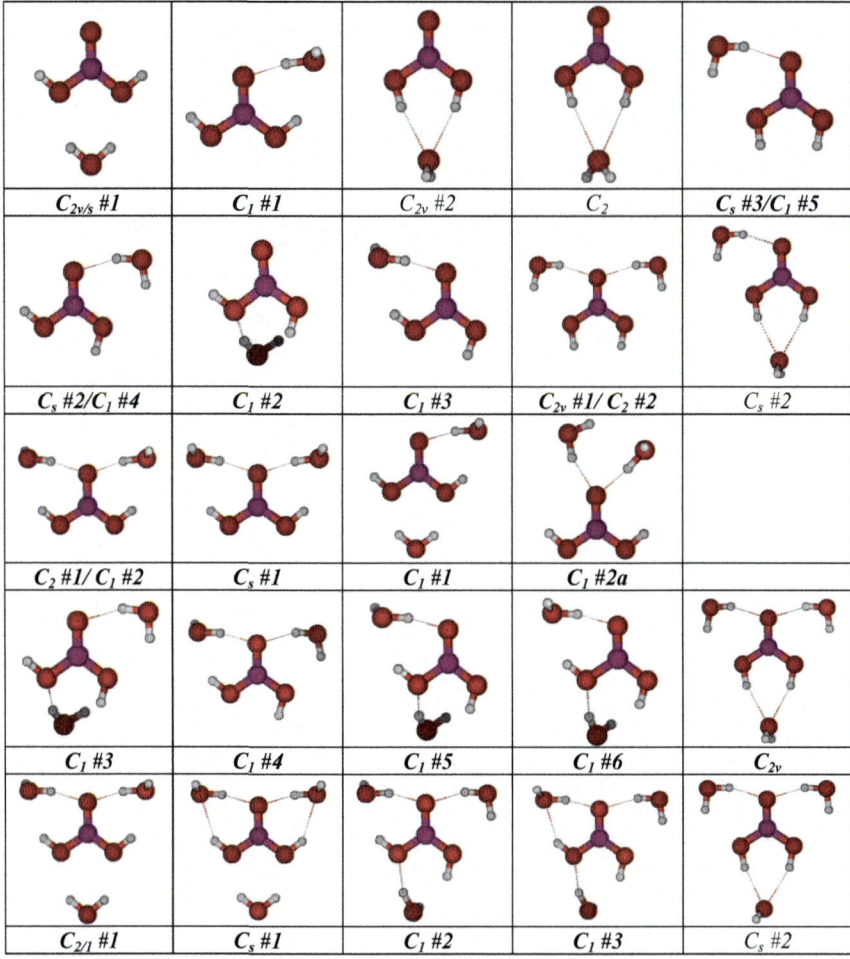

$C_{2v/s}$ #1	C_1 #1	C_{2v} #2	C_2	C_s #3/C_1 #5
C_s #2/C_1 #4	C_1 #2	C_1 #3	C_{2v} #1/ C_2 #2	C_s #2
C_2 #1/ C_1 #2	C_s #1	C_1 #1	C_1 #2a	
C_1 #3	C_1 #4	C_1 #5	C_1 #6	C_{2v}
$C_{2/1}$ #1	C_s #1	C_1 #2	C_1 #3	C_s #2

Fig. 13 Structure of hydrated oxodihydroxoborate, $BO(OH)_2^-$ · nH_2O

observe it spectroscopically. The vibrational frequencies of the most stable form are given in Table 12. In the isotropic Raman spectra, one would predict the observation of the BO_3 deformation mode at around 415 cm^{-1}, the BO(H) stretching mode at around 800 cm^{-1}, the BOH deformation mode at around 1100 cm^{-1}, and the BO stretching mode at about 1530 cm^{-1}. There is a fair amount of coupling between the BO stretching and BOH deformation modes.

A monohydrate can be based on any of the three stable anhydrous forms. From the most stable anhydrous C_{2v} #1 form, both the C_{2v} #1 and the more stable C_1 #1 forms can be derived (Fig. 13). The C_{2v} #1 is only stable at the HF levels, and at B3LYP/6-31G*, converting to the C_s #1 form at the other levels. From the next most stable anhydrous C_s form, the C_s #2, C_1 #2 and C_1 #3 forms can be derived.

The C_1 #2 form is higher in energy than the nearly isoenergetic C_s #2 and C_1 #3 forms. At MP2/6-311+G*, C_s #2 converts to C_1 #4. From the high-energy anhydrous C_{2v} #2 form, the unstable C_{2v} #2 and C_s #3 forms arise. The lower-energy C_s #3 form converts to C_1 #5. The C_{2v} #2 form could convert to a C_2 form (unstable) or one of two C_s forms. In all of these, the water molecule moves (C_s) or would move (C_2) to a structure close to C_1 #5. In all cases, conformers possessing a hydrogen bond to the oxo group are the most stable within each grouping based on the naked anion.

For the dihydrate, we can construct several structures based on the three anhydrous forms. From C_{2v} #1, the C_2 #1, C_s #1, and C_1 #1 forms may be constructed. The C_2 #1 form is unstable at HF/6-311+G* and MP2/6-311+G*, giving C_1 #2a and C_1 #2 forms, respectively. From C_s #1, the stable C_1 #3–6 forms can be constructed. From C_{2v} #2, the C_{2v} #1 form can be constructed (only stable at HF/6-31+G*), as well as C_s #2 (stable at MP2/6-311+G*). The C_{2v} #1 form converts to the stable C_2 #2 form (C_1 #8 at MP2/6-311+G*), whereas most attempts to obtain C_s #2 result in migration of the double acceptor water molecule towards the double donor water molecule.

For the trihydrate, two structures exist for each of the three anhydrous forms. Both a C_2 and a C_s #1 exist for the C_{2v} #1 anhydrous form, although at B3LYP/6-31G* and the MP2 levels, the C_2 structure desymmetrizes to C_1 #1. Both C_1 #2 and C_1 #3 can exist for the C_s #1 anhydrous form. The C_{2v} form could exist for the C_{2v} #2 anhydrous form, but it has imaginary frequencies. Nearly all attempts to desymmetrize (C_s #2,3) result in the double acceptor water molecule moving toward the other double donor water molecules, and at the only level where such a structure exists (C_s #3, MP2/6-31+G*), an imaginary frequency would desymmetrize by moving the double acceptor water molecule towards the double donor water molecules.

In the presence of additional water molecules, the BOH twisting vibrations can increase to 600–800 cm^{-1} (if hydrogen bonded), the in-plane BO_3 deformations increase to 480–600 cm^{-1}, the out-of-plane BO_3 deformation is hardly affected, the B-O(H) and B-O symmetric stretching frequency increases slightly, and the coupled antisymmetric B-OH stretch and BOH deformations all increase in frequency to about 1100–1300 cm^{-1} (Fig. 14). For water molecules that are hydrogen-bonded to the lone oxygen of this strong base, the OH frequency is lowered to around 3000 cm^{-1}.

3.9 Boric Acid-Hydroxide Complex and Its Hydrates, $B(OH)_3 \cdot OH^- \cdot nH_2O$

The boric acid-boric acid complex was investigated next (Fig. 15). All attempts to locate the complex between hydroxide and the C_{3h} form of boric acid resulted in deprotonation to form the oxodihydroxoborate anion-water complex. The C_s

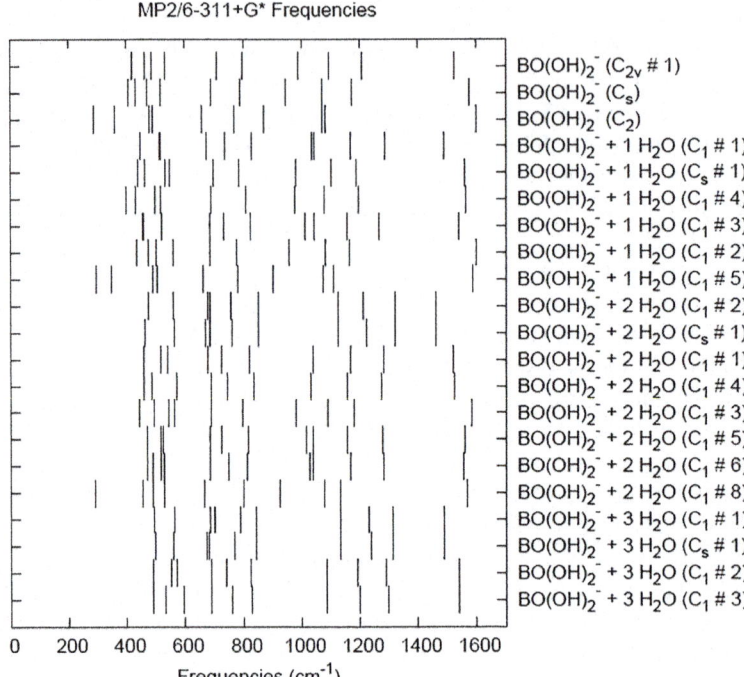

MP2/6-311+G* Frequencies

BO(OH)$_2^-$ (C$_{2v}$ # 1)
BO(OH)$_2^-$ (C$_s$)
BO(OH)$_2^-$ (C$_2$)
BO(OH)$_2^-$ + 1 H$_2$O (C$_1$ # 1)
BO(OH)$_2^-$ + 1 H$_2$O (C$_s$ # 1)
BO(OH)$_2^-$ + 1 H$_2$O (C$_1$ # 4)
BO(OH)$_2^-$ + 1 H$_2$O (C$_1$ # 3)
BO(OH)$_2^-$ + 1 H$_2$O (C$_1$ # 2)
BO(OH)$_2^-$ + 1 H$_2$O (C$_1$ # 5)
BO(OH)$_2^-$ + 2 H$_2$O (C$_1$ # 2)
BO(OH)$_2^-$ + 2 H$_2$O (C$_s$ # 1)
BO(OH)$_2^-$ + 2 H$_2$O (C$_1$ # 1)
BO(OH)$_2^-$ + 2 H$_2$O (C$_1$ # 4)
BO(OH)$_2^-$ + 2 H$_2$O (C$_1$ # 3)
BO(OH)$_2^-$ + 2 H$_2$O (C$_1$ # 5)
BO(OH)$_2^-$ + 2 H$_2$O (C$_1$ # 6)
BO(OH)$_2^-$ + 2 H$_2$O (C$_1$ # 8)
BO(OH)$_2^-$ + 3 H$_2$O (C$_1$ # 1)
BO(OH)$_2^-$ + 3 H$_2$O (C$_s$ # 1)
BO(OH)$_2^-$ + 3 H$_2$O (C$_1$ # 2)
BO(OH)$_2^-$ + 3 H$_2$O (C$_1$ # 3)

0   200   400   600   800   1000 1200 1400 1600

Frequencies (cm^{-1})

Fig. 14 Variation of BO(OH)$_2^-$ vibrational frequencies (0–1700 cm^{-1}) as a function of hydration and structure

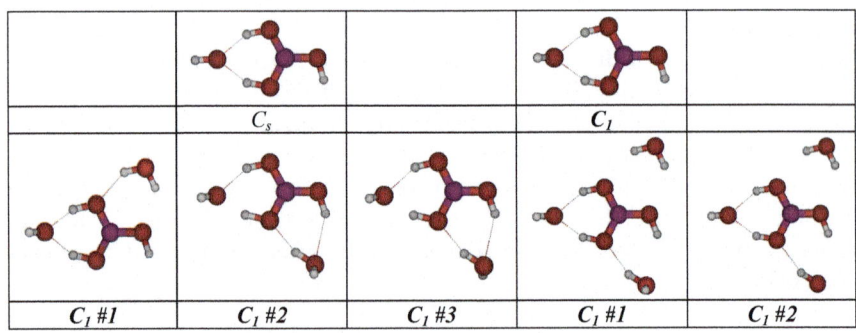

| | C$_s$ | | C$_1$ | |
| C$_1$ #1 | C$_1$ #2 | C$_1$ #3 | C$_1$ #1 | C$_1$ #2 |

Fig. 15 Structure of boric acid-hydroxide complex, B(OH)$_3$+OH$^-$+nH$_2$O

complex between the C$_s$ #2 form of boric acid acting as a double hydrogen bond donor to hydroxide was not stable at B3LYP/6-31G* and MP2/6-31G*, reverting instead to a oxodihydroxoborate-water complex via proton transfer. At the other levels, an imaginary frequency gave rise to the stable C$_1$ structure, which was 10–25 kJ/mol less stable than the corresponding oxodihydroxoborate form. Upon hydration to give one of three possible C_1 structures, the B3LYP/6-31G*, B3LYP/6-31+G*, and MP2/6-31G* converge to the oxodihydroxo form, and with the exception of C_1 #1, this also happens at the HF/6-31G* and B3LYP/6-311+G* levels. The forms that actually exist are 15–45 kJ/mol less stable than the corresponding oxodihydroxoborate dihydrate. Addition of the second water molecule only gives stable structures at HF/6-31+G* and HF/6-311+G* levels, reverting to the oxodihydroxoborate at the other levels. When both can exist, the oxodihydroxoborate is 45–50 kJ/mol more stable. Hydrating the boric acid part of the complex stabilizes the corresponding anion form.

3.10   Transition State Connecting Tetrahydroxoborate and Oxodihydroxoborate-Water Complex

The transition states connecting hydrated tetrahydroxoborate and the hydrated oxodihydroxoborate anions are shown in Fig. 16. The transition state essentially looks like a C_s boric acid interacting with a hydroxyl anion. It is stabilized somewhat by the interaction between the empty p-orbital of the boron and the hydroxyl non-bonding electron pairs, and also hydrogen bonding between the hydroxyl oxygen and the *syn*-hydrogens of boric acid, which tilt towards the hydroxyl. The electronic barrier (from tetrahydroxoborate) is between 84 and 114 kJ/mol, depending on level. The transition state could conceivably connect to either the van der Waals complex between boric acid and hydroxide, or to either the C_1 #1 or #3 forms of the monohydrated oxodihydroxoborate anion.

A water molecule can stabilize the transition state in one of two ways. It can stabilize the boric acid portion of the transition state (C_1 #1, #3, and #5) in one of the three possible sites (as a hydrogen bond donor-acceptor and two donor-donor types, respectively). Alternatively, it can stabilize the hydroxyl portion of the molecule (C_1 #2 and #4) where the free boric acid hydrogen is pointing either toward or away from the water molecule, respectively. The stabilization of the hydroxide lowers the energy more. When compared with the unhydrated transition state, addition of the water to the boric acid part of the molecule actually increases the barrier by between 10 and 25 kJ/mol, whereas stabilization of the hydroxide part of the molecule lowers the barrier by between 5 and 15 kJ/mol.

Waton and coworkers used the temperature-jump method to study the equilibrium between boric acid and borate [77]. The kinetics did not fit a simple equilibrium, but was analyzed by postulating an intermediate which they called [B(OH)$_3$,OH$^-$]. Our analysis of their published rate constants (k_{23}) at 4 and 20 °C suggests an activation barrier of 42 kJ/mol, giving an enthalpy of activation of 39 kJ/mol. We hypothesize that this intermediate is actually [BO(OH)$_2$]$^-$ · H$_2$O. Our calculated electronic barriers are much too high (101–117 kJ/mol), but these are lowered upon addition of an extra one (90–103 kJ/mol) or two (69–86 kJ/mol) water molecules. This leads one to think that additional water molecules might

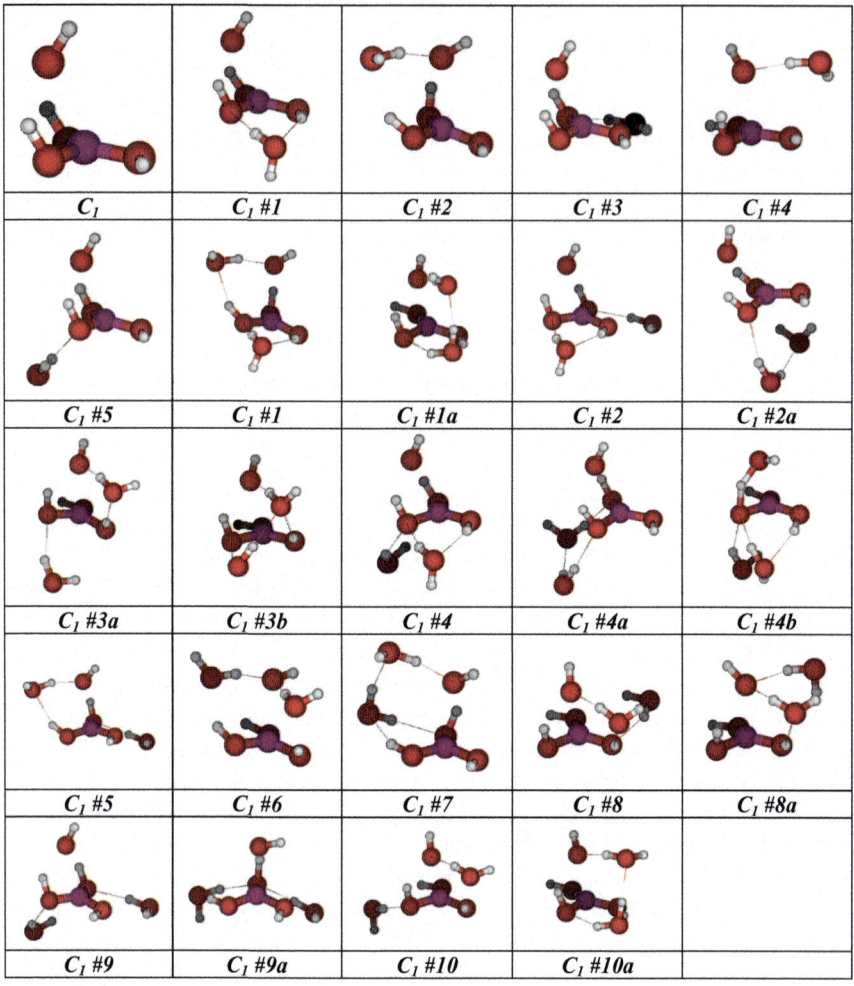

Fig. 16 Structure of boric acid-hydroxide transition structure + nH$_2$O

stabilize the transition state even more to bring the results into even better agreement with experiment, assuming that the postulated intermediate is correct.

3.11   Trioxoborate, BO_3^{3-}, and Dioxohydroxoborate $BO_2(OH)^{2-}$

While trioxoborate and dioxohydroxoborate are not expected to exist in aqueous solution owing to their strong basicity, these ions do exist as molecular entities in

Table 13 Geometrical parameters of BO_3^{3-} and $BO_2(OH)^{2-}$

Level	B-O distances (Å)			
	BO_3^{3-}	$BO_2(OH)^{2-}$	$s\text{-}BO_2(OH)^{2-}$	$a\text{-}BO_2(OH)^{2-}$
HF/6-31G*	1.4186	1.5941	1.3347	1.3188
HF/6-31+G*	1.4135	1.5556	1.3440	1.3251
HF/6-311+G*	1.4127	1.5613	1.3405	1.3209
B3LYP/6-31G*	1.4411	1.6307	1.3551	1.3332
B3LYP/6-31+G*	1.4234	1.5731	1.3637	1.3413
B3LYP/6-311+G*	1.4211	1.5812	1.3587	1.3359
MP2/6-31G*	1.4461	1.6296	1.3616	1.3401
MP2/6-31+G*	1.4363	1.5781	1.3755	1.3513
MP2/6-311+G*	1.4299	1.5741	1.3679	1.3437

Table 14 Vibrational frequencies (cm^{-1}) of the D_{3h} form of BO_3^{3-}

Level	BO_3 i.p def E'	BO str. A_1'	BO_3 o.o.p. def A_2''	BO str. E'
HF/6-31G*	581	848	876	1216
HF/6-31+G*	536	829	851	1085
HF/6-311+G*	535	826	844	1062
B3LYP/6-31G*	530	764	773	1107
B3LYP/6-31+G*	271	791	679	897
B3LYP/6-311+G*	227	785	680	867
MP2/6-31G*	533	769	788	1124
MP2/6-31+G*	358	767	659	843
MP2/6-311+G*	344	768	654	824

their sodium salts (Table 1). The calculated B-O bond distance of BO_3^{3-} (Table 13) is slightly longer than what is observed experimentally in the sodium salt, and this is partly due to the neglect of the sodium counterion and crystal packing. For $BO_2(OH)^{2-}$, the B-O distances for the unprotonated oxygens are in reasonable agreement with experiment, whereas the B-O(H) distance is quite long. It is expected that hydrogen bonding present in the crystal structure would shorten the corresponding B-O(H) distance considerably.

The calculated vibrational frequencies of BO_3^{3-} are shown in Table 14. There is a quite surprising dependence of the level of theory and basis set. The stretching frequencies decrease when going from Hartree-Fock to the correlated levels, as expected, as well as when going from nondiffuse to diffuse basis sets. It is more surprising that the deformation modes are affected even more when going from nondiffuse to diffuse basis sets at the correlated levels. This may be an artifact of using a correlated level on a system, which, in the gas-phase, is likely unbound with respect to electron detachment.

The calculated vibrational frequencies of $BO_2(OH)^{2-}$ are shown in Table 15. The dependence of the frequency on the level of theory and basis set is not as

Table 15 Vibrational frequencies (cm^{-1}) of the C_s #1 form of $BO_2(OH)^{2-}$

Level	BO₃ i.p def	BO (H) str	OHtors	BO₃ i.p def	BO₃ o.o. p def	OH def	BO str	BO str
	A'	A'	A''	A'	A''	A'	A'	A'
HF/6-31G*	453	491	547	723	812	1052	1134	1661
HF/6-31+G*	471	523	572	719	812	1080	1104	1557
HF/6-311+G*	472	515	560	704	811	1086	1093	1551
B3LYP/6-31G*	391	426	523	656	724	958	1047	1563
B3LYP/6-31+G*	416	476	543	635	712	986	1014	1426
B3LYP/6-311+G*	417	467	528	615	710	990	1007	1424
MP2/6-31G*	397	445	537	668	737	973	1050	1572
MP2/6-31+G*	418	478	547	643	708	991	1001	1406
MP2/6-311+G*	425	480	518	639	703	995	1001	1408

pronounced as for trioxoborate. The B-O and B-OH stretching frequencies decrease when going from Hartree-Fock to the correlated levels, as expected. When going from nondiffuse to diffuse basis sets, the B-O stretching frequencies decrease, but the B-OH frequency increases. The other frequencies do not show any surprising trends.

4 Conclusions

The calculated bond lengths and vibrational frequencies of boric acid and borate agree fairly well with that observed experimentally and with previous calculations, where available, when the comparison is appropriate. The oxodihydroxoborate ion is much more stable than the boric acid-hydroxide complex, when the latter exists. The oxodihydroxoborate ion, if it can be observed, should have a strong vibrational band at approximately 1400–1600 cm^{-1}. A transition state that links the tetrahydroxoborate to the hydrated oxodihydroxoborate ion has been found. The addition of water molecules lowers the barrier significantly, bringing the activation energy to closer agreement with experiment (assuming an oxodihydroxoborate intermediate).

Acknowledgements The author acknowledges the Atlantic Computational Excellence Network (ACEnet) for computational support.

References

1. Richens DT (1997) The chemistry of aqua ions. Wiley, Chichester
2. Oi T (2000) Calculations of reduced partition function ratios of monomeric and dimeric boric acids and borates by the ab initio molecular orbital theory. J Nucl Sci Tech 37(2):166–172

3. Oi T (2000) Ab initio molecular orbital calculations of reduced partition function ratios of polyboric acids and polyborate anions. Z Naturforsch A 55:623–628
4. Oi T, Yanase S (2001) Calculations of reduced partition function ratios of hydrated monoborate anion by the ab initio molecular orbital theory. J Nucl Sci Tech 38:429–432
5. Zeebe RE (2005) Stable boron isotope fractionation between dissolved $B(OH)_3$ and $B(OH)_4^-$. Geochim Cosmochim Acta 69:2753–2766
6. Liu Y, Tossell JA (2005) Ab initio molecular orbital calculations for boron isotope fractionation on boric acids and borates. Geochim Cosmochim Acta 69:3995–4006
7. Tossell JA (2005) Boric acid, "carbonic" acid, and N-containing oxyacids in aqueous solution: ab initio studies of structure, pKa, NMR shifts, and isotopic fractionation. Geochim Cosmochim Acta 69:5647–5658
8. Rustad JR, Bylaska EJ (2007) Ab initio calculation of isotopic fractionation in $B(OH)_3$(aq) and $B(OH)_4^-$(aq). J Am Chem Soc 129:2222–2223
9. Rustad JR, Bylaska EJ, Jackson VE, Dixon DA (2010) Calculation of boron-isotope fractionation between $B(OH)_3$(aq) and $B(OH)_4$-(aq). Geochim Cosmochim Acta 74:2843–2850
10. Zeebe RE, Sanyal A, Ortiz JD, Wolf-Gladrow DA (2001) A theoretical study of the kinetics of the boric acid-borate equilibrium in seawater. Marine Chem 73:113–124
11. Kracek FC, Morey GW, Merwin HE (1938) The system, water-boron oxide. Am J Sci A 35:143–171
12. Blasdale WC, Slansky CM (1939) The solubility curves of boric acid and the borates of sodium. J Am Chem Soc 61:917–920
13. Berger SV (1953) The crystal structure of boron oxide. Acta Chem Scand 7:611–622
14. Strong SL, Kaplow R (1968) The structure of crystalline B_2O_3. Acta Crystallogr B 24:1032–1036
15. Gurr GE, Montgomery PW, Knutson CD, Gorres BT (1970) The crystal structure of trigonal diboron trioxide. Acta Crystallogr B 26:906–915
16. Effenberger H, Lengauer CL, Parthe E (2001) Trigonal B_2O_3 with higher space-group symmetry: results of a reevaluation. Monat Chem 132:1515–1517
17. Prewitt CT, Shannon RD (1968) Crystal structure of a high-pressure form of B_2O_3. Acta Crystallogr B 24:869–874
18. Zachariasen WH (1934) The crystal lattice of boric acid, BO_3H_3. Z Kristallogr 88:150–161
19. Cowley JM (1953) Structure analysis of single crystals by electron diffraction. II. Disordered boric acid structure. Acta Crystallogr 6:522–529
20. Zachariasen WH (1954) The precise structure of orthoboric acid. Acta Crystallogr 7:305–310
21. Shuvalov RR, Burns PC (2003) A new polytype of orthoboric acid, H_3BO_3-3T1. Acta Crystallogr C 59:i47–i49
22. Tazaki H (1940) Single crystals of metaboric acid. J Sci Hiroshima Univ A 10:37–54
23. Tazaki H (1940) The structure of orthorhombic metaboric acid, HBO_2(a). J Sci Hiroshima Univ A 10:55–61
24. Peters CR, Milberg ME (1964) The refined structure of orthorhombic metaboric acid. Acta Crystallogr 17:229–234
25. Zachariasen WH (1952) A new analytical method for solving complex crystal structures. Acta Crystallogr 5:68–73
26. Zachariasen WH (1963) The crystal structure of monoclinic metaboric acid. Acta Crystallogr 16:385–389
27. Freyhardt CC, Wiebcke M, Felsche J (2000) The monoclinic and cubic phases of metaboric acid (precise redeterminations). Acta Crystallogr C 56:276–278
28. Zachariasen WH (1963) The crystal structure of cubic metaboric acid. Acta Crystallogr 16:380–384
29. Konig H, Hoppe R (1977) Zur Kenntnis von Na_3BO_3. Z Anorg Allg Chem 434:225–232
30. Menchetti S, Sabelli C (1982) Structure of hydrated sodium borate $Na_2[BO_2(OH)]$. Acta Crystallogr B 38:1282–1284

31. Block S, Perloff A (1963) The direct determination of the crystal structure of $NaB(OH)_4 2H_2O$. Acta Crystallogr 16:1233–1238
32. Csetenyi LJ, Glasser FP, Howie RA (1993) Structure of sodium tetrahydroxyborate. Acta Crystallogr C 49:1039–1041
33. Touboul M, Betourne E, Nowogrocki G (1995) Crystal structure and dehydration process of $Li(H_2O)_4B(OH)_4.2H_2O$. J Solid State Chem 115:549–553
34. Zachariasen WH (1964) The crystal structure of lithium metaborate. Acta Crystallogr 17:749–751
35. Hohne E (1964) Die Kristallstruktur des $LiB(OH)_4$. Z Chem 4:431–432
36. Fronczek FR, Aubry DA, Stanley GG (2001) Refinement of lithium tetrahydroxoborate with low-temperature CCD data. Acta Crystallogr E 57:i62–i63
37. Onak TP, Landesman H, Williams RE, Shapiro I (1959) The B11 nuclear magnetic resonance chemical shifts and spin coupling values for various compounds. J Phys Chem 63:1533–1535
38. Momii RK, Nachtrieb NH (1967) Nuclear magnetic resonance study of borate-polyborate equilibria in aqueous solution. Inorg Chem 6:1189–1192
39. How MJ, Kennedy GR, Mooney EF (1969) The pH dependence of the boron-11 chemical-shift of borate-boric acid solutions. J Chem Soc D Chem Commun 267–268
40. Smith HDJ, Wiersema RJ (1972) Boron-11 nuclear magnetic resonance study of polyborate ions in solution. Inorg Chem 11:1152–1154
41. Covington AK, Newman KE (1973) Base dissociation constant of the borate ion from 11B chemical shifts. J Inorg Nucl Chem 35:3257–3262
42. Henderson WG, How MJ, Kennedy GR, Mooney EF (1973) The interconversion of aqueous boron species and the interaction of borate with diols: a 11B N.M.R. study. Carbohydrate Res 28:1–12
43. Janda R, Heller G (1979) 11B–NMR-spektroskopische Untersuchungen an waessrigen Polyboratloesungen. Z Naturforsch B 34:1078–1083
44. Epperlein BW, Lutz O, Schwenk A (1975) Fourier-Kernresonanzuntersuchungen an 10B und 11B in Waessriger Loesung. Z Naturforsch A 30:955–958
45. Salentine CG (1983) High-field 11B NMR of alkali borates. Aqueous polyborate equilibria. Inorg Chem 22:3920–3924
46. Frisch MJ et al (2004) Gaussian 03, Revision D.02. Gaussian Inc., Wallingford, CT
47. Gupta A, Tossell JA (1981) A theoretical study of bond distances, X-ray spectra and electron density distributions in borate polyhedra. Phys Chem Miner 7:159–164
48. Gupta A, Tossell JA (1983) Quantum mechanical studies of distortions and polymerization of borate polyhedra. Am Miner 68:989–995
49. Zhang ZG, Boisen MBJ, Finger LW, Gibbs GV (1985) Molecular mimicry of the geometry and charge density distribution of polyanions in borate minerals. Am Miner 70:1238–1247
50. Zaki K, Pouchan C (1995) Vibrational analysis of orthoboric acid H_3BO_3 from ab initio second-order perturbation calculations. Chem Phys Lett 236:184–188
51. Tian SX, Xu KZ, Huang M-B, Chen XJ, Yang JL, Jia CC. Theoretical study on infrared vibrational spectra of boric-acid in gas-phase using density functional methods. J Mol Struct (Theochem) 459:223–227, 459
52. Tachikawa M (2004) A density functional study on hydrated clusters of orthoboric acid, B $(OH)_3(H_2O)n$ (n = 1–5). J Mol Struct (Theochem) 710:139–150
53. Stefani D, Pashalidis I, Nicolaides AV (2008) A computational study of the conformations of the boric acid $(B(OH)_3)$, its conjugate base $((HO)_2BO^-)$ and borate anion $(B(OH)_4^-)$. J Mol Struct (Theochem) 853:33–38
54. Zhou Y, Fang C, Fang Y, Zhu F (2011) Polyborates in aqueous borate solution: a Raman and DFT theory investigation. Spectrochim Acta A 83:82–87
55. Ananthakrishnan R (1936) The Raman spectra of some boron compounds (methyl borate, ethyl borate, boron tri-bromide and boric acid). Proc Indian Acad Sci A 4:74–81
56. Ananthakrishnan R (1937) The Raman spectra of crystal powders. IV. Some organic and inorganic compounds. Proc Indian Acad Sci A 5:200–221
57. Hibben JH (1938) The constitution of some boric oxide compounds. Am J Sci A 35:113–125

58. Mitra SM (1938) Raman effect in boric acid and in some boron compounds. Ind J Phys 12:9–14
59. Kahovec L (1938) Studien zum Raman-Effekt. Mitteilung LXXXV. Borsauere und Derivate. Z Phys Chem 40:135–145
60. Miller FA, Wilkins CH (1952) Infrared spectra and characteristic frequencies of inorganic ions. Anal Chem 24:1253–1294
61. Bethell DE, Sheppard N (1955) The infra-red spectrum and structure of boric acid. Trans Faraday Soc 51:9–15
62. Servoss RR, Clark HM (1957) Vibrational spectra of normal and isotopically labeled boric acid. J Chem Phys 26:1175–1178
63. Maya L (1976) Identification of polyborate and fluoropolyborate ions in solution by Raman spectroscopy. Inorg Chem 15:2179–2184
64. Maeda M, Hirao T, Kotaka M, Kakihana H (1979) Raman spectra of polyborate ions in aqueous solution. J Inorg Nucl Chem 41:1217–1220
65. Janda R, Heller G (1979) Ramanspektroskopische Untersuchungen an festen und in Wasser geloesten Polyboraten. Z Naturforsch B 34:585–590
66. Ogden JS, Young NA (1988) The characterisation of molecular boric acid by mass spectrometry and matrix isolation infrared spectroscopy. J Chem Soc Dalton Trans 1645–1652
67. Gilson TR (1991) Characterization of ortho- and meta-boric acids in the vapour phase. J Chem Soc Dalton Trans 2463–2466
68. Andrews L, Burkholder TR (1992) Infrared spectra of molecular $B(OH)_3$ and HOBO in solid argon. J Chem Phys 97:7203–7210
69. Gupta A, Swanson DK, Tossell JA, Gibbs GV (1981) Calculation of bond distances, one-electron properties and electron density distributions in first-row tetrahedral hydroxy and oxyanions. Am Miner 66:601–609
70. Hess AC, McMillan PF, O'Keeffe M (1988) Torsional barriers and force fields in H_4TO_4 molecules and molecular ions (T = C, B, Al, Si). J Phys Chem 92:1785–1791
71. Nielsen JR, Ward NE (1937) Raman spectrum and structure of the metaborate ion. J Chem Phys 5:201
72. Edwards JO, Morrison GC, Ross VF, Schultz JW (1955) The structure of the aqueous borate ion. J Am Chem Soc 77:266–268
73. Oertel RP (1972) Raman study of aqueous monoborate-polyol complexes. Equilibria in the monoborate-1,2-ethanediol system. Inorg Chem 11:544–549
74. Liu Z, Gao B, Hu M, Li S, Xia S (2003) FT-IR and Raman spectroscopic analysis of hydrated cesium borates and their saturated aqueous solution. Spectrochim Acta A 59:2741–2745
75. Zhu FY, Fang CH, Fang Y, Zhou YQ, Ge HW, Liu HY (2014) Structure of aqueous potassium metaborate solution. J Mol Struct 1070:80–85
76. Attina M, Cacace F, Occhiucci G, Ricci A (1992) Gaseous borate and polyborate anions. Inorg Chem 31:3114–3117
77. Waton G, Mallo P, Candau SJ (1984) Temperature-jump rate study of the chemical relaxation of aqueous boric acid solutions. J Phys Chem 88:3301–3305

Construction of a Potential Energy Surface Based on a Diabatic Model for Proton Transfer in Molecular Pairs

Yuta Hori, Tomonori Ida and Motohiro Mizuno

Abstract We propose a simple construction method of the potential energy surface based on diabatic model for proton transfer in molecular pairs. Assuming two-state valence bond electronic wave functions as a diabatic basis, the diagonal and non-diagonal matrix elements in diabatic potential of water, ammonia, and imidazole pairs were obtained. The validity of the construction procedure was confirmed by comparing two adiabatic potentials: one was transformed from the obtained diabatic potential and another was calculated by DFT calculation. Diabatic potentials were also obtained using fewer reference points than conventional methods at various intermolecular distances. Finally, we discuss the resulting diabatic potential and non-diagonal elements in detail.

Keywords Potential energy surface · Proton transfer · Diabatic

1 Introduction

One of the important steps in the theoretical treatment of chemical reactions is representation of the potential energy surface (PES) [1, 2]. Most approaches to chemical reactions analyze the PES by quantum chemical calculations derived under the Born-Oppenheimer approximation, also known as the adiabatic PES. Once the adiabatic potential is obtained, the scattering cross section, reaction constant, and reaction path, which are important for understanding chemical reactions, can be obtained from the potential [3]. Though ab initio quantum chemical calculations are becoming possible for large molecular systems, however, accurate PES calculations for understanding chemical reaction tend to be unfeasible. In addition, though analytical function for PES requires to analyze the reaction, the global function has not been known.

Y. Hori (✉) · T. Ida · M. Mizuno
Chemistry Course, Division of Material Chemistry, Graduate School of Natural
Science and Technology, Kanazawa University, Kanazawa 920-1192, Japan
e-mail: yu.hori59@gmail.com

© Springer International Publishing AG, part of Springer Nature 2018

179

Y. A. Wang et al. (eds.), *Concepts, Methods and Applications of Quantum Systems in Chemistry and Physics*, Progress in Theoretical Chemistry and Physics 31,
https://doi.org/10.1007/978-3-319-74582-4_9

Rather than examine the adiabatic PES, another approach to chemical reactions is to analyze the diabatic PES. In contrast to the adiabatic potential, the diabatic potential presents electronic states that change constantly to confine the eigenstates of the electronic Hamiltonian. The approaches using diabatic picture have been utilized various area in chemistry and physics where the coupling between nuclei and electrons such as vibronic coupling [4–6]. There are some approaches to describing the diabatic potential [7], constructing using some valence bond (VB) electronic wave functions [8–11]. Especially, empirical valence bond (EVB) [12] or multistate empirical valence bond (MS-EVB) [13] approach extended EVB is used the molecular mechanical functions to construct the PES and applied to the molecular dynamics (MD) simulations for many proton transfer systems [12, 14–29]. Furthermore, quantum dynamical approach using molecular mechanical functions (double Morse potential) for proton transfer was also performed [30]. However, to construct PES using diabatic potentials corresponding to reactant and product states, which is based on VB picture, is important for understanding chemical reactions in terms on chemical bond character. Although VB structures are usually not orthogonal, in this study we consider orthonormal VB structures. To basic idea, consider a two-state VB electronic wave function as the diabatic basis:

$$|\psi\rangle = c_1|\phi_1\rangle + c_2|\phi_2\rangle, \tag{1}$$

where $|\phi_1\rangle$ and $|\phi_2\rangle$ are VB wave functions that describes the electronic structure of the reactant and product states, respectively. The lowest adiabatic potential energy V^{ad} is then given by the lower root of the 2×2 secular equation; specifically:

$$V^{\mathrm{ad}} = \frac{V_{11}^{\mathrm{di}} + V_{22}^{\mathrm{di}}}{2} - \sqrt{\left(\frac{V_{11}^{\mathrm{di}} - V_{22}^{\mathrm{di}}}{2}\right)^2 + V_{12}^{\mathrm{di}\,2}}, \tag{2}$$

where

$$V_{11}^{\mathrm{di}} = \langle\phi_1|H|\phi_1\rangle, \tag{3}$$

$$V_{22}^{\mathrm{di}} = \langle\phi_2|H|\phi_2\rangle, \tag{4}$$

$$V_{12}^{\mathrm{di}} = \langle\phi_1|H|\phi_2\rangle. \tag{5}$$

V_{11}^{di} and V_{22}^{di} are the potential energies for the two VB structures of the reactant and product states, respectively. In this approach V_{11}^{di}, V_{22}^{di} and V_{12}^{di} function forms including parameters can be obtained to fit in experimental or ab initio data. In these works, V_{11}^{di} and V_{22}^{di} are related to use of molecular mechanics potential functions, especially, which are taken as the harmonic normal-mode potential or Morse potential etc. [14, 15, 17, 18, 25, 26]. On the other hand, although the selection of V_{12}^{di} is less obvious, Gaussian function as V_{12}^{di} proposed by Chang and Millar [14] has been widely used. However, they has not been confirmed whether the obtained diabatic potentials produce the reliable adiabatic potentials or not, although these functions are obtained to fit in some data. Additionally, to obtain these analytical functions with parameters uniquely or using fewer data is important to construct the PES for various chemical reactions. Therefore, a proposal of simple method for light or more uniquely construction of the diabatic potentials (V_{11}^{di} and V_{22}^{di}) and non-diagonal matrix element (V_{12}^{di}) using the analytical functions is important to analyze the chemical reaction and it can be widely applied to describing the large molecular systems such as proteins.

In this study, we focus on one-dimensional proton transfer models and suggest a simple construction method of global PES for intermolecular proton transfer by use of Morse potential as V_{11}^{di} and V_{22}^{di} and Gaussian function as V_{12}^{di} in the diabatic potential matrix and confirm the validity to use these potential functions. In addition, we investigate whether it is possible to apply to the proton transfer for various intermolecular distance. Here, we focused on the proton-bonded four bimolecular models for ammonia (AmH$^+$-Am) and imidazole (ImH$^+$-Im) pairs as symmetrical homo-molecular proton transfer systems and for imidazole-ammonia (ImH$^+$-Am) and ammonia-water (AmH$^+$-Wat) pairs as asymmetrical hetero-molecular systems, the four model structures of which were shown in Fig. 1. We investigate the portability to use the Morse potential and Gaussian function as the diabatic potential matrix elements by comparison the transformed adiabatic potential from the diabatic one with the calculated by DFT calculation in these systems. Finally, we discuss the proton transfer characters using obtained diabatic potential (V_{11}^{di} and V_{22}^{di}) and non-diagonal elements (V_{12}^{di}) for homo- and hetero-molecular pairs.

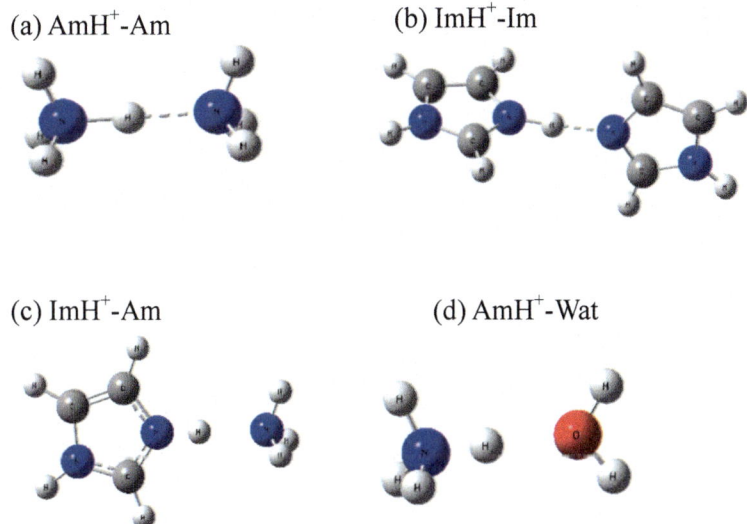

(a) AmH$^+$-Am   (b) ImH$^+$-Im

(c) ImH$^+$-Am   (d) AmH$^+$-Wat

Fig. 1 Four model structures of the proton-bonded **a** AmH$^+$-Am, **b** ImH$^+$-Im, **c** AmH$^+$-Wat, and **d** ImH$^+$-Am pairs

2 Theoretical and Calculation Methods

2.1 Calculation Models

In the construction of PES for proton transfer, we focused on the four proton transfer models: (a) AmH$^+$-Am, (b) ImH$^+$-Im as homo-molecular pairs, and (c) ImH$^+$-Am, (d) AmH$^+$-Wat as hetero-molecular pairs (Fig. 1). Figure 2 shows also these models and potential energy coordinates. For homo-molecular pairs, coordinate R denotes the intermolecular distance and x is the translating proton position that defines as a displacement from the center of the intermolecular distance. For hetero-molecular pairs, r is used as the displacement between a proton-bonded atom and the proton.

2.2 Diabatic Potential Functions (V_{11}^{di}, V_{22}^{di} and V_{12}^{di})

In this study, we constructed a PES for proton transfer by using diabatic picture. Diabatic potentials can be constructed using a variety of valence bond (VB) configurations [7–11]. Here, we considered a two-state VB electronic wave function as the diabatic basis corresponding to the reactant and product states, i.e., Equation (1). In particular, the EVB approach can be related to use of molecular mechanical potential functions describing the molecular vibration, which is rational for understanding chemical reaction. In most cases, the diagonal matrix elements (V_{11}^{di} and V_{22}^{di}) are taken as the harmonic normal-mode or Morse potentials [14, 15, 17, 18, 25, 26]. On the other hand, Chang and Millar [14] suggested the use of a generalized Gaussian function as the non-diagonal matrix element (V_{12}^{di}).

Therefore, we selected the Morse function as the V_{11}^{di} and V_{22}^{di}, and Gaussian function as the V_{12}^{di} to construct PES for proton transfer. For homo-molecular pairs, V_{11}^{di}, V_{22}^{di} and V_{12}^{di} are explicitly defined as

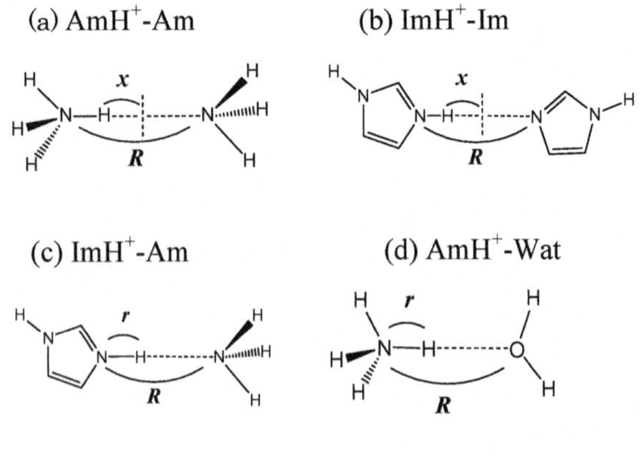

Fig. 2 Proton transfer model and potential energy coordinates for **a** AmH$^+$-Am, **b** ImH$^+$-Im, **c** AmH$^+$-Wat, and **d** ImH$^+$-Am

(a) AmH$^+$-Am

(b) ImH$^+$-Im

(c) ImH$^+$-Am

(d) AmH$^+$-Wat

$$V_{11}^{di}(x;R) = D\left(1 - e^{-k(x+x_0)}\right)^2 + c, \tag{6}$$

$$V_{22}^{di}(x;R) = D\left(1 - e^{k(x-x_0)}\right)^2 + c, \tag{7}$$

$$V_{12}^{di}(x;R) = A\exp\left(-bx^2\right). \tag{8}$$

For hetero-molecular pairs, V_{11}^{di}, V_{22}^{di} and V_{12}^{di} are defined as

$$V_{11}^{di}(r;R) = D_1\left(1 - e^{-k_1(r-r_0)}\right)^2 + c, \tag{9}$$

$$V_{22}^{di}(r;R) = D_2\left(1 - e^{-k_2(R-r-r_0')}\right)^2 + c + D_3, \tag{10}$$

$$V_{12}^{di}(r;R) = A\exp\left(-b(r-r_c)^2\right). \tag{11}$$

We constructed the PES by optimization of the potential parameters in these functions (V_{11}^{di}, V_{22}^{di} and V_{12}^{di}). For V_{11}^{di} and V_{22}^{di}, parameters D, D_1, and D_2 are binding energies; x_0, r_0, r_0' are the equilibrium bond lengths; k, k_1, and k_2 are the decay constants. For V_{12}^{di}, parameter A is the amplitude of the Gaussian function; parameter b corresponds to the spread of the function; r_c is the point of the maximum value of the function.

2.3 Optimization of Potential Parameters

Optimization of the potential parameters was conducted according to the following procedures.

First, we estimated the parameters of V_{11}^{di} and V_{22}^{di}. Parameters D, D_1, and D_2 were estimated by dissociation energy of the proton corresponding to one side of the molecular pairs, i.e., ammonium, imidazolium, and oxonium ions. x_0, r_0, r_0' were used the equilibrium bond lengths of these molecules. The estimated values of these parameters are shown in Table 1. For hetero-molecular pairs, parameter D_3 was obtained from the difference between $V^{ad}(r_0;R)$ and $V^{ad}(r_0';R)$:

$$D_3 = V^{ad}\left(r_0';R\right) - V^{ad}(r_0;R). \tag{12}$$

The remaining parameters k, k_1, and k_2 included in V_{11}^{di} and V_{22}^{di} are considered to be dependent on the R coordinate. These parameters were estimated by comparison with the adiabatic potential obtained from DFT calculations according to Eq. (2). V_{11}^{di} and V_{22}^{di} describes molecular vibration for the reactant and product states, respectively. Therefore, we assumed that V_{11}^{di} and V_{22}^{di} reproduced the adiabatic

Table 1 Diabatic potential energy parameters for (a) AmH$^+$-Am, (b) ImH$^+$-Im, (c) ImH$^+$-Am, and (d) AmH$^+$-Wat

(a) AmH$^+$-Am		(b) ImH$^+$-Im	
D(kJ mol^{-1})	r_0(Å)	D(kJ mol^{-1})	r_0(Å)
664.758	1.027	668.81	1.0148
(c) ImH$^+$-Am			
D_1(kJ mol^{-1})	D_2(kJ mol^{-1})	r_0(Å)	r_0'(Å)
668.81	664.758	1.015	1.027
(d) AmH$^+$-Wat			
D_1(kJ mol^{-1})	D_2(kJ mol^{-1})	r_0(Å)	r_0'(Å)
725.69	593.731	1.027	0.977

potential, when the proton position was closer to a binding atom than a stable point ($r < r_0$). Here, we estimated the parameter k by comparing them with the V_{22}^{di} and adiabatic potential at $x = 2x_0$ for homo-molecular pairs for fixed R. In the same way, for hetero-molecular pairs, k_1 and k_2 were determined by comparing with the adiabatic potential at $r = 2r_0 - R/2 \equiv r_{k_1}$ and $r = 3R/2 - 2r_0' \equiv r_{k_2}$, respectively, for fixed R. These relations are explicitly defined as

$$V_{22}^{\text{di}}(2x_0) = V^{\text{ad}}(2x_0), \tag{13}$$

$$V_{11}^{\text{di}}(r_{k_1}) = V^{\text{ad}}(r_{k_1}), \tag{14}$$

$$V_{22}^{\text{di}}(r_{k_2}) = V^{\text{ad}}(r_{k_2}). \tag{15}$$

for homo- and hetero-molecular pairs, respectively. Parameters k, k_1, and k_2 are then denoted as

$$k = \frac{1}{x_0} \ln\left(1 + \sqrt{\frac{V^{\text{ad}}(2x_0) - c}{D}}\right), \tag{16}$$

$$k_1 = \frac{1}{r_0 - r_{k_1}} \ln\left(1 + \sqrt{\frac{V^{\text{ad}}(r_{k_1}) - c}{D_1}}\right), \tag{17}$$

$$k_2 = \frac{1}{r_{k_2} + r_0' - R} \ln\left(1 + \sqrt{\frac{V^{\text{ad}}(r_{k_2}) - c - D_3}{D_2}}\right). \tag{18}$$

Subsequently, parameters A and b of V_{12}^{di} depending on the coordinate R were estimated. These parameters were also estimated by comparison with the adiabatic potential obtained from DFT calculations and using obtained diabatic potentials (V_{11}^{di} and V_{22}^{di}). Determination of these parameters was conducted according to the following steps for each pairs.

First, parameter A was estimated. Because this parameter represents the amplitude of the Gaussian type function, parameter b vanishes at $x=0$ for homo-molecular pairs. Thus, the relationship between the adiabatic and diabatic potentials at $x=0$ and fixed R is given by

$$\frac{V_{11}^{di}(0) + V_{22}^{di}(0)}{2} - \sqrt{\left(\frac{V_{11}^{di}(0) - V_{22}^{di}(0)}{2}\right)^2 + V_{12}^{di\,2}(0)} = V^{ad}(0). \qquad (19)$$

Parameter A was then estimated by using obtained V_{11}^{di} and V_{22}^{di}, which was expressed as

$$A = V_{11}^{di}(0) - V^{ad}(0), \qquad (20)$$

for homo-molecular pairs. For hetero-molecular pair, parameter A was estimated at the cross point $(r=r_c)$ of V_{11}^{di} and V_{22}^{di} substituted $D_3=0$. Parameter A was then expressed as

$$A = V_{11}^{di}(r_c) - V^{ad}(r_c), \qquad (21)$$

for hetero-molecular pairs.

Next, the remaining parameter b of V_{12}^{di} was estimated. This parameter corresponds to the spread of the Gaussian type function. For homo-molecular pair, the midpoint between the local minimum (x_0) and local maximum $(x=0)$ is assumed to define the spread of V_{12}^{di}. Therefore, the relationship between the adiabatic and diabatic potentials at $x=x_0/2$ and fixed R is also given by

$$\frac{V_{11}^{di}\left(\frac{x_0}{2}\right) + V_{22}^{di}\left(\frac{x_0}{2}\right)}{2} - \sqrt{\left(\frac{V_{11}^{di}\left(\frac{x_0}{2}\right) - V_{22}^{di}\left(\frac{x_0}{2}\right)}{2}\right)^2 + V_{12}^{di\,2}\left(\frac{x_0}{2}\right)} = V^{ad}\left(\frac{x_0}{2}\right). \qquad (22)$$

According to this relation, parameter b is expressly defined by

$$b = \frac{2}{x_0^2}\left[\ln A^2 - \ln\left\{\left(\frac{V_{11}^{di}\left(\frac{x_0}{2}\right) + V_{22}^{di}\left(\frac{x_0}{2}\right)}{2} - V^{ad}\left(\frac{x_0}{2}\right)\right)^2 - \left(\frac{V_{11}^{di}\left(\frac{x_0}{2}\right) - V_{22}^{di}\left(\frac{x_0}{2}\right)}{2}\right)^2\right\}\right]. \tag{23}$$

In the same way, for hetero-molecular pairs, b was determined by comparing with the adiabatic potential at $r = (r_c + r_0)/2 \equiv r_b$ for fixed R, which was defined by

$$b = \frac{1}{(r_b - r_c)^2}\left[\ln A^2 - \frac{1}{2}\ln\left\{\left(\frac{V_{11}^{di}(r_b) + V_{22}^{di}(r_b)}{2} - V^{ad}(r_b)\right)^2 - \left(\frac{V_{11}^{di}(r_b) - V_{22}^{di}(r_b)}{2}\right)^2\right\}\right]. \tag{24}$$

2.4   Adiabatic Potential Energy Calculation

To obtain the adiabatic PES $V^{ad}(x, R)$ or $V^{ad}(r, R)$ for the proton transfer reaction, quantum chemical calculations were performed with stepwise movement of the proton and intermolecular distances by 0.02 and 0.05 Å, respectively. The geometries of the molecules were then kept in the minimum energy structures, except for the proton position (x or r) and the intermolecular distance (R). Minimum energy structures, as shown in Fig. 1, were determined by geometrical optimization. Finally, the adiabatic potential energies were fitted using a polynomial series function of the fourth order with respect to x or r and of the third order with respect to R [31, 32]. In previous work [33, 34], the proton transfer of ImH$^+$-Im systems was discussed with the B3LYP approach. Therefore, all DFT calculations were performed at the B3LYP/aug-cc-pVDZ level using the Gaussian-09 package [35].

3   Results and Discussion

We first estimated the potential parameters k, k_1, k_2, and D_3 of V_{11}^{di} and V_{22}^{di} using Eqs. (12) and (16)–(18) and obtained parameters (Table 1) for (a) AmH$^+$-Am, (b) ImH$^+$-Im, (c) ImH$^+$-Am, and (d) AmH$^+$-Wat. The parameters A and b were then estimated using Eqs. (20), (21), (23), and (24). Figure 3 shows the computed V_{11}^{di}, V_{22}^{di}, and V_{12}^{di} values using obtained the potential parameters at some intermolecular distance for (a) AmH$^+$-Am, (b) ImH$^+$-Im, (c) ImH$^+$-Am and (d) AmH$^+$-Wat, respectively. To compare with the adiabatic potential by DFT calculation, Figs. 4 and 5 show the transformed adiabatic potential derived from the diabatic potential matrix elements (V_{11}^{di}, V_{22}^{di}, and V_{12}^{di}) using Eq. (2) and one obtained from DFT. Although the potentials using diabatic model not reproduce the ones using DFT around the local minimum of the potentials, however, the figures show that the transformed adiabatic potentials are qualitatively in good agreement with those calculated by DFT calculations for all proton transfer systems at various intermolecular distance R. Furthermore, to compare with the present work and DFT data Table 2 shows the coefficient of determination (R-squared). Because the values of R-squared for all models were close to unity, we confirmed the validity of our approach. Thus, it is indicates that PES for various proton transfer systems can qualitatively reproduce by using V_{11}^{di} and V_{22}^{di} with Morse potential described the vibrational motion and V_{12}^{di} with the Gaussian function by assumption of two-state VB wave functions as a diabatic basis.

In our approach, parameters k, A, and b can be uniquely determined not necessary to fit the potential energy. Especially, the construction procedures of the

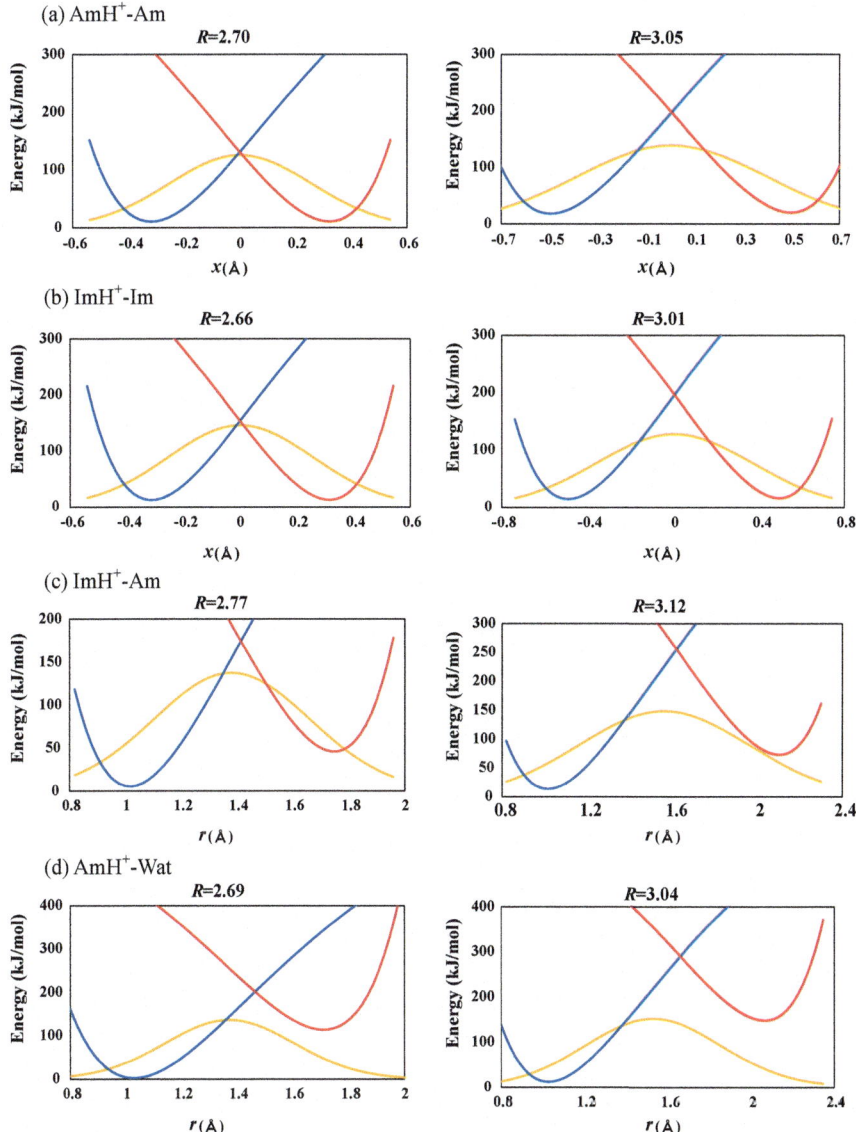

Fig. 3 Diagonal matrix elements, V_{11}^{di} (blue line) and V_{22}^{di} (red line), and non-diagonal matrix element V_{12}^{di} (yellow line) in diabatic potential matrix computed using obtained potential parameters at some intermolecular distance R for **a** AmH$^+$-Am, **b** ImH$^+$-Im, **c** ImH$^+$-Am, and **d** AmH$^+$-Wat

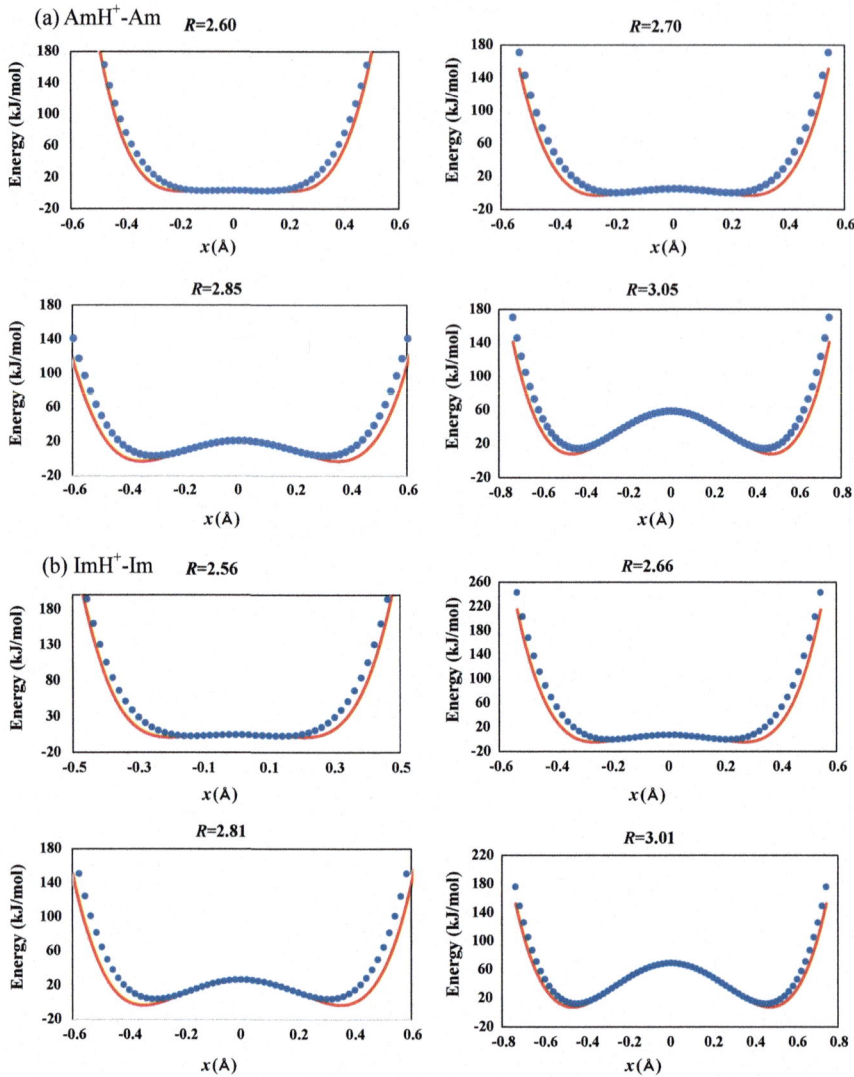

Fig. 4 Transformed adiabatic potential derived from the diabatic potentials using Eq. (2) (red line) and from DFT (blue dots) at some intermolecular distance R for **a** AmH$^+$-Am and **b** ImH$^+$-Im

diabatic potential employed four reference points ($x = x_0$, $2x_0$, 0 and $x_0/2$) at fixed R for homo-molecular pairs, while five reference points ($r = r_0$, r_{k_1}, r_{k_2}, r_c and r_b) for hetero-molecular pairs. Thus, the whole potentials of proton transfer are described by using energies of 40 or 50 reference points, while the number of DFT data points to describe the whole adiabatic potential required approximately 500 points. Therefore, the PES describing the entire proton transfer system for diabatic picture can be obtained using less than one-tenth of the reference points required for

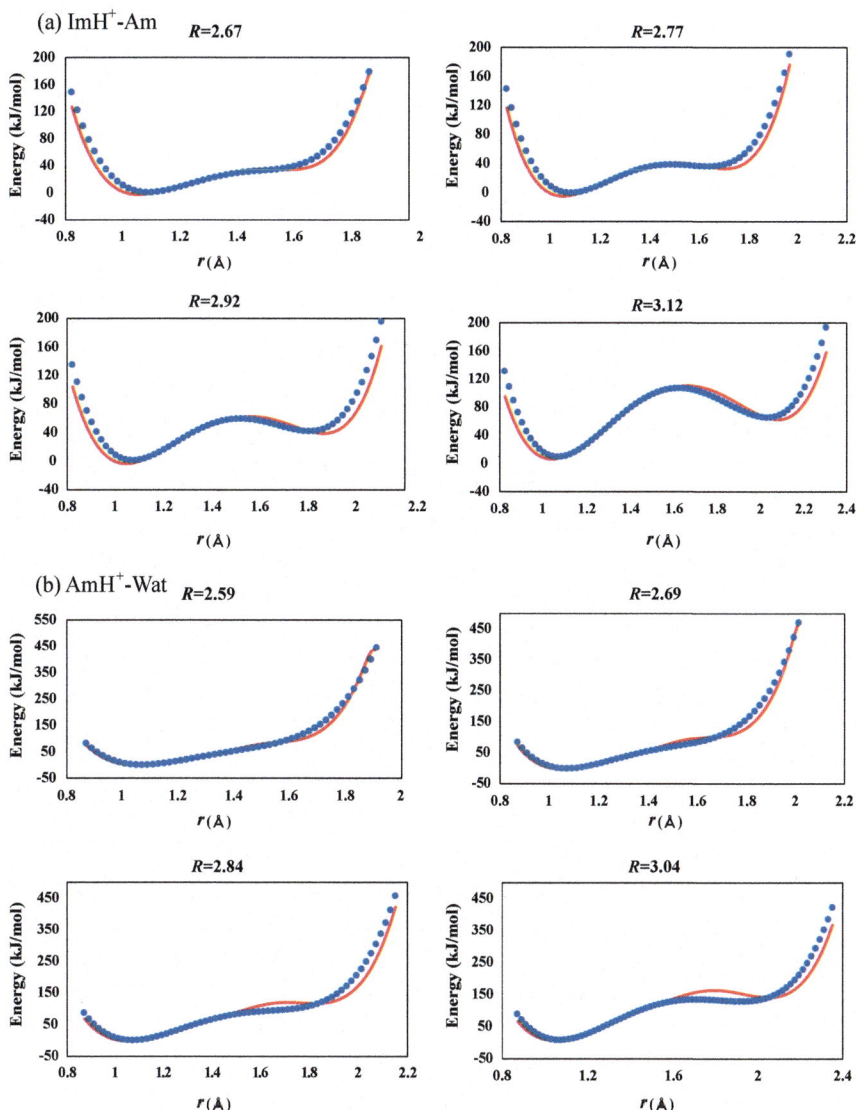

Fig. 5 Transformed adiabatic potential derived from the diabatic potentials using Eq. (2) (red line) and from DFT (blue dots) at some intermolecular distance R for **c** ImH$^+$-Am and **d** AmH$^+$-Wat

Table 2 Coefficient of determination (R-squared) between adiabatic potential derived from the diabatic potentials and from DFT

	(a) AmH$^+$-Am	(b) ImH$^+$-Im	(c) ImH$^+$-Am	(d) AmH$^+$-Wat
R-squared	0.966	0.974	0.932	0.965

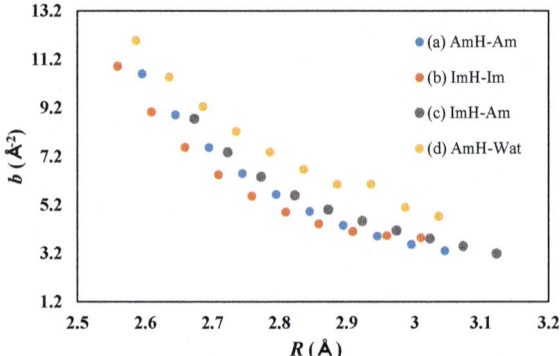

Fig. 6 Intermolecular dependence of potential parameter b of V_{12}^{di} for **a** AmH$^+$-Am, **b** ImH$^+$-Im, **c** ImH$^+$-Am, and **d** AmH$^+$-Wat

the adiabatic picture. Furthermore, our method can be applied to proton transfer system even when the transition state (TS) cannot be calculated, although the information of TS requires the construction of potential in the previous work. Therefore, it is concluded that our procedures are useful for PES construction by using the diabatic picture for proton transfer systems and can be applied to large molecular systems such as proteins.

Finally, we discuss the obtained potential parameters. Figure 6 shows that intermolecular dependence of potential parameter b of V_{12}^{di} for (a) AmH$^+$-Am, (b) ImH$^+$-Im, (c) ImH$^+$-Am, and (d) AmH$^+$-Wat. The values of parameter b, which describe the spread of the Gaussian function, decreased as R increased. This result indicates that the non-diagonal matrix element (V_{12}^{di}) is broadly distributed along the proton transfer coordinate and the bond mixture between the reactant and the product states occurs over a wide range, not only at the TS. In addition, because V_{12}^{di} is broadly distributed along the intermolecular distance, the proton can be formed mixture between reactant and product states and transferred at the location formed hydrogen bond. To clarify the effect of non-diagonal matrix element V_{12}^{di}, the ratio of amplitude for the V_{12}^{di} (parameter A) divided by the crossing point energy of the V_{11}^{di} and V_{22}^{di} was estimated, the results of which were shown in Fig. 7. According to

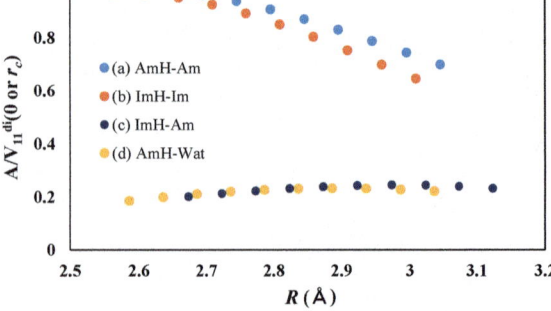

Fig. 7 Ratio of amplitude for non-diagonal matrix element divided by the crossing point energy of the diabatic potential at $x = 0$ for homo-molecular pairs or $r = r_c$ for hetero-molecular pairs, i.e., $A/V_{11}^{di}(x = 0 \text{ or } r = r_c)$

Eq. (2), V_{12}^{di} contributes to the stability of the adiabatic potential V^{ad} and the $A/V_{11}^{di} (x=0$ or $r=r_c)$ is assumed to be determined the rate of contribution to the possibility of proton transfer. For both homo-molecular pairs, the value of A/V_{11}^{di} decreased as R increased and was approximately 50 to 80% at various inter-molecular distance. Thus, the proton can easily transfer at the location formed hydrogen bond for homo-molecular pairs. On the other hand, for both hetero-molecular pairs, the value of A/V_{11}^{di} was constantly about 20% over wide range intermolecular distance, which was lower than homo-molecular pairs for all intermolecular distance. It indicates that the proton for homo-molecular pairs can easily transfer than for hetero-molecular pairs. From the above discussion, we find that the obtained potentials give the qualitative information about proton transfer even if a simple two-state diabatic model is used.

4 Conclusion

Potential energy surface (PES) is an important theoretical approach for under-standing chemical reactions. Diabatic potentials are used to understand proton transfer reactions. Especially, EVB approach based on the diabatic picture is used to the molecular mechanical function to construct PES and applied to many appli-cations including molecular dynamics simulation. In this paper, we constructed the PES based on diabatic model for proton transfer models: (a) AmH^+-Am, (b) ImH^+-Im, (c) ImH^+-Am and (d) AmH^+-Wat. We confirmed that Morse potentials as the diagonal matrix element (V_{11}^{di} and V_{22}^{di}) and Gaussian function as the non-diagonal matrix element (V_{12}^{di}), which are important to apply widely to understanding the chemical reaction including the classical or quantum dynamics simulations, described the proper PES for proton transfer. In addition, we proposed a simple method to uniquely construct the diabatic potentials using these analytical func-tions. The diabatic potentials at various intermolecular distance were obtained using fewer reference points than the adiabatic potentials to describe an entire proton transfer system. Therefore, our construction method is useful and can be applied to the large molecular systems such as proteins.

From the values of estimated the potential parameters, V_{12}^{di} was broadly dis-tributed and the proton-bonded mixture between the reactant and product states occurred over a wide range of reaction coordinates, and not only at the TS. Fur-thermore according to the relation between diagonal and non-diagonal matrix ele-ments, it is found that the proton for homo-molecular pairs can transfer easily than for hetero-molecular pairs.

Acknowledgements This work was supported by a Grant-in-Aid for Scientific Research (23310063, 26286002) from the Ministry of Education, Science and Technology, Government of Japan.

References

1. Truhlar DG, Steckler R, Gordon MS (1987) Chem Rev 87:217–236
2. Schatz GC (1989) Rev Mod Phys 61:669–688
3. Levine RD (2009) Molecular reaction dynamics. Cambridge University Press
4. Worth GA, Cederbaum LS (2004) Annu Rev Phys Chem 55:127–158
5. Köppel H, Domcke W, Cederbaum LS (1984) Adv Chem Phys 57:59–246
6. Nicolas G, Gadea FX (1999) J Chem Phys 111:10537
7. Voorhis TV et al (2010) Annu Rev Phys Chem 61:149–170
8. Åqvist J, Warshel A (1993) Chem Rev 93:2523–2544
9. Mo Y, Gao J (2000) J Phys Chem A 104:3012–3020
10. Truhlar DG (2007) J Comp Chem 28:73–86
11. Song L, Gao J (2008) J Phys Chem A 112:12925–12935
12. Warshel A, Weiss RM (1980) J Am Chem Soc 102:6218–6226
13. Lobaugh J, Voth GA (1996) J Chem Phys 104:2056
14. Chang Y-T, Miller WH (1990) J Phys Chem 94:5884–5888
15. Hinsen K, Roux BJ (1997) Comp Chem 18:368–380
16. Minichino C, Voth GA (1997) J Phys Chem B 101:4544–4552
17. Sagnella DE, Tuckerman ME (1998) J Chem Phys 108:2073–2083
18. Vuilleumier R, Borgis D (1998) Chem Phys Lett 284:71–77
19. Wang Y, Gunn JR (1999) Int J Quantum Chem 73:357–367
20. Schmitt UW, Voth GA (1999) J Chem Phys 111:9361–9381
21. Čuma M, Schmitt UW, Voth GA (2000) Chem Phys 258:187–199
22. Čuma M, Schmitt UW, Voth GA (2001) J Phys Chem A 105:2814–2823
23. Brancato G, Tuckerman ME (2005) J Chem Phys 122:224507
24. Hong G, Rosta E, Warshel AJ (2006) Phys Chem B 110:19570–19574
25. Lammers S, Lutz S, Meuwly MJ (2008) Comp Chem 29:1048–1063
26. Chen H, Yan T, Voth GA (2009) J Phys Chem A 113:4507–4517
27. Feng S, Voth GA (2011) J Phys Chem B 115:5903–5912
28. Kamerlin SCL, Warshel A (2011) IREs Comput Mol Sci 1:30–45
29. Li A, Cao Z, Li Y, Yan T, Shen P (2012) J Phys Chem B 116:12793–12800
30. Marinica DC, Gaigeot M-P, Borgis D (2006) Chem Phys Lett 423:390–394
31. Jaroszewski L, Lesyng B, Tanner JJ, McCammon JA (1990) Chem Phys Lett 175:282–288
32. Jaroszewski L et al (1993) J Mol Struct THEOCHEM 283:57–62
33. Tatara W et al (2003) J Phys Chem A 107:7827–7831
34. Mangiatordi GF et al (2011) J Phys Chem A 115:2627–2634
35. Frisch MJ, Trucks GW, Schlegel HB, Scuseria GE, Robb MA, Cheeseman JR, Scalmani G, Barone V, Mennucci B, Petersson GA, Nakatsuji H, Caricato M, Li X, Hratchian HP, Izmaylov AF, Bloino J, Zheng G, Sonnenberg JL, Hada M, Ehara M, Toyota K, Fukuda R, Hasegawa J, Ishida M, Nakajima T, Honda Y, Kitao O, Nakai H, Vreven T, Montgomery Jr. JA, Peralta JE, Ogliaro F, Bearpark M, Heyd JJ, Brothers E, Kudin KN, Staroverov VN, Kobayashi R, Normand J, Raghavachari K, Rendell A, Burant JC, Iyengar SS, Tomasi J, Cossi M, Rega N, Millam JM, Klene M, Knox JE, Cross JB, Bakken V, Adamo C, Jaramillo J, Gomperts R, Stratmann RE, Yazyev O, Austin AJ, Cammi R, Pomelli C, Ochterski JW, Martin RL, Morokuma K, Zakrzewski VG, Voth GA, Salvador P, Dannenberg JJ, Dapprich S, Daniels AD, Farkas Ö, Foresman JB, Ortiz JV, Cioslowski J, Fox DJ (2009) Gaussian 09, (Revision C.01). Gaussian, Inc., Wallingford CT

Ab Initio Investigations of Stable Geometries of the Atmospheric Negative Ion NO_3^- $(HNO_3)_2$ and Its Monohydrate

Atsuko Ueda, Yukiumi Kita, Kanako Sekimoto
and Masanori Tachikawa

Abstract The possible stable geometries of the atmospheric negative core ion NO_3^- $(HNO_3)_2$ and its monohydrate were theoretically investigated with the second order Møller-Plesset perturbation theory (MP2) in consideration of the effect of electron correlation. For both ionic clusters, we obtained the different stable geometries from the previous study by Drenck and coworkers (Int J Mass Spectrom 273:126–131, 2008) [1] with the density functional theory of Becke 3-parameters hybrid functional (B3LYP). The non-planar geometry with two hydrogen-bondings between one oxygen atom on NO_3^- and each hydrogen atom of two HNO_3 fragments is found as the most stable structure of the core ion at 0 K. For the monohydrate, the most stable geometry at 0 K is found as the *H₂O-embedded* form in which one water molecule is located at the center of the cluster with hydrogen-bondings to NO_3^- and HNO_3 fragments. Our results show that the hydrogen bond network of the core ion can be strongly perturbed by a single water molecule. We also discussed the relative abundance of conformers of these ionic clusters under a finite temperature.

1 Introduction

Atmospheric ions are led by multistage reactions, e.g. a charge-transfer reaction, an ion induced nucleation, etc. [2]. Atmospheric ions affect an atmospheric environment by generating atmospheric aerosols [3, 4]. Atmospheric ion clusters consisting of organic compounds, water molecule, etc. have a crucial role in physical and chemical processes in the atmosphere. Negative ion clusters in the atmosphere lead to acid rain which has been known to affect the environment, e.g. forest trees, plants, animals, and buildings [5]. Especially H_2SO_4 and HNO_3 have the dominant contribution to acid rain, because strong acids release H^+ when water vapor exists

A. Ueda · Y. Kita (✉) · K. Sekimoto · M. Tachikawa
Quantum Chemistry Division, Yokohama City University,
Seto 22-2, Kanazawa-ku, Yokohama 236-0027, Japan
e-mail: ykita@yokohama-cu.ac.jp

© Springer International Publishing AG, part of Springer Nature 2018
Y. A. Wang et al. (eds.), *Concepts, Methods and Applications of Quantum Systems in Chemistry and Physics*, Progress in Theoretical Chemistry and Physics 31,
https://doi.org/10.1007/978-3-319-74582-4_10

193

[6]. Experimental results observed in situ mass spectrometry show the existence of atmospheric negative ion $NO_3^-(HNO_3)_2$ in the upper troposphere (altitude between 9 and 12 km) [7, 8]. This negative ion is also known as a dominant negative ion at the upper troposphere region [7].

Recently, Sekimoto and Takayama established the atmospheric pressure corona discharge ionization (APCDI) technique, which enables us to reproducibly generate negative ions [9]. They reported the existence of stable negative ion water clusters, $NO_3^-(HNO_3)_2(H_2O)_n$, and the specific stability referred to as a *magic number* for $n = 8$ [10]. Geometric structures of these water clusters are, however, still unclear even for the core ion and its monohydrate ($n = 1$).

From theoretical points of view, Drenck and coworkers reported stable structures of $NO_3^-(HNO_3)_2(H_2O)_n$ ($n = 0$–4) obtained at B3LYP/6-31++G** level of density functional theory (DFT) calculations [1]. They also reported the stable structures of $NO_3^-(HNO_3)_m(H_2O)_n$ up to $n + m = 6$, and assessed the validity of theoretical calculations by comparing theoretical dissociation energies of the $NO_3^-(HNO_3)_m(H_2O)_n$ cluster into NO_3^-, mHNO_3, and nH_2O fragments to the experimental results with mass-analyzed ion kinetic energy (MIKE) spectra measurement [11]. Their theoretical results agree with the experiments reasonably, but assumed only one kind of geometry for each $NO_3^-(HNO_3)_m(H_2O)_n$ clusters despite that a water cluster should generally have various kinds of conformers. In order to elucidate the stable geometries of $NO_3^-(HNO_3)_2$ and its hydrates, a more comprehensive geometry searching with first-principles calculations must be indispensable.

In this study, to elucidate stable geometries of these ionic clusters in details, we theoretically analyzed stable geometries of the negative core ion $NO_3^-(HNO_3)_2$ and its monohydrate in consideration of a lot of possible conformers with the post Hartree-Fock ab initio method. We also discussed the relative abundance of conformers of these ionic clusters under a finite temperature.

2   Computational Details

We employed the second order Møller-Plesset perturbation theory (MP2) with 6-31 ++G** Gaussian type basis sets in ab initio calculations of the negative core ion, and its monohydrate. The basis set superposition error (BSSE) is not corrected because of the less convergence in BSSE corrected geometrical optimization procedure. The harmonic approximation was used to evaluate the zero-point vibration energy (ZPE) and Gibbs free energy. Natural Population Analysis (NPA) [12] was used to analyze electronic populations on each atom. All calculations were performed with *GAUSSIAN 09* program package [13].

In the comformational searching, we picked up the initial geometries to be a molecular cluster consisting of one NO_3^-, two HNO_3's, and one H_2O fragments, and optimized all the geometric degrees of freedom of the cluster simultaneously.

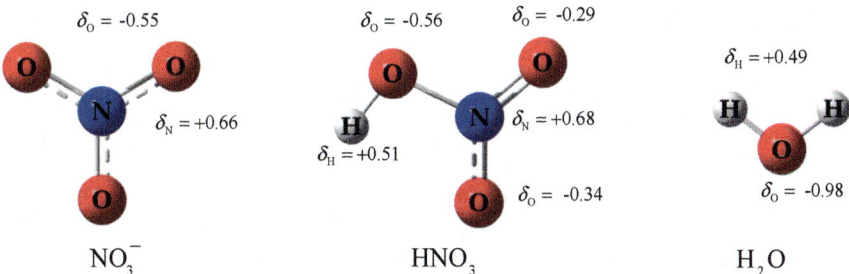

Fig. 1 Geometries of molecular fragments, NO_3^-, HNO_3, and H_2O obtained with MP2/6-31+ +G** calculations. The δ_X means the NPA charge on the element X

Table 1 Intermolecular interaction energies with zero-point vibration correction (E_{Int}, unit in kcal/mol) between two fragments obtained with MP2/6-31++G** calculation. The E_{Int} value for the complex $X \cdots Y$ is calculated as $E_{Int} = E_{ZPE}(X) + E_{ZPE}(Y) - E_{ZPE}(X \cdots Y)$, where $E_{ZPE}(X)$ is the sum of electronic total energy and the zero-point energy (ZPE) of the system X

Complex	E_{Int}	Exptl. [11]
$NO_3^- \cdots H_2O$	14.9	14.6 ± 0.2
$NO_3^- \cdots HNO_3$	30.6	
$HNO_3 \cdots HNO_3$	7.9	
$HNO_3 \cdots H_2O$	9.2	

The stable structures of these molecular fragments are shown in Fig. 1. The energetic stabilities of $NO_3^- (HNO_3)_2$ and its monohydrate can be mainly determined by the $O \cdots H - O$ type intermolecular hydrogen-bonding (HB) between the fragments, where the ionic HB between NO_3^- and HNO_3 fragments is the most strong interaction among possible two fragments as shown in Table 1. We note here that MP2/ 6-31 ++G** level of ab initio calculations reasonably reproduce the corresponding experimental intermolecular interaction energy between NO_3^- and H_2O with MIKE spectra [11]. In the geometry optimizations of $NO_3^- (HNO_3)_2$, initial geometries were chosen to have HBs between the fragments as much as possible, where the total number of initial geometries is 45 including conformers having a different angle between molecular planes of NO_3^- and HNO_3. In the case of the monohydrate, all the possible 218 hydrogen-bonded structures between NO_3^-, HNO_3, and H_2O fragments were chosen as initial geometries.

3   Results and Discussion

3.1   Negative Core Ion NO_3^- $(HNO_3)_2$

Fifteen different geometries were obtained in the geometry searching of NO_3^- $(HNO_3)_2$. We classified them into two groups, α and β, according to the number of oxygen atom(s) on NO_3^- making HBs with two HNO_3's. In the group α containing three kinds of conformers having a different angle between molecular planes of NO_3^- and HNO_3, one oxygen atom on NO_3^- has two HBs with two HNO_3's. In the group β containing twelve kinds of conformers, two different oxygen atoms on NO_3^- have one HBs with different HNO_3's. The stable geometries of all conformers and their relative energies at 0 K (ΔE_{ZPE}) are shown in Fig. 2, where the most stable structure at 0 K is the structure $\alpha 1$. Drenck and coworkers reported the stable structure similar to the structure $\beta 9$ at B3LYP/6-31++G** level of DFT calculations [1]. At MP2/6-31++G** level of ab initio calculations, however, the total energy of the structure $\beta 9$ is 1.6 kcal/mol higher than that of the most stable structure $\alpha 1$.

The present ab initio calculations suggested that the negative core ion has the structure of $\alpha 1$ as the most stable structure at 0 K. It is, however, strongly expected that the relative abundance of these conformers depends on a thermal condition, because most of the relative energies at 0 K are within a few kcal/mol. The temperature dependence of relative abundance of conformers is shown in Fig. 3, where the relative abundance of ith conformer (p_i) at the temperature T is estimated in the standard way as

$$p_i(T) = \exp\{-\Delta G_i/k_B T\}/Z, \quad Z = \sum_j^{all} \exp\{-\Delta G_j/k_B T\}, \qquad (1)$$

where ΔG_i is the relative Gibbs energy including the enthalpic (H: the sum of the electronic energy, ZPE, thermal corrections, etc.) and the entropic terms ($-TS$). In Fig. 3, the temperature dependences of p_i values for all conformers are shown in the upper figure (a), and those of the sum of p_i values over conformers in each group are given in the lower figure (b). Figure 3 clearly indicates that the relative abundance of NO_3^- $(HNO_3)_2$ conformers depends on the temperature, and the inversion of the total abundance of the groups α and β is found at around 250 K. We note here that the most stable structure $\alpha 1$ at 0 K is a dominant conformer in the low temperature region below 90 K, while the structure of $\beta 9$ has the largest relative abundance in high temperature region above 350 K due to the large entropic contribution from the lowest twisting mode of two HNO_3 fragments (the harmonic vibrational frequency $\omega_e = 10$ cm^{-1}). Such temperature dependence of the relative abundances as well as the inversion of the abundance is arising from the entropic

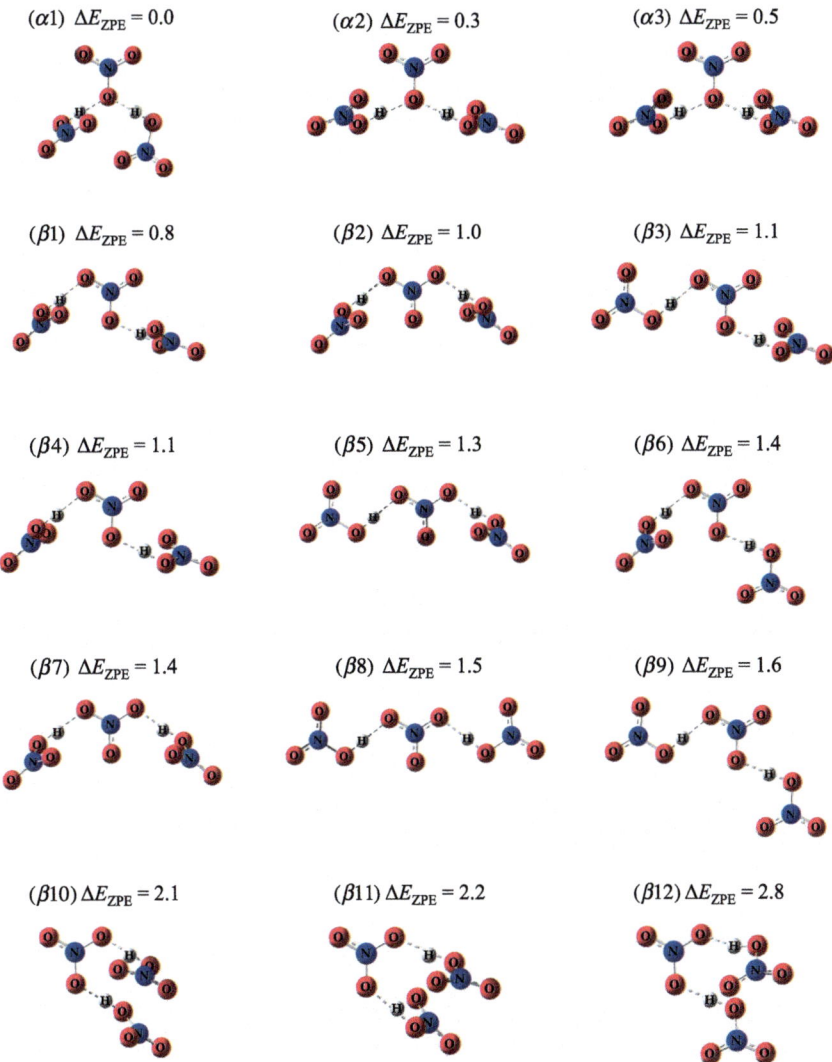

(α1) $\Delta E_{ZPE} = 0.0$ (α2) $\Delta E_{ZPE} = 0.3$ (α3) $\Delta E_{ZPE} = 0.5$

(β1) $\Delta E_{ZPE} = 0.8$ (β2) $\Delta E_{ZPE} = 1.0$ (β3) $\Delta E_{ZPE} = 1.1$

(β4) $\Delta E_{ZPE} = 1.1$ (β5) $\Delta E_{ZPE} = 1.3$ (β6) $\Delta E_{ZPE} = 1.4$

(β7) $\Delta E_{ZPE} = 1.4$ (β8) $\Delta E_{ZPE} = 1.5$ (β9) $\Delta E_{ZPE} = 1.6$

(β10) $\Delta E_{ZPE} = 2.1$ (β11) $\Delta E_{ZPE} = 2.2$ (β12) $\Delta E_{ZPE} = 2.8$

Fig. 2 Stable geometries of NO_3^- $(HNO_3)_2$ obtained with MP2/6-31++G** calculations. The definitions of each group are given in the text. The ΔE_{ZPE} (unit in kcal/mol) means the relative energy with zero-point vibration correction from the most stable structure α1

contributions and the difference in the number of conformers in each group (*numerical* contribution). Since the conformer α2 has the largest p_l value in the temperature range from 90 to 350 K as shown in Fig. 3a, at around the inversion temperature, the *numerical* advantage of the group β conformers plays a dominant role in the inversion of the relative abundance.

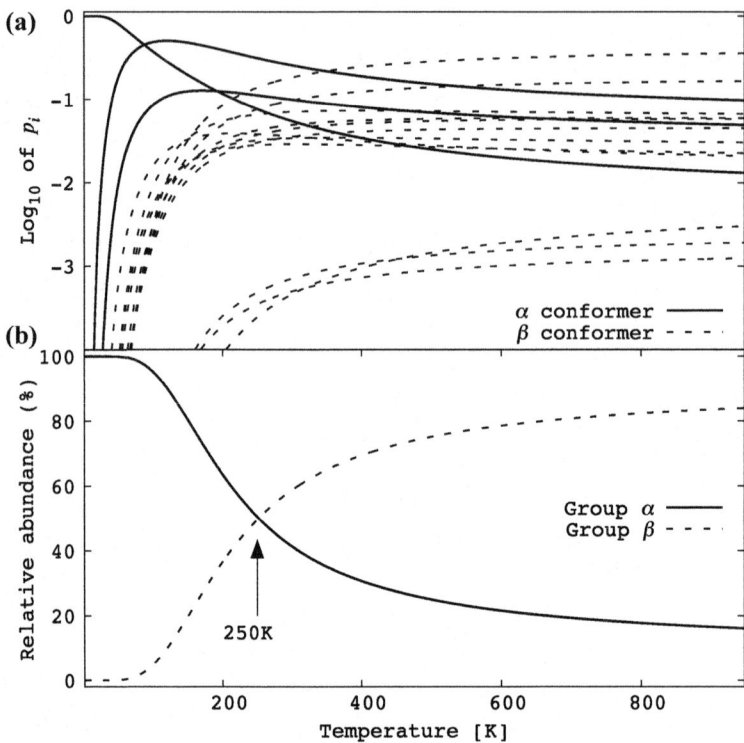

Fig. 3 The temperature dependence of relative abundances of $NO_3^-(HNO_3)_2$ conformers: **a** p_i values defined in Eq. (1) of all conformers (in the common logarithmic scale), **b** the sum of p_i values in each group (in percent figures). The numbers of conformers in the group α and β are 3 and 12, respectively

3.2 Monohydrated Core Ion $NO_3^-(HNO_3)_2H_2O$

Ninety different structures were obtained in the conformational searching for $NO_3^-(HNO_3)_2H_2O$. We classified these conformers into four groups, A–D, according to the structure of $NO_3^-(HNO_3)_2$ and/or the hydrogen-bonded structure between H_2O and other fragments. The group A (B) contains 23 (56) kinds of conformers which have a hydrogen-bonded structure between H_2O and the core ions of the group α (β). In these two groups, H_2O is located outside of the cluster (*H_2O-attached* form). In the group C containing ten conformers, the H_2O molecule is located at the center of a cluster (*H_2O-embedded* form), and has two or three HBs with NO_3^- and HNO_3 fragments. The group D contains only one conformer in which the proton transfer is occurred from H_2O to NO_3^- as $OH^-(HNO_3)_3$ having C_3 symmetry. The lowest energy geometries in each group and those relative energies at 0 K are shown in Fig. 4. As seen in Fig. 4, the most stable structure of $NO_3^-(HNO_3)_2H_2O$ at 0 K is the H_2O-embedded form of the group C. Drenck and

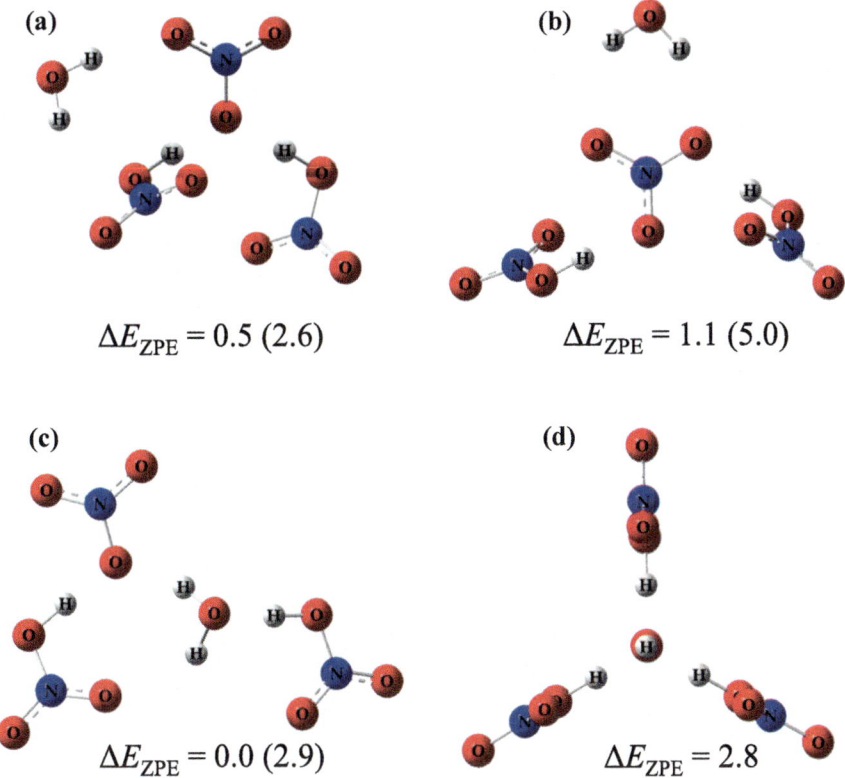

(a) $\Delta E_{\mathrm{ZPE}} = 0.5\ (2.6)$

(b) $\Delta E_{\mathrm{ZPE}} = 1.1\ (5.0)$

(c) $\Delta E_{\mathrm{ZPE}} = 0.0\ (2.9)$

(d) $\Delta E_{\mathrm{ZPE}} = 2.8$

Fig. 4 The most stable structures of NO_3^- $(HNO_3)_2 H_2O$ in the group A–D obtained with MP2/6-31 ++G** calculations. The definitions of each group are given in the text. The ΔE_{ZPE} (unit in kcal/mol) means the relative energy with zero-point vibration correction from the most stable structure. The value in the parenthesis is the maximum ΔE_{ZPE} value in each group

coworkers reported the structure classified into the group B at B3LYP/6-31++G** level of DFT calculations [1]. We address here that their structure has a high relative energy, $\Delta E_{\mathrm{ZPE}} = 2.1$ kcal/mol, at MP2 level of ab initio calculations.

Our results clearly show that it is not necessary for the most stable geometry in the monohydrate to maintain the most stable one of the core ion itself. To our knowledge, the energetic stability of H_2O-embedded form has not been reported so far for NO_3^- $(HNO_3)_2 H_2O$. Such energetic stability of the group C conformers can be reasonably explained as resulting from the enhancement of polarization of the embedded water molecule. In the structures (C) in Fig. 4, for instance, the NPA charges on two hydrogen (δ_H) atoms of the embedded water molecule are +0.52 and +0.54, while those of the outer water molecule are +0.50 and +0.50 in the structures (A) and (B) in Fig. 4. Such increment of the net charges on two hydrogen atoms can strengthen the intermolecular HBs between H_2O and other fragments.

Since the relative energy of most conformers are within a few kcal/mol (ΔE_{ZPE} values of 76 conformers (84%) are within 3 kcal/mol), the relative abundances of conformers should depend on a thermal condition as in the case of the core ion. Figure 5 shows the theoretical temperature dependence of the relative abundances of conformers: Fig. 5a the temperature dependences of p_i values for all conformers, and Fig. 5b those of the sum of p_i values over conformers in each group. As shown in Fig. 5b, the H_2O-embedded conformers of the group C (red solid line) have the largest relative abundance among all groups in the low-temperature region below 330 K. For the H_2O-attached conformers, the conformers of the group A (blue dashed lines) have a small relative abundance over whole temperature range, but the group B conformer (green dotted line) has the largest relative abundance above 330 K. The relative abundance of $OH^-(HNO_3)_3$ conformer is less than 1% in this

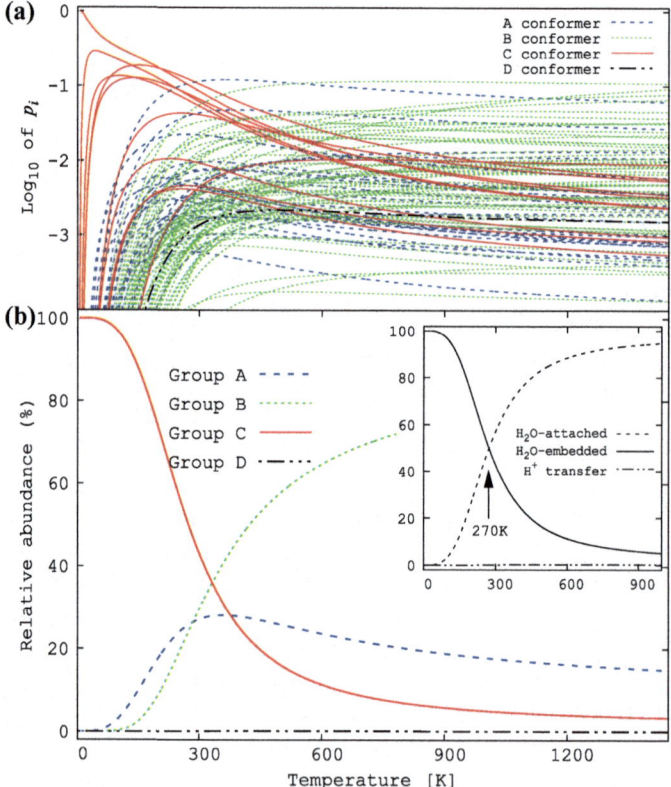

Fig. 5 The temperature dependence of relative abundances of $NO_3^-(HNO_3)_2H_2O$ conformers: **a** p_i values of all conformers (in the common logarithmic scale), **b** the sum of p_i values in each group (in percent figures). The inset figure in (**b**) is the relative abundance of H_2O-attached conformers (the sum of the group A and B), H_2O-embedded conformers (the group C), and the proton transfer structure (the group D). The numbers of conformers in the group A–D are 23, 56, 10 and 1, respectively

temperature range. As shown in the inset figure in Fig. 5b, the inversion of the relative abundance of the H_2O-embedded and attached conformers occurs at around 270 K. At around this inversion temperature, the numerical advantage of H_2O-attached conformers contributes to the inversion dominantly, because the relative abundances of H_2O-embedded conformers are larger than that of H_2O-attached conformers. The large relative abundances of H_2O-attached conformers in the high temperature region are due to the entropic contributions from slow molecular vibrations relevant to the attached H_2O molecule ranging from 20 to 30 cm^{-1}, e.g. H_2O rocking mode.

4 Conclusion

We theoretically analyzed the stable geometries of the atmospheric negative core ion, $NO_3^-(HNO_3)_2$, and its monohydrate, $NO_3^-(HNO_3)_2H_2O$, at MP2/6-31++G** level of ab initio calculations, and discussed the relative energetic stability of conformers for both ionic clusters. We found a total of 15 and 90 kinds of different conformers for $NO_3^-(HNO_3)_2$ and its monohydrate, respectively.

Unlike the previous DFT study by Drenck and coworkers [1], in the most stable geometry at 0 K, the core ion has the hydrogen-bonded structure in which one oxygen atom on NO_3^- has two hydrogen bonds with each HNO_3. For the monohydrate, the most stable geometry is the hydrogen-bonded structure in which the H_2O molecule is located at the center of a cluster (*H_2O-embedded* form) rather than the structure in which H_2O is located outside of the cluster (*H_2O-attached* form). This means that the hydrogen-bonding network of the core ion can be strongly perturbed by a single water molecule. Analyzing the theoretical temperature dependence of relative abundances of conformers, we also confirmed that the conformers having the H_2O-attached form become dominant at high temperature region above 270 K.

Acknowledgements The present study was supported by Grant-in-Aid for Scientific Research and for Priority Areas by Ministry of Education, Culture, Sports, Science and Technology, Japan (KAKENHI). A part of the present computations were performed using Research Center for Computational Science, Okazaki, Japan.

References

1. Drenck K, Hvelplund P, Nielsen SB, Panja S, Støchkel K (2008) Int J Mass Spectrom 273:126–131
2. Yu F, Turco RP (2000) Geophys Res Lett 27:883–886
3. Fend J, Möller D (2004) J Atmos Chem 48:217–233
4. Harrison RG, Carslaw KS (2003) Rev Geophys 41:1012–1027
5. Singh A, Agrawal M (2008) J Environ Biol 29:15–24

6. Jacob DJ (1999) Introduction to atmospheric chemistry, Princeton University
7. Heitmann H, Arnold F (1983) Nature 306:747–751
8. Arnold F (1980) Nature 284:610–611
9. Sekimoto K, Takayama M (2007) Int J Mass Spectrom 261:38–44
10. Sekimoto K, Takayama M (2011) J Mass Spectrom 46:50–60
11. Lee N, Keesee RG, Castleman AW Jr (1980) J Chem Phys 72:1089–1094
12. Reed AE, Weinstock RB, Weinhold F (1985) J Chem Phys 83:735–746
13. Frisch MJ, et al (2010) Gaussian 09, revision C.01, Gaussian, Inc., Wallingford CT

A Theoretical Study of Covalent Bonding Formation Between Helium and Hydrogen

Taku Onishi

Abstract In order to investigate chemical bonding between helium and hydrogen in the He–H model, coupled-cluster calculations were performed. In this study, three different hydrogen formal charges (positive, neutral and negative) were considered. In the case of positive hydrogen, it has been concluded that covalent bonding is formed between helium 1s orbital and hydrogen 1s orbital. Zero-point vibration energy was smaller than dissociation energy. It has been concluded that positive hydrogen is kept fixed at optimized structure.

Keywords Helium · Hydrogen · Molecular orbital · Covalent bonding Chemical bonding rule

1 Introduction

It has been recognized that helium cannot form covalent bonding with other atoms and molecules. It is because helium has the stable closed shell configuration. Recently, Helgaker et al. demonstrated that helium clusters are not dispersed under the strong magnetic field [1]. Several types of helium clusters such as He_3, He_4 and He_6 were optimized under the circumstance [2]. On the other hand, without magnetic field, many quantum chemical calculations were performed, from the interest of van der Waals interaction [3–5] and potential energy [6–8]. However, molecular orbital analysis was not performed for helium clusters and helium-including clusters. In our previous study, the chemical bonding character of helium dimer

T. Onishi (✉)
Center of Ultimate Technology on Nano-electronics, Mie University, Mie, Japan
e-mail: taku@chem.mie-u.ac.jp; taku.onishi@kjemi.uio.no

T. Onishi
Centre for Theoretical and Computational Chemistry (CTCC), Department of Chemistry, University of Oslo, Oslo, Norway

T. Onishi
Department of Applied Physics, Osaka University, Osaka, Japan

© Springer International Publishing AG, part of Springer Nature 2018   203
Y. A. Wang et al. (eds.), *Concepts, Methods and Applications of Quantum Systems in Chemistry and Physics*, Progress in Theoretical Chemistry and Physics 31, https://doi.org/10.1007/978-3-319-74582-4_11

(He–He) [9, 10] was investigated by coupled cluster calculations. At a local minimum, bonding and anti-bonding molecular orbitals (MOs) were obtained. By the use of our chemical bonding rule [11, 12], it was concluded that covalent bonding is formed at optimized structure, due to orbital overlap between helium 1s orbitals.

Bond order is approximate equation to investigate a stability of chemical bonding in two-atom system:

$$N = \frac{N_a - N_b}{2}$$

where N_a and N_b denote the number of electrons in bonding and its anti-bonding MOs. In the case of helium dimer, bonding and anti-bonding MOs are formed between helium 1s orbitals. Since two electrons occupy both MOs ($N_a = N_b = 2$), N becomes zero. It was understood that covalent bonding is formed, even if bond order is zero [10].

In this study, we have investigated chemical bonding character between helium and hydrogen. The coupled cluster calculations have been performed for He–H model. Here, three different hydrogen formal charges have been assumed: (1) positive (H$^+$); (2) neutral (H); (3) negative (H$^-$).

2   Computations

2.1   Calculation Method and Model

The calculations presented here were performed using the coupled cluster singles and doubles (CCSD) method. It is because CCSD method accurately reproduces accurate interatomic distance, electronic state and molecular spectra for simple molecule [13, 14]. We used aug-cc-pVTZ basis set for helium and hydrogen [15]. All calculations were performed with the Gaussian program [16]. To investigate chemical bonding character between helium and hydrogen, the simple He–H model was constructed. Three types of He–H$^+$, He–H and He–H$^-$ models were constructed. Potential energy curves were obtained, changing the interatomic distance. In addition, zero point vibrational energy has been also obtained at the optimized structure.

2.2   Chemical Bonding Rule

In order to judge chemical bonding character such as covalency and ionicity in the calculated MOs, chemical bonding rule is very useful and applicable (see Fig. 1).
How to utilize chemical bonding rule

1. In MOs including outer shell electrons, check whether the orbital overlap between helium and hydrogen exists or not.

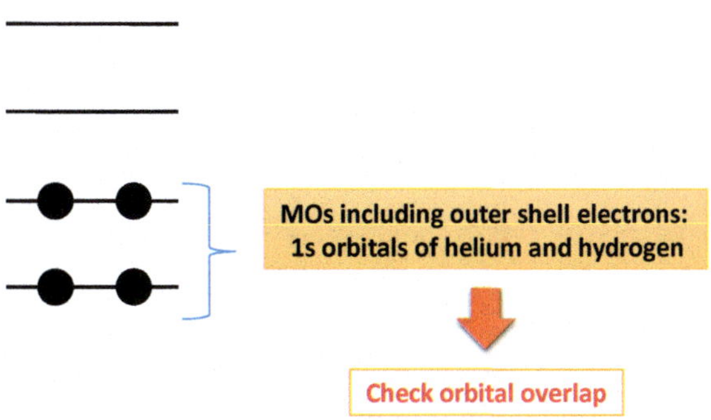

Fig. 1 Schematic drawing of chemical bonding rule

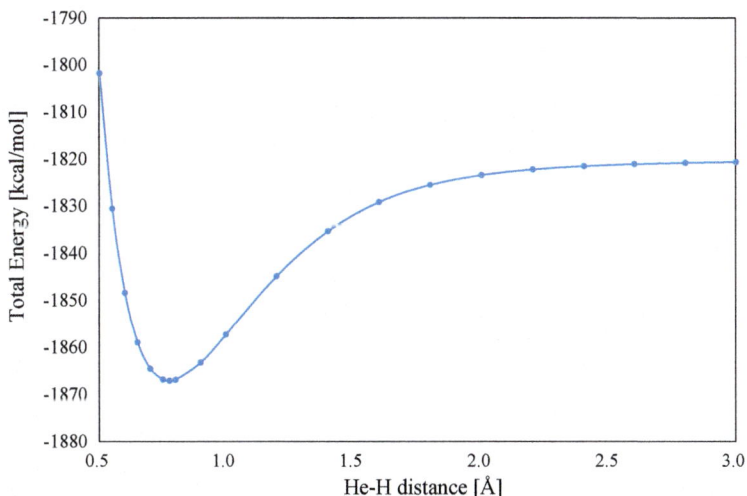

MOs including outer shell electrons:
1s orbitals of helium and hydrogen

Check orbital overlap

2. With orbital overlap, bonding character is covalent. Without orbital overlap, bonding character is ionic.

3 Results and Discussion

3.1 He–H⁺ Model

Figure 2 shows potential energy curve of He–H⁺ model, changing the interatomic He–H distance. A local minimum is given at 0.776 Å. Two electrons occupy MO1, which is only one occupied MO. The wave-function of MO1 at a local minimum is

Fig. 2 Potential energy curve of He–H⁺ model, changing the interatomic He–H distance

Fig. 3 Schematic drawing of electrons and orbitals in He–H$^+$ model

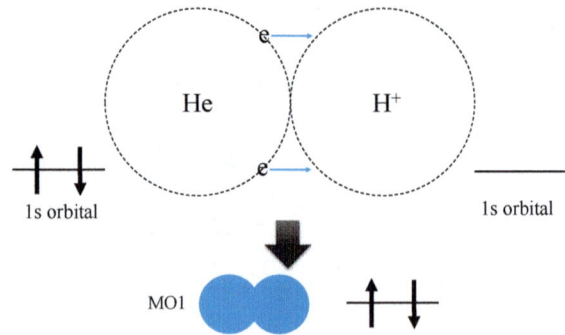

$$\psi_{MO1}(He - H^+) = 0.13\,\phi_{H(1s')} + 0.12\,\phi_{H(1s'')} + 0.35\,\phi_{He(1s')} + 0.45\,\phi_{He(1s'')} + 0.16\,\phi_{He(1s''')}$$

There is orbital overlap between helium 1s orbital and hydrogen 1s orbital. From chemical bonding rule, it is concluded that helium forms covalent bonding with hydrogen. Mulliken charge densities of helium and hydrogen are 0.315 and 0.685, respectively.

Figure 3 depicts the schematic drawing of electrons and orbitals in He–H$^+$ model. Helium 1s electrons are shared by both helium and hydrogen, through covalent bonding formation between helium and hydrogen. Hence, the interatomic He–H$^+$ distance becomes smaller.

3.2 He–H Model

Figure 4 shows potential energy curve of He–H model, changing the interatomic He–H distance. A local minimum is given at 3.577 Å. It is much larger than the interatomic He–H$^+$ distance (0.776 Å). Alpha and beta electrons are occupied in different MO1α and MO1β, respectively. The wave-function of MO1α is

$$\psi_{MO1\alpha}(He - H) = 0.35\,\phi_{He(1s')} + 0.48\,\phi_{He(1s'')} + 0.30\,\phi_{He(1s''')}$$

On the other hand, the wave-function of MO1β is

$$\psi_{MO1\beta}(He - H) = 0.35\,\phi_{He(1s')} + 0.48\,\phi_{He(1s'')} + 0.30\,\phi_{He(1s''')}$$

It is found that MO1α and MO1β are paired. Orbital energies of MO1α and MO1β are slightly different: −0.91792 au for MO1α; −0.91787 au for MO1β. It is due to broken spin symmetry. MO2α has no paired beta MO. The wave-function of MO2α is

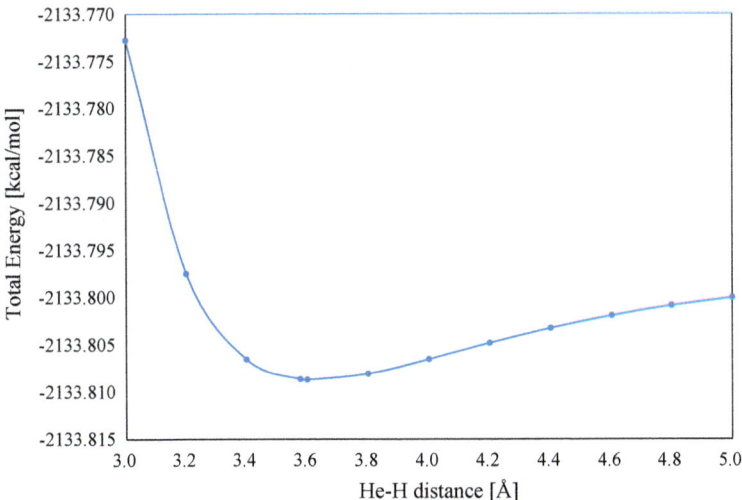

Fig. 4 Potential energy curve of He–H model, changing the interatomic He–H distance

$$\psi_{MO2\alpha}(He - H) = 0.24\,\phi_{H(1s')} + 0.51\,\phi_{H(1s'')} + 0.38\,\phi_{H(1s''')}$$

There is no orbital overlap between helium and neutral hydrogen. From chemical bonding rule, no covalent bonding is formed in He–H model. Mulliken spin densities of helium and hydrogen are 0.00 and 1.00, respectively. MO2α is responsible for the spin density.

Figure 5 depicts the schematic drawing of electrons and orbitals in He–H model. Electron repulsion between two helium 1s electrons and hydrogen 1s electron is dominative, though Coulomb interaction exists between helium 1s electrons and positive hydrogen atomic nucleus. It is considered that the repulsion lets the interatomic He–H distance elongated.

3.3 He–H⁻ Model

Figure 6 shows potential energy curve of He–H⁻ model, changing the interatomic He–H distance. A local minimum is given at 6.452 Å. It is much larger than He–H⁺ and He–H models. It is found that He–H⁻ is weakly bounded. Four electrons occupy MO1 and MO2. The wave-function of MO1 is

$$\psi_{MO1}(He - H^-) = 0.35\,\phi_{He(1s')} + 0.48\,\phi_{He(1s'')} + 0.30\,\phi_{He(1s''')}$$

On the other hand, the wave-function of MO2 is

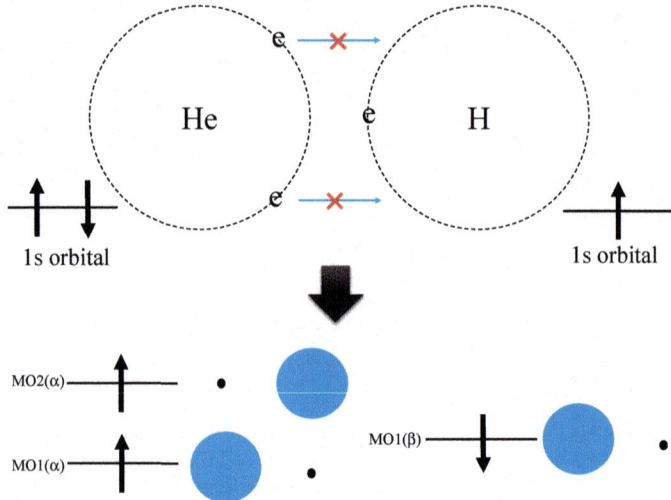

Fig. 5 Schematic drawing of electrons and orbitals in He–H model

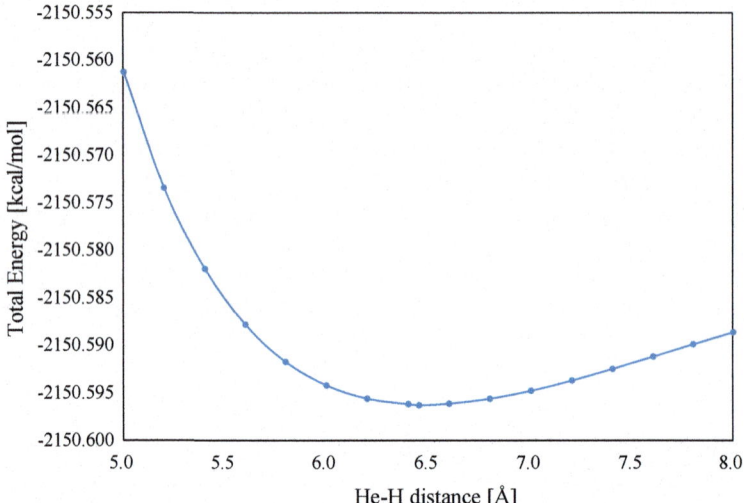

Fig. 6 Potential energy curve of He–H⁻ model, changing the interatomic He–H distance

$$\psi_{MO2}(He - H^-) = 0.16\,\phi_{H(1s')} + 0.27\,\phi_{H(1s'')} + 0.41\,\phi_{H(1s''')} + 0.37\,\phi_{H(2s')}$$

There is no orbital overlap between helium and hydrogen in He–H⁻ model. From chemical bonding rule, it is found that no covalent boning is formed. MO1 is for helium 1s orbital. In MO2, two electrons are delocalized over not only hydrogen

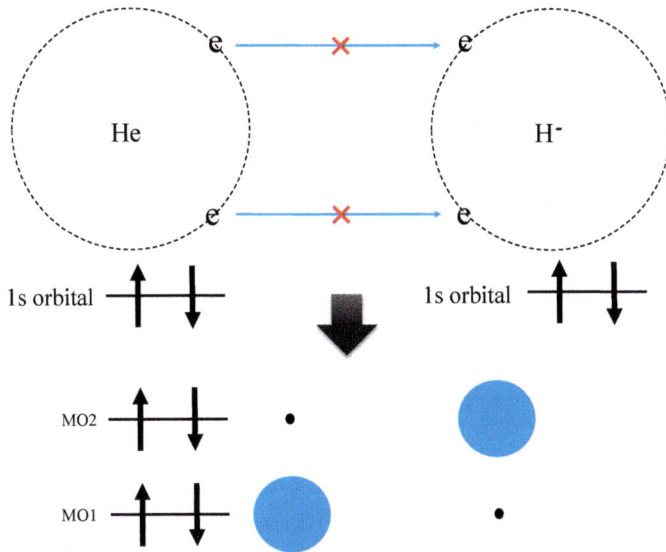

Fig. 7 Schematic drawing of electrons and orbitals in He–H⁻ model

1s orbital but also hydrogen 2s orbital. Mulliken charge densities of helium and hydrogen are 0.00 and −1.00, respectively.

Figure 7 depicts the schematic drawing of electrons and orbitals in He–H⁻ model. There is electron repulsion between two helium 1s electrons and two hydrogen 1s electrons, though Coulomb interaction exists between helium 1s electrons and positive hydrogen atomic nucleus. It is considered that this effect is stronger than He–H⁻ model, due to existence of two hydrogen 1s electrons.

3.4 Zero-Point Vibration Energy in the He–H⁺, He–H and He–H⁻ Models

The dissociation energies for He–H⁺, He–H and He–H⁻ models can be estimated from the total energy difference between local minimum and completely dissociated point. They were 46.9 kcal/mol, 0.0117 kcal/mol and 0.0191 kcal/mol for He–H⁺, He–H and He–H⁻ models, respectively. Except He–H⁺ model, the values are very small. It is because the stable covalent bonding between helium and hydrogen is formed only in He–H⁺ model. In order to investigate the effect of quantum vibration on the dissociation, we obtain zero point vibration energy. When zero point vibration energy is smaller than dissociation energy, hydrogen is kept fixed at optimized position. On the other hand, when it is larger, the dissociation may be caused by the external factor.

Zero point vibration energies for He–H$^+$, He–H and He–H$^-$ moleds, were 4.58, 0.0467 and 0.0208 kcal/mol, respectively. In He–H$^+$ model, it is much smaller than dissociation energy. It shows that optimized structure is much stabilized, because hydrogen is kept fixed at optimized structure. It is concluded that helium can be strongly bounded with positive hydrogen. On the other hand, zero point vibration energies are larger than dissociation energies in He–H and He–H$^-$ models. In addition, the dissociation energies are very small and no covalent bonding is formed.

4 Conclusions

From chemical bonding rule, it was found that helium 1s orbital and hydrogen 1s orbital forms the stable covalent bonding in He–H$^+$ model. Zero point vibration energy was estimated to be 4.58 kcal/mol. It is much smaller than dissociation energy (46.9 kcal/mol). On the other hand, no covalent bonding is formed, and the small dissociation energy is given in He–H and He–H$^-$ models. Helium is weakly bound with hydrogen.

Acknowledgements This work was supported by the Research Council of Norway (RCN) through CoE Grant No. 179568/V30 (CTCC) and through NOTUR Grant No. NN4654 K for HPC resources. The author would like to thank Prof. Trygve Helgaker and Prof. Josef Paldus for valuable comments.

References

1. Lange KK, Tellgren EI, Hoffmann MR, Helgaker T (2012) Science 337:327
2. Tellgren EI, Reine SS, Helgaker T (2012) Phys Chem Chem Phys 14:9492
3. Tao J, Perdew JP (2005) J Chem Phys 122:114102
4. Zhao Y, Truhlar DG (2006) J Phys Chem A 110:5121
5. Kamiya M, Tsuneda T, Hirao K (2002) J Chem Phys 117:6010
6. Lotrich VF, Bartrett RJ, Grabowski I (2005) Chem Phys Lett 405:43
7. Snook I, Per MC, Russo SP (2008) J Chem Phys 129:164109
8. Allen MJ, Tozer DJ (2002) J Chem Phys 117:11113
9. Onishi T (2016) J Chin Chem Soc 63:83
10. Onishi T (2016) AIP Conf Proc 1790:02002
11. Onishi T (2012) Adv Quant Chem 64:31
12. Onishi T (2015) Adv Quant Chem 70:31
13. Helgaker T, Jorgensen P, Olsen J (2000) Molecular electronic-structure theory. Wiley, p 648
14. Bartlett R, Musial M (2007) Rev Mod Phys 79:291
15. Woon DE, Dunning TH Jr (1994) J Chem Phys 100:2975
16. Gaussian 09, Frisch MJ, Trucks GW, Schlegel HB, Scuseria GE, Robb MA, Cheeseman JR, Scalmani G, Barone V, Mennucci B, Petersson GA, Nakatsuji H, Caricato M, Li X, Hratchian HP, Izmaylov AF, Bloino J, Zheng G, Sonnenberg JL, Hada M, Ehara M, Toyota K, Fukuda R, Hasegawa J, Ishida M, Nakajima T, Honda Y, Kitao O, Nakai H, Vreven T, Montgomery JA Jr, Peralta JE, Ogliaro F, Bearpark M, Heyd JJ, Brothers E,

Kudin KN, Staroverov VN, Kobayashi R, Normand J, Raghavachari K, Rendell A, Burant JC, Iyengar SS, Tomasi J, Cossi M, Rega N, Millam JM, Klene M, Knox JE, Cross JB, Bakken V, Adamo C, Jaramillo J, Gomperts R, Stratmann RE, Yazyev O, Austin AJ, Cammi R, Pomelli C, Ochterski JW, Martin RL, Morokuma K, Zakrzewski VG, Voth GA, Salvador P, Dannenberg JJ, Dapprich S, Daniels AD, Farkas Ö, Foresman JB, Ortiz JV, Cioslowski J, Fox DJ (2009) Gaussian, Inc., Wallingford CT

Small Rhodium Clusters: A HF and DFT Study–III

M. A. Mora and M. A. Mora-Ramírez

Abstract Small neutral and ionic Rhodium clusters Rh_n (n = 6, 8, 13) are investigated by ab initio molecular orbital calculations with full optimization at the Restricted Open Shell Hartree-Fock (ROHF) level with a LANL2DZ basis set, and with the methods based on Density Functional Theory, B3LYP/MWB, B3LYP/PBE. The clusters are found favor close-packed icosahedron structures in contrast to previous theoretical predictions that rhodium clusters should favor cubic motifs. A range of spin multiplicities are investigated for each cluster and we present the minimum energy conformation along with the vertical and adiabatic ionization potentials.

Keywords Rhodium clusters · ROHF calculations · Transition metal Ionization potential

1 Introduction

It is well known that small-sized Rhodium clusters develop a magnetic moment [1, 2] while larger clusters and Rh-bulk are non-magnetic. Both basic and applied science researchers have been attracted to this behavior, because of the implications in applications such as magnetic recording [3]. In fact, while the structural characterization and hence magnetism of Rh clusters is an open problem as its potential for use as a high-density storage media, Rhodium also has applications in catalysis [4, 5]. In 2012, 81% of the 30 tons corresponding to the annual world production was used to produce three-way catalytic converters [6]. Rhodium catalysts are used in

M. A. Mora (✉)
Depto. de Química, Universidad Autónoma Metropolitana, campus Iztapalapa,
Av. Sn. Rafael Atlixco 186, 09340 Mexico, D. F., Mexico
e-mail: mam@xanum.uam.mx

M. A. Mora-Ramírez
Depto. Fisicomatemáticas, Facultad de C. Químicas, Benemérita Universidad
Autónoma de Puebla, Sn. Claudio y Sur 22 Col. Sn. Manuel, 72570 Puebla, Mexico
e-mail: marco.x.mora@gmail.com

© Springer International Publishing AG, part of Springer Nature 2018   213
Y. A. Wang et al. (eds.), *Concepts, Methods and Applications of Quantum Systems in Chemistry and Physics*, Progress in Theoretical Chemistry and Physics 31,
https://doi.org/10.1007/978-3-319-74582-4_12

several chemical processes such as manufacturing of certain silicon rubbers [7] and the reduction of benzene to cyclohexane [8]. Rhodium also finds use in the jewelry industry and as an agent for hardening and improving corrosion resistance [9].

The main problem to obtain a deep atomic-level understanding of the physical and chemical properties of clusters relies on an accurate determination of their equilibrium atomic structure, which is not as simple as it might appear. A direct identification of the equilibrium atomic structure by experimental techniques is very difficult and only indirect measurements can provide few clues about the atomic structure. Thus, the combinations of experimental techniques with first-principles calculations have been used. For example, vibrational spectroscopies combined with theoretical calculations have lead to important insights into the atomic structure of small Rh clusters [10, 11]. Isolated metal clusters have also been investigated by Stern-Gerlach molecular-beam deflection experiments [2, 12–17].

However, there are difficulties in the direct identification of the atomic structure of clusters by experimental techniques. Thus, most of the structural studies have been based on theoretical calculations, which can directly determine the atomic structure of clusters using several well-defined algorithms. Several calculations based on density functional theory (DFT) have focused on these clusters, for example, on metal particles containing 13 atoms, Rh_{13} [18–29]. Furthermore, it is important to mention that few studies have focused on the search for the lowest-energy structures with most studies assuming predefined structures. Sophisticated algorithms have been employed in the search for the lowest-energy structure, namely, generic algorithms (GA) [30], basin-hopping Monte Carlo (BHMC) [31–34], Monte Carlo (MC) [35], conformational space annealing [29], taboo search in descriptor space (TSDE) [23, 36], high-temperature molecular Dynamics (high-T-MD) [22]. Almost all the studies with these algorithms have been used in combination with empirical pair potentials. These potentials have difficulties in providing a correct description of the atomic structure [37–39], and hence, the ground state structures might not be correct.

In this paper, we present HF and DFT calculations on clusters with 6 and 8 rhodium atoms for comparison with theoretical and experimental results. We also present results for the 13-atom cluster since it is one of the most studied clusters and to the best of our knowledge there are no experimental results.

2 Method and Computational Details

It is well known that the method of calculation and the chosen basis set are the two most important factors in determining the accuracy of results. Ab initio methods must represent all the electrons in some manner. However, for heavy atoms it is desirable to reduce the amount of computation burden. One way to do this is by replacing the core electrons and their basis functions in the wave function by a potential term in the Hamiltonian. These are called core potentials, effective core potentials (ECP) or relativistic effective core potentials (RECP). In this work, we use the ECP LANL2DZ [40–42] potential, which is one of the most widely used for heavy elements.

Another factor to consider in selecting the method of calculation is the spin contamination. A high spin contamination can affect the geometry and population analysis and significantly affect the spin density. The error introduced by spin contamination is unacceptable when systems with transition metals are investigated. The restricted open-shell Hartree-Fock (ROHF) ab initio method is one of the best ways to avoid spin contamination and obtain a reliable wave function.

We performed ab initio molecular orbital calculations with full optimization at the ROHF level with a LANL2DZ basis set. Also in this research we employed the B3LYP hybrid functional with the small-core quasirelativistic approach of Wood and Boring [43] MWB ECP, and, the General Gradient Approximation (GGA) formulated by Perdew, Burcke, and Hernzerhof (PBE) [44, 45] since it has been shown that these hybrid functional can yield reliable energetics and structural results for other metal compounds. The geometries were adjusted until a stationary point on the potential energy surface was found, using the Berny algorithm [46, 47] for the minimization.

3   Results and Discussion

Figure 1 shows the Rh_6 optimized geometry calculated with the B3LYP/MWB chemical model obtained from different initial geometries; Fig. 1a shows the final geometry obtained from an initial octahedron, Fig. 1b is a final triangular prism conformer, Fig. 1c is a deformed pentagonal pyramid which converges to an distorted trigonal prism or capped tetrahedron as in Fig. 1d.

Table 1 shows the relative energy (eV) of the Rh_6 cluster obtained with the B3LYP/MWB chemical model; we searched the spin multiplicity up to 23. The lowest energy isomer is an octahedron with a multiplicity equal to 9, followed by the distorted pentagonal pyramid with a multiplicity of 11. Finally the trigonal prism isomer has $M = 9$. The minimum energy isomer is in complete agreement with the experimental results [10, 48], and with previous theoretical results [21, 24, 25, 48–56]. The relative energy of ROHF-isomers with different geometry is 1.039, 3.973 eV for pentagonal pyramid and octahedron respectively. For the B3LYP-isomers with different energy the relative energy is 0.218 for pentagonal pyramid, and 0.490 eV for trigonal prism. With the ROHF/LANL2DZ method, the triangular prism has a minimum energy conformation with multiplicity equal to 15. The more stable pentagonal pyramid is when $M = 17$, and the square bi-pyramid with $M = 11$. A geometry formed by two perpendicular squares (an incomplete cube) with $M = 15$ is the fundamental state for this particular geometry. Among these four geometries, the minimum energy is the triangular prism. All calculations in this series were performed with the ROHF/LANL2DZ methods and with multiplicities from 1 to 23 and full optimization.

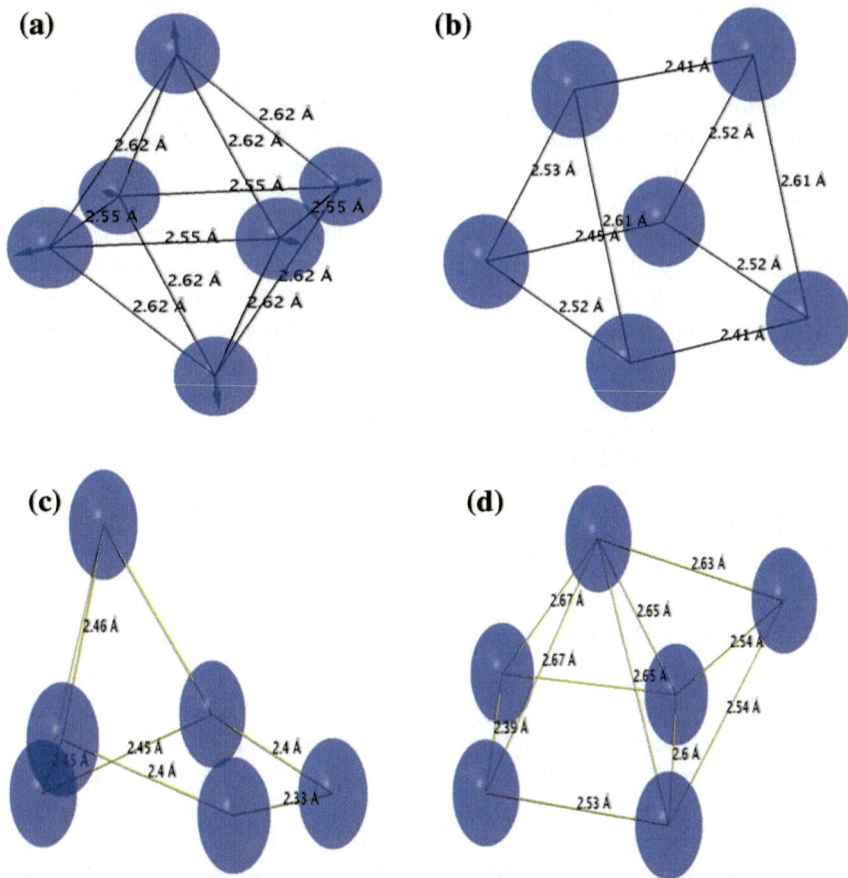

Fig. 1 Isomers of Rh_6 geometry optimized with the B3LYP/MWB model chemistry, **a** tetrahedron, **b** triangular prism, **c** deformed pentagonal pyramid, **d** capped tetrahedron

The relative energy (a.u.) calculated with the ROHF/LANL2DZ method for Rh_8 clusters are shown in Table 2. We searched the spin multiplicity up to 21. The relative energy of isomers with different geometry is 23.76, 1.023, and 1.249 eV for the cubic, bctp, and incomplete-ico respectively. In Fig. 2, we show the optimized structures of the Rh_8 clusters. Figure 2a is the lowest energy isomer, a bicapped octahedron, bcoh, with a multiplicity equal to 19. Figure 2b corresponds to the bicapped triangular prism isomer, bctp, with a multiplicity of 19. Figure 2c shows an isomer with the incomplete icosahedron structure. Finally, a cubic isomer with M = 13 is presented in Fig. 2d. The octahedron isomer with the minimum energy is in complete agreement with the experimental results reported by D. J. Harding et al. [10, 11], and by M. R. Beltran et al. [48], and with theoretical results obtained with different methods such as local spin density [24, 57], molecular dynamics [26], effective core potential [57], B3LYP [48, 49], which unlike of the reported here

Table 1 Rh_6 relative energy (eV) obtained from B3LYP/MWB calculations for isomers of Rh_6 in different states of spin multiplicity, in bold the octahedron isomer

Multiplicity	Triangular prism	Pentagonal pyramid	**Square bi-pyramid**
1	0.346	2.503	1.309
3	0.354	1.037	0.786
5	0.275	0.786	0.522
7	0.139	0.340	0.375
9	0	0.340	**0**
11	0.272	0	0.250
13	0.381	0.609	0.675
15	1.167	1.162	1.178
17	3.570	3.763	3.891
19	6.620	nc	7.132
21	nc	nc	nc
23	14.528	14.749	nc

nc not converged

Table 2 Rh_8 relative energy (eV) obtained from ROHF/ LanL2DZ calculations for isomers of Rh_8 in different states of spin multiplicity, in bold the bi-capped octahedron isomer

Multiplicity	Incomplete-ico	**bcoh**	bctp	cubic
1	17.82	21.52	15.29	3.92
3	16.19	16.14	16.57	0.65
5	11.40	24.81	15.16	1.93
7	12.00	13.09	10.45	nc
9	10.58	11.24	nc	2.64
11	7.48	9.03	6.94	1.01
13	4.82	5.93	5.06	0
15	1.09	7.05	nc	5.90
17	1.20	2.37	1.63	5.63
19	0	**0**	0	1.93
21	nc	2.93	1.63	11.51

nc not converged

were performed by keeping the symmetry, i.e. only optimized bond distances [19, 56].

For the cluster formed by 13 rhodium atoms, taking as the initial geometry that presented in Fig. 3a, we performed two series of calculations using density functional theory. The first one using the Becke-Lee-Yang-Parr [58, 59] functional and a double-Z basis set with effective core potential to represent the electrons close to the nucleus, and the second using the PBE functional [44, 45]. Both sets of calculations were carried out by varying the multiplicity from 2 to 26, performed with full optimization, without symmetry constraints, and following the Berny algorithm for minimization as it is implemented in the Gaussian 03 computer package.

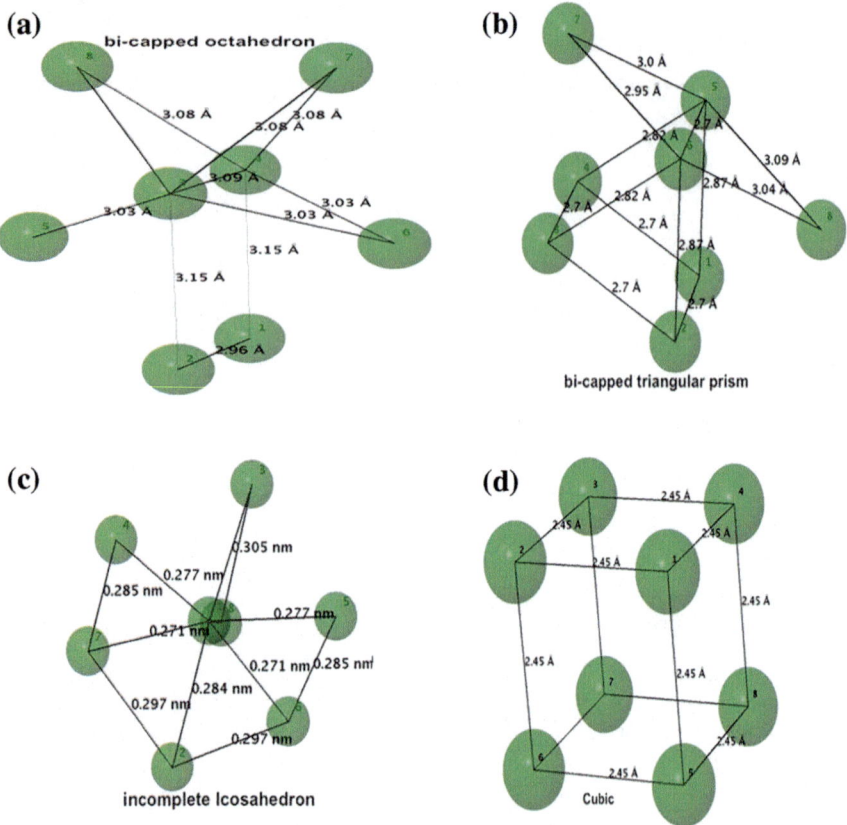

Fig. 2 Isomers of Rh_8 geometry in its minimum energy conformation obtained with the ROHF/ LANL2DZ. **a** bi-capped octahedron, **b** bi-capped triangular prism, **c** incomplete icosahedron, **d** cube

The B3LYP/LANL2DZ optimized geometry for Rh_{13} with M = 2 is shown in Fig. 3b. We notice that the initial geometry is not retained. The positions of the atoms 9 and 10 are considerably modified. The plane containing the atoms 3, 4, 7, 8, 11, 12 is slightly modified in the optimization process. The calculations made for the other multiplicities converge to a geometry similar to that found for M = 2, which is basically maintained until the cluster with M = 26. The conformation of minimum energy for this series corresponds to a multiplicity equal to 22, and is presented in Fig. 3c, where we can see that the plane of the six atoms has been distorted. Table 3 shows the energy calculated for these clusters as well as the eigenvalues of the S^2 operator. Note that the spin contamination for these calculations is large. As expected, during the optimization process the initial geometry of the cluster is not equal to the final geometry, i.e., during the optimization process the geometry evolves. Geometries known as non-icosahedral or low symmetry have

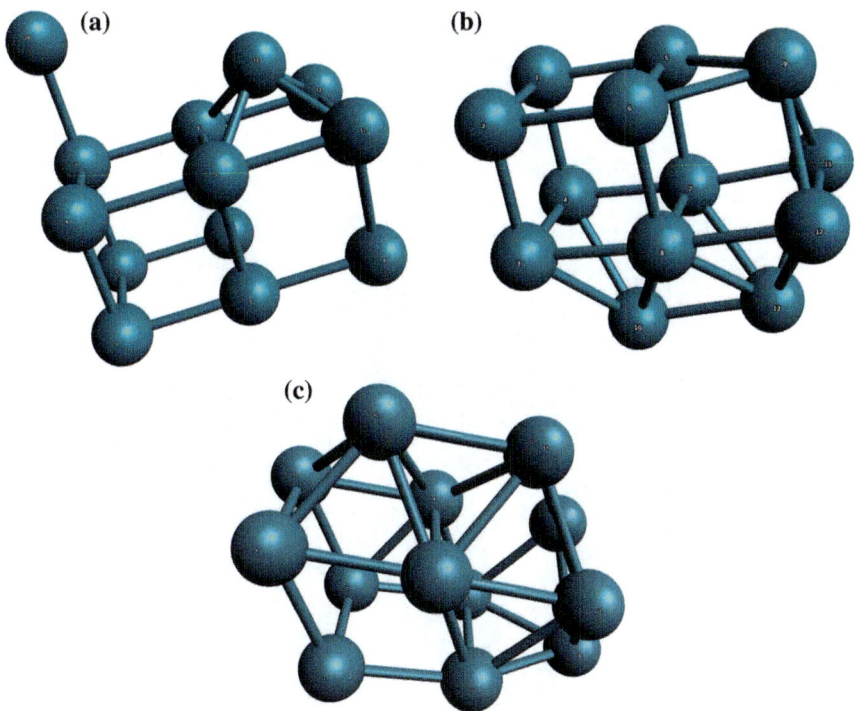

Fig. 3 Rh$_{13}$ with the functional B3LYP **a** configuration initial **b** optimized with M = 2 **c** the minimum energy configuration for M = 22

Table 3 Rh$_{13}$ relative energy obtained from B3LYP/MWB calculations for in different states of spin multiplicity and the eigenvalue of $<S^2>$ operator

Multiplicity	Relative energy (eV)	$<S^2>$
4	0.095	7.179
6	0.248	9.75
8	0.093	17.53
10	0.003	25.6556
12	0.095	36.364
14	0.035	49.373
16	0.0	64.166
18	0.218	81.228
20	0.299	100.073
22	0.250	121.073
24	1.045	144.030
26	2.041	168.985
28	2.985	195.932

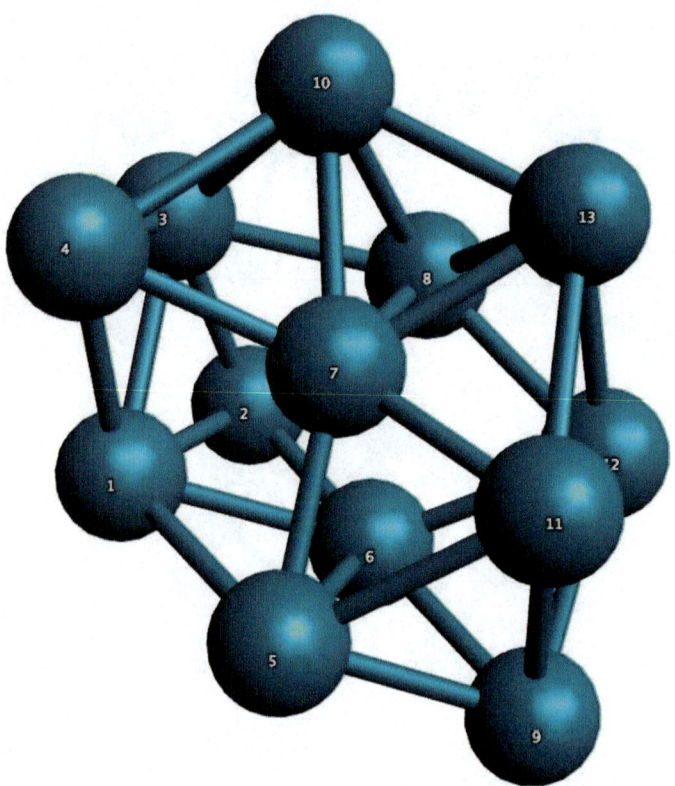

Fig. 4 Final geometry obtained with the Perdew-Burke-Ernzerhof exchange-correlation functional for Rh_{13}

been reported as optimized [27, 55, 57], and are markedly different from those obtained by us using the same method of calculation. It seems that in those calculations, geometries reported as the minimum energy were not optimized at the DFT level, probably assuming predefined structures obtained by molecular dynamic or similar methods.

The series of PBE calculations was also initiated with the geometry presented in Fig. 3a, without any restrictions during the optimization process. The final geometry, Fig. 4, is similar to that found with the B3LYP functional. The difference is that the plane formed by the atoms 3, 4, 7, 8, 11, 12 is now more deformed, and the cube formed by the atoms 1–8 is markedly deformed. Again, the electronic state of minimum energy corresponds to M = 22.

This same type of calculation was performed earlier [22]. In that report, it is mentioned that the minimum energy geometry is the so-called non-icosahedral one. Since our initial geometry is not exactly that considered in those studies, we take as the initial conformation a double single cube geometry, DSC, previously reported [19, 28] as the minimum energy conformer. This initial conformation was used to

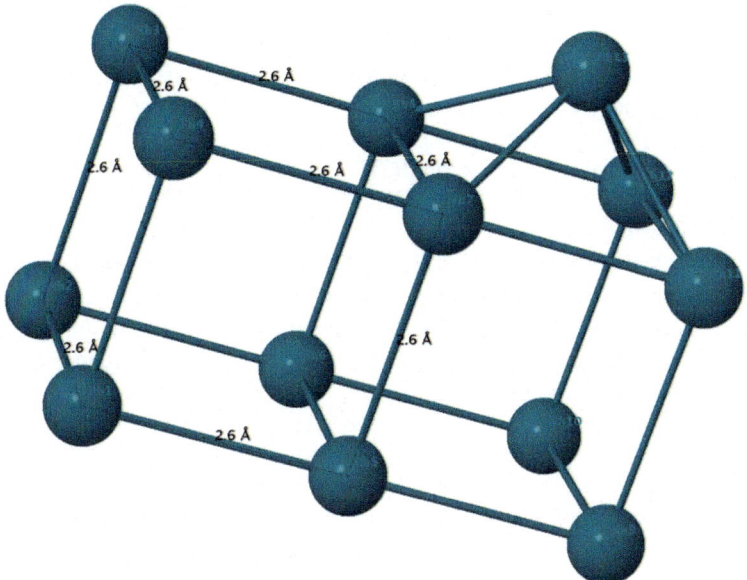

Fig. 5 Double-simple-cubic initial geometry for PBE and B3LYP calculations

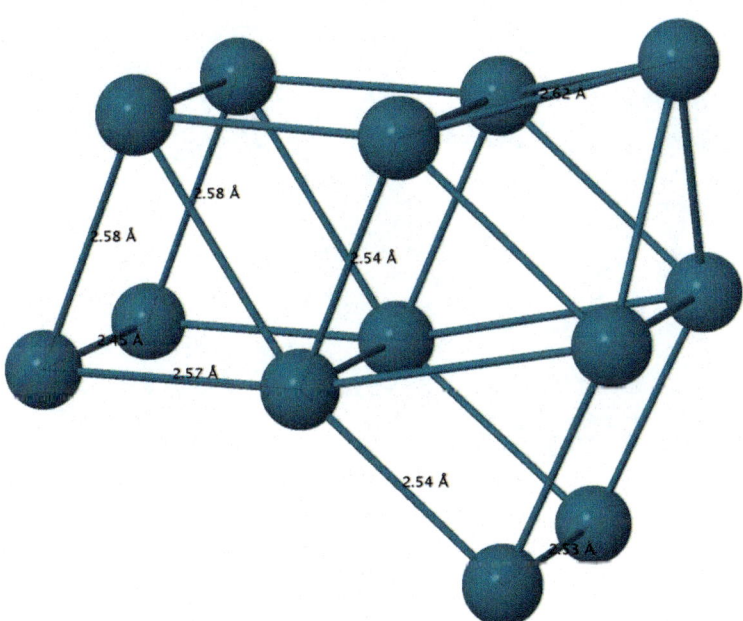

Fig. 6 Final geometry obtained for Rh_{13} with full optimization. The geometry obtained for both PBE and B3LYP functionals are very similar

perform two other sets of calculations with the same B3LYP and PBE functionals. Figure 5 shows the initial geometry, the final geometry obtained without any kind of restriction for both functionals. Figure 6 is very similar. It is a simple double distorted cube formed by oblique parallelepipeds. Although the geometries are very similar, that obtained with the functional PBE shows shorter distances than the corresponding one calculated with the B3LYP functional. The angle formed by the distorted cubes is 140° and 124° for B3LYP and PBE respectively. The initial geometry is not preserved during the optimization process. The geometry optimized by us is 0.13 eV more stable with PBE, and 0.58 eV more stable with B3LYP, than the capped double cube conformation previously reported. The energies are not the only differences. Reference [56] reported a magnetic moment of 9 μ_B for the double simple cubic, DSC, geometry with the PBE method while we obtain a magnetic moment of 13 μ_B with the same method (PBE), and 21 μ_B using B3LYP, equal to that obtained by Reddy et al. [18] using the von Barth-Hedin form of the exchange-correlation contributions in the discrete variational method. It is

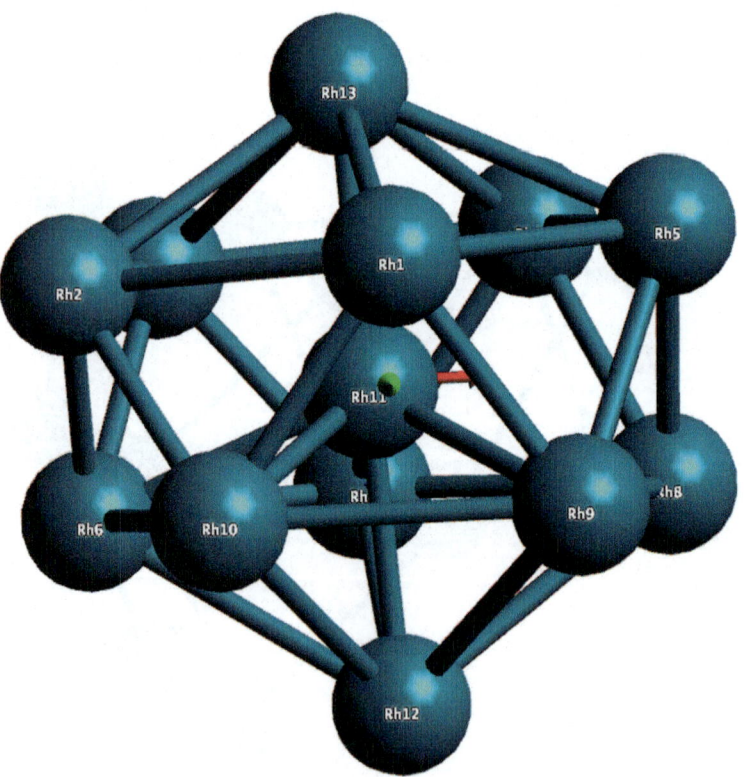

Fig. 7 Rh_{13} cluster in its minimum energy conformation, a centered icosahedron, with multiplicity 16, obtained with the ROHF method

important to emphasize that the DSC geometry reported as the most stable is not retained once it is subjected to a process of optimization without any kind of restrictions.

Rh_{13} is one of the most studied [18–29, 38, 51, 60] clusters because it is considered as the seed for different cluster growth patterns. A wide dispersion in the calculated multiplicities is present in the literature.

The ROHF calculations for Rh_{13} converge to an interesting structure presented in Fig. 7. It is an interlocking series of pentagons, an irregular icosahedron, with a rhodium atom in the center. The icosahedron is composed of interlocking pentagonal "caps". Every vertex of the icosahedron is the top of a pentagonal cap. We performed calculations for multiplicities in the range from 4 to 30. Table 4 shows that the energy has an oscillating behavior. Three minimum energy states are present at multiplicities of 10, 16 and 28. There is a small energy difference of 6.17 and 2.17 eV in relation to the multiplicity equal to 28.

In Fig. 7, the pentagons formed by the Rh_1-Rh_5, and the Rh_6-Rh_{10} atoms have different bond distances. These are 2.8 Å and 2.66 Å in average respectively. The average distance from the vertex, Rh_{11} to the Rh_1-Rh_5 pentagon is 2.89 Å and the distance from the vertex Rh_{13} to the pentagon Rh_6-Rh_{10} is 3.8 Å. Also shown in Table 4 is the dipole moment of the cluster in different electronic states.

There are two possibilities for the formation of the anion or the cation. The cation could be formed when the neutral cluster loses an electron. It may be a previously paired electron in which case the multiplicity of the cation is increased by one unit. If the loss is that of a previously unpaired electron, the multiplicity decreases by one unit in relation to the neutral cluster. On the other hand, the

Table 4 Rh13, energy (a.u.), relative energy (eV), dipolar moment (D), energy of the HOMO, energy of the LUMO, and the band gap values obtained from ROHF calculations for different states of spin multiplicity

Mult.	Energy	R. Energy	D	HOMO	LUMO	GAP
4	−1411.243	11.97	2.712	−0.031	−0.010	0.021
6	−1411.314	10.04	2.143	−0.065	−0.003	0.062
8	−1411.317	9.982	1.460	−0.051	−0.014	0.037
10	−1411.457	6.177	1.409	−0.035	−0.010	0.026
12	−1411.416	7.265	2.021	−0.037	0.002	0.039
14	−1411.423	7.075	1.978	−0.032	−0.004	0.028
16	−1411.604	2.176	0.995	−0.044	−0.002	0.042
18	−1411.588	2.585	1.132	−0.045	−0.001	0.041
20	−1411.579	2.830	2.872	−0.024	−0.004	0.020
22	−1411.618	1.769	1.082	−0.031	0.000	0.031
24	−1411.624	1.633	2.618	−0.025	−0.002	0.023
26	−1411.672	0.299	0.968	−0.026	−0.011	0.015
28	−1411.684	0.000	2.254	−0.036	−0.009	0.025
30	−1411.646	1.034	3.048	−0.033	−0.015	0.019

Table 5 Rh13 selected parameters ROHF/LANL2DZ for RH13 neutral with multiplicity 16 and its ionic species

Specie	Charge, multiplicity	Energy (a. u.)	E. A.	I. P.	HOMO	LUMO	GAP	Dip. Mom.
Rh13	+1, 15	−1411.399		0.206	−0.127	−0.115	0.012	0.85
	+1, 17	−1411.516		0.087	−0.135	−0.120	0.016	1.05
	0, 16	−1411.604			−0.444	−0.003	0.014	0.99
	−1, 17	−1411.615	0.012		0.052	0.102	0.049	0.84
	−1, 15	−1411.596	0.007		0.599	0.098	0.500	0.62
	+1, 27	NC						
	+1, 29	NC						
	0, 28	−1411.684			−0.036	−0.009	0.027	2.254
	−1, 27	−1411.654	−0.03		−0.046	0.096	0.05	2.656
	−1, 29	−1411.641	−0.04		−0.036	−0.009	0.027	2.005

formation of the anion occurs through the gain of an electron. The new electron may be unpaired in which case the multiplicity increases, or it might pair with a previously unpaired electron in the neutral cluster thus decreasing the multiplicity.

Table 5 reports the Restricted Open shell Hartree Fock energies of Rh13 $M = 16$ and $M = 28$, and its ionic clusters, the charge and multiplicity, the electron affinities (EA) and the ionization potential (IP). The electron affinities were calculated as the energy difference between the neutral and the anionic clusters while the ionization potential was calculated as the energy difference between the cation and the neutral molecule. On the other hand the Ionization Potential calculated from the HOMO following Koopman's theorem, is 0.116 a.u. = 3.15 eV for $M = 16$.

4 Conclusions

We report a study of rhodium clusters using ROHF and DFT methods with full optimization.

Our results agree with those of other researchers concerning the equilibrium geometry, but disagree on the magnetic ground state

- The equilibrium structure of the isomers is unambiguously determined with ROHF methods. For Rh_6 and Rh_8, the ROHF minimum energy conformation is in excellent agreement with experiment.
- Different spin states are quite close in energy. All of them have essentially the same equilibrium structure.
- With different XC functionals, different spin states are obtained for the same conformation.

- The Growth behavior follows the icosahedral structure route.
- In the formation of ions, a larger number of unpaired electrons lead to greater stability.

References

1. Cox AJ, Louderback JG, Bloomfield LA (1993) Experimental observation of magnetism in rhodium clusters. Phys Rev Lett 71:923–926
2. Cox AJ, Louderback JG, Apsel SE, Bloomfield LA (1994) Magnetism in 4d-transition metal clusters. Phys Rev B 4:12295–12298
3. Schmid G (2004) Nanoparticles from theory to applications. Wiley-VCH. ISBN 3527305076
4. Wei J, Iglesia E (2004) Structural requirements and reaction pathways in methane activation and chemical conversion catalized by rhodium. J Catal 225:116–127
5. Nolte P, Stierle A, Jin-Phillipp NY, Kasper N, Schulli TU, Dosch H (2008) Shape changes of supported Rh nanoparticles during oxidation and reduction cycles. Science 321:1654–1658
6. Loferty PJ (2013) Commodity report: platinum group metals. United States Geological Survey
7. Heldings FM, Capka M (2003) Rhodium complexes as catalyst for hydrosilylation crosslinking of silicone ruber. J Appl Polymer Sci 30(5):1837
8. Halligudi SB et al (1992) Hydrogenation of bencene to cycloexene catalized by Rhodium(I) complex supported on montmorillonite clay. React Kinet Catal Lett 48(2):547–552
9. Cramer Stepen S Jr, Covino, Bernard (1990) ASM handbook materials park OH: ASM international report pp 393–396
10. Harding DJ, Gruene P, Haertelt M, Meijer G, Fielicke A, Hamilton SM, Hopkins WS, Mackenzie RS, Neville SP, Walsh TR (2010) Probing the structures of gas-phase rhodium cluster cation by far-infrared spectroscopy. J Chem Phys 133:214304–214313
11. Harding DJ, Walsh TR, Hamilton SM, Hopkins WS, Mackenzie SR, Gruene P, Haertelt M, Maijer G, Fielicke F (2010) Comunications: the structure of Rh_8^+ in the gas phase. J Chem Phys 132:011101–011104
12. Knickelbin MB (2005) Phys Rev B 71:18444
13. Apsel SE, Emmert JW, Deng J, Bloomfield LA (1996) Surface-enhanced magnetism in nickel clusters. Phys Rev Lett 76:1441
14. Douglas DC, Bucher JP, Bloomfield LA (1992) Magnetic studies of free ferromagnetic clusters. Phys Rev B 45:6341–6344
15. Knickelbein MB (2001) Experimental observation of superparamagnetism in manganese clusters. Phys Rev Lett 86:5255–5257
16. Knickelbein MB (2004) Magnetic ordering in manganese clusters. Phys Rev B 70:014424
17. Xu X, Yin S, Moro R, de Herr WA (2005) Magnetic moments and adiabatic magnetization of free cobalt clusters. Phys Rev Lett 95:237209
18. Reddy BV, Khanna SN, Dunlap BI (1993) Giant magnetic moments of 4d clusters. Phys Rev Lett 70:3323–3326
19. Bae YC, Kumar V, Osanai H, Kawazoe Y (2005) Cubic magic clusters of rhodium stabilized with eight-center bonding: magnetism and growth. Phys Rev B 72:115427–115432
20. Chang CM, Chou MY (2004) Alternative low-symmetry structure for 13 atoms metal clusters. Phys Rev Lett 93:133401–133404
21. Aguilera-Granja F, Rodríguez-López JL, Michaelian K, Berlanga-Ramírez EO, Vega A (2002) structure and magnetism of small rhodium clusters. Phys Rev B 66:224410–224419
22. Wang LL, Johnson DD (2007) Density functional study of structural trends for late-transition-metal 13-atoms clusters. Phys Rev B 75:235405–235409

23. Sun Y, Zhang M, Fournier R (2008) Periodic trends in the geometric structures of 13-atoms metal clusters. Phys Rev B 77:0754435
24. Jinlong Y, Toigo F, Kelin W (1994) Structural electronic, and magnetic properties of small rhodium clusters. Phys Rev B 50:7915–7924
25. Reddy BV, Nayak SK, Khanna SN, Rao BK, Jena P (1999) Electronic structure and magnetism of Rh_n (n = 2-13) clusters. Phys Rev B 59:5214–5222
26. Guirado-López R, Villaseñor-González P, Dorantes-Dávila J, Pastor GM (2000) Magnetism of Rh_n clusters. J Appl Phys 87:4906
27. Aguilera-Granja F, Montejano-Carrizales JM, Guirado-López RA (2006) Magnetic properties of small 3d and 4d transition metal clusters: the role of a noncompact growth. Phys Rev B 73:115422
28. Bae YC, Osansi H, Kumar V, Kawazone Y (2004) Nonicosahedral growth and magnetism behavior of rhodium clusters. Phys Rev B 70:195413–195419
29. Rogan J, García G, Loyola C, Orellana W, Ramírez R, Kiwi M (2006) Alternative search strategy for minimal energy nanocluster structures: the case of rhodium, palladium, and silver. J Chem Phys 125:214708–214712
30. Joswig JO, Springborg M (2003) Generic algorithms search for global minima of aluminium clusters using Sutton-Chen potential. Phys Rev B 68:085408
31. Zhan L, Chen JZY, Montejano-Carrizalez JM, Guirado-López RA (2006) Phys Rev B 73:115422
32. Aprá E, Ferrando R, Fontunelli A (2006) Density-functional global optimization of gold clusters. Phys Rev B 73:205414
33. Wales DJ, Doye JPK (1997) Global optimization by basin-hopping and the lowest energy structures of Lennard-Jones clusters containing up to 110 atoms. J Phys Chem A 101: 5111–5116
34. Kim HG, Choi SK, Lee HM (2008) New algorithm in the basin-hopping Monte Carlo to find the global minimum structure of unary and binary metallic nano clusters. J Chem Phys 128:144702
35. Erkoc S, Saltaf R (1999) Monte Carlo computer simulations of copper clusters. Phys Rev A 60:3053
36. Cheng J, Fournier R (2004) Structural optimization of atomic clusters by Tabu search in descriptor spaces. Theor Chem Acc 112:7–15
37. Baletto F, Ferrando R (2005) Structural properties of nanoclusters: energetic, thermodynamic, and kinetic effects. Rev Mod Phys 77:371
38. Piotrowsky MJ, Piquini P, Da Silva JLF (2010) Density functional theory investigation of 3d, 4d, and 5d 13-atom metal clusters. Phys Rev B 81:155446–155459
39. Piotrowski MJ, Piquini P, Odashima MM, DaSilva JLF (2011) Transition-metal 13-atom clusterts assessed with solid and surface-biased functionals. J Chem Phys 134:134105–134110
40. Hay PJ, Wadt WR (1985) Ab initio effective core potentials for molecular calculations. Potentials for main group elements Na to Bi. J Chem Phys 82:270
41. Wadt WR, Hay PJ (1985) Ab initio effective core potentials for molecular calculations. Potentials for the transition metal atoms Sc to Hg. J Chem Phys 82:284
42. Hay PJ, Wadt WR (1985) Ab initio effective core potentials for molecular calculations. Potentials for K to Au including the outermost core orbitals. J Chem Phys 82:299
43. Wood JH, Boring AM (1978) Improved Pauli Hamiltonian for local-potential problems. Phys Rev B 18:2701
44. Perdew JP, Burke K, Ernzerhof M (1996) General gradient approximation made simple. Phys Rev Lett 77:3865
45. Perdew JP, Burke K, Ernzerhof M (1997) General gradient approximation made simple [Phys Rev Lett 77:4884(1996)] Phys Rev Lett 78:1396
46. Peng C, Ayala PY, Schelegel HB, Frish MJ (1996) Using redundant internal coordinates to optimize equilibrium geometries and transition states. J Comput Chem 17:49

47. Peng C, Schelegel HB (1994) Combining synchronous transit and quasi-Newton methods to find transit states. Israel J Chem 33:449

48. Beltrán MR, Buendía Zamudío F, Chanhan V, Sen P, Wang H, Ko YJ, Bowen K (2013) Ab initio and anion photoelectron studies of Rh_n (n = 1-9) clusters. Eur Phys J D 67:63–70

49. Bertin V, Lopez-Rendón R, del Angel G, Poulain E, Avilés R, Uc-Rosas V (2010) Comparative theoretical study of small Rh_n nanoparticles ($2 \leq n \leq 8$) using DFT methods. Int J Quantum Chem 110:1152–1164

50. Harding DJ, Davies RDL, Mackenzie SR, Walsh TR (2008) Oxides of small rhodium clusters: theoretical investigation of experimental reactivities. J Chem Phys 129:124304–124310

51. Mora MA, Mora-Ramírez MA, Rubio-Arroyo Manuel F (2010) Structural and electronic study of neutral, positive and negative small rhodium clusters [Rh_n, Rh_n^+, Rh_n^-]. Int J Quantum Chem 110:2541–2547

52. Li ZQ, Yu JZ, Ohno K, Kawazoe Y (1999) Calculations on the magnetic properties of rhodium clusters. J Phys Condens Matter 7:47–53

53. Hang TD, Hung HM, Thiem LN, Nguyen HMT (2015) Electronic structure and thermochemical properties of neutral and anionic rhodium clusters Rhn, n = 2-13. Evolution of structures and stabilities of binary clusters RhmM (M = Fe Co, Ni; m = 1-6). Comput Theor Chem 1068:30–41

54. Chien CH, Blaisten-Barojas E, Pedersen MR (1998) Magnetic and electronic properties of rhodium clusters. Phys Rev A 58:2196–2202

55. Harding D, Mackenzie SR, Walsh TR (2006) Structural isomers and reactivity for Rh_6, Rh_6^+. J Phys Chem B 110:18272–18277

56. Da Silva JLF, Piotrowski MJ, Aguilera-Granja F (2012) Phys Rev B 86:125430–125435

57. Sun Y, Fournier R, Zhang M (2009) Structural and electronic properties of 13-atom 4d transition-metal clusters. Phys Rev B 79:043202–043211

58. Lee C, Yang W, Parr RG (1988) Development Colle-Salvetti correlation-energy formula into a functional of the electron density. Phys Rev B 37:785

59. Miehlich B, Sabin A, Stoll H, Preus H (1989) Results obtained with the correlation energy density functional of Becke and Lee, Yang and Parr. Chem Phys Lett 157:200–206

60. Futschek T, Marsman M, Hafner J (2005) Structural and magnetic isomers of small Pd and Rh clusters: an ab initio functional study. J Phys Condens Matter 17:5927–5963

Spectroscopy of Radiative Decay Processes in Heavy Rydberg Alkali Atomic Systems

Valentin B. Ternovsky, Alexander V. Glushkov, Olga Yu. Khetselius, Marina Yu. Gurskaya and Anna A. Kuznetsova

Abstract We present the results of studying the radiation decay processes and computing the probabilities and oscillator strengths of radiative transitions in spectra of heavy Rydberg alkali-metal atoms. All calculations of the radiative decay (transitions) probabilities have been carried out within the generalized relativistic energy approach (which is based on the Gell-Mann and Low S-matrix formalism) and the relativistic many-body perturbation theory with using the optimized one-quasiparticle representation and an accurate accounting for the critically important exchange-correlation effects as the perturbation theory second and higher orders ones. The precise data on spectroscopic parameters (energies, reduced dipole transition matrix elements, amplitude transitions) of the radiative transitions $nS_{1/2} \rightarrow n'P_{1/2,3/2}$ ($n = 5, 6$; $n' = 10$–70), $nP_{1/2,3.2} \rightarrow n'D_{3/2,5/2}$ ($n = 5, 6$; $n' = 10$–80) in the Rydberg Rb, Cs spectra and the transitions $7S_{1/2}$-$nP_{1/2,3/2}$, $7P_{1/2,3.2}$-$nD_{3/2,5/2}$ ($n = 20$–80) in the Rydberg francium spectrum are presented. The obtained results are analyzed and discussed from viewpoint of the correct accounting for the relativistic and exchange-correlation effects. It has been shown that theoretical approach used provides an effective accounting of the multielectron exchange-correlation effects, including effect of essentially non-Coulomb grouping of Rydberg levels and others.

Keywords Radiation decay processes · Multielectron atoms and multicharged ions · Relativistic energy approach

V. B. Ternovsky (✉) · A. V. Glushkov · O. Yu. Khetselius · M. Yu. Gurskaya
A. A. Kuznetsova
Odessa State Environmental University, L'vovskaya Str., 15,
Odessa-9 65016, Ukraine
e-mail: ternovskyvb@gmail.com

© Springer International Publishing AG, part of Springer Nature 2018   229
Y. A. Wang et al. (eds.), *Concepts, Methods and Applications of Quantum Systems in Chemistry and Physics*, Progress in Theoretical Chemistry and Physics 31,
https://doi.org/10.1007/978-3-319-74582-4_13

1 Introduction

Accurate radiative decay widths and probabilities, oscillator strengths of radiative transitions in spectra of the Rydberg atomic systems (atoms in the highly excited states with large values of the principal quantum number n $\gg$ 1) are of a great interest for astrophysical analysis, laboratory, thermonuclear plasma diagnostics, fusion research etc. (see, for example [1–60]). In recent years intensive theoretical and experimental investigations of spectroscopic properties of the Rydberg atoms are also stimulated by a great number of their possible important applications in atomic and molecular optics and spectroscopy, quantum electronics, laser physics (for example, speech is about new lasing schemes in the short-wave range with using the Rydberg systems), quantum informatics and computing, astrophysics etc. It is well known that the Rydberg atoms make the contribution into interstellar clouds absorption spectrum (Rydberg states with n $\sim$ 300–700). The unique properties of the Rydberg atoms are associated with too small ionization potentials, sufficiently large size, enough long lifetime compared to conventional atomic states, finally, unprecedented sensitivity to external fields. Really, it is well known that the Rydberg atomic systems are very sensitive to electromagnetic fields and can be used for the detection and sensing static and AC electric and magnetic fields. Strongly interacting Rydberg systems have unique photon emission properties. These facts stimulate more intensive research of the Rydberg atoms, in particular, on the basis of new experimental methods of laser spectroscopy, beam-foil spectroscopy, using magneto-optical traps, synchrotron radiation sources, cryogenic devices and so on. It is worth to remind about such unique and interesting physical objects and phenomena such as the Rydberg matter, Bose-condensate in vapors of the Rydberg alkali-metal atoms, fountains of cold Rydberg atoms etc.

The well-known quasiclassical and quantum-mechanical approaches such as the Hartree-Fock (HF) and Dirac-Fock (DF) methods, quantum defect and the Coulomb approximations, the model potential and pseudopotential methods etc. have been used to calculate the spectroscopic properties of different light and middle Rydberg atoms. In a case of the heavy Rydberg atoms in a free state or in an external electromagnetic field a modern level of description of the Rydberg atoms is not sufficiently satisfactory. A precise accounting for the relativistic and exchange-correlation (XC) effects, including an effect of the non-Coulomb grouping levels in the Rydberg spectra (the effect, which, as a rule, is not considered within simplified Coulomb and quantum defect models) is of a great interest and importance.

The purpose of this work is to present the results of studying the radiation decay processes and computing probabilities and oscillator strengths of the radiative transitions in the spectra of heavy Rydberg atoms of alkali-metal elements. The precise data on spectroscopic parameters (energies, reduced dipole transition matrix elements, amplitude transitions) of the radiative transitions $nS_{1/2} \rightarrow n'P_{1/2,3/2}$ (n = 5, 6; n' = 10–70), $nP_{1/2,3.2} \rightarrow n'D_{3/2,5/2}$ (n = 5, 6; n' = 10–80) in the Rydberg Rb, Cs spectra and the transitions $7S_{1/2}\text{-}nP_{1/2,3/2}$, $7P_{1/2,3.2}\text{-}nD_{3/2,5/2}$ (n = 20–80) in the

Rydberg francium spectrum are listed. The data are discussed from the viewpoint of the correct accounting for the relativistic and exchange-correlation effects. It has been shown that the theoretical approach used provides a precise accounting for the important exchange-correlation effects, including the effect of essentially non-Coulomb grouping of Rydberg levels, continuum pressure etc.

All calculations of the radiative decay (transitions) probabilities (matrix elements) in the studied atomic systems have been performed with using the generalized relativistic energy approach and the relativistic many-body perturbation theory (PT) with using the optimized one-quasiparticle representation and an accurate accounting of the exchange-correlation effects, including the effect of essentially non-Coulomb grouping of Rydberg levels [61–63].

Let us remind that the theoretical fundamentals of an energy approach in a case of the one-electron ions have been considered by Labzovsky et al. [57, 58]. Originally the energy approach to radiative and autoionization processes in multielectron atoms and ions has been developed by Ivanova-Ivanov et al. [59–62, 64–67]. More accurate, advanced version of the relativistic energy approach has been further developed in Refs. [63, 68–72]. The energy approach is based on the Gell-Mann and Low S-matrix formalism combined with the relativistic perturbation theory. In relativistic case the Gell-Mann and Low formula expressed an energy shift ΔE through the electrodynamical scattering matrix including interaction with as the photon vacuum field as a laser field. The first case is corresponding to determination of radiative decay characteristics for atomic systems. Earlier we have applied the corresponding generalized versions of the energy approach to many problems of atomic, nuclear and even molecular spectroscopy, including, cooperative electron-gamma-nuclear "shake-up" processes, electron-muon-beta-gamma-nuclear spectroscopy, spectroscopy of atoms in a laser field etc. [73–98].

2 Relativistic Energy Approach and Many-Body Perturbation Theory with the Dirac-Kohn-Sham Zeroth Approximation

Let us describe in brief the key moments of our theoretical approach (look for more details in Refs. [63, 65–69, 73–76]). As usually, the wave functions zeroth basis is found from the Dirac equation solution with self-consistent total potential.

The bare Hamiltonian is as follows:

$$H = \sum_{i} \left\{ \alpha c p - \beta m c^2 + U(r_i|Z) \right\} + + \sum_{i>j} exp\left(i\omega_{ij}r_{ij}\right) \cdot \frac{\left(1 - \alpha_i \alpha_j\right)}{r_{ij}}, \qquad (1)$$

where α_i, α_j—the Dirac matrices, ω_{ij}—the transition frequency, c—the light velocity, Z is a charge of the atomic nucleus. Within relativistic perturbation theory [3, 4] we introduce the zeroth–order Hamiltonian as:

$$H_0 = \sum_i \{\alpha c p_i - \beta m c^2 + [-Z/r_i + U_{MF}(r_i|b) + V_{XC}(r_i)]\}, \tag{2}$$

where $V_{XC}(r_i)$—one-particle exchange-correlation potential, $U_{MF}(r_i|b)$—a self-consistent Coulomb-like mean-field potential (b is the potential parameter, which is further determined within ab initio procedure), that potential interaction "quasiparticles-core" in the case of atomic system consisting of closed electron shells and external quasiparticles.

The relativistic wave functions are calculated by solution of the Dirac equation with the potential, which includes the Coulomb potential of the closed electron shells core of an alkali atomsplus the exchange Kohn-Sham potential and correlation Lundqvist-Gunnarsson potential (see details in Refs. [63, 70, 73–76]).

In order to provide the construction of the optimized one-quasiparticle representation and improve an effectiveness of the numerical code we have used special ab initio procedure within relativistic energy approach [68] (see also [69, 70]). It reduces to accurate treating the lowest order multielectron effects, in particular, the gauge dependent radiative contribution into imaginary part of the electron system energy $\mathrm{Im}\,\delta E_{ninv}$ for the certain class of the photon propagator calibrations and minimization of the corresponding density functional $\mathrm{Im}\,\delta E_{ninv}$. Some known alternative approaches to construction of an optimized one-quasiparticle representation for multielectron atom can be found in Refs. [11–22].

Within the relativistic energy approach [61, 62, 64, 65] an imaginary part of the electron energy shift of an atom is directly connected with the radiation decay possibility (transition amplitude). An approach, using the Gell-Mann and Low formula with the QED scattering matrix, is used in treating the relativistic atom. The total energy shift of the state is usually presented in the form:

$$\Delta E = \mathrm{Re}\Delta E + i\Gamma/2 \tag{3}$$

where Γ is interpreted as the level width, and the decay possibility $P = \Gamma$.

The imaginary part of an electron energy of the atomic system can be determined in the lowest second order of perturbation theory as:

$$\mathrm{Im}\Delta E(B) = -\frac{e^2}{4\pi} \sum_{\begin{bmatrix} \alpha > n > f \\ \alpha > n \leq f \end{bmatrix}} V^{|\omega_{\alpha n}|}_{\alpha n \alpha n}, \tag{4}$$

where $(\alpha > n > f)$ for electron and $(\alpha < n < f)$ for vacancy. The matrix element is determined as follows:

$$V^{|\omega|}_{ijkl} = \iint dr_1 dr_2 \Psi_i^*(r_1)\Psi_j^*(r_2) \frac{\sin|\omega|r_{12}}{r_{12}} (1 - \alpha_1\alpha_2)\Psi_k^*(r_2)\Psi_l^*(r_1) \tag{5}$$

When calculating the matrix elements (5), one should use the angle symmetry of the task and write the corresponding expansion for $\sin|\omega|r_{12}/r_{12}$ on spherical harmonics as follows [62]:

$$\frac{\sin |\omega|r_{12}}{r_{12}} = \frac{\pi}{2\sqrt{r_1 r_2}} \sum_{\lambda=0}^{\infty} (\lambda) J_{\lambda+1/2}(|\omega|r_1) J_{\lambda+1/2}(|\omega|r_2) P_{\lambda}(\cos \widehat{r_1 r_2}) \tag{6}$$

where J is the Bessel function of first kind and $(\lambda) = 2\lambda + 1$.

This expansion is corresponding to usual multipole one for probability of radiative decay (an amplitude approach of quantum mechanics). Substitution of the expansion (5) to matrix element allow to get the following expression:

$$V_{1234}^{\omega} = [(j_1)(j_2)(j_3)(j_4)]^{1/2} \sum_{\lambda\mu} (-1)^{\mu} \begin{pmatrix} j_1 j_3 & \lambda \\ m_1 - m_3 & \mu \end{pmatrix} \times \mathrm{Im} Q_{\lambda}(1234) , \tag{7}$$

$$Q_{\lambda} = Q_{\lambda}^{\mathrm{Qul}} + Q_{\lambda}^{\mathrm{Br}}$$

where j_i is the total single electron momentums, m_i—the projections; Q^{Qul} is the Coulomb part of interaction, Q^{Br}—the Breit part.

The total radiation width of the one-quasiparticle state can be presented in the following form:

$$\Gamma(\gamma) = -2 \operatorname{Im} M^1(\gamma) = -2 \sum_{\lambda n l j} (2j+1) \operatorname{Im} Q_{\lambda}(n_{\gamma} l_{\gamma} j_{\gamma} n l j)$$

$$Q_{\lambda} = Q_{\lambda}^{Cul} + Q_{\lambda}^{Br}. \tag{8}$$

$$Q_{\lambda}^{Br} = Q_{\lambda, \lambda-1}^{Br} + Q_{\lambda, \lambda}^{Br} + Q_{\lambda, \lambda+1}^{Br}$$

The individual terms of the Σ_{nlj} sum correspond to the partial contribution of the $n_{\lambda} l_{\lambda} j_{\lambda} \rightarrow nlj$ transitions; Σ_{λ} is a sum of the contributions of the different multiplicity transitions. The detailed expressions for the Coulomb and Breit parts can be found in Refs. [62–66].

The imaginary parts of the Coulomb part Q_{λ}^{Cul} and the Breit part contain the radial R_{λ} and angular S_{λ} integrals as follows (in the Coulomb units) [65]:

$$\operatorname{Im} Q_{\lambda}^{Cul}(12;43) = Z^{-1} \operatorname{Im} \left\{ R_{\lambda}(12;43) S_{\lambda}(12;43) + R_{\lambda}\left(\widetilde{12};4\widetilde{3}\right) S_{\lambda}\left(\widetilde{12};4\widetilde{3}\right) + \right.$$
$$\left. + R_{\lambda}\left(1\widetilde{2};4\widetilde{3}\right) S_{\lambda}\left(1\widetilde{2};4\widetilde{3}\right) + R_{\lambda}\left(\widetilde{1}\widetilde{2};4\widetilde{3}\right) S_{\lambda}\left(\widetilde{1}\widetilde{2};4\widetilde{3}\right) \right\}. \tag{9}$$

$$\operatorname{Im} Q_{\lambda, l}^{Br} = \frac{1}{Z} \operatorname{Im} \left\{ R_{\lambda}\left(12;\widetilde{4}\widetilde{3}\right) S_{\lambda}^{l}\left(12;\widetilde{4}\widetilde{3}\right) + R_{\lambda}\left(\widetilde{1}\widetilde{2};43\right) S_{\lambda}^{l}\left(\widetilde{1}\widetilde{2};43\right) + \right.$$
$$\left. + R_{\lambda}\left(\widetilde{12};\widetilde{4}3\right) S_{\lambda}^{l}\left(\widetilde{12};\widetilde{4}3\right) + R_{\lambda}\left(\widetilde{1}\widetilde{2};4\widetilde{3}\right) S_{\lambda}^{l}\left(\widetilde{1}\widetilde{2};4\widetilde{3}\right) \right\}. \tag{10}$$

Here $\{\lambda l_1 l_3\}$ means that λ, l_1 and l_3 must satisfy the triangle rule and the sum $\lambda + l_1 + l_3$ must be an even number. The rest terms in (9), (10) include the small components of the Dirac functions. The tilde designates that the large radial component f must be replaced by the small one g, and instead of l_i, $\tilde{l}_i = l_i - 1$ should be taken for $j_i < l_i$ and $\tilde{l}_i = l_i + 1$ for $j_i > l_i$. The detailed definitions for the radial R_λ and angular S_λ integrals can be found in Refs. [59, 64–67].

The total probability of a λ—pole transition is usually represented as a sum of the electric P_λ^E and magnetic P_λ^M parts. The electric (or magnetic) λ—pole transition $\gamma \to \delta$ connects two states with parities which by λ (or $\lambda + 1$) units. In our designations (the radiative $\gamma \to \delta$ transition) one could write:

$$
\begin{aligned}
P_\lambda^E(\gamma \to \delta) &= 2(2j+1)Q_\lambda^E(\gamma\delta;\gamma\delta) \quad Q_\lambda^E = Q_\lambda^{Cul} + Q_{\lambda,\lambda-1}^{Br} + Q_{\lambda,\lambda+1}^{Br} \\
P_\lambda^M(\gamma \to \delta) &= 2(2j+1)Q_\lambda^M(\gamma\delta;\gamma\delta) \quad Q_\lambda^M = Q_{\lambda,\lambda}^{Br}
\end{aligned}
\tag{11}
$$

The adequate, precise computation of radiative parameters of the heavy Rydberg alkali-metal atoms within relativistic perturbation theory requires an accurate accounting for the multi-electron exchange-correlation effects (including polarization and screening effects, a continuum pressure etc.). These effects within our approach are treated as the effects of the perturbation theory second and higher orders. Using the standard Feynman diagram technique one should consider two kinds of diagrams (the polarization and ladder ones), which describe the polarization and screening exchange-correlation effects. The detailed description of the polarization diagrams and the corresponding analytical expressions for matrix elements of the polarization interelectron interaction (through the polarizable core of an alkali atom) potential is presented in Refs. [63, 73–76].

An effective approach to accounting of the polarization diagrams contributions is in adding the effective two-quasiparticle polarizable operator into the perturbation theory first order matrix elements. In Ref. [65] the corresponding non-relativistic polarization functional has been derived. More correct relativistic expression has been presented in the Refs. [34, 35] and used in our theory. The corresponding two-quasiparticle polarization potential looks as follows:

$$
V_{pol}^d(r_1 r_2) = X \left\{ \int \frac{dr'\left(\rho_c^{(0)}(r')\right)^{1/3}\theta(r')}{|r_1 - r'|\cdot|r' - r_2|} \right.
$$
$$
\left. - \int \frac{dr'\left(\rho_c^{(0)}(r')\right)^{1/3}\theta(r')}{|r_1 - r'|} \int \frac{dr''\left(\rho_c^{(0)}(r'')\right)^{1/3}\theta(r'')}{|r'' - r_2|} \Big/ \left\langle \left(\rho_c^{(0)}\right)^{1/3}\right\rangle \right\}
\tag{12a}
$$

$$
\left\langle \left(\rho_c^{(0)}\right)^{1/3}\right\rangle = \int dr \left(\rho_c^{(0)}(r)\right)^{1/3}\theta(r),
\tag{12b}
$$

$$
\theta(r) = \left\{1 + \left[3\pi^2 \cdot \rho_c^{(0)}(r)\right]^{2/3} / c^2\right\}^{1/2}
\tag{12c}
$$

where ρ_c^0 is the core electron density (without account for the quasiparticle), X is numerical coefficient, c is the light velocity. The contribution of the ladder diagrams (these diagrams describe the immediate interparticle interaction) is summarized by a modification of the perturbation theory zeroth approximation mean-field central potential (look [35, 65]), which include the screening (anti-screening) of the core potential of each particle by the two others. The details of calculating this contribution can be found in Refs. [63, 65–76]. All computing was performed with using the modified PC code "Superatom-ISAN" (version 93).

3 Results and Conclusions

In Tables 1 and 2 we present the experimental and theoretical values (in atomic units: a.u.) of the reduced dipole transition matrix elements for the Fr and Cs atoms: experimental data—Exp; theoretical data: perturbation theory (PT)-DF^{SD}—PT with the Dirac-Hartree-Fock zeroth approximation (single-double SD approximation in which single and double excitations of Dirac-Hartree-Fock wave functions are

Table 1 The reduced dipole transition matrix elements for Fr (see text)

Transition/ method	PT-DF^{SD}	PT-DF^{SD} (corr)	EMP	PT-RHF (corr)	PT-RHF	DF	RPT-EA	Exp.
$7p_{1/2}$-7s	4.256	–	–	4.279	4.304	4.179	4.272, 4.274	4.277
$8p_{1/2}$-7s	0.327	0.306	0.304	0.291	0.301	–	0.339	
$9p_{1/2}$-7s	0.110	0.098	0.096	–	–	–	0.092	
$10p_{1/2}$-7s	–	–	–	–	–	–	0.063	
$7p_{3/2}$-7s	5.851	–	–	5.894	5.927	5.791	5.891	5.898
$8p_{3/2}$-7s	0.934	0.909	0.908	0.924	–	–	0.918	–
$9p_{3/2}$-7s	0.436	0.422	0.420	–	–	–	0.426	–
$10p_{3/2}$-7s	–	–	–	–	–	–	0.284	–
$7p_{1/2}$-8s	4.184	4.237	4.230	4.165	4.219	4.196	4.228	–
$8p_{1/2}$-8s	10.02	10.10	10.06	10.16	10.00		10.12	–
$9p_{1/2}$-8s	0.985	–	0.977	–	–	–	0.972	–
$10p_{1/2}$-8s	–	–	–	–	–	–	0.395	–
$7p_{3/2}$-8s	7.418	7.461	7.449	7.384	7.470	7.472	7.453	–
$8p_{3/2}$-8s	13.23	13.37	13.32	13.45	13.26		13.35	–
$9p_{3/2}$-8s	2.245	–	2.236	–	–	–	2.232	–
$10p_{3/2}$-8s	–	–	–	–	–	–	1.058	
$7p_{1/2}$-9s	1.016	–	1.010	–	–	–	1.062	–
$8p_{1/2}$-9s	9.280	–	9.342	–	–	–	9.318	–
$9p_{1/2}$-9s	17.39	–	17.40	–	–	–	17.42	–

(continued)

Table 1 (continued)

Transition/ method	PT-DF^{SD}	PT-DF^{SD} (corr)	EMP	PT-RHF	PT-RHF (corr)	DF	RPT-EA	Exp.
$10p_{1/2}$-9s	–	–	–	–	–	–	1.836	–
$7p_{3/2}$-9s	1.393	–	1.380	–	–	–	1.41	–
$8p_{3/2}$-9s	15.88	–	15.92	–	–	–	15.96	–
$9p_{3/2}$-9s	22.59	–	22.73	–	–	–	22.68	–
$10p_{3/2}$-9s	–	–	–	–	–	–	3.884	–

Table 2 The reduced dipole transition matrix elements for Cs (see text)

Transition	PT-DF^{SD}	PT-DF^{SD} (corr)	DF	PT-RHF	QDA	EF-RMP	Exp.
$6p_{1/2}$-6s	4.482	4.535	4.510	–	4.282	4.489	4.4890(7)
$6p_{3/2}$-6s	6.304	6.382	6.347	–	5.936	6.323	6.3238(7)
$7p_{1/2}$-6s	0.297	0.279	0.280	0.2825	0.272	0.283	0.284(2)
$7p_{3/2}$-6s	0.601	0.576	0.576	0.582	0.557	0.583	0.583(9)
$8p_{1/2}$-6s	0.091	0.081	0.078	–	0.077	0.088	–
$8p_{1/2}$-6s	0.232	0.218	0.214	–	0.212	0.228	–
$6p_{1/2}$-7s	4.196	4.243	4.236	4.237	4.062	4.234	4.233(22)
$6p_{3/2}$-7s	6.425	6.479	6.470	6.472	6.219	6.480	6.479(31)
$7p_{1/2}$-7s	10.254	10.310	10.289	10.285	9.906	10.309	10.309 (15)
$7p_{3/2}$-7s	14.238	14.323	14.293	14.286	13.675	14.323	14.325 (20)

included to all PT orders); PT-DF^{SD}—PT with the Dirac-Hartree-Fock zeroth approximation (plus compilation), EMP—empirical model potential method; PT-RHF—PT with the relativistic Hartree-Fock (RHF) zeroth approximation and PT-RHF(corr)—corrected version with using empirical data; DF-PT with the Dirac-Fock zeroth approximation; RPT-EA—our method (relativistic PT with the Dirac-Kohn-Sham zeroth approximation combined with an energy approach (EA)). All data have been taken from Refs. [1–6, 10, 30, 31]).

In Fig. 1 we present a dependence of the calculated reduced dipole matrix elements upon a principal quantum number for different states of the Rydberg atom Rb: $5P_{3/2}$-$nD_{5/2}$ (n ∼ 70): available experimental data—the circles; Theory: continuous line—our data; the dotted line- data by Piotrowicz et al., obtained within the quasiclassical Dyachkov-Pankratov model [4, 10, 12, 30, 31]. In Figs. 2 and 3 we present the same dependences for the Rydberg states of the Cs atom: $6P_{3/2} \rightarrow nD_{5/2}$ and the Fr atom: $7P_{3/2} \rightarrow nD_{5/2}$, n = 10–70.

The detailed analysis of the computation data shows very important role of the relativistic and interelectron exchange-correlation effects (for example the contribution due to the perturbation theory second and higher orders, including the

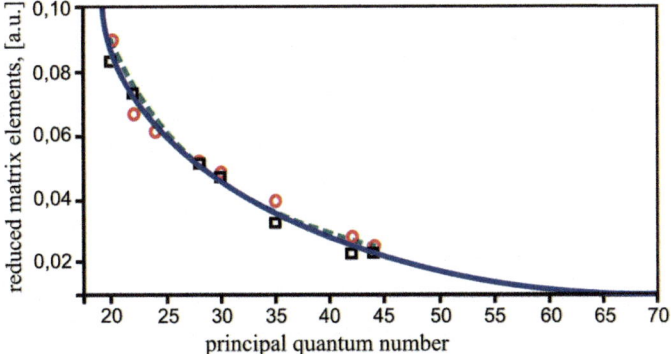

Fig. 1 A dependence of the calculated reduced dipole matrix elements upon principal quantum number for Rydberg atom Rb: $5P_{3/2} \rightarrow nD_{5/2}$ ($n \sim 70$). The available experimental data are listed as a circle; Theory: continuous line—our data, dotted line- data by Piotrowicz et al. within the quasi-classical Dyachkov-Pankratov model (see text)

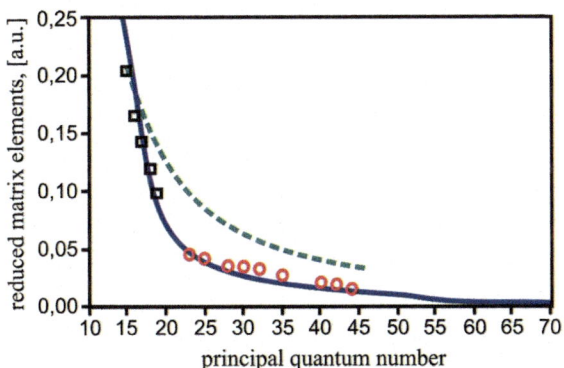

Fig. 2 A dependence of the calculated reduced dipole matrix elements upon principal quantum number for Rydberg atom Cs: $6P_{3/2} \rightarrow nD_{5/2}$ ($n \sim 70$). The available experimental data are listed as a circle; Theory: continuous line—our data, dotted line- data by Piotrowicz et al. within the quasi-classical Dyachkov-Pankratov model (see text)

interelectron polarization interaction and mutual screening ones reaches ~40%), as well as the effect of the non-Coulomb grouping levels in the Rydberg spectra.

To conclude, we have presented the results of studying the radiation decay processes and computing reduced dipole matrix elements (radiative amplitudes) of transitions in spectra of heavy Rydberg atoms of alkali elements (Rb, Cs, Fr; for the states with the principal quantum number n = 10–80) on the basis of the generalized relativistic energy approach and the relativistic many-body perturbation theory with the optimized one-quasiparticle representation. A critically important

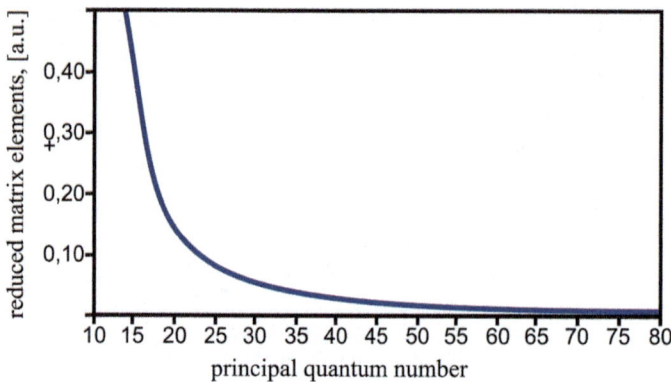

Fig. 3 A dependence of the calculated reduced dipole matrix elements upon principal quantum number for Rydberg atom Fr: $7P_{3/2} \rightarrow nD_{5/2}$, n = 10–80 (our data)

role of the many-body exchange-polarization effects and effect of the non-Coulomb grouping levels in the Rydberg spectra has been found. The detailed numerical data on the dipole matrix elements and transition probabilities are listed in Refs. [99, 100].

Acknowledgements The authors are very much thankful to Prof. J. Maruani and Dr. Y. A. Wang for invitation to make contributions on the QSCP-XVI workshop (Vancouver, Canada). The useful comments of the anonymous referees are very much acknowledged too.

References

1. Martin W (2004) NIST spectra database, version 2.0. NIST, Washington. http://physics.nist.gov.asd
2. Moore C (1987) NBS spectra database. NBS, Washington
3. Weiss A (1977) J Quant Spectrosc Radiat Transf 18:481; Phys Scripta T65:188 (1993)
4. Nadeem A, Haq SU (2010) Phys Rev A 81:063432; (2011) Phys Rev A 83:063404
5. Simsarian JE, Orozco LA, Sprouse GD, Zhao WZ (1998) Phys Rev A 57:2448
6. Curtis L (1995) Phys Rev A 51:4574
7. Li Y, Pretzler G, Fill EE (1995) Phys Rev A 52:R3433–R3435
8. Feng Z-G, Zhang L-J, Zhao J-M, Liand C-Y, Jia S-T (2009) J Phys B At Mol Opt Phys 42:145303
9. Marinescu M, Vrinceanu D, Sadeghpour HR (1998) Phys Rev A 58:R4259
10. Safronova UI, Johnson W, Derevianko A (1999) Phys Rev A 60:4476; Dzuba V, Flambaum V, Safranova MS (2006) Phys Rev A 73:02211
11. Grant IP (2007) Relativistic quantum theory of atoms and molecules, theory and computation. In: Springer series on atomic, optical, and plasma physics, vol 40. Springer, Berlin, pp 587–626
12. Glushkov AV (2008) Relativistic quantum theory. Quantum mechanics of atomic systems. Astroprint, Odessa, p 700

13. Wilson S (2007) Recent advances in theoretical physics and chemistry systems. In: Maruani J, Lahmar S, Wilson S, Delgado-Barrio G (eds) Series: Progress in theoretical chemistry and physics, vol 16. Springer, Berlin, pp 11–80
14. Feller D, Davidson ER (1981) J Chem Phys 74:3977
15. Froelich P, Davidson ER, Brändas E (1983) Phys Rev A 28:2641
16. Rittby M, Elander N, Brändas E (1983) Int J Quantum Chem 23:865
17. Wang YA, Yam CY, Chen YK, Chen GH (2011) J Chem Phys 134:241103
18. Maruani J (2016) J Chin Chem Soc 63:33
19. Pavlov R, Mihailov L, Velchev C, Dimitrova-Ivanovich M, Stoyanov Z, Chamel N, Maruani J (2010) J Phys Conf Ser 253:012075
20. Dietz K, Heβ BA (1989) Phys Scripta 39:682
21. Kohn JW, Sham LJ (1964) Phys Rev A 140:1133
22. Gross EG, Kohn W (2005) Exchange-correlation functionals in density functional theory. Plenum, New York
23. Froese Fischer C, Tachiev G, Irimia A (2006) GAtom Data Nucl Data Tables 92:607
24. Cheng K, Kim Y, Desclaux J (1979) At Data Nucl Data Tables 24:11
25. Safranova UI, Safranova MS, Johnson W (2005) Phys Rev A 71:052506
26. Ternovsky EV, Antoshkina OA, Florko TA, Tkach TB (2017) Photoelectronics 26:139
27. Indelicato P, Desclaux JP (1993) Phys Scripta T 46:110
28. Dzuba VA, Flambaum VV, Sushkov OP (1995) Phys Rev A 51:3454
29. Sapirstein J (1998) Rev Modern Phys 70:55
30. Piotrowicz M, MacCormick C, Kowalczyk A et al (2011) arXiv:1103.0109v2 [quant-ph]; Beterov I, Mansell CW, Yakshina EA et al (2012) arXiv:1207.3626v1 [physics.atom-ph]
31. Dyachkov LG, Pankratov PM (1994) J Phys B 27(3):461; Piotrowicz M, MacCormick C, Kowalczyk A, Bergamini S, Yakshina EA (2011) New Journ Phys 13:093012
32. Quinet P, Argante C, Fivet V, Terranoval C, Yushchenko AV, Biémont É (2007) Astrophys Astron 474:307
33. Biémont É, Fivet V, Quinet P (2004) J Phys B At Mol Opt Phys 37:4193
34. Glushkov AV (1991) Opt Spectrosc 70:555
35. Glushkov AV (1990) Soviet Phys J 33(1):1
36. Khetselius OY (2009) Int J Quantum Chem 109:3330
37. Khetselius OY (2009) Phys Scripta T135:014023
38. Malinovskaya SV, Glushkov AV, Khetselius OY, Svinarenko AA, Mischenko EV, Florko TA (2009) Int J Quant Chem 109(4):3325
39. Glushkov AV, Loboda AV, Gurnitskaya EP, Svinarenko AA (2009) Phys Scripta T135:014022
40. Glushkov AV, Khetselius OY, Svinarenko AA (2013) Phys Scripta T153:014029
41. Svinarenko AA (2014) J Phys Conf Ser 548:012039
42. Glushkov AV, Mansarliysky VF, Khetselius OY, Ignatenko AV, Smirnov A, Prepelitsa GP (2017) J Phys Conf Ser 810:012034
43. Glushkov AV, Loboda AV (2007) J Appl Spectrosc 74:305. (Springer)
44. Malinovskaya SV, Glushkov AV, Khetselius OY, Loboda AV, Lopatkin YuM, Svinarenko AA, Nikola LV, Perelygina TB (2011) Int J Quantum Chem 111:288
45. Glushkov AV, Khetselius OY, Loboda AV, Ignatenko AV, Svinarenko AA, Korchevsky DA, Lovett L (2008) Spectral line shapes. AIP Conf Proc 1058:175
46. Khetselius OY (2012) J Phys Conf Ser 397:012012
47. Khetselius OY, Florko TA, Svinarenko AA, Tkach TB (2013) Phys Scripta T153:014037
48. Khetselius OY (2008) Spectral line shapes. AIP Conf Proc 1058:363
49. Khetselius OY, Glushkov AV, Gurnitskaya EP, Loboda AV, Mischenko EV, Florko T, Sukharev D (2008) Spectral line shapes. AIP Conf Proc 1058:231
50. Khetselius OY (2007) Photoelectronics 16:129
51. Khetselius OY, Gurnitskaya EP (2006) Sens Electron Microsyst Technol N 3:35
52. Khetselius OY, Gurnitskaya EP (2006) Sens Electron Microsyst Technol N 2:25
53. Johnson WR, Lin CD, Cheng KT (1980) Phys Scr 21:409

54. Johnson WR, Sapistein J, Blundell S (1988) Phys Rev A 37:307
55. Khetselius OY (2011) Quantum structure of electroweak interaction in heavy finite Fermi-systems. Astroprint, Odessa
56. Hibbert A, Hansen JE (1994) J Phys B At Mol Opt Phys 27:3325
57. Braun MA, Dmitriev YuYu, Labzovsky LN (1969) JETP 57:2189
58. Tolmachev VV (1969) Adv Quantum Chem 4:331
59. Ivanov LN, Ivanova EP (1979) At Data Nucl Data Tables 24:95
60. Driker MN, Ivanova EP, Ivanov LN, Shestakov AF (1982) J Quant Spectrosc Radiat Transf 28:531
61. Ivanov LN, Letokhov VS (1985) Com Mod Phys D At Mol Phys 4:169
62. Vidolova-Angelova E, Ivanov LN, Ivanova EP, Angelov DA (1986) J Phys B At Mol Opt Phys 19:2053
63. Glushkov AV (2006) Relativistic and correlation effects spectra of atomic systems. Astroprint, Odessa
64. Ivanov LN, Ivanova EP, Aglitsky EV (1988) Phys Rep 166:315
65. Ivanova EP, Ivanov LN, Glushkov AV, Kramida AE (1985) Phys Scripta 32:513
66. Ivanova EP, Glushkov AV (1986) J Quant Spectrosc Radiat Transf 36:127
67. Ivanov LN, Ivanova EP, Knight L (1993) Phys Rev A 48:4365
68. Glushkov AV, Ivanov LN (1992) Phys Lett A 170:33
69. Glushkov AV, Ivanov LN, Ivanova EP (1986) Autoionization phenomena in atoms. Moscow University Press, Moscow, pp 58–160
70. Glushkov AV (2012) Quantum systems in chemistry and physics: progress in methods and applications. In: Nishikawa K, Maruani J, Brändas E, Delgado-Barrio G, Piecuch P (eds) Series: Progress in theoretical chemistry and physics, vol 26. Springer, pp 231–252
71. Glushkov AV, Ivanov LN (1993) J Phys B At Mol Opt Phys 26:L379
72. Glushkov AV (1992) JETP Lett 55:97
73. Khestelius OYu (2008) Hyperfine structure of atomic spectra. Astroprint, Odessa, 210p
74. Svinarenko AA, Glushkov AV, Khetselius OY, Ternovsky VB, Dubrovskaya YV, Kuznetsova AA, Buyadzhi VV (2017) In: Orjuela JEA (ed) Rare earth element. InTech, pp 83–104
75. Glushkov AV, Khetselius OY, Svinarenko AA, Buyadzhi VV, Ternovsky VB, Kuznetsova, Bashkarev PG (2017) In: Uzunov DI (ed) Recent studies in perturbation theory. InTech, pp 131–150
76. Glushkov AV, Ambrosov SV, Loboda AV, Chernyakova Yu G, Svinarenko AA, Khetselius OY (2004) Nucl Phys A Nucl Hadr Phys 734:21
77. Glushkov AV, Malinovskaya SV, Sukharev DE, Khetselius OY, Loboda AV, Lovett L (2009) Int J Quantum Chem 109:1717
78. Glushkov AV (2013) Advances in quantum methods and applications in chemistry, physics, and biology. In: Hotokka M, Maruani J, Brändas E, Delgado-Barrio G (ed) Series: Progress in theoretical chemistry and physics, vol 27B. Springer, pp 161–178
79. Glushkov AV, Khetselius OY, Svinarenko AA, Prepelitsa GP (2010) In: Duarte FJ (ed) Coherence and ultrashort pulsed emission. InTech, Rijeka, pp 159–186
80. Buyadzhi VV, Glushkov AV, Lovett L (2014) Photoelectronics 23:38
81. Buyadzhi VV, Glushkov AV, Mansarliysky VF, Ignatenko AV, Svinarenko AA (2015) Sens Electron Microsyst Technol 12(4):27
82. Glushkov AV, Khetselius OY, Malinovskaya SV (2008) Frontiers in quantum systems in chemistry and physics. In: Wilson S, Grout PJ, Maruani J, Delgado-Barrio G, Piecuch P (eds) Series: Progress in theoretical chemistry and physics, vol 18. Springer, pp 525–541
83. Glushkov AV, Khetselius OY, Malinovskaya SV (2008) Eur Phys J Spec Top 160:195
84. Glushkov AV, Khetselius OY, Malinovskaya SV (2008) Mol Phys 106:1257
85. Glushkov AV, Khetselius OY, Svinarenko AA (2012) Advances in the theory of quantum systems in chemistry and physics. In: Hoggan P, Maruani J, Brandas E, Delgado-Barrio G, Piecuch P (eds) Series: Progress in theoretical chemistry and physics, vol 22. Springer, pp 51–68

86. Glushkov AV, Khetselius OY, Lovett L (2010) Advances in the theory of atomic and molecular systems dynamics, spectroscopy, clusters, and nanostructures. In: Piecuch P, Maruani J, Delgado-Barrio G, Wilson S (eds) Series: Progress in theoretical chemistry and physics, vol 20. Springer, pp 125–152

87. Glushkov AV, Khetselius OY, Loboda AV, Svinarenko AA (2008) Frontiers in quantum systems in chemistry and physics. In: Wilson S, Grout PJ, Maruani J, Delgado-Barrio G, Piecuch P (eds) Series: Progress in theoretical chemistry and physics, vol 18. Springer, pp 543–560

88. Glushkov AV, Khetselius O, Gurnitskaya E, Loboda A, Florko T, Sukharev D, Lovett L (2008) Frontiers in quantum systems in chemistry and physics. In: Wilson S, Grout P, Maruani J, Delgado-Barrio G, Piecuch P (eds) Series: Progress in theoretical chemistry and physics, vol 18. Springer, pp 507–524

89. Glushkov AV, Ambrosov SV, Loboda AV, Gurnitskaya EP, Khetselius OY (2006) Recent advances in theoretical physics and chemistry systems. In: Julien JP, Maruani J, Mayou D, Wilson S, Delgado-Barrio G (eds) Series: Progress in theoretical chemistry and physics, vol 15. Springer, pp 285–299

90. Malinovskaya SV, Glushkov AV, Dubrovskaya YV, Vitavetskaya LA (2006) Recent advances in theoretical physics and chemistry systems. In: Julien J-P, Maruani J, Mayou D, Wilson S, Delgado-Barrio G (eds) Series: Progress in theoretical chemistry and physics, vol 15. Springer, pp 301–307

91. Khetselius OY (2012) Quantum systems in chemistry and physics: progress in methods and applications. In: Nishikawa K, Maruani J, Brandas E, Delgado-Barrio G, Piecuch P (eds) Series: Progress in theoretical chemistry and physics, vol 26. Springer, pp 217–229

92. Khetselius OY (2015) Frontiers in quantum methods and applications in chemistry and physics. In: Nascimento M, Maruani J, Brändas E, Delgado-Barrio G (eds) Series: Progress in theoretical chemistry and physics, vol 29. Springer, pp 55–76

93. Glushkov AV, Svinarenko AA, Khetselius OY, Buyadzhi VV, Florko TA, Shakhman AN (2015) Frontiers in quantum methods and applications in chemistry and physics. In: Nascimento M, Maruani J, Brändas E, Delgado-Barrio G (eds) Series: Progress in theoretical chemistry and physics, vol 29. Springer, pp 197–217

94. Khetselius OY, Zaichko PA, Smirnov AV, Buyadzhi VV, Ternovsky VB, Florko TA, Mansarliysky VF (2017) Quantum systems in physics, chemistry, and biology. In: Tadjer A, Pavlov R, Maruani J, Brändas E, Delgado-Barrio G (eds) Series: Progress in theoretical chemistry and physics, vol 30. Springer, pp 271–281

95. Glushkov AV, Rusov VD, Ambrosov SV, Loboda AV (2003) New projects and new lines of research in nuclear physics. In: Fazio G, Hanappe F (eds). World Scientific, Singapore, pp 126–132

96. Glushkov AV, Malinovskaya SV, Loboda AV, Shpinareva IM, Prepelitsa GP (2006) J Phys Conf Ser 35:420

97. Glushkov AV, Malinovskaya SV, Chernyakova YG, Svinarenko AA (2004) Int J Quantum Chem 99:889

98. Glushkov AV, Ambrosov SV, Ignatenko AV, Korchevsky DA (2004) Int J Quantum Chem 99:936

99. Glushkov AV, Ignatenko AV, Khetselius OY, Ternovsky VB (2017) Spectroscopy of Rydberg atoms and relativistic quantum chaos, OSENU, Odessa, p 152

100. Ternovsky VB, Buyadzhi VV, Gurskaya MY, Kuznetsova AA (2015) Preprint OSENU, NAM-3 (Odessa, 2015), p 32; Preprint OSENU, OSENU, NAM-4 (Odessa, 2015), p 36

Enhancement Factors for Positron Annihilation on Valence and Core Orbitals of Noble-Gas Atoms

D. G. Green and G. F. Gribakin

Abstract Annihilation momentum densities and vertex enhancement factors for positron annihilation on valence and core electrons of noble-gas atoms are calculated using many-body theory for s, p and d-wave positrons of momenta up to the positronium-formation threshold. The enhancement factors parametrize the effects of short-range electron-positron correlations which increase the annihilation probability beyond the independent-particle approximation. For all positron partial waves and electron subshells, the enhancement factors are found to be relatively insensitive to the positron momentum. The enhancement factors for the core electron orbitals are also almost independent of the positron angular momentum. The largest enhancement factor ($\sim$10) is found for the 5p orbital in Xe, while the values for the core orbitals are typically $\sim$1.5.

Keywords Positron annihilation · Annihilation momentum density Many-body theory · Enhancement factors · Noble-gas atoms

1 Introduction

Low-energy positrons annihilate in atoms and molecules forming two γ rays whose Doppler-broadened spectrum is characteristic of the electron velocity distribution in the states involved, and thus of the electron environment. This makes positrons a unique probe in materials science. For example, vacancies and defects in semiconductors and other industrially important materials can be studied [1–6]. Positron-induced Auger-electron spectroscopy (PAES) [7–11] and time-resolved PAES [11, 12] enable studies of surfaces with extremely high sensitivity, including dynamics of catalysis, corrosion, and surface alloying [13]. The γ spectra are also sensitive to the positron momentum at the instant of annihilation. This is exploited in Age-

D. G. Green (✉) · G. F. Gribakin
Centre for Theoretical Atomic, Molecular and Optical Physics,
Queen's University Belfast, Belfast, Northern Ireland BT71NN, UK
e-mail: d.green@qub.ac.uk

G. F. Gribakin
e-mail: g.gribakin@qub.ac.uk

© Springer International Publishing AG, part of Springer Nature 2018
Y. A. Wang et al. (eds.), *Concepts, Methods and Applications of Quantum Systems in Chemistry and Physics*, Progress in Theoretical Chemistry and Physics 31,
https://doi.org/10.1007/978-3-319-74582-4_14

MOmentum Correlation (AMOC) experiments (see, e.g., [14–16]), in which the γ spectra are measured as a function of the positron "age" (i.e., time after emission from source). AMOC enables study of positron and positronium cooling (and, more generally, transitions between positron states, e.g., for different trapping states, or via chemical reactions) [14–16].

Interpretation of the experiments relies heavily on theoretical input, e.g., in PAES one requires accurate relative annihilation probabilities for core electrons of various atoms [17]. Such quantities, however, are not easy to calculate, as the annihilation process is strongly affected by short-range electron-positron and long-range positron-atom correlations. These effects significantly enhance the annihilation rates [18, 19] and alter the shape and magnitude of the annihilation γ spectra [20–24], compared to independent-particle approximation (IPA) calculations.

A powerful method that allows for systematic inclusion of the correlations in atomic systems is many-body theory (MBT). MBT enables one to calculate the so-called *enhancement factors* (EF), which quantify the increase of the electron density at the positron due to the effect of correlations. The EF can be used to correct the IPA annihilation probabilities and γ-spectra [2, 17]. They are particularly large (~ 10) for the valence electrons, but are also significant for the core electrons [25].[1] EF were introduced in early MBT works involving positron annihilation in metals that were based on considering positrons in a homogeneous electron gas [28, 29]. Subsequently, density functional theories (DFT) were developed to describe positron states and annihilation in a wider class of condensed-matter systems [30, 31]. These methods usually rely on some input in the form of the correlation energy and EF for the positron in electron gas from MBT [32]. When applied to real, inhomogeneous systems, position-dependent EF can lead to spurious effects in the spectra [5], and show deficiencies when benchmarked against more accurate calculations [33].

A recent study of a model system of eight electrons and one positron confined in a harmonic potential [34] highlighted significant discrepancies between the annihilation momentum densities (AMD) and EF obtained using exact diagonalization and those found using common DFT approaches. The best agreement for the shapes of AMD was observed for position-dependent EF [35, 36] calculated in the Kahana formalism [36, 37], though there was a factor-of-two difference for the total annihilation rate. It was also suggested in [34] (see also [38]) that the EF could be defined rigorously using natural geminals. These quantities are electron-positron pair wavefunctions which diagonalise the two-body reduced density matrix, and which can be extracted from the accurate many-particle wavefunction. It would be very useful if the natural geminals could be used without the knowledge of the total wavefunction, and it is possible that this can be done using MBT.

In the context of the positron-atom problem, the MBT calculations provided an accurate and essentially complete picture of low-energy positron interaction with

---

[1] Positron annihilation with core electrons is also affected by exchange-assisted tunnelling [26, 27]. This is a manifestation of electron exchange, which increases the wavefunctions of inner electrons in the range of distances of the valence electrons. For this effect to be properly included in a calculation, one needs to use true nonlocal exchange potentials, e.g., at the Hartree-Fock level, as is the case in the present calculations.

noble-gas atoms [19], with excellent agreement between the theoretical results and experimental scattering cross sections and annihilation rates. The MBT work was extended recently [24] to the γ-spectra for (thermal) positron annihilation on noble-gas atoms. It provided an accurate description of the measured spectra for Ar, Kr and Xe [3] and firmly established the relative contributions of various atomic orbitals to the spectra. The calculations also yielded "exact" ab initio EF $\bar{\gamma}_{nl}$ for individual electron orbitals nl, and found that they follow a simple scaling with the orbital ionization energy [24].

In this work we provide a more detailed analysis and report EF for annihilation of s-, p- and d-wave positrons with momenta up to the positronium formation threshold. We demonstrate that the EF for a given electron orbital and positron partial wave are insensitive to the positron momentum (in spite of the strong momentum dependence of the annihilation probability [19]). Moreover, we show that whilst the EF for the core orbitals are almost independent of the positron angular momenta, those for the valence subshells vary between the positron s, p and d waves. In addition to their use in correcting IPA calculations of positron annihilation with core electrons in condensed matter, the positron-momentum dependent EF calculated here can be used to determine accurate pick-off annihilation rates for positronium in noble gases [39].

2 Theory of Positron Annihilation in Many-Electron Atoms

2.1 Basics

Consider annihilation of a low-energy ($\varepsilon \sim 1$ eV) positron with momentum $\mathbf{k}$ in a many-electron system, e.g., an atom. In the dominant process, the positron annihilates with an electron in state n to form two γ-ray photons of total momentum $\mathbf{P}$ [40]. In the centre-of-mass frame, where the total momentum $\mathbf{P}$ is zero, the two photons are emitted in opposite directions and have equal energies $E_\gamma = p_\gamma c = mc^2 + \frac{1}{2}(E_i - E_f) \simeq mc^2 \simeq 511$ keV, where E_i and E_f denote the energy of the initial and final states of the system (excluding rest mass). When $\mathbf{P}$ is non-zero, however, the two photons no longer propagate in exactly opposite directions and their energy is Doppler shifted. For example, for the first photon $E_{\gamma_1} = E_\gamma + mcV \cos \theta$, where θ is the angle between the momentum of the photon and the centre-of-mass velocity of the electron-positron pair $\mathbf{V} = \mathbf{P}/2m$ (assuming that $V \ll c$, and $p_{\gamma_1} = E_{\gamma_1}/c \approx mc$). The Doppler shift of the photon energy from the centre of the line then is

$$\epsilon = E_{\gamma_1} - E_\gamma = mc \, V \cos \theta = \frac{Pc}{2} \cos \theta. \tag{1}$$

The typical momenta of electrons bound with energy ε_n determine the characteristic width of the annihilation spectrum $\epsilon \sim Pc \sim \sqrt{|\varepsilon_n| mc^2} \gg |\varepsilon_n|$. Hence the shift $\varepsilon_n/2$ of the line centre E_γ from $mc^2 = 511$ keV can usually be neglected, even for the core electrons. The γ spectrum averaged over the direction of emitted photons (or that of the positron momentum $\mathbf{k}$) takes the form (see, e.g., [20])

$$w_n(\epsilon) = \frac{1}{c} \int \int_{2|\epsilon|/c}^{\infty} |A_{n\mathbf{k}}(\mathbf{P})|^2 \frac{PdPd\Omega_{\mathbf{P}}}{(2\pi)^3}, \tag{2}$$

where $A_{n\mathbf{k}}(\mathbf{P})$ is the annihilation amplitude, whose calculation using MBT is described below. The quantity $|A_{n\mathbf{k}}(\mathbf{P})|^2$ is the annihilation momentum density.[2]

The annihilation rate λ for a positron in a gas of atoms or molecules with number density n_m is usually parametrized by

$$\lambda = \pi r_0^2 c n_m Z_{\mathrm{eff}}, \tag{3}$$

where $r_0 = e^2/mc^2$ is the classical radius of the electron (in CGS units) and Z_{eff} is the effective number of electrons per target atom or molecule that contribute to annihilation [42, 43]. It is found as a sum over electron states $Z_{\mathrm{eff}} = \sum_n Z_{\mathrm{eff},n}$, where

$$Z_{\mathrm{eff},n} = \int w_n(\epsilon)\, d\epsilon = \int |A_{n\mathbf{k}}(\mathbf{P})|^2 \frac{d^3\mathbf{P}}{(2\pi)^3} \tag{4}$$

is the partial contribution due to positron annihilation with electron in state n, and where it is assumed that the incident positron wavefunction used in the calculation of $A_{n\mathbf{k}}(\mathbf{P})$ is normalized to a plane wave. In general, the parameter Z_{eff} is different from the number of electrons in the target atom Z. In particular, positron-atom and electron-positron correlations can make $Z_{\mathrm{eff}} \gg Z$ [19, 24, 44–46].

2.2   Many-Body Theory for the Annihilation Amplitude

The incident positron wavefunction is taken in the form of a partial-wave expansion[3]

$$\psi_{\mathbf{k}}(\mathbf{r}) = \frac{4\pi}{r}\sqrt{\frac{\pi}{k}} \sum_{\ell m} i^\ell e^{i\delta_\ell} Y_{\ell m}^*(\hat{\mathbf{k}}) Y_{\ell m}(\hat{\mathbf{r}}) P_{\varepsilon\ell}(r), \tag{5}$$

where δ_ℓ is the scattering phaseshift [47], $Y_{\ell m}$ is the spherical harmonic, and where the radial function with orbital angular momentum ℓ is normalized by its asymptotic behaviour $P_{\varepsilon\ell}(r) \simeq (\pi k)^{-1/2} \sin(kr - \pi\ell/2 + \delta_\ell)$. In the simplest approximation the radial wavefunctions are calculated in the static field of the ground-state (Hartree-Fock, HF) atom. This approximation is very inaccurate for the positron-atom problem. It fails to describe the scattering cross sections and grossly

---

[2]Alternatively to the Doppler-shift spectrum, experiments measure the one-dimensional angular correlation of annihilation radiation (1D-ACAR), i.e., the small angle Θ between the direction of one photon and the plane containing the other. The corresponding distribution can be obtained from $w(\varepsilon)$ using $\Theta = 2\varepsilon/mc^2$. Not also that if the positron wavefunction is constant, then the annihilation momentum density is proportional to the electron momentum density, and the γ spectrum becomes similar to the Compton profile [22, 23, 41].

[3]In this and subsequent sections we make wide use of atomic units (a.u.).

underestimates the annihilation rates. Much more accurate positron wavefunctions (Dyson orbitals) are obtained by solving the Dyson equation which includes the non-local, energy-dependent positron-atom correlation potential [19, 48] (Sect. 2.2.2).

In the lowest-order approximation the annihilation amplitude is given by

$$A_{n\mathbf{k}}(\mathbf{P}) = \int e^{-i\mathbf{P}\cdot\mathbf{r}} \psi_{\mathbf{k}}(\mathbf{r}) \varphi_n(\mathbf{r}) d^3\mathbf{r}, \tag{6}$$

where $\varphi_n(\mathbf{r}) \equiv \varphi_{nlm}(\mathbf{r}) = \frac{1}{r} P_{nl}(r) Y_{lm}(\hat{\mathbf{r}})$ is the wavefunction of electron in subshell nl. Equation (6) is equivalent to IPA. After integration over the directions of $\mathbf{P}$ in the spectrum (2), all positron partial waves contribute to the AMD incoherently. This means that the annihilation amplitude can be calculated independently for each ℓ, replacing $\psi_{\mathbf{k}}(\mathbf{r})$ in Eq. (6) by the corresponding positron partial wave orbital $\psi_\varepsilon(\mathbf{r})$. Omitting the index ℓ, we denote such amplitude $A_{n\varepsilon}(\mathbf{P})$.

As described below, the main corrections to the zeroth-order amplitude originate from the electron-positron Coulomb interaction which increases the probability of finding the electron and positron at the same point in space.

2.2.1 The Annihilation Vertex

Figure 1 shows the amplitude $A_{n\varepsilon}(\mathbf{P})$ in diagrammatic form [20, 21, 24, 49].[4] The total amplitude is depicted on left-hand side of the diagrammatic equation, with the double line (ε) corresponding to incoming positron that annihilates an electron in orbital n, producing two γ-rays (double-dashed line), and the circle with a cross denoting the full annihilation vertex. The main contributions to the amplitude are shown on the right-hand side of the equation. Diagram (a) is the zeroth-order amplitude [IPA, Eq. (6)], diagram (b) is the first-order correction and diagram (c) is the nonperturbative 'virtual-positronium' correction. This correction contains the shaded 'Γ-block' which represents the sum of an infinite series of electron-positron ladder diagrams shown in the lower part of Fig. 1.

The ladder diagrams represented by the Γ-block are important because the electron-positron Coulomb attraction supports bound states of the positronium (Ps) atom. To form Ps, the energy of the incident positron needs to be greater than the Ps formation threshold $E_{\mathrm{Ps}} = I - 6.8$ eV, where I is the ionization potential of the atom. However, even at lower energies where the Ps can only be formed virtually, this process gives a noticeable contribution. In practice, the Γ-block is found from the linear equation $\Gamma = V + V\chi\Gamma$, shown diagrammatically in the lower part of Fig. 1, where V is the electron-positron Coulomb interaction and χ is the propagator of the intermediate electron-positron state. Discretizing the electron and positron continua by confining the system in a spherical cavity reduces this to a linear matrix equation, which is easily solved numerically (see [19, 21, 48, 50] for further details).

---

[4]It is also possible to develop a diagrammatic expansion for Z_{eff} [19, 20, 44, 45, 48] that enables one to calculate the annihilation rate directly, rather than from Eq. (4).

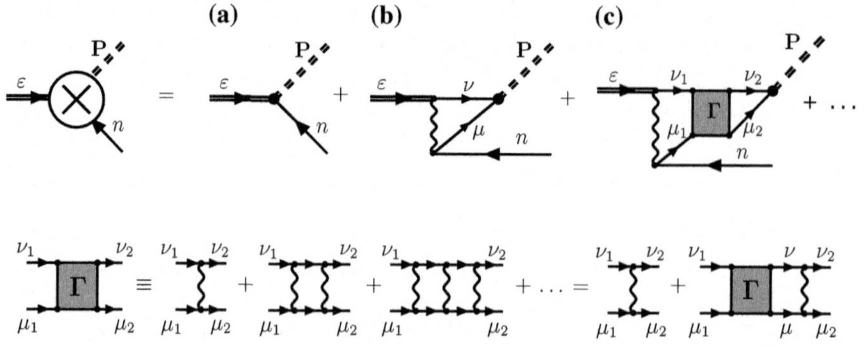

Fig. 1 Amplitude of positron annihilation with an electron in state n: **a** zeroth-order, **b** first-order, and **c** with virtual-positronium corrections. Double lines labelled ε represent the incident positron; single lines labelled ν (μ) represent positron (excited electron) states; lines labelled n represent holes in the atomic ground state; wavy lines represent the electron-positron Coulomb interaction, and double-dashed lines represent the two γ-ray photons. The Γ-block is the sum of the electron-positron ladder diagram series. Summation over all intermediate positron, electron, and hole states is assumed.

The total amplitude takes the form

$$A_{n\varepsilon}(\mathbf{P}) = \int e^{-i\mathbf{P}\cdot\mathbf{r}} \left\{ \psi_\varepsilon(\mathbf{r})\varphi_n(\mathbf{r}) + \tilde{\Delta}_\varepsilon(\mathbf{r};\mathbf{r}_1,\mathbf{r}_2)\psi_\varepsilon(\mathbf{r}_1)\varphi_n(\mathbf{r}_2)d^3\mathbf{r}_1 d^3\mathbf{r}_2 \right\} d^3\mathbf{r}. \quad (7)$$

Here, the first term, corresponding to the diagram Fig. 1a, is simply the Fourier transform of the product of electron and positron wavefunctions, taken at the same point. The second term, involving the non-local annihilation kernel $\tilde{\Delta}_\varepsilon$ (of non-trivial form), describes the vertex corrections. Note that $A_{n\varepsilon}(\mathbf{P})$ is the Fourier transform of the correlated pair wavefunction (the term in the braces[5]). References [49, 50] present the partial-wave analysis and corresponding working analytic expressions for the matrix elements involving the vertex corrections.

2.2.2 Dyson Equation for the Positron Wavefunction

As mentioned above, accurate annihilation rates and γ-spectra can be obtained only by taking into account the positron-atom correlation potential. This potential is described by another class of diagrams that "dress" the positron wavefunction. The corresponding positron *quasiparticle* wavefunction (or Dyson orbital, double line in Fig. 1) is calculated from the Dyson equation (see, e.g., [51–53])

---

[5]The term in braces can also be compared with the expression for the natural geminal corresponding to the positron state ε and electron orbital n, $\alpha_{\varepsilon n}(\mathbf{r},\mathbf{r}) = \sqrt{\gamma_{\varepsilon n}(\mathbf{r})}\psi_\varepsilon(\mathbf{r})\varphi_n(\mathbf{r})$ (cf. Eq. (9) in Ref. [34]), which can be used to determine the position dependent EF $\gamma_{\varepsilon n}(\mathbf{r})$, see Sect. 4.

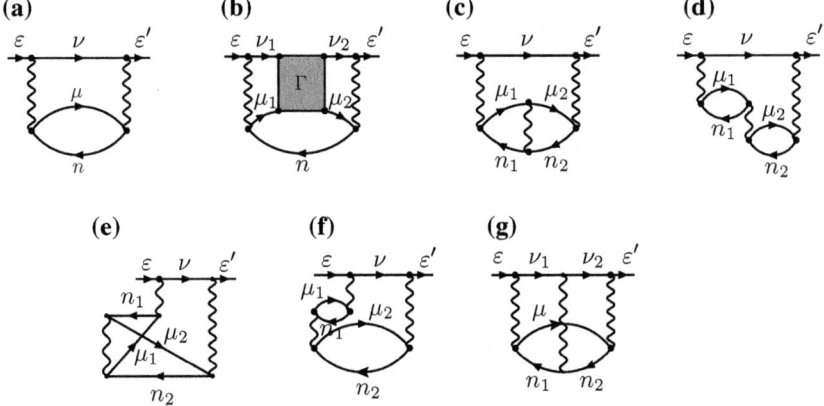

Fig. 2 Main contributions to the positron self-energy matrix $\langle \varepsilon' | \hat{\Sigma}_E | \varepsilon \rangle$. The lowest, second-order diagram **a** describes the effect of polarization; diagram **b** accounts for virtual Ps formation represented by the Γ-block. Diagrams **c–g** represent leading third-order corrections not included in (**b**). Top lines in the diagrams describe the positron. Other lines with the arrows to the right are excited electron states, and to the left, holes, i.e., electron states occupied in the target ground state. Wavy lines represent Coulomb interactions. Summation over all intermediate states is assumed

$$\left(\hat{H}_0 + \hat{\Sigma}_\varepsilon \right) \psi_\varepsilon(\mathbf{r}) = \varepsilon \psi_\varepsilon(\mathbf{r}). \tag{8}$$

Here $\hat{H}_0$ is the Hamiltonian of the positron in the static field of the N-electron atom in the ground state (described at the HF level). The nonlocal, energy-dependent correlation potential $\hat{\Sigma}_\varepsilon$ is equal to the self-energy of the positron Green's function [54], and acts as an integral operator $\hat{\Sigma}_\varepsilon \psi_\varepsilon(\mathbf{r}) = \int \Sigma_\varepsilon(\mathbf{r}, \mathbf{r}') \psi_\varepsilon(\mathbf{r}') d^3\mathbf{r}'$.

The main contributions to $\hat{\Sigma}_\varepsilon$ are shown in Fig. 2. At large positron-atom distances the correlation potential reduces to the local polarization potential $\Sigma_\varepsilon(\mathbf{r}, \mathbf{r}') \simeq -\alpha_d \delta(\mathbf{r} - \mathbf{r}')/2r^4$, where α_d is the dipole polarizability of the atom. If only diagram Fig. 2a is included, the polarizability is given by the HF approximation

$$\alpha_d = \frac{2}{3} \sum_{n,\mu} \frac{|\langle \mu | \mathbf{r} | n \rangle|^2}{\epsilon_\mu - \epsilon_n}. \tag{9}$$

Diagrams Fig. 2c–f are third-order corrections to the polarization diagram (a) of the type described by the random-phase approximation with exchange [55]. Including these gives asymptotic behaviour of $\Sigma_\varepsilon(\mathbf{r}, \mathbf{r}')$ with a more accurate value of α_d. The diagram Fig. 2b describes the virtual Ps-formation contribution. Adding it to diagram Fig. 2a nearly doubles the strength of the correlation potential in heavier noble-gas atoms (Ar, Kr and Xe). The diagram Fig. 2g describes the positron-hole repulsion. Including the diagrams of Fig. 2 in the positron-atom correlation potential provides accurate scattering phaseshifts and cross sections for all noble-gas atoms [19].

The positron self-energy diagrams and the annihilation amplitude contain sums over the intermediate excited electron and positron states. In practice we calculate them numerically using sets of electron and positron basis states constructed using 40 B-splines of order 6, in a spherical box of radius 30 a.u. We use an expotential knot sequence for the B-splines, which provides for an efficient spanning of the electron and positron continua in the sums over intermediate states [48]. The maximum angular momentum of the intermediate states is $l_{max}=15$, and we extrapolate to $l_{max} \to \infty$ as in [19, 21, 48, 50].

3 Annihilation Momentum Densities for Valence and Core Electron Orbitals in Noble Gases

Figures 3 and 4 show the AMD $|A_{n\varepsilon}(\mathbf{P})|^2$ [spherically averaged, as in Eq. (2)] for thermal ($k = 0.04$ a.u.) s-wave positrons annihilating on individual core and valence subshells of the noble gas atoms, calculated using different approximations for the annihilation amplitude and positron wavefunction. The range of two-γ momenta $P = 0$–6 a.u. corresponds to the maximum Doppler energy shift $\epsilon \approx 11$ keV.

The simplest approximation shown uses the zeroth-order (IPA) annihilation amplitude (6) with positron wavefunctions in the static field of the HF atom. Better approximations involve using the full annihilation vertex of Fig. 1 [Eq. (7)], or the best (Dyson) positron wavefunction, or both. In general, including correlations of either types increases the AMD and the annihilation probability.

General trends are observed throughout the noble-gas sequence. The AMD are broader for the core orbitals for which the typical electron momenta are greater, leading to greater Doppler shifts. The core AMD (and the core annihilation probabilities [24, 50]) also have noticeably smaller magnitudes compared with those of the valence electrons. For all electron orbitals whose radial wavefunctions have nodes (e.g., 2s in Ne, 2s, 3s and 3p in Ar, etc.) the AMD display deep minima related to the nodes of the annihilation amplitude $A_{n\varepsilon}(\mathbf{P})$. Their number and positions are related to the number and positions of the nodes in the orbital's radial wavefunction (i.e., the radial nodes that occur closer to the nucleus result in the nodes of $A_{n\varepsilon}(\mathbf{P})$ at higher momenta). This behaviour is easy to understand from the zeroth-order amplitude (6), which is the Fourier transform of the product of the electron and positron wavefunctions. For low positron energies, its wavefunction inside the atom decreases monotonically towards the nucleus (suppressed by the repulsive electrostatic potential at smaller distances), and has no nodes. Hence, the nodal structure of the annihilation amplitude is determined by the behaviour of the electron wavefunction. Inclusion of the correlation corrections to the annihilation vertex, as described by Eq. (7), leads only to a small shift in the positions of the nodes.

For a given approximation for the annihilation vertex, the AMD calculated using the positron Dyson orbitals (red curves) are larger than those calculated using the HF positron wavefunction (blue curves). The corresponding increase is nearly the

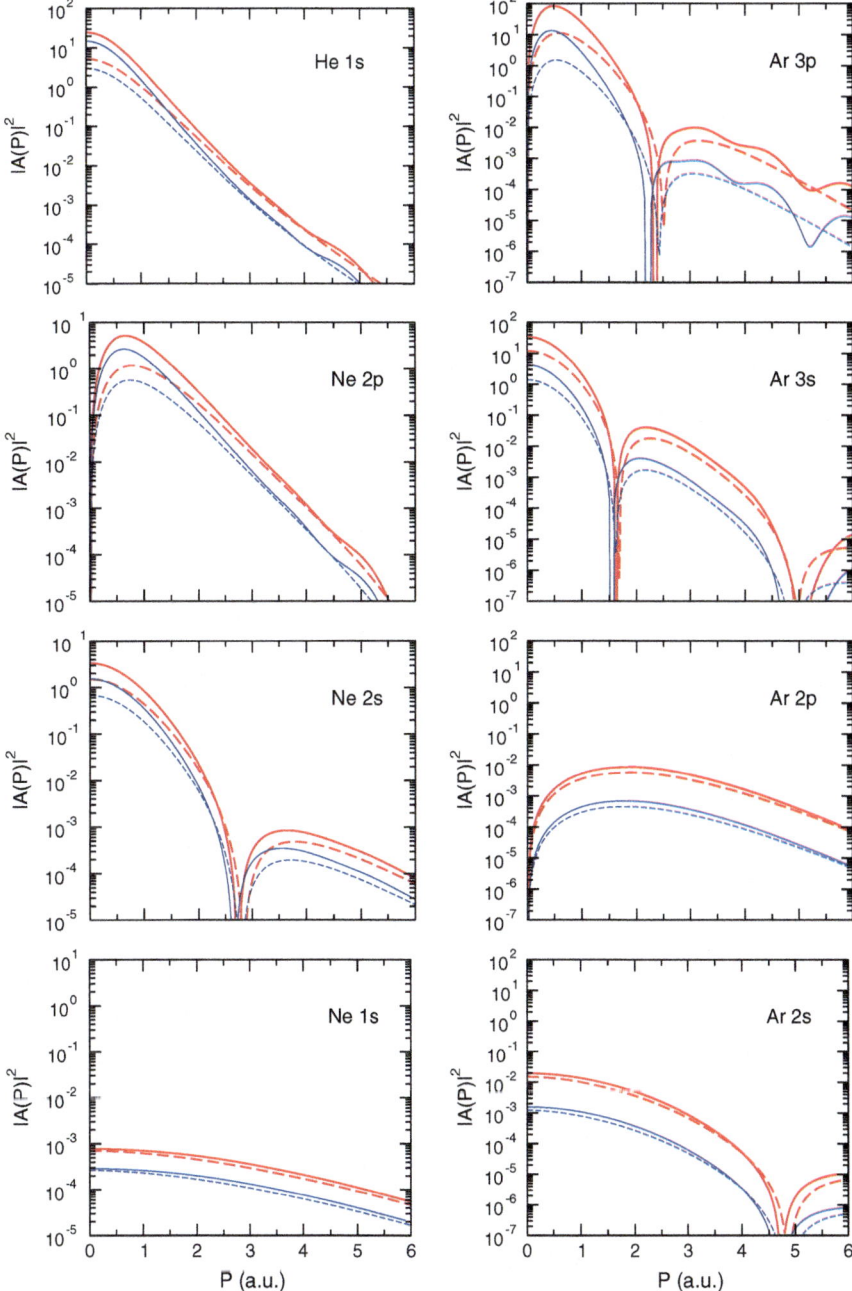

Fig. 3 Annihilation momentum density as a function of the total 2γ momentum P, for s-wave positron annihilation in He, Ne and Ar, calculated in different approximations for the annihilation vertex: zeroth-order vertex (dashed lines); full vertex $(0 + 1 + \Gamma)$ (solid lines), for HF (thin blue lines) and Dyson (thick red lines) positron of momentum $k = 0.04$ a.u. (For a given approximation for the vertex, the lines for the calculation with Dyson positron lie above those for HF positron)

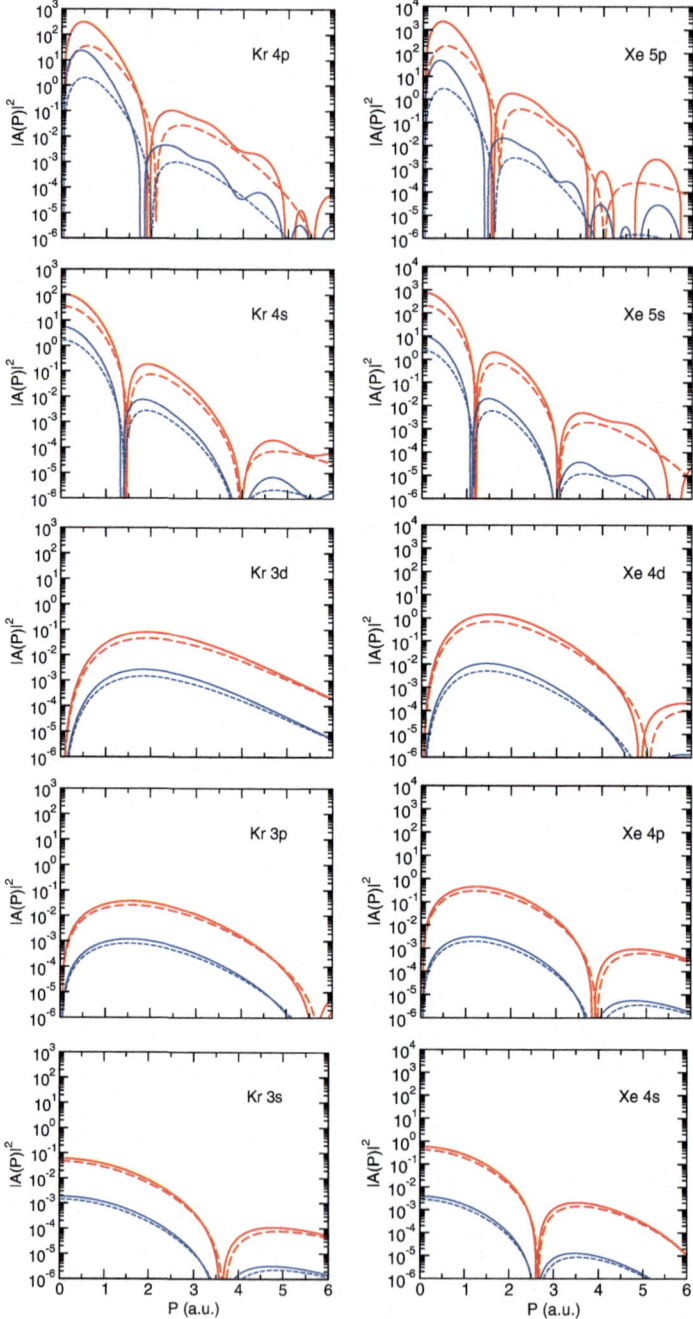

Fig. 4 Annihilation momentum densities as a function of the total 2γ momentum P, for s-wave positron annihilation in Kr and Xe. Various lines as described in Fig. 3

same for the valence and core orbitals in a given atom. It ranges from a factor of ~ 2 in Helium to ~ 100 in Xe, and is only weakly dependent on P. (Note that the AMD are plotted on the logarithmic scale.) This increase is due to the action of the attractive correlation potential $\hat{\Sigma}_\varepsilon$ on the positron (Dyson) wavefunction. It leads to a build-up of the positron density in the vicinity of the atom and greater overlap with the atomic electron density. This effect is stronger for the heavier, more polarizable atoms, leading to much greater Z_{eff} values. In Ar, Kr and Xe the attractive positron-atom potential supports low-lying virtual s levels, leading to a characteristic resonant growth of the annihilation rates at low positron energies [19, 44, 45, 56].

When the vertex corrections are included in the annihilation amplitude (solid curves), the AMD is enhanced above the zeroth-order (IPA) result (dashed curves). This is due to the Coulomb attraction within the annihilating pair, which increases the probability of finding the electron and positron at the same point in space. The size of the enhancement is similar for the HF and Dyson positron wavefunctions. At the same time, the enhancement is much greater for the valence electrons than for the core electrons, as the former are more easily perturbed by the positron's Coulomb field. For the core electrons, the vertex correction is dominated by the first-order diagram Fig. 1b (similar to the case of hydrogen-like ions [21]). For the valence electrons, the nonperturbative Γ-block contribution Fig. 1c is also very important.

From Figs. 3 and 4 one can also see that the vertex enhancement is significantly stronger at low momenta P of the electron-positron pair (which leads to narrowing of the γ-ray spectra in comparison with those obtained with the zeroth-order amplitude [20, 24]). This can be seen most clearly in the AMD of the valence electrons. Their high-P content is due to positron annihilation with the electron when the latter is closer to the nucleus, and where its local velocity is higher, making it less suscep-tible to the positron's attraction. (Note that for large P the calculated Γ-block vertex corrections contain numerical errors which manifest themselves as extra oscillations visible in AMD for valence electrons. This, however, has a negligible effect on the annihilation spectra, since the AMD for the valence electrons at such momenta are very small.) The vertex enhancement is considered in more detail below.

We conclude this section by noting that calculations of the corresponding γ spec-tra were reported in [24, 50]. They showed excellent agreement with the measured spectra for Ar, Kr and Xe [3], and firmly established the fraction of core annihilation for these atoms.

4 Vertex Enhancement Factors

The enhancement of the AMD and annihilation rates due to the correlation cor-rections to the vertex with respect to those obtained using the zeroth-order (IPA) approximation [cf. Eqs. (7) and (6)] can be parameterized through so-called *vertex enhancement factors*. The MBT enables a direct *ab initio* calculation of 'exact' EF: one simply compares the results obtained using the annihilation amplitude calculated in different approximations, as described here.

The EF were introduced originally to correct the IPA annihilation rates for positrons in condensed matter (see, e.g., [31] and references therein),

$$\lambda = \pi r_0^2 c \int n_-(\mathbf{r})n_+(\mathbf{r})\gamma(\mathbf{r})d^3\mathbf{r}, \tag{10}$$

where $n_-(\mathbf{r})$ and $n_+(\mathbf{r})$ are the electron and positron densities, respectively, and $\gamma(\mathbf{r})$ is the EF. The latter is typically computed for a uniform electron gas (e.g., using MBT [32, 57]) and parameterized in terms of the electron density, e.g., as $\gamma = 1 + 1.23r_s - 0.0742r_s^2 + \frac{1}{6}r_s^3$, where $r_s = (3/4\pi n_-)^{1/3}$ [58] (see also [30, 59]). An approximation commonly used to account for the vertex enhancement of the IPA annihilation amplitude (6) is [5, 60]

$$A_{n\varepsilon}(\mathbf{P}) = \int e^{-i\mathbf{P}\cdot\mathbf{r}}\psi_\varepsilon(\mathbf{r})\psi_n(\mathbf{r})\sqrt{\gamma(\mathbf{r})}d^3\mathbf{r}. \tag{11}$$

However, this method is known to give spurious effects in the high-momentum regions of the γ spectra [5].

The general MBT expression for the annihilation amplitude in a finite rather than infinite and homogeneous system, Eq. (7), shows that the correlation contribution to the vertex is nonlocal, i.e., it involves the positron and electron wavefunctions $\psi_\varepsilon(\mathbf{r}_1)$ and $\psi_n(\mathbf{r}_2)$ at different points in space. The corresponding enhancement is described by the three-point function $\tilde{\Delta}_\varepsilon(\mathbf{r};\mathbf{r}_1,\mathbf{r}_2)$. This allows one to formally define the EF for the electron in orbital n and positron of energy ε by

$$\sqrt{\gamma_{n\varepsilon}(\mathbf{r})} \equiv 1 + \frac{\iint \tilde{\Delta}_\varepsilon(\mathbf{r};\mathbf{r}_1\mathbf{r}_2)\psi_\varepsilon(\mathbf{r}_1)\varphi_n(\mathbf{r}_2)d^3\mathbf{r}_1 d^3\mathbf{r}_2}{\psi_\varepsilon(\mathbf{r})\varphi_n(\mathbf{r})}. \tag{12}$$

However, the presence of nodes in the wavefunctions in the denominator renders this quantity of limited use and we must opt for a more pragmatic approach.

It is clear from Figs. 3 and 4 that the vertex enhancement of the AMD $|A_{n\varepsilon}(\mathbf{P})|^2$, i.e., full-vertex results compared with zeroth-order, has a weak dependence on the momentum P (except near the nodes of the amplitude). This momentum dependence of the vertex enhancement has little effect on the annihilation γ spectra for the noble-gas atoms, especially for the core orbitals [50]. It is thus instructive to define a two-γ momentum-averaged vertex EF as the ratio of the full-vertex partial annihilation rate to that calculated using the zeroth-order (IPA) vertex:

$$\bar{\gamma}_{nl}(k) = \frac{Z_{\text{eff},nl}^{(0+1+\Gamma)}(k)}{Z_{\text{eff},nl}^{(0)}(k)}, \tag{13}$$

where the superscript denotes the vertex order (see Fig. 1) and nl labels the subshell of the electron that the positron of momentum k annihilates with. Analogous EF are commonly used to analyse and predict the annihilation rates and γ spectra in solids (see, e.g., [5, 61–66]). The true spectrum for annihilation on a given subshell for a given positron momentum can then be approximated by

$$w_{nl}(\epsilon) \approx \bar{\gamma}_{nl} w_{nl}^{(0)}(\epsilon) \tag{14}$$

where $w_{nl}^{(0)}$ is the γ spectrum calculated using the zeroth-order vertex. Accurate reconstruction of the true spectra for s-wave thermal positrons using Eq. (14) has been demonstrated for noble-gas atoms in [50].

In a recent paper [24] we showed that for thermal s-wave positrons ($k = 0.04$ a.u.) $\bar{\gamma}_{nl}$ follows a near-universal scaling with the orbital ionization energy I_{nl},

$$\bar{\gamma}_{nl} = 1 + \sqrt{A/I_{nl}} + (B/I_{nl})^{\beta}, \tag{15}$$

where A, B and β are constants.[6] The second term on the RHS of Eq. (15) describes the effect of the first-order correction, Fig. 1b. Its scaling with I_{nl} was motivated by the $1/Z$ scaling for positron annihilation in hydrogen-like ions [21]. The third term is phenomenological and describes the effect of the Γ-block correction that is particularly important for the valence subshells.

Here we extend the calculations of the enhancement factors to s-, p- and d-wave positrons with momenta up to the positronium-formation threshold. At small, e.g., room-temperature, thermal positron momenta $k \sim 0.04$ a.u., the contributions of the positron p and d waves to the annihilation rates are very small, owing to $Z_{\mathrm{eff}}(k) \propto k^{2\ell}$ low-energy behaviour. (This is a manifestation of the suppression of the positron wavefunction in the vicinity of the atom by the centrifugal potential $\ell(\ell + 1)/2r^2$.) However, for higher momenta close to the Ps-formation threshold, the s-, p- and d-wave contributions to the annihilation rates become of comparable magnitude (see Fig. 16 and Tables III–VII in Ref. [19]).

Figures 5–9 show the enhancement factors for positron annihilation with electrons in the valence (np and ns) and core [$(n-1)$s, $(n-1)$p, $(n-1)$d, as applicable] orbitals of He, Ne, Ar, Kr, and Xe, as functions of the positron momentum.

The EF for positron annihilation with 1s electrons in He (Fig. 5) are 2.6–3.0 for the s-wave, 3.8–4.1 for the p-wave, and 5.2–5.9 for the d-wave. They show only a weak dependence on the positron momentum, which is a typical feature of all the data. There is also little difference between the EF obtained with the static-field (HF) positron wavefunctions (dashed lines) and those found using the positron Dyson orbitals (solid lines). This is in spite of the fact that the use of the correlated Dyson positron wavefunctions increases the AMD (and the annihilation rates [19]) by almost an order of magnitude for s-wave positrons (Fig. 3).

The weak dependence of the EF on the positron energy and the type of positron wavefunction used is related to the nature of the vertex enhancement. The intermediate electron and positron states in diagrams Fig. 1b and c that describe the

---

[6]For HF positron wavefunctions the values of the parameters are $A = 1.54$ a.u. $= 42.0$ eV, $B = 0.92$ a.u. $= 24.9$ eV, and $\beta = 2.54$. For Dyson positron wavefunctions the values are $A = 1.31$ a.u. $= 35.7$ eV, $B = 0.83$ a.u. $= 22.7$ eV, and $\beta = 2.15$ [24].

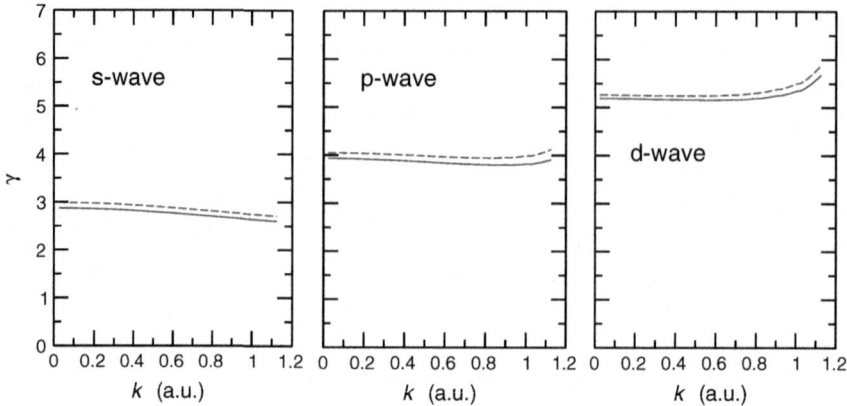

Fig. 5 Enhancement factors for s-, p- and d-wave positrons annihilating on the 1s electrons in He, obtained with HF (dashed lines) and Dyson (solid lines) positron wavefunctions

short-range vertex enhancement (v, μ, v_1, μ_1, etc.) are highly virtual, i.e., have relatively large energies. For example, the energy denominator of diagram Fig. 1b is $\varepsilon - \varepsilon_v - \varepsilon_\mu + \varepsilon_n$ (see Refs. [20, 50]). Estimating the typical electron and positron energies as $\varepsilon_{v,\mu} \sim |\varepsilon_n|$ (the ionization energy of electron orbital n), we see that for few-electronvolt positrons, the positron energy ε can be neglected. For the same reason, the vertex correction function $\tilde{\Delta}_\varepsilon(\mathbf{r}; \mathbf{r}_1, \mathbf{r}_2)$ is only weakly nonlocal, i.e., it is large only for $|\mathbf{r}_1 - \mathbf{r}_2| \ll |\mathbf{r}_{1,2}| \sim |\mathbf{r}|$ (see the "annihilation maps" in Figs. 4.14–4.16 of Ref. [67]). The situation becomes different at large momenta close to the Ps formation threshold. Here the p- and d-wave EF show an upturn related to the virtual Ps formation becoming "more real" [68]. This is also seen in $\bar{\gamma}_{np}$ for heavier atoms.

The increase of the EF with the positron orbital angular momentum ℓ seen in Fig. 5 can be related to the behaviour of the low-energy positron wavefunctions near the atom. Due to the action of the centrifugal potential, the p- and d-wave radial wavefunctions are suppressed as $(kr)^\ell$ with $\ell = 1$ and 2, compared with the s wave. The nonlocal correlation corrections Fig. 1b and c "help" the positron to pull the atomic electron towards larger distances, which has a greater advantage for the higher partial waves.

It is interesting to compare the values of $\bar{\gamma}_{1s}$ for He with the EF for positron annihilation with atomic hydrogen: 6–7, 10–12, and 15–17, for the s-, p-, and d-wave positrons, respectively, with $k \leq 0.4$ a.u. (see Fig. 13 in Ref. [48]). The greater values of the EF for hydrogen are related to the smaller binding energy of the 1s electron in hydrogen (13.6 eV) compared with that in He (24.6 eV). The vertex corrections are generally greater for the more weakly bound electrons that have more diffuse orbitals and are more easily perturbed by the positron's Coulomb interaction. The same trend will be seen throughout the noble-gas-atom sequence, with more strongly bound electron orbitals, in particular those in the core, displaying smaller EF [cf. Eq. (15)].

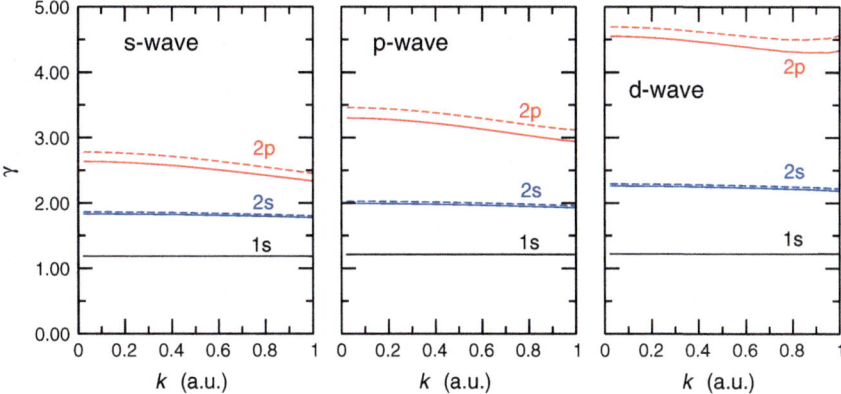

Fig. 6 Enhancement factors for s-, p- and d-wave positrons annihilating on the 1s, 2s and 2p sub-shells in Ne, obtained with HF (dashed lines) and Dyson (solid lines) positron wavefunctions.

Turning to Ne (Fig. 6), we observe that the EF for the outer valence 2p subshell are slightly smaller than those for 1s in He, in spite of the binding energy of the 2p electrons (21.6 eV) being lower than that of He 1s. Ne also has the broadest γ ray spectrum of all the noble gases (see AMD in Fig. 3, and the data for the calculated and measured spectra [50, 69]). The latter indicates that the 2p electrons in Ne have large typical momenta, which makes the correlation correction to the annihilation vertex relatively small. The EF for the inner valence 2s subshell is around 2, while for the deeply bound 1s electrons, $\bar{\gamma}_{1s} \approx 1.2$. We also note that for the core orbitals, the values of the EF for the positron s, p and d waves are quite close. This is in fact a general trend observed for all atoms that the *relative difference* between the values of $\bar{\gamma}_{nl} - 1$ for the positron s, p and d waves is becoming small with the increase in the binding energy. The smaller effect of the orbital angular momentum of the positron on the EF for core orbitals is due to the vertex correction becoming "more local", and hence, less sensitive to the variation of the positron radial wavefunction.

The EF in Ar, Kr and Xe (Figs. 7, 8 and 9) become progressively larger, for both the valence and core electrons. For example, the vertex EF for s-wave positron annihilation with the outer valence np electrons increases from $\bar{\gamma}_{3p} = 5.2$ (Ar), to $\bar{\gamma}_{4p} - 6.6$ (Kr), to $\bar{\gamma}_{5p} = 9.2$ (Xe) (for the HF positron wavefunction at low momenta $k \lesssim 0.1$ a.u.). The EF for the $(n-1)l$ core orbitals also increase to $\bar{\gamma}_{(n-1)l} \sim 1.5$–2, with the values for the 3d and 4d orbitals being noticeably larger than those of the 3s/3p and 4s/4p orbitals, for Kr and Xe, respectively.

Another feature of the data is the growing difference between the EF for the np electrons obtained with the Dyson positron wavefunction (solid lines) and those found using the static-field (HF) positron wavefunction (dashed lines). This effect is related to the increase in the strength of the positron-atom correlation potential $\hat{\Sigma}_\varepsilon$ for the heavier noble-gas atoms [19, 44, 45]. For s-wave positrons it results in the creation of positron-atom virtual levels [47] whose energies $\varepsilon = \kappa^2/2$ become lower for heavier atoms, with values of $\kappa = -0.23, -0.10$ and -0.012 a.u. for Ar,

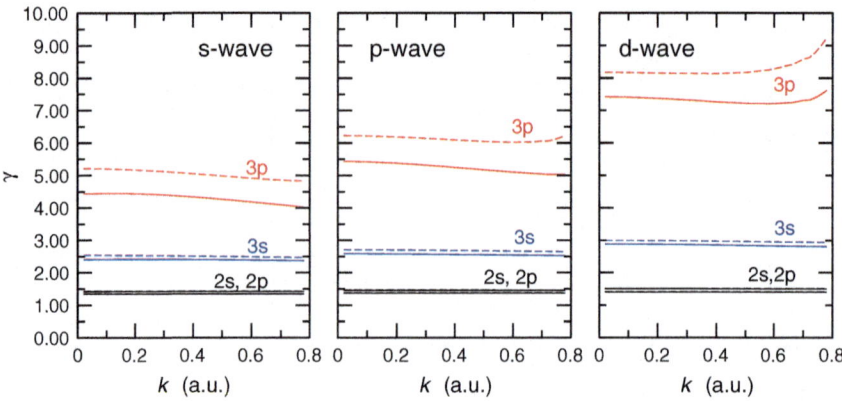

Fig. 7 Enhancement factors for s-, p- and d-wave positrons annihilating on the 2s, 2p, 3s and 3p subshells in Ar, obtained with HF (dashed lines) and Dyson (solid lines) positron wavefunctions

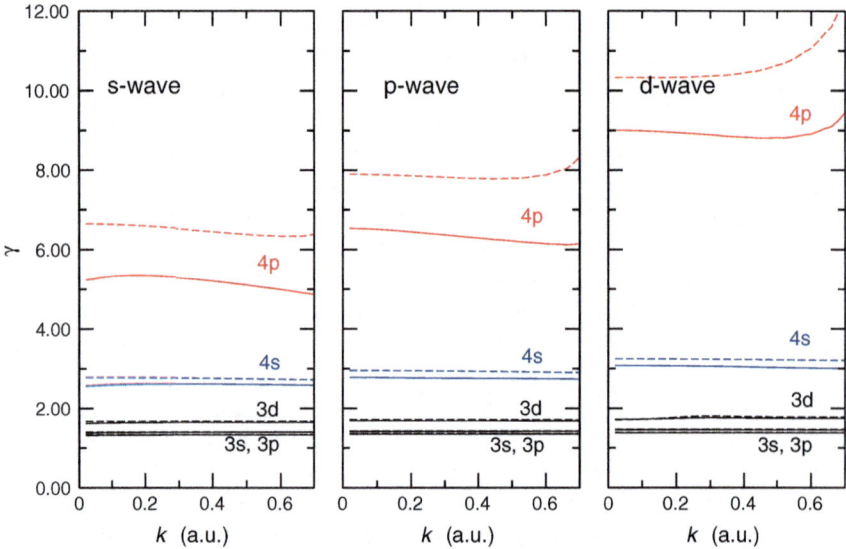

Fig. 8 Enhancement factors for s-, p- and d-wave positrons annihilating on the 3s, 3p, 3d, 4s and 4p subshells in Kr, obtained with HF (dashed lines) and Dyson (solid lines) positron wavefunctions

Kr and Xe, respectively [19]. This is accompanied by a rapid growth of the positron wavefunction near the atom, with the Dyson orbitals being enhanced by a factor $\sim 1/|\kappa|$ compared to the static-field positron wavefunctions at low energies. Hence, the inclusion of the correlation potential makes the radial dependence of the positron wavefunction more vigorous. This is evidenced by some broadening of the γ spectra obtained with the Dyson rather than the HF positron wavefunction [50]. This also

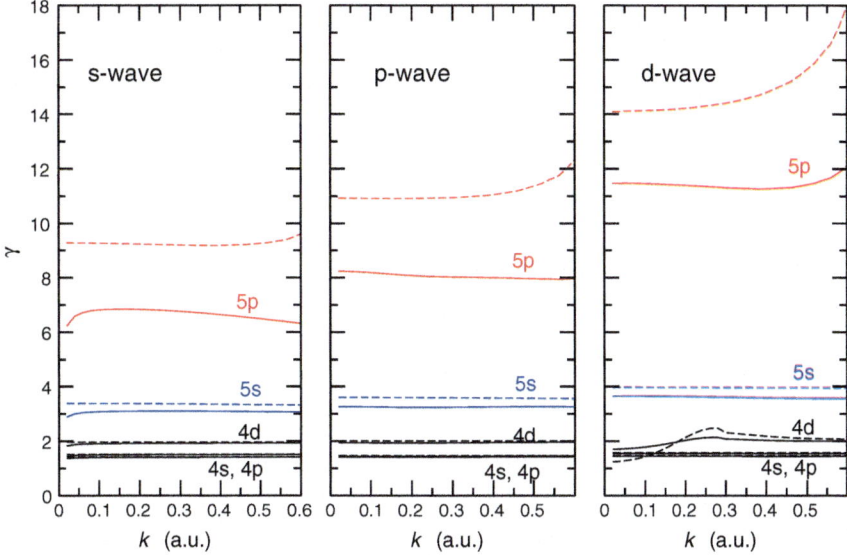

Fig. 9 Enhancement factors for s-, p- and d-wave positrons annihilating on the 4s, 4p, 4d, 5s and 5p subshells in Xe, obtained with HF (dashed lines) and Dyson (solid lines) positron wavefunctions

results in a reduction of the EF, which is most noticeable for the valence electrons, and is largest in Xe, which has the strongest correlation potential for the positron.

Besides the rise in the valence EF for high positron momenta (related to the proximity of the Ps-formation threshold), one other exception from the weak momentum-dependence of the EF is seen at low momenta for d-wave positron annihilating on the 3d and 4d orbitals in Kr and Xe. The enhancement factors in this case are approximately constant from the Ps-formation threshold down to $\sim$0.3 a.u., but then deviate at lower momenta, especially in Xe. Both the zeroth-order and full-vertex Z_{eff} values for these orbitals are calculated to be smooth functions of k. However, they obey the $\sim k^4$ behaviour and become very small at low k (e.g., for Xe, using the Dyson wavefunction we find $Z_{\mathrm{eff},4d} \sim 10^{-3}$ at $k \sim 0.3$ a.u., decreasing to $\sim 10^{-7}$ for $k \sim 0.03$ a.u. It appears that for such small k numerical inaccuracies arise in the calculation of the Γ-block contribution, leading to errors when evaluating the ratio in Eq. (13).

5 Conclusions

We used many-body theory methods to calculate the annihilation momentum densities and vertex enhancement factors for s-, p- and d-wave positrons annihilating on valence and core electrons in noble-gas atoms. The general trends of the EF is their weak dependence on the positron momentum and decrease with the increasing electron binding energy. We find that the type of the positron wavefunction used, i.e., Dyson orbital which accounts for the positron-atom correlation attraction vs repulsive static-field (HF) wavefunction, has relatively little effect on the EF, except for the valence orbitals in most polarisable targets. We also find a relatively weak dependence of the EF for the core and inner-valence electrons on the positron's orbital angular momentum.

The weak momentum-dependence of the EF obtained in positron-atom calculations suggests that they can be used to improve the calculations of positron annihilation in more complex environments. One such system is positronium colliding with noble-gas atoms, where calculations of Ps-atom pick-off annihilation rates that neglect the short-range vertex enhancement strongly underestimate the measured rates [70]. Another context where similar EF can be used is positron annihilation in molecules. Here there is a sharp contrast between the large amount of experimental information, including γ-spectra, for a wide range of molecule [69] and paucity of credible theoretical data [23, 71]. The positron-molecule problem is particularly interesting because the Z_{eff} values for most polyatomic molecules show orders-of-magnitude increases due to resonant positron annihilation [71]. In such molecules the positron annihilates from a temporarily formed weakly-bound state. Attempts to calculate such states using standard quantum-chemistry methods have been numerous but not very successful [71] (i.e., there is only a small number of systems where theory and experiment can be compared, and the agreement is mostly qualitative [72]).

The calculations presented in this paper could be extended to other atoms, including those with open shells. Partial filling of electron shells can be taken into account in the many-body-theory sums using fractional occupation numbers (cf. Ref. [73]), and the positron wavefunction is insensitive to such details of the electronic structure at the static (HF) level. (At the level of Dyson orbitals, one will need to calculate the positron self-energy using fractional electron-shell occupancies.) Calculations are particularly straightforward for annihilation with core electrons, as their enhancement factor is described well by the first-order correction to the annihilation vertex.

Acknowledgements DGG is supported by a United Kingdom Engineering and Physical Sciences Research Council Fellowship, grant number EP/N007948/1.

References

1. Asoka-Kumar P, Alatalo M, Ghosh V, Kruseman A, Nielsen B, Lynn K (1996) Phys Rev Lett 77:2097. https://doi.org/10.1103/PhysRevLett.77.2097
2. Lynn KG, MacDonald JR, Boie RA, Feldman LC, Gabbe JD, Robbins MF, Bonderup E, Golovchenko J (1977) Phys Rev Lett 38:241. https://doi.org/10.1103/PhysRevLett.38.241
3. Iwata K, Gribakin GF, Greaves RG, Surko CM (1997) Phys Rev Lett 79:39. https://doi.org/10.1103/PhysRevLett.79.39

4. Lynn K, Dickman J, Brown W, Robbins M, Bonderup E (1979) Phys Rev B 20:3566. https://doi.org/10.1103/PhysRevB.20.3566
5. Alatalo M, Barbiellini B, Hakala M, Kauppinen H, Korhonen T, Puska M, Saarinen K, Hautojärvi P, Nieminen R (1996) Phys Rev B 54:2397. https://doi.org/10.1103/PhysRevB.54.2397
6. Tuomisto F, Makkonen I (2013) Rev Mod Phys 85:1583. https://doi.org/10.1103/RevModPhys.85.1583
7. Weiss A, Mayer R, Jibaly M, Lei C, Mehl D, Lynn KG (1988) Phys Rev Lett 61:2245. https://doi.org/10.1103/PhysRevLett.61.2245
8. Ohdaira T, Suzuki R, Mikado T, Ohgaki H, Chiwaki M, Yamazaki T (1997) Appl Surf Sci 116:177. https://doi.org/10.1016/S0169-4332(96)01049-5
9. Weiss AH, Fazleev NG, Nadesalingam MP, Mukherjee S, Xie S, Zhu J, Davis BR (2007) Radiat Phys Chem 76:285. https://doi.org/10.1016/j.radphyschem.2006.03.053
10. Mayer J, Hugenschmidt C, Schreckenbach K (2010) Surf Sci 604:1772. https://doi.org/10.1016/j.susc.2010.07.003
11. Hugenschmidt C (2016) Surf Sci Rep 71:547. https://doi.org/10.1016/j.surfrep.2016.09.002
12. Hugenschmidt C, Lwe B, Mayer J, Piochacz C, Pikart P, Repper R, Stadlbauer M, Schreckenbach K (2008) Nucl Instrum Methods A 593:616. https://doi.org/10.1016/j.nima.2008.05.038
13. Mayer J, Hugenschmidt C, Schreckenbach K (2010) Phys Rev Lett 105:207401. https://doi.org/10.1103/PhysRevLett.105.207401
14. Stoll H, Koch KMM, Major J (1991) Nucl Instrum Methods B 582:56
15. Coleman P (ed) (2000) Positron beams and their applications. World Scientific
16. Sano Y, Kino Y, Oka T, Sekine T (2015) J Phys Conf Ser 618:012010. http://stacks.iop.org/1742-6596/618/i=1/a=012010
17. Jensen KO, Weiss A (1990) Phys Rev B 41:3928. https://doi.org/10.1103/PhysRevB.41.3928
18. Iwata K, Greaves RG, Murphy TJ, Tinkle MD, Surko CM (1995) Phys Rev A 51:473. https://doi.org/10.1103/PhysRevA.51.473
19. Green DG, Ludlow JA, Gribakin GF (2014) Phys Rev A 90:032712. https://doi.org/10.1103/PhysRevA.90.032712
20. Dunlop LJM, Gribakin GF (2006) J Phys B 39:1647. https://doi.org/10.1088/0953-4075/39/7/008
21. Green DG, Gribakin GF (2013) Phys Rev A 88:032708. https://doi.org/10.1103/PhysRevA.88.032708
22. Green DG, Saha S, Wang F, Gribakin GF, Surko CM (2010) Mater Sci Forum 666:21. https://doi.org/10.4028/www.scientific.net/MSF.666.21
23. Green DG, Saha S, Wang F, Gribakin GF, Surko CM (2012) New J Phys 14:035021. http://stacks.iop.org/1367-2630/14/i=3/a=035021
24. Green DG, Gribakin GF (2015) Phys Rev Lett 114:093201. https://doi.org/10.1103/PhysRevLett.114.093201
25. Bonderup E, Andersen JU, Lowy DN (1979) Phys Rev B 20:883. https://doi.org/10.1103/PhysRevB.20.883
26. Flambaum VV (2009) Phys Rev A 79:042505. https://doi.org/10.1103/PhysRevA.79.042505
27. Kozlov MG, Flambaum VV (2013) Phys Rev A 87:042511. https://doi.org/10.1103/PhysRevA.87.042511
28. Kahana S (1963) Phys Rev 129:1622. https://doi.org/10.1103/PhysRev.129.1622
29. Carbotte JP (1967) Phys Rev 155:197. https://doi.org/10.1103/PhysRev.155.197
30. Boroński E, Nieminen R (1986) Phys Rev B 34:3820. https://doi.org/10.1103/PhysRevB.34.3820
31. Puska MJ, Nieminen RM (1994) Rev Mod Phys 66:841. https://doi.org/10.1103/RevModPhys.66.841
32. Arponen J, Pajanne E (1979) Ann Phys 121:343. https://doi.org/10.1016/0003-4916(79)90101-5
33. Mitroy J, Barbiellini B (2002) Phys Rev B 65:235103. https://doi.org/10.1103/PhysRevB.65.235103

34. Makkonen I, Ervasti MM, Siro T, Harju A (2014) Phys Rev B 89:041105. https://doi.org/10.1103/PhysRevB.89.041105
35. Daniuk S, Kontrym-Sznajd G, Rubaszek A, Stachowiak H, Mayers J, Walters PA, West RN (1987) J Phys F: Metal Phys 17:1365. https://doi.org/10.1088/0305-4608/17/6/011
36. Jarlborg T, Singh AK (1987) Phys Rev B 36:4660. https://doi.org/10.1103/PhysRevB.36.4660
37. Rubaszek A , Stachowiak H (1984) Physica Status Solidi (B) 124:159. https://doi.org/10.1002/pssb.2221240117
38. Zubiaga A, Ervasti MM, Makkonen I, Harju A, Tuomisto F, Puska MJ (2016) J Phys B: At Mol Opt Phys 49:064005
39. Swann AR, Green DG, Gribakin GF arXiv: 1709.00394
40. Berestetskii VB, Lifshitz EM, Pitaevskii LP (1982) Quantum electrodynamics, 2nd edn. Pergamon, Oxford
41. Kaijser P, Smith Jr VH (1977) Adv Quantum Chem 10:37. https://doi.org/10.1016/S0065-3276(08)60578-X
42. Fraser PA (1968) Adv At Mol Phys 4:63
43. Pomeranchuk I, Eksp Zh (1949) Teor Fiz 19:183
44. Dzuba VA, Flambaum VV, King WA, Miller BN, Sushkov OP (1993) Phys Scripta T46:248. https://doi.org/10.1088/0031-8949/1993/T46/039
45. Dzuba VA, Flambaum VV, Gribakin GF, King WA (1996) J Phys B 29:3151. https://doi.org/10.1088/0953-4075/29/14/024
46. Surko CM, Gribakin GF, Buckman SJ (2005) J Phys B 38:R57. https://doi.org/10.1088/0953-4075/38/6/R01
47. Landau LD, Lifshitz EM (1977). Quantum mechanics (Non-relativistic theory). In: Course of theoretical physics, 3rd edn, vol 3. Pergamon, Oxford
48. Gribakin GF, Ludlow J (2004) Phys Rev A 70:032720. https://doi.org/10.1103/PhysRevA.70.032720
49. Green DG (2011) PhD thesis, Queen's University Belfast
50. Green DG, Gribakin GF (2015) arXiv:1502.08045
51. Abrikosov AA, Gorkov LP, Dzyalonshinkski IE (1975) Methods of quantum field theory in statistical physics. Dover, New York
52. Fetter AL, Walecka JD (2003) Quantum theory of many-particle systems. Dover, New York
53. Dickhoff WH, Neck DV (2008) Many body theory exposed!—Propagator description of quantum mechanics in many-body systems, 2nd edn. World Scientific, Singapore
54. Bell JS, Squires EJ (1959) Phys Rev Lett 3:96. https://doi.org/10.1103/PhysRevLett.3.96
55. Amusia MY, Cherepkov NA (1975) Case studies in atomic physics 5:47
56. Goldanski VI, Sayasov YS (1968) Phys Lett 13:300
57. Arponen J, Pajanne E (1979) J Phys F 9:2359. https://doi.org/10.1088/0305-4608/9/12/009
58. Barbiellini B, Puska MJ, Torsti T, Nieminen RM (1995) Phys Rev B 51:7341. https://doi.org/10.1103/PhysRevB.51.7341
59. Stachowiak H, Lach J (1993) Phys Rev B 48:9828. https://doi.org/10.1103/PhysRevB.48.9828
60. Daniuk S, Kontrym-Sznajd G, Rubaszek A, Stachowiak H, Mayers J, Walters PA, West RN (1987) J Phys F 17:1365. https://doi.org/10.1088/0305-4608/17/6/011
61. Jensen KO (1989) J Phys Condens Matter 1:10595. https://doi.org/10.1088/0953-8984/1/51/027
62. Alatalo M, Kauppinen H, Saarinen K, Puska MJ, Mäkinen J, Hautojärvi P, Nieminen RM (1995) Phys Rev B 51:4176. https://doi.org/10.1103/PhysRevB.51.4176
63. Barbiellini B, Puska MJ, Alatalo M, Hakala M, Harju A, Korhonen T, Siljamäki S, Torsti T, Nieminen RM (1997) Appl Surf Sci 116:283. https://doi.org/10.1016/S0169-4332(96)01070-7
64. Barbiellini B, Hakala M, Puska MJ, Nieminen RM, Manuel AA (1997) Phys Rev B 56:7136. https://doi.org/10.1103/PhysRevB.56.7136
65. Makkonen I, Hakala M, Puska MJ (2006) Phys Rev B 73:035103. https://doi.org/10.1103/PhysRevB.73.035103

66. Kuriplach J, Morales AL, Dauwe C, Segers D, Šob M (1998) Phys Rev B 58:10475. https://doi.org/10.1103/PhysRevB.58.10475
67. Ludlow J (2003) PhD thesis, Queen's University Belfast
68. Gribakin GF, Ludlow J (2002) Phys Rev Lett 88:163202. https://doi.org/10.1103/PhysRevLett.88.163202
69. Iwata K, Greaves RG, Surko CM (1997) Phys Rev A 55:3586. https://doi.org/10.1103/PhysRevA.55.3586
70. Mitroy J, Ivanov IA (2001) Phys Rev A 65:012509. https://doi.org/10.1103/PhysRevA.65.012509
71. Gribakin GF, Young JA, Surko CM (2010) Rev Mod Phys 82:2557. https://doi.org/10.1103/RevModPhys.82.2557
72. Tachikawa M (2014) J Phys Conf Ser 488:012053. https://doi.org/10.1088/1742-6596/488/1/012053
73. Dzuba VA, Kozlov A, Flambaum VV (2014) Phys Rev A 89:042507. https://doi.org/10.1103/PhysRevA.89.042507

Geometric Phase and Interference Effects in Ultracold Chemical Reactions

N. Balakrishnan and B. K. Kendrick

Abstract Electronically non-adiabatic effects play an important role in many chemical reactions and light induced processes. Non-adiabatic effects are important, when there is an electronic degeneracy for certain nuclear geometries leading to a conical intersection between two adiabatic Born-Oppenheimer electronic states. The geometric phase effect arises from the sign change of the adiabatic electronic wave function as it encircles the conical intersection between two electronic states (e.g., a ground state and an excited electronic state). This sign change requires a corresponding sign change on the nuclear motion wave function to keep the overall wave function single-valued. Its effect on bimolecular chemical reaction dynamics remains a topic of active experimental and theoretical interrogations. However, most prior studies have focused on high collision energies where many angular momentum partial waves contribute and the effect vanishes under partial wave summation. Here, we examine the geometric phase effect in cold and ultracold collisions where a single partial wave, usually the s-wave, dominates. It is shown that unique properties of ultracold collisions, including isotropic scattering and an effective quantization of the scattering phase shift, lead to large geometric phase effects in state-to-state reaction rate coefficients. Illustrative results are presented for the hydrogen exchange reaction in the fundamental $H+H_2$ system and its isotopic counterparts.

Keywords Geometric phase · Ultracold molecules · Ultracold chemistry Ultracold collisions

N. Balakrishnan (✉)
Department of Chemistry, University of Nevada, Las Vegas, NV 89154, USA
e-mail: naduvala@unlv.nevada.edu

B. K. Kendrick
Los Alamos National Laboratory, Theoretical Division (T-1, MS B221),
Los Alamos, NM 87545, USA

© Springer International Publishing AG, part of Springer Nature 2018   265
Y. A. Wang et al. (eds.), *Concepts, Methods and Applications of Quantum Systems in Chemistry and Physics*, Progress in Theoretical Chemistry and Physics 31,
https://doi.org/10.1007/978-3-319-74582-4_15

1 Introduction

The Born-Oppenheimer approximation is the basis of much of the development in electronic structure theory, quantum dynamics of nuclear motion and molecular spectroscopy. The approach exploits the vast difference in timescale for electronic and nuclear motion due to the small electron/nuclei mass ratio. In this approach, the Schrödinger equation for electronic motion is solved for various fixed nuclear configurations and the resulting electronic energy as a function of the nuclear degrees of freedom is called the electronic potential energy surface (PES). The nuclei evolve under the influence of this electronic PES (or PESs) and subsequent solution of the nuclear Schrödinger equation yields energy levels for rotational, vibrational and translational motion of the nuclei. This two-step procedure for quantum chemical dynamics has been wildly successful for many elementary chemical reactions. However, when there is an electronic degeneracy for certain nuclear configurations, i.e., a conical intersection between two electronic PESs, this adiabatic solution of the Schrödingier equation for nuclear motion breaks down and a fully non-adiabatic treatment is desirable. Such non-adiabatic treatments which include the coupling between the ground and excited electronic PESs are computationally challenging and not practical for the vast majority of chemical reactions.

An important consequence of the electronic degeneracy is that the real-valued ground state electronic wave function changes sign when the nuclear motion encircles the conical intersection (CI) between two electronic states. The sign change requires a corresponding sign change on the nuclear motion wave function to keep the overall wave function single-valued. In other words, the nuclear motion Schrödinger equation acquires a vector potential as originally pointed out by Mead and Truhlar [1]. The effect of the vector potential is equivalent to a magnetic solenoid centered at the conical intersection [2–4]. Flux of this magnetic field through the surface enclosed by the CI yields a phase shift. This phase shift, due to its geometric origin, is referred to as the geometric phase (GP) or the Berry phase [5]. The geometric phase resulting from the vector potential is analogous to the Aharonov-Bohm effect [6] and Mead initially referred to this as the molecular Aharonov-Bohm effect [7]. There have been numerous attempts in the literature to include the geometric phase in both bound state [8–10] and scattering calculations of triatomic systems [11–23]. While bound state studies of alkali metal trimers such as Li_3 [24], Na_3 [25] and transition metal systems like Cu_3 [26] showed much better agreement with experimental results when the GP effect is included, experimental verification of the GP effects in a bimolecular chemical reaction has not been successful yet [27–33].

Almost all of the experimental studies of GP effects in bimolecular chemical reactions have so far been limited to H or D atom exchange reactions in H+HD/D+HD systems at energies close to the conical intersection [27–33]. At these high collision energies, many angular momentum partial waves contribute and any small GP effect present in a partial wave resolved cross section washes out when a summation over all partial waves is carried out to evaluate the total differential or integral cross sections [14–23]. Thus, it appears that an energy regime where only a single partial wave contributes is the most relevant regime to explore GP effects in a chemical reaction. This regime, referred to as the cold and ultracold regime, has gained much interest in recent years, thanks to the dramatic progress in cooling and trapping of molecules in the mK and μK regimes. Here, we will focus on our recent studies of the GP effect in chemical reactions in the ultracold regime taking the hydrogen exchange reaction as an illustrative example.

The ultracold regime [34–38] provides a fascinating domain to explore quantum effects in chemical reactions. Because s-wave scattering dominates at ultracold temperatures (for bosons and distinguishable particles), only the $l = 0$ partial wave contributes and the GP effect is not smeared out by partial wave summation. Furthermore, isotropic scattering in the s-wave regime allows for maximum constructive or destructive interference between wave functions along alternative paths around the CI (direct and looping/exchange paths). These properties combined with an effective quantization of the scattering phase-shift in the ultracold regime (Levinson's theorem [39] $\delta(0) = n\pi$ where δ is the phase shift and n is the number of bound states supported by the potential well) entail maximum constructive or destructive interference between the direct and exchange/looping scattering amplitudes. This leads to a large enhancement or suppression of reactivity, as recently demonstrated for $O+OH(v = 0, 1) \rightarrow H+O_2(v', j')$ [40, 41] and the hydrogen exchange processes in $H+H_2(v = 4, j = 0)$, $H+HD(v = 4, j = 0)$ and $D+HD(v = 4, j = 0)$ reactions [42–45]. The $H+H_2$ reaction has an energy barrier for vibrational levels $v < 3$ but becomes barrierless for $v > 3$ [43, 46–48]. Indeed, vibrationally adiabatic potentials for the $H+H_2$ reaction for $v = 4$ and higher vibrational levels depict an effective potential well. The bound state structure of this potential well has a dramatic effect on the scattering process at ultracold temperatures as discussed below. Also, barrierless reactions occur with much larger rate coefficients at ultracold temperatures and are more amenable to experiments than barrier reactions that proceed via tunneling.

The chapter is organized as follows. In Sect. 2 we briefly discuss the mechanism of the GP effect in ultracold reactions. Section 3 outlines the coupled channel method employed in the scattering calculations. Illustrative results of GP effects in $H+H_2/H+HD/D+HD$ reactions are presented in Sect. 4 followed by conclusions in Sect. 5.

2 Mechanism of the GP Effect in the Ultracold Regime

In our previous work [40, 42] we showed that to observe the GP effect in reactive and inelastic collisions two criteria should be satisfied: (i) the adiabatic PES must exhibit a conical intersection; (ii) the scattering amplitudes along the two scattering pathways (direct and exchange/looping) must have comparable magnitude and scatter into the same angular region. Isotropic scattering in the ultracold regime and the effective quantization of the scattering phase shift as required by Levinson's theorem provide the criterion for maximum constructive and destructive interference between

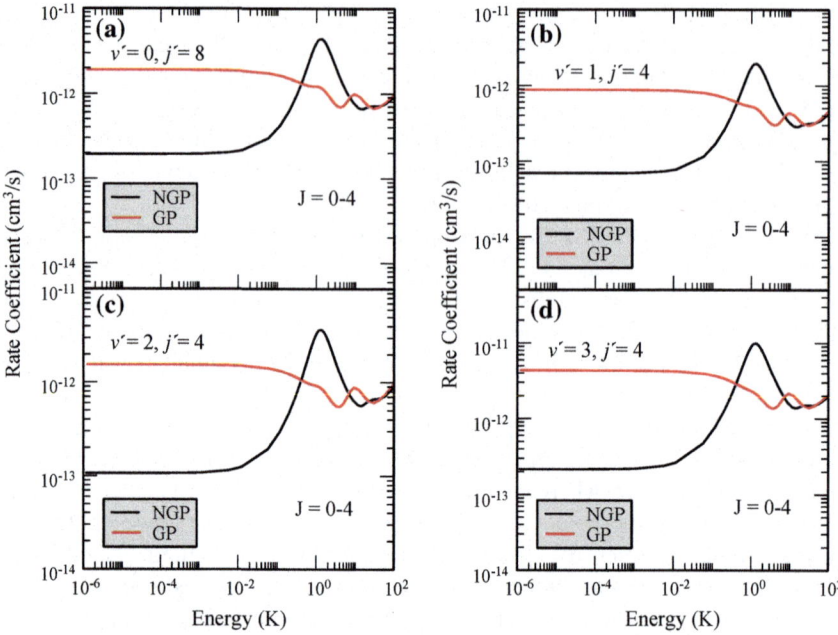

Fig. 1 Rotationally resolved reaction rate coefficients for the $H + H_2(v = 4, j = 0) \rightarrow H + H_2(v', j')$ (para-para) as functions of the incident collision energy: **a** $v' = 0, j' = 8$, **b** $v' = 1, j' = 4$, **c** $v' = 2$, $j' = 4$, and **d** $v' = 3, j' = 4$. In all panels the red curves include the geometric phase (GP) and the black curves do not (NGP). Contributions from all values of total angular momentum $J = 0$–4 are included in the rates. Reproduced with permission from [45]

the two scattering amplitudes [40–42]. The details of the interference mechanism have been discussed in prior works [40–42] and only a brief description is given here.

In H+HD collisions, due to the presence of two identical H atoms, a correct treatment of the H+HD channel should include both purely non-reactive collisions such as $H_a + H_b D(v, j) \rightarrow H_a + H_b(v', j')D$ (where a and b labels the two hydrogen atoms for illustrative purpose) and exchange collisions such as $H_a + H_b D(v, j) \rightarrow H_b + H_a D(v', j')$ where the two identical H atoms exchange with one another. The same applies to the D+HD system where the identical D atoms are exchanged in the exchange scattering amplitude. Both reactions also include purely reactive channels leading to D+H$_2$ and H+D$_2$ products. These reactive pathways may involve a "direct" path (traverses over one transition state) or a "looping path" (traverses over two transition states). For a schematic illustration of these reaction pathways, see Fig. 1a, b of Kendrick et al. [42]. Our recent studies have shown large GP effects in the H+HD and D+HD channels due to strong interference between the inelastic and exchange components of the scattering amplitudes. The GP effect was found to be not significant for the purely reactive channels due to small values of the scattering amplitudes for the looping pathway resulting in negligible interference with the dominant "direct" pathway.

Here, we focus on the exchange channel that shows the largest GP effect. Our theoretical description is based on the exchange pathways depicted in Fig. 1a of Ref. [42] and earlier works of Althorpe and collaborators [20–23].

For H+HD and D+HD systems (or in general, A+AB collisions), the GP and NGP (no geometric phase) scattering amplitudes can be written in terms of the "inelastic" and "exchange" scattering amplitudes, f_{inel} and f_{ex}, respectively,

$$f_{NGP/GP} = \frac{1}{\sqrt{2}}(f_{inel} \pm f_{ex}). \tag{1}$$

The square modulus of the scattering amplitudes for the NGP and GP calculations may be written as

$$|f_{NGP/GP}|^2 = \frac{1}{2}(|f_{inel}|^2 + |f_{ex}|^2 \pm 2|f_{inel}| \, |f_{ex}| \cos \Delta), \tag{2}$$

where the complex scattering amplitudes $f_{inel} = |f_{inel}|e^{i\delta_{inel}}$ and $f_{ex} = |f_{ex}|e^{i\delta_{ex}}$ and $\Delta = \delta_{ex} - \delta_{inel}$ is the phase difference between the exchange and inelastic pathways. For comparable values of the two scattering amplitudes, i.e., $|f_{ex}| = |f_{inel}| = |f|$, Eq. (2) becomes $|f_{NGP/GP}|^2 = |f|^2(1 \pm \cos \Delta)$. Furthermore, if $\cos \Delta = +1$ then maximum (constructive) interference occurs for the NGP case and $|f_{NGP}|^2 \sim 2|f|^2$ and $|f_{GP}|^2 \sim 0$. On the other hand, if $\cos \Delta = -1$ then maximum (constructive) interference occurs for the GP case and $|f_{GP}|^2 \sim 2|f|^2$ and $|f_{NGP}|^2 \sim 0$. Recalling that $\Delta = n\pi$ can occur in the ultracold regime (Levinson's theorem) where n is an integer, the reaction can be turned on or off depending simply on the sign of the interference term (since $| \cos \Delta | \sim 1$). In contrast, if one of the scattering amplitudes is much greater than the other, $|f_{ex}|^2 \gg |f_{inel}|^2$ or $|f_{inel}|^2 \gg |f_{ex}|^2$, then Eq. (2) becomes $|f_{NGP/GP}|^2 \sim |f_{ex}|^2/2$ or $|f_{NGP/GP}|^2 \sim |f_{inel}|^2/2$. The GP effect vanishes in this case and the interference term containing $| \cos \Delta |$ plays no role. In the high partial wave limit (high collision energies), the interference term averages out to zero ($\cos \Delta \sim 0$) and there is no GP effect. This description is also valid for the pure reactive case, e.g., H+HD→D+H$_2$ except the two scattering amplitudes for the different paths are replaced by $|f_{ex}| = |f_{loop}|$ and $|f_{inel}| = |f_{direct}|$. In our previous work, we have shown that the phase quantization of $\Delta = n\pi$ can be understood by considering scattering in a simple spherical well potential for the different pathways (i.e., Levinson's theorem $\delta_{ex} = n_{ex}\pi$ and $\delta_{inel} = n_{inel}\pi$ but with a different number of bound states n_{ex} and n_{inel} for the spherical well potentials traversed by the two pathways) [40].

3 Quantum Scattering Method

The reactive scattering calculations were carried out using hyperspherical coordinates. Two sets of hyperspherical coordinates are employed: the adiabatically-adjusting principle axis hyperspherical (APH) coordinates of Pack and Parker

[49, 50] in the inner hyper-radial region where the three-body interaction is strong and the Delves hyperspherical coordinates in the outer region where the three-body forces vanish and different atom-diatom configurations emerge. The APH coordinates are independent of the different atom-diatom arrangement channels and allow an evenhanded description of all three arrangement channels in an A+BC system compared to the Delves hyperspherical coordinates. The method accurately treats the body-frame Eckart singularities [50] associated with non-zero total angular momentum quantum number J and includes the geometric phase using the general vector potential approach [15, 16, 18]. The geometric phase is included only in the APH coordinates as it is relevant only in the region of three-body interaction where the CI is located. Regardless of the choice of the hyperspherical coordinates, the basic numerical approach involves a sector-adiabatic formalism. The hyper radius (ρ) is divided into a large number of sectors and at the center of each sector, the total wave function is expanded in terms of five-dimensional hyperspherical surface functions. The surface functions are in turn expanded in primitive angular functions. Convergence is sought with respect to the number of primitive functions included in the expansion. A sequential truncation/diagonalization procedure is used to reduce the size the surface function matrix. The expansion coefficients depend on the hyper radius but within a sector they are assumed to be independent of ρ. Coupled channel equations resulting from the Schrödinger equation with this expansion of the total wave function in terms of hyperspherical surface functions are solved from sector-to-sector. Asymptotic boundary conditions are applied in Jacobi coordinates at the last sector in ρ to evaluate the reactance and scattering matrices from which cross sections and rate coefficients are computed using standard expressions [49].

4 Results

In a series of papers [42–45], we have carried out a detailed analysis of geometric phase effects in the H+H$_2$, H+HD and D+HD reactions with the H$_2$/HD molecule excited to the $v = 4$ vibrational level. As discussed previously, the vibrationally adiabatic potential curves display a barrierless path for $v > 3$ with a small potential well compared to $v = 0$ which proceeds through an energy barrier. For the symmetric H+H$_2$ reaction, the geometric phase can be accounted for by properly symmetrizing the scattering amplitude and including a phase factor. For para-para transition (even j to even j' transitions) the properly symmetrized differential cross section is given by [18, 51]

$$\left.\frac{d\sigma}{d\Omega}\right|_{vjm\to v'j'm'} = \frac{\bar{k}_{v'j'}}{\bar{k}_{vj}}|f^N_{vjm\to v'j'm'} - (-1)^{i_{gp}}f^R_{vjm\to v'j'm'}|^2, \tag{3}$$

Fig. 2 Total reaction rate coefficients for the H + $H_2(v = 4, j = 0) \rightarrow H + H_2$ (para-para) reaction as a function of the collision energy: **a** summed over all values of total angular momentum $J = 0$–4, and **b** individual contributions from each J. Solid curves $J = 0$, dashed curves $J = 1$, dot dashed $J = 2$, dotted $J = 3$, and double-dot dashed $J = 4$. The red curves include the geometric phase (GP) and the black curves do not (NGP). Reproduced with permission from [45]

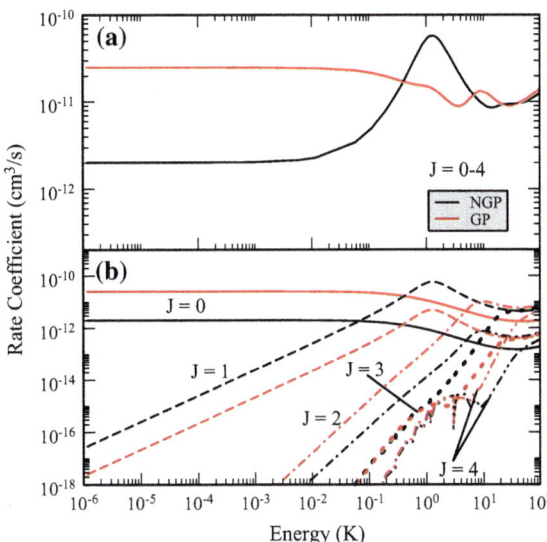

where f^N and f^R are the scattering amplitudes for the non-reactive and reactive channels, $\bar{k}_{vj}$ are the appropriately normalized wave vector magnitudes, and $i_{gp} = 1$ or 0 for calculations which include or do not include the GP, respectively (see Ref. [18] for details). As demonstrated in previous [18] studies using the vector potential approach [1] the GP effect is captured almost entirely by the sign change (i.e., $i_{gp} = 1$) given in Eq. (3). Thus, the GP can be accurately included for H + H_2 by performing calculations *without* the vector potential [18, 51]. The computed f^N and f^R are then properly combined using Eq. (3) to include ($i_{gp} = 1$) or not include ($i_{gp} = 0$) the GP. We used this approach for the H + H_2 calculations reported in this work. For H+HD and D+HD reactions, the GP effect was included using the vector potential approach. The H_3 PES of Boothroyd et al. [52] is used in the calculations reported here. We have verified that the H_3 PES of Mielke et al. [53] yields comparable results [42].

Figure 1 shows the rotationally resolved reaction rate coefficients for the H+$H_2(v = 4, j = 0) \rightarrow$ H+$H_2(v', j')$ reaction for $v' = 0, j' = 8$, $v' = 1, j' = 4$, $v' = 2, j' = 4$ and $v' = 3, j' = 4$ [45]. It is seen that the GP rates dominate the NGP rates for all four rotational levels shown in Fig. 1. Indeed, a similar trend is found for all the state-to-state rotational transitions leading to even rotational levels of H_2 for all v' levels (see Table I of Kendrick et al. [45]). As a result, a similar GP effect is found for vibrationally resolved rate coefficients for para-para transitions when summed over all rotational levels in a given vibrational state. The trend also prevails in the total rate coefficients as illustrated in Fig. 2. The upper panel of Fig. 2 shows the total rate coefficient and the lower panel displays contributions from different partial waves (orbital angular momentum $l = J$ for initial rotational level $j = 0$).

Figure 3 shows a comparison between GP and NGP rate coefficients for the H+HD($v = 4, j = 0) \rightarrow$ HD(v', j')+ H reaction for $v' = 0, j' = 3$, $v' = 1, j' = 2$, and

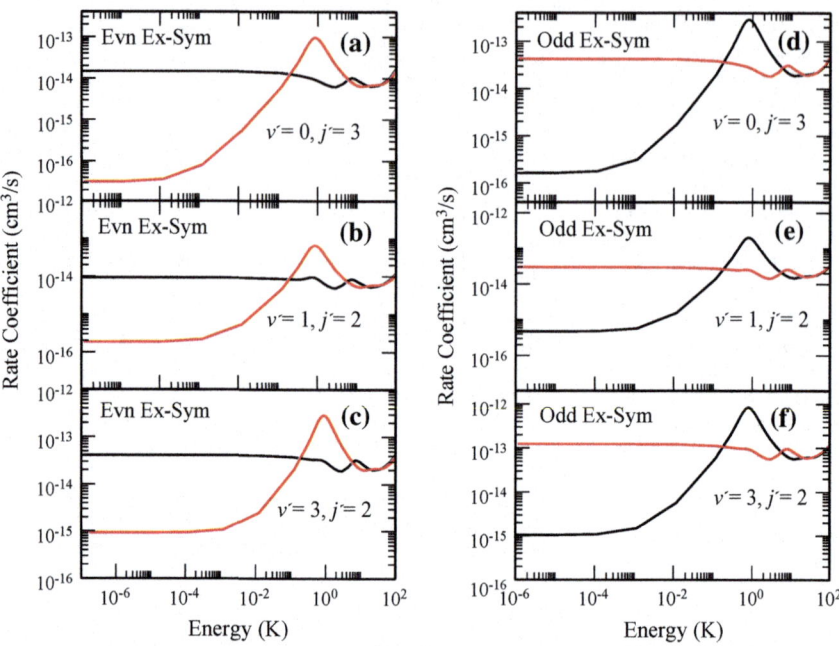

Fig. 3 Rotationally resolved rate coefficients (cross section times the relative velocity) for the H+HD($v = 4, j = 0$) → H+HD(v', j') reaction for $v' = 0, j' = 3$, $v' = 1, j' = 2$, and $v' = 3, j' = 2$ as functions of the collision energy. Even and odd exchange symmetry results are given in the left and right panels, respectively. In each panel, the red curves show the GP results and the black curves denote the NGP results. The results include all values of total angular momentum $J = 0$–4. Reproduced with permission from [43]

$v' = 3, j' = 2$ as a function of the incident collision energy. Total angular momentum quantum numbers $J = 0$–4 have been included the calculations to yield converged results up to 20 K though results are presented for energies up to 100 K. Results for even exchange symmetry are shown in the left panel and those for odd exchange symmetry are given in the right panel. It is seen that the GP and NGP results differ dramatically in the ultracold regime but they merge and show little difference for energies above 20 K. The GP/NGP effect is sensitive to the final rovibrational levels of the HD molecule and the exchange parity symmetry. Even and odd exchange symmetry results exhibit opposite GP/NGP effects and the overall GP effect becomes smaller when the total rate is computed by combining even/odd exchange symmetry contributions with appropriate nuclear spin statistics factors. This is illustrated in Fig. 4 for the H+HD($v = 4, j = 0$) → HD+ H reaction. A similar trend is observed for the corresponding D+HD reaction.

The dominance of the GP/NGP effect for a given state-to-state rate coefficient can be explained based on the sign and magnitude of $\cos \Delta$ and the scattering amplitudes for the direct and exchange pathways. This can be extracted from the GP and NGP scattering amplitudes (see Eqs. (1) and (2)). Figure 5 shows the ratio of the squares

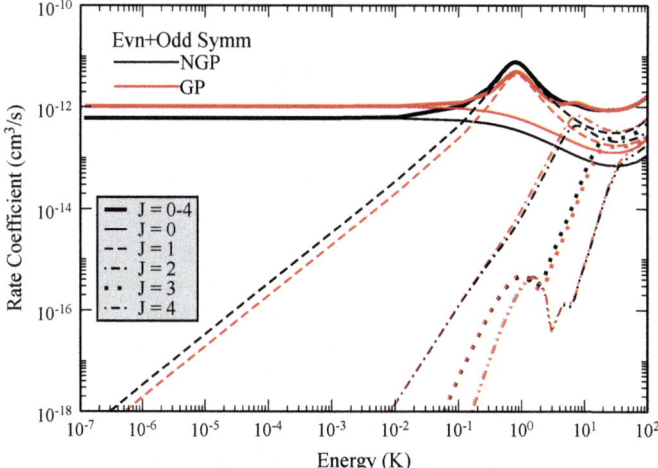

Fig. 4 Total angular momentum resolved (partial wave resolved since $l = J$ for $j = 0$) rate coefficients for the H+HD($v = 4, j = 0$) → H+HD reaction summed over all energetically accessible rotational and vibrational levels of the product HD molecule as well as even and odd exchange symmetry contributions weighted by appropriate nuclear spin statistics factors. Reproduced with permission from [43]

of the scattering amplitudes for the exchange and direct paths (upper panel) and the $\langle \cos \Delta \rangle$ values (lower panel) as a function of the collision energy for the $v' = 0, j' = 3$ transition shown in Fig. 3 for the H+HD reaction. It is seen that the NGP rates dominate when $\langle \cos \Delta \rangle = +1$ and the GP rates dominate when $\langle \cos \Delta \rangle = -1$. This is most clearly seen at energies below 1 mK where s-wave scattering dominates. At higher collision energies where non-zero partial waves contribute, the $\langle \cos \Delta \rangle$ values oscillate around zero making the contribution from the interference term in Eq. (2) less significant leading to a negligible GP effect. This trend is observed in all state-to-state rotationally resolved rate coefficients for H+HD and D+HD reactions [43, 44].

The large GP effect in state-to-state cross sections when isotropic scattering dominates in the s-wave regime is illustrated in Fig. 6 where differential cross sections (DCSs) for the $v' = 3, j' = 0$ final state in the D+HD($v = 4, j = 0$) reaction are plotted as a function of the collision energy and the scattering angle. It is seen that the NGP results are enhanced for even exchange symmetry while the GP results are enhanced for the odd exchange symmetry through constructive interference between the scattering amplitudes for the direct and exchange pathways. The interference is destructive when the DCSs for these cases are suppressed. The interference pattern changes when resonances are present as seen at energies near 1 K where a $l = 2$ shape resonance occurs for the D+HD reaction. A detailed discussion of the resonances and parameters characterizing them (position, width and lifetimes) are given in Hazra et al. [43] and Kendrick et al. [44] for the H+HD and D+HD reactions.

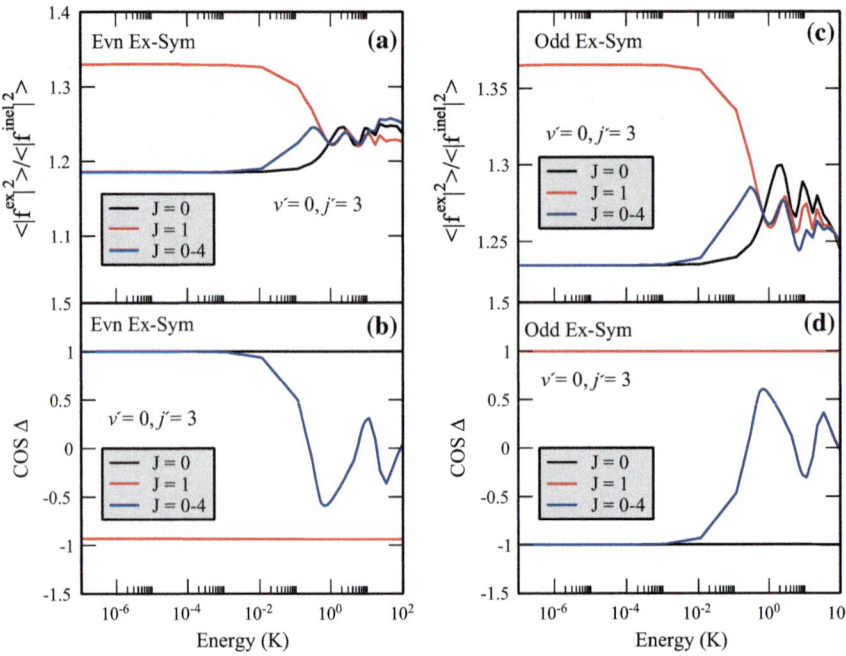

Fig. 5 Ratio of the square of the scattering amplitudes for the exchange (reactive) and inelastic (non-reactive) pathways (upper panels) and average value of cos Δ (lower panels) for both even (left panels) and odd (right panels) exchange symmetries as functions of the collision energy. Results are presented for the H+HD($v' = 0, j' = 3$) product channel. Reproduced with permission from [43]

5 Summary and Conclusions

We have discussed the importance of the geometric phase effect in ultracold hydrogen exchange reactions in collisions of H and D atoms with vibrationally excited H_2 and HD molecules. For vibrational levels $v > 3$ these reactions occur through a barrierless path. Results presented for the $v = 4$ vibrational levels of the H_2 and HD molecules illustrate strong interference between the direct and exchange components of the scattering amplitudes leading to enhancement or suppression of the reactivity. Isotropic scattering in the ultracold s-wave regime allows maximum constructive/destructive interference leading to large GP effects in state-to-state reaction rate coefficients. The effect persists but to a lesser extent in the total reaction rate, due in part, to the cancellation of the GP effects when even and odd exchange symmetry results are added to yield the total rates for H+HD and D+HD reactions. The results presented here illustrate that the GP effect may be experimentally observable by the selection of a particular nuclear spin-state of the HD molecule in H+HD/D+HD collisions.

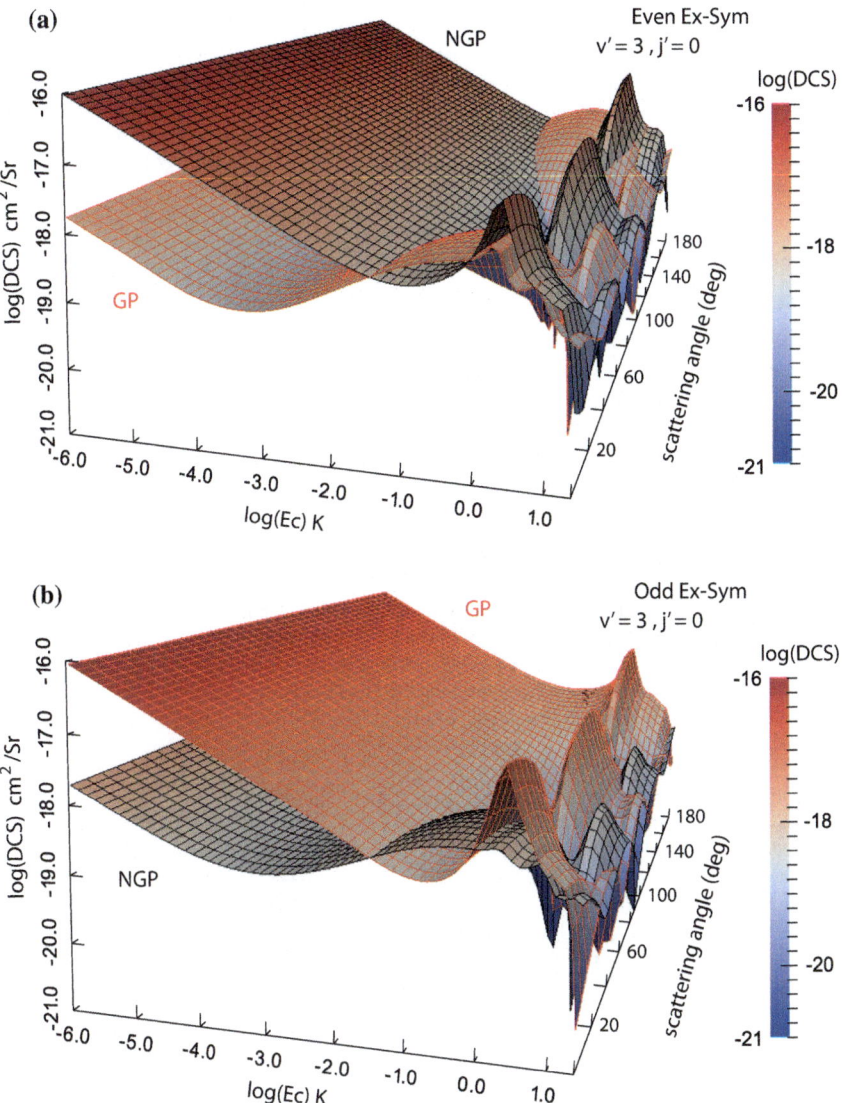

Fig. 6 The DCS is plotted as a function of collision energy and scattering angle for the D + HD($v = 4, j = 0$) → D + HD($v' = 3, j' = 0$) reaction. Panel **a** is even exchange symmetry and **b** is odd exchange symmetry. The DCS plotted with the red mesh includes the geometric phase (GP) while the one with the black mesh does not (NGP). The results include all values of total angular momentum $J = 0$–4. Reproduced with permission from [44]

Acknowledgements N. B. acknowledges support from the Army Research Office, MURI grant No. W911NF-12-1-0476 and the National Science Foundation, grant No. PHY-1505557. B. K. K. acknowledges that part of this work was done under the auspices of the US Department of Energy under Project No. 20140309ER of the Laboratory Directed Research and Development Program at Los Alamos National Laboratory. Los Alamos National Laboratory is operated by Los Alamos National Security, LLC, for the National Security Administration of the US Department of Energy under contract DE-AC52-06NA25396.

References

1. Mead CA, Truhlar DG (1979) J Chem Phys 70:2284; 78:6344E (1983)
2. Simon B (1983) Phys Rev Lett 51:2167
3. Bohm A, Boya LJ, Kendrick B (1991) Phys Rev A 43:1206
4. Mead CA (1992) Rev Mod Phys 64:51
5. Berry MV (1984) Proc R Soc Lon Ser A 392:45
6. Aharonov Y, Bohm D (1959) Phys Rev 115:485
7. Mead CA (1980) Chem Phys 49:23
8. Kendrick BK (1997) Phys Rev Lett 79:2431
9. Kendrick BK (1997) Int J Quantum Chem 64:581
10. Babikov D, Kendrick BK, Zhang P, Morokuma K (2005) J Chem Phys 122:044315
11. Kuppermann A, Wu YSM (1993) Chem Phys Lett 205:577; 213:636E (1993)
12. Wu YSM, Kuppermann A (1995) Chem Phys Lett 235:105
13. Kuppermann A, Wu YSM (1995) Chem Phys Lett 241:229
14. Kendrick BK (2001) J Chem Phys 114:8796
15. Kendrick B, Pack RT (1996) J Chem Phys 104:7475
16. Kendrick B, Pack RT (1996) J Chem Phys 104:7502
17. Kendrick BK, (2000) J Chem Phys 112:5679; 114:4335E (2001)
18. Kendrick BK (2003) J Phys Chem 107:6739
19. Kendrick BK (2003) J Chem Phys 118:10502
20. Juanes-Marcos C, Althorpe SC (2005) J Chem Phys 122:204324
21. Juanes-Marcos JC, Althorpe SC, Wrede E (2005) Science 309:1227
22. Althorpe SC (2006) J Chem Phys 124:084105
23. Althorpe SC, Stecher T, Bouakline F (2008) J Chem Phys 129:214117
24. von Busch H, Eckel V, Dev H, Kasahara S, Wang J, Demtröder W, Sebald P, Meyer W (1998) Phys Rev Lett 81:4584
25. Keil M, Krämer H-G, Kudell A, Baig MA, Zhu J, Demtröder W, Meyer W (2000) J Chem Phys 113:7414
26. Rohlfing EA, Valentini JJ (1986) Chem Phys Lett 126:113
27. Kliner DAV, Adleman DE, Zare RN (1991) J Chem Phys 95:1648
28. Adelman DE, Shafer NE, Kliner DAV, Zare RN (1992) J Chem Phys 97:7323
29. Kitsopoulos TN, Buntine MA, Baldwind DP, Zare RN, Chandler DW (1993) Science 260:1605
30. Jankunas J, Sneha M, Zare RN, Bouakline F, Althorpe SC (2013) J Chem Phys 139:144316
31. Jankunas J, Sneha M, Zare RN, Bouakline F, Althorpe SC, Phys Z (2013) Chemistry 227:1281
32. Jankunas J, Sneha M, Zare RN, Bouakline F, Althorpe SC, Herráez-Aguilar D, Aoiz FJ (2014) PNAS 111:15
33. Sneha M, Gao H, Zare RN, Jambrina PG, Menéndez M, Aoiz FJ (2016) J Chem Phys 145:024308
34. Krems RV, Stwalley WC, B. Friedrich B (eds) (2009) Cold molecules: theory, experiment, applications. CRC Press, Taylor & Francis Group
35. Carr LD, DeMille D, Krems RV, Ye J (2009) New J Phys 11:055049

36. Ospelkaus S, Ni K-K, Wang D, de Miranda MHG, Neyenhuis B, Quéméner G, Julienne PS, Bohn JL, Jin DS, Ye J (2010) Science 327:853
37. Knoop S, Ferlaino F, Berninger M, Mark M, Nägerl H-C, Grimm R, D'Incao JP, Esry BD (2010) Phys Rev Lett 104:053201
38. Balakrishnan N (2016) J Chem Phys 145:150901
39. Levinson N (1949) Kgl. Danske Videnskab Selskab Mat Fys Medd 25:9
40. Kendrick BK, Hazra J, Balakrishnan N (2015) Nat Commun 6:7918
41. Hazra J, Kendrick BK, Balakrishnan N (2015) J Phys Chem A 119:12291
42. Kendrick BK, Hazra J, Balakrishnan N (2015) Phys Rev Lett 115:153201
43. Hazra J, Kendrick BK, Balakrishnan N (2016) J Phys B At Mol Opt Phys 49:194004
44. Kendrick BK, Hazra J, Balakrishnan N (2016) New J Phys 18:123020
45. Kendrick BK, Hazra J, Balakrishnan N (2016) J Chem Phys 145:164303
46. Simbotin I, Ghosal S, Côté R (2011) Phys Chem Chem Phys 13:19148
47. Simbotin I, Ghosal S, Côté R (2014) Phys Rev A 89:040701
48. Simbotin I, Côté R (2015) N J Phys 17:065003
49. Pack RT, Parker GA (1987) J Chem Phys 87:3888
50. Kendrick BK, Pack RT, Walker RB, Hayes EF (1999) J Chem Phys 110:6673
51. Mead CA (1980) J Chem Phys 72:3839
52. Boothroyd AI, Keogh WJ, Martin PG, Peterson MR (1996) J Chem Phys 104:7139
53. Mielke SL, Garrett BC, Peterson KA (2002) J Chem Phys 116:4142

Part III
Biochemistry and Biophysics

Adducts of Arzanol with Explicit Water Molecules: An Ab Initio and DFT Study

Liliana Mammino

Abstract Arzanol ($C_{22}H_{26}O_7$) is a naturally occurring acylphloroglucinol present in *Helichrysum italicum*. It is the major responsible of its medicinal properties, which include anti-oxidant properties. In the arzanol molecule, the R of the COR group characterising acylphloroglucinols is a methyl group, and the two substituents in *meta* to COR are an α-pyrone ring, bonded to the benzene ring through a methylene bridge, and a prenyl chain. The high number of hydrogen bond donor and acceptor sites in the molecule entails an investigation taking into account solute-solvent hydrogen bonds in an explicit manner. The current work considers adducts of arzanol with explicit water molecules for a representative selection of its conformers. Adducts with one water molecule attached in turn to each of the H-bond donors or acceptors were calculated to estimate the strength with which each site can bind a water molecule. Adducts with varying numbers of water molecules were calculated to identify preferred arrangements of the water molecules around the various sites and around the molecule as a whole. These adducts also suggest possible geometries for the first solvation layer. All the adducts were calculated at the HF/6-31G(d, p) and the DFT/B3LYP/6-31+G(d, p) levels, with fully relaxed geometry.

Keywords Acylphloroglucinols · Adducts with explicit water molecules Arzanol · Hydrogen bonding · Solute-solvent interactions

Electronic supplementary material The online version of this chapter (https://doi.org/10.1007/978-3-319-74582-4_16) contains supplementary material, which is available to authorized users.

L. Mammino (✉)
Department of Chemistry, University of Venda, P/Bag X5050, Thohoyandou 0950, South Africa
e-mail: sasdestria@yahoo.com

© Springer International Publishing AG, part of Springer Nature 2018　　　　281
Y. A. Wang et al. (eds.), *Concepts, Methods and Applications of Quantum Systems in Chemistry and Physics*, Progress in Theoretical Chemistry and Physics 31, https://doi.org/10.1007/978-3-319-74582-4_16

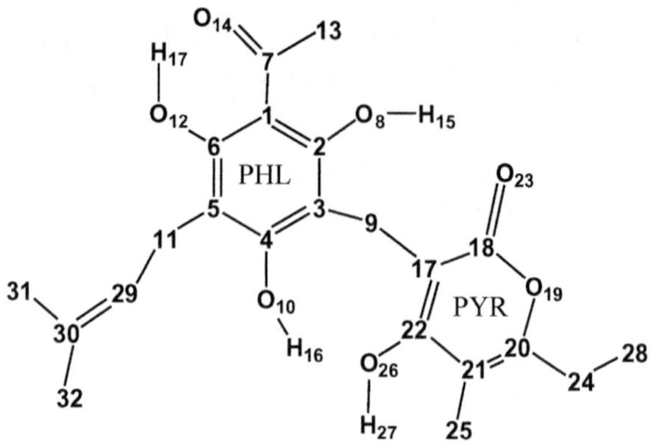

Fig. 1 Structure of the arzanol molecule and atom numbering utilized in this work [5]. The C atoms are denoted by their numbers. The figure shows the carbon skeleton of the molecule, the O atoms, and the H atoms pertaining to OH groups. The other H atoms are hidden, to better highlight the molecular structure

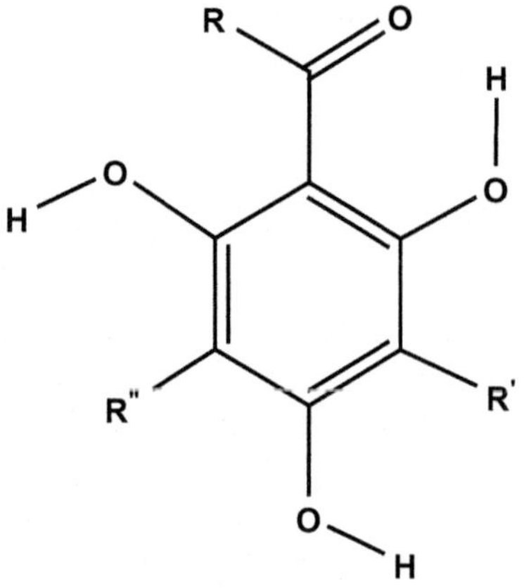

Fig. 2 General structure of acylphloroglucinols. The molecules are characterised by the presence of three equally spaced OH and a COR group

1   Introduction

Arzanol ($C_{22}H_{26}O_7$, Fig. 1) is a naturally-occurring acylphloroglucinol, and is the major responsible of the anti-inflammatory, anti-oxidant, antibiotic and antiviral activities of *Helichrysum italicum* [1–3]. Acylphloroglucinols (ACPL, Fig. 2, [4]) are derivatives of phloroglucinol (1,3,5-trihydroxybenzene) characterised by the presence of a COR group. In the arzanol molecule, R is a methyl group, R′ is an α-pyrone ring attached to the phloroglucinol moiety through a methylene bridge and R″ is a prenyl chain (Fig. 1).

Figure 1 shows the atom numbering utilized in this work, which is consistent with the numbering utilised in a thorough conformational study of the molecule [5]; for the phloroglucinol moiety, it is also consistent with the numbering utilized in previous works on ACPLs [6–10], to facilitate cross-references and comparisons. For the sake of conciseness, the two moieties and the molecule are denoted by the following acronyms in the rest of the text: PHL for the acylphloroglucinol moiety (comprising the prenyl chain attached to C5), PYR for the α-pyrone moiety, and ARZ for arzanol.

The computational study of the ARZ molecule [5] showed that its conformational preferences are influenced by the patterns of intramolecular hydrogen bonds (IHB), which are the dominant stabilising factor, by the mutual orientation of the PHL and PYR moieties (which also determines part of the IHB patterns), by the orientation of the phenol OHs (as is true for ACPLs in general [6–10]) and by the orientation of the prenyl chain (which is generally true for prenylated ACPLs [8]). The IHBs comprise the IHB formed by O14 and either H15 or H16 (here termed "first IHB" [6–10]), the IHBs between the two moieties (which will be categorised as IMHB, for 'intermoiety H-bonds', when it is relevant to underline this role, [5]) and the O10-H16···π or O12-H17···π interactions, when either O10-H16 or O12-H17 and the prenyl chain have favourable orientations. The distribution of donor and acceptor sites in the ARZ molecule enables the formation of two simultaneous IMHBs, one on either side of the methylene bridge, and all the lowest energy conformers are characterised by the presence of the first IHB and two IMHBs [5]; when the first IHB engages H15, it is cooperative with the IMHB engaging O8.

The computational study of ARZ [5] included calculations in three solvents (chloroform, acetonitrile and water) utilising the Polarizable Continuum Model (PCM, [11–13]). In general, "continuum solvation models are the ideal conceptual framework to describe solvent effects within the QM approach" [13]. However, PCM does not take into explicit account directional solute-solvent interactions such as hydrogen bonding [14] (except implicitly for some effects [15]). On the other hand, solute-solvent H-bonding is important for solute molecules containing H-bond donors or acceptors and solvent molecules capable of forming H-bonds. The most important of these solvents is water, which constitutes the highest proportion of the mass of living organisms. The consideration of adducts with explicit water molecules is the most informative option on solute-solvent H-bonding utilising QM approaches. It can provide information about preferential arrangements of water molecules in the vicinity of the various donor or acceptor sites. It can also contribute information on the outcome of the competition between intramolecular H-bonding and intermolecular solute-water H-bonding through energetics comparisons (by comparing an adduct maintaining a certain IHB and an adduct in which its donor or acceptor is engaged in a solute-solvent H-bond), and also through the optimisation itself, which may 'open' (break) specific IHBs, as verified, e.g., in the study of adducts of caespitate [16] or other ACPLs [17] with explicit water molecules.

This work considers adducts of various conformers of ARZ with explicit water molecules, trying to identify patterns for the energy of the solute-solvent interaction at different binding sites and for preferred arrangements of water molecules around different sites of the ARZ molecule. The study appears to be particularly interesting because of the high number of H-bond donors and acceptors in the ARZ molecule and because of the presence of IHBs (including cooperative ones), which influences the way in which water molecules approach the corresponding regions. The work pertains to an ongoing study of antioxidant ACPLs which, for ARZ, has so far produced the results reported in [5, 18].

2 Calculation Details

2.1 Selection of Adducts

ARZ can have a high number of conformers, with different mutual orientations of the PHL and PYR moieties and different IHB patterns. More than 90 conformers were considered in [5] (which included all the lowest energy ones and disregarded only some very high energy ones without the first IHB). Four mutual orientations of the two moieties were identified and denoted with the numbers 1, 2, 3 and 4 [5]. Corresponding conformers of the #1 and #2 series differ only by the fact that the orientation of the PYR ring is symmetrical with respect to the plane identified by the PHL ring and have very close energies; the same is true for corresponding conformers of the #3 and #4 series. Therefore, it is sufficient to select one series from each pair to ensure the consideration of the different IHMB patterns. The #1 and #3 series were selected, to include the minimum energy conformers of each pair (Fig. 3). It can be expected that adducts of conformers of the #2 series with the same number and input-arrangement of water molecules as for corresponding conformers of the #1 series will be very similar, as the PYR-related portion is symmetrical in the two cases and the IHB and steric patterns are the same. Equivalently, adducts of corresponding conformers of the #3 and #4 series are expected to be very similar. A few adducts of conformers of the #2 and #4 series were considered to verify this expectation.

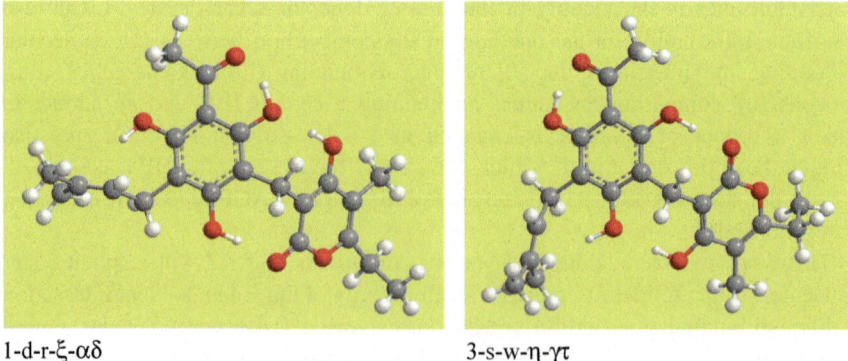

1-d-r-ξ-αδ   3-s-w-η-γτ

Fig. 3 The two lowest energy conformers with different mutual orientations of the two ring systems, as identified in [5], and the. Acronyms denoting them

The variety of possible adducts with explicit water molecules is extremely high for a solute molecule like ARZ, which contains a high number of H-bond donor or acceptor sites and can form a variety of conformers in which one or another site may be more or less available to form H-bonds with water molecules. It was opted to consider all the lower energy conformers of the #1 and #3 series and representative higher energy ones, to ensure that the most interesting arrangements of water molecules around the various sites of ARZ are captured.

Two sets of adducts were considered. One set involves most of the conformers of the #1 and #3 series, and entails adducts in which one water molecules is attached in turn to each donor or acceptor site of each conformer. These adducts enable a comparison of the energy with which each site can bind a water molecule, and also offer indications about how close a water molecule can approach the given site. In the real situation within the solvent, this is determined also by the interactions between water molecules and, therefore, a water molecule attached to a given site might remain at a greater distance than in the models with only one water molecule. On the other hand, it may happen that one (or, sometimes, more) water molecules remain attached to the solute molecule when it enters the active site of the biological target and may contribute to the binding between the molecule and the target; in such cases, the water molecule will likely remain as close as possible to the ARZ site to which it binds. The knowledge of the strength with which a water molecule binds to a certain donor or acceptor site of ARZ may thus be useful also for a better understanding of its permanence (when it occurs) when ARZ binds to its target, or its role in such binding.

The second set comprises adducts with several explicit water molecules, attempting to approximate a first solvation layer or portions of it. Like in previous studies on adducts of ACPLs with explicit water molecules [16, 17, 19], the 'first solvation layer' concept is expanded to include not only the water molecules directly H-bonded to suitable sites of the solute molecule, but also water molecules that might bridge them (the presence of a third water molecule bridging two molecules directly H-bonded to the solute often has a stabilizing effect [17]). The distribution and spacing of the several H-bond donors or acceptors in ARZ enable the possibility of considering adducts in which the water molecules attached to ARZ, and those bridging them, approximate a continuous layer in the vicinity of extensive portions of ARZ.

The inputs were prepared placing water molecules in the vicinity of H-bond donor or acceptor sites of the selected conformers of ARZ. Different numbers and arrangements of water molecules were considered for each conformer, also taking into account the resulting arrangements of already optimised outputs. For instance, when one or more water molecules 'moved' into a second solvation layer on optimization (out of contact with the ARZ molecule and with no bridging role between water molecules attached to it), those molecules were removed and the resulting input was optimised as a new adduct.

The selection of the number/s of water molecule in the adducts likely to better contribute the desired information is a rather delicate issue. Too small a number would not enable the incorporation of the effects of water-water interactions relevant for the first solvation layer, such as the stabilising effect of a water molecule bridging two water molecules H-bonded to ARZ. On the other hand, the tendency of water molecules to cluster together limits the number of water molecules in an adduct, if one wishes them to 'remain' in the first solvation layer on optimisation. For the case of ARZ, it was found that when more than 9–10 water molecules are present, their tendency to cluster becomes dominant and several of them may move away from their initial binding sites of ARZ, yielding arrangements in which they 'crowd' in the vicinity of only a portion of ARZ, and one or more of them may move beyond the first solvation layer.

2.2   Computational Approaches

All the adducts were calculated in vacuo, performing optimisation with fully relaxed geometry at the same levels utilised in all the previous calculations on ACPLs [6–10, 16, 17, 19–25], i.e., Hartree Fock (HF) with the 6-31G(d,p) basis set and Density Functional Theory (DFT) with the B3LYP functional [26–28] and the 6-31+G(d,p) basis set. The reasons for the selection of the two levels of theory and basis sets are explained in the previous works [6–10, 16–24]; it is considered important to maintain them in this and further studies involving ACPLs, to enable informative and straightforward comparisons.

In the previous studies [5–10], DFT calculations have mostly been performed as post-HF calculations; random testing had shown that the same inputs optimise to the same conformers with the two methods; thus, treating DFT as post-HF calculations was expedient to decrease computational costs. In the case of the adducts of ARZ considered here, HF and DFT calculations were performed independently for each input because the presence of several H-bond donor and acceptor sites and the non-covalent nature of the solute-solvent H-bonds suggests the possibility that the two methods may lead to different arrangements of water molecules. In most cases, the same input optimised to similar arrangements, but, in a number of cases, the optimised adducts differed substantially. Such outputs were then utilised as inputs for the other method, what enabled the consideration of additional geometries that had not been envisaged on the initial input-preparation.

The interaction energy ($\Delta E_{arz\text{-}n\cdot aq}$) between the ARZ molecule and the water molecules bonded to it was calculated for each adduct. The general equation is [29]

$$\Delta E_{arz-n\cdot aq} = E_{adduct} - \left(E_{arz} + n\,E_{aq}\right) - \Delta E_{aq-aq} \tag{1}$$

where E_{adduct} is the energy of the adduct, n is the number of water molecules in the adduct, E_{arz} is the energy of the isolated ARZ conformer, E_{aq} is the energy of an isolated water molecule and $\Delta E_{aq\text{-}aq}$ is the overall interaction energy between water molecules, resulting mainly from water-water H-bonds.

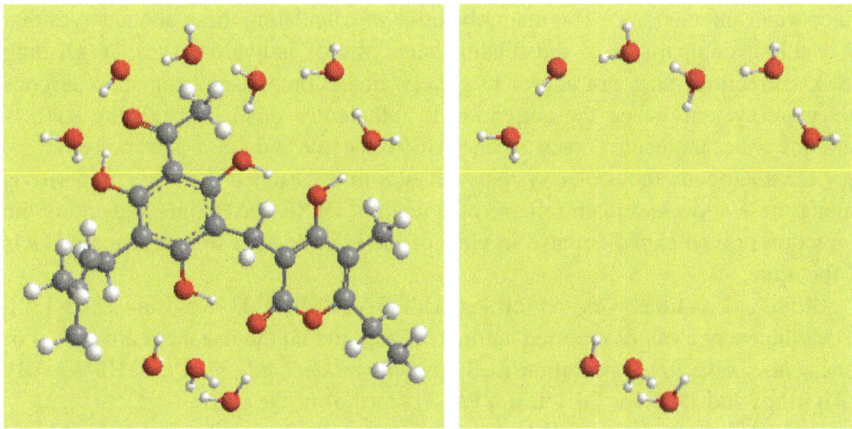

Fig. 4 Example of an adduct of arzanol with 9 explicit water molecules and the system of the sole water molecules in the same arrangement as in the adduct, used to calculate $\Delta E_{aq\text{-}aq}$

For adducts with only one water molecule, the equation becomes simply

$$\Delta E_{arz-aq} = E_{adduct} - E_{arz} - E_{aq} \tag{2}$$

For adducts with n non-interacting water molecules, the equation would be

$$\Delta E_{arz-n \cdot aq} = E_{adduct} - \left(E_{arz} + n\, E_{aq}\right) \tag{3}$$

For adducts with interacting water molecules (which is the most common case when there are several water molecules), $\Delta E_{aq\text{-}aq}$ is evaluated through a single point (SP) calculation on a group of water molecules arranged exactly as in the adduct, but without the ARZ molecule ([29], Fig. 4). If $E_{aq\text{-}set}$ is the energy of this set of water molecules, then

$$\Delta E_{aq-aq} = E_{aq-set} - n\, E_{aq}$$

and Eq. (1) becomes

$$\Delta E_{arz-n \cdot aq} = E_{adduct} - E_{arz} - E_{aq-set} \tag{4}$$

Both E_{adduct} and $E_{aq\text{-}set}$ have been corrected for basis set superposition error (BSSE), using the counterpoise method [30], when the adduct contains more than one water molecule. The BSSE correction was not applied to adducts containing only one water molecule because it would be small for these adducts and because these calculations are meant for comparisons. The values for the calculated adducts with four or more water molecules show that the BSSE correction increases substantially as the number of water molecules in the adduct increases; therefore, it will be smallest for adducts with one water molecule (also in view of the absence of water-water interactions). The main objective of calculating these adducts was that of enabling comparison of the binding strengths of individual sites. Neglecting BSSE corrections does not appear to greatly affect comparisons among analogous molecular systems when the correction is sufficiently small [31] (all the adducts with one water molecule consist of one ARZ molecule and one water molecule, so, they are analogous molecular systems). It was thus assumed that the comparisons among these adducts remain reliable also without BSSE correction. Neglecting the correction proved expedient also in view of the high number of calculated adducts of this type.

All the calculations were performed with GAUSSIAN 03, Revision D 01 [32].

All the energy values reported are in kcal/mol and all the distances are in Å. For conciseness sake, the calculation methods will be denoted simply as HF for HF/6-31G(d,p) and DFT for DFT/B3LYP/6-31+G(d,p) in the text.

Detailed information, including tables with all energy values and H-bond parameters for the calculated adducts, and figures showing the geometries of the calculated adducts, is provided in the Supplementary Information.

3   Results

3.1   Naming of Conformers and Adducts

Following a practice introduced since the initial studies of ACPLs [6–10], the conformers are denoted by acronyms which provide information about their characteristics, to enable easy tracking of geometric characteristics and energy-influencing features. The acronyms for the ARZ conformers start with the number denoting the mutual orientation of the moieties. The other geometry features are denoted by letters, whose meanings are listed in Table 1 [5]. The two lowest energy conformers and their acronyms are shown in Fig. 3.

In order to identify them in a straightforward way, the adducts are denoted by the acronym of the ARZ conformer at their centre, followed by information about the water molecules. For the adducts with one water molecule, the name of the conformer is followed by '1aq' and by the position to which the water molecule is attached. Thus, 1-d-r-ξ-αδ-1aq-O14 denotes the adduct of the 1-d-r-ξ-αδ conformer in which one water molecule is attached to O14; 1-d-r-αδ-1aq-H17-π1 denotes the adduct of 1-d-r-αδ in which a water molecule is attached to H17 and is also interacting with the π bond of the prenyl chain; and so on. The adducts containing more than one water molecule are denoted with the name of the conformer followed by the number of molecules in the adduct. For instance, 1-d-r-ξ-αδ-5aq denotes an adduct of conformer 1-d-r-ξ-αδ with 5 water molecules. Since different arrangements of the water molecules are possible for adducts of the same conformer and with the same number of water molecules, letters are added at the end of the acronym to distinguish them from one another. However, the information on the site to which each water molecule is attached would be too bulky to summarise it in an acronym, and 3D models showing the arrangements of the water molecules are

Table 1 Symbols utilised to specify geometrical characteristics in the acronyms denoting the conformers. The symbols d, s, r, w, u, η and ξ have the same meanings as in other studies on ACPLs [6–10] and the symbols α, β, γ, δ, ε and τ had been introduced in [5]

Symbol	Meaning
d	The H15⋯O14 first IHB is present
s	The H17⋯O14 first IHB is present
r	H16 is oriented towards the α-pyrone ring
w	H16 is oriented towards the prenyl chain
u	H15 or H17, not engaged in the first IHB, is oriented toward the COR group ('upwards')
η	Presence of O-H⋯π interaction between H16 and the C29=C30 double bond
ξ	Presence of O-H⋯π interaction between H17 and the C29=C30 double bond
a	No O-H⋯π interaction is present, and the prenyl chain is oriented 'upwards'
b	No O-H⋯π interaction is present, and the prenyl chain is oriented 'downwards'
α	The H27⋯O8 intermonomer hydrogen bond is present
β	The H15⋯O26 intermonomer hydrogen bond is present
γ	The H15⋯O23 intermonomer hydrogen bond is present
δ	The H16⋯O23 intermonomer hydrogen bond is present
ε	The H16⋯O26 intermonomer hydrogen bond is present
τ	The H27⋯O10 intermonomer hydrogen bond is present
π1	The C29=C30 double bond, when acting as binding site for a water molecule

the only option to convey a clear image of each adduct (the 3D models are included in the Supplementary Information).

3.2 Results for the Adducts of Arzanol with One Water Molecule

Given the high number of conformers of the ARZ molecule, and the high number of sites to which a water molecule can bind (including the possibility of simultaneous binding to two geometrically suitable sites), considering all the possible adducts of this type would be unaffordable. It was opted to calculate a selection of sufficiently representative adducts for each binding site. While some sites (e.g., O14 or O23) are accessible in all conformers, other sites are not available in some conformers, and the combinations for simultaneous binding to two sites depend on the type of conformer; therefore, only a limited number of adducts may be obtainable for certain combinations. Changes during optimisation reduce the number of adducts for some sites while increasing it for others (which informs that the former sites or site-combinations are less favourable than others).

A total of more than 200 adducts were calculated. Figure 5 shows the main geometries obtained, according to the type of conformer of the ARZ molecule and to the site to which the water molecule binds. All the geometries are shown in the

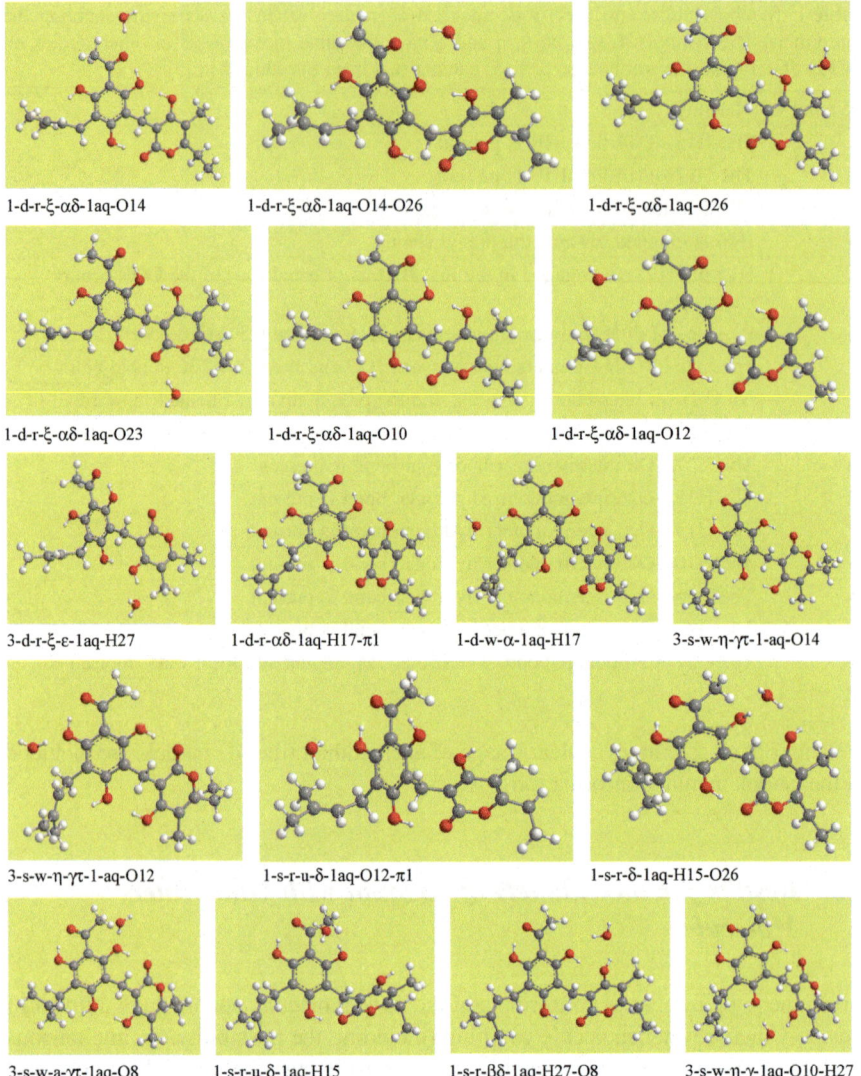

1-d-r-ξ-αδ-1aq-O14   1-d-r-ξ-αδ-1aq-O14-O26   1-d-r-ξ-αδ-1aq-O26

1-d-r-ξ-αδ-1aq-O23   1-d-r-ξ-αδ-1aq-O10   1-d-r-ξ-αδ-1aq-O12

3-d-r-ξ-ε-1aq-H27   1-d-r-αδ-1aq-H17-π1   1-d-w-α-1aq-H17   3-s-w-η-γτ-1-aq-O14

3-s-w-η-γτ-1-aq-O12   1-s-r-u-δ-1aq-O12-π1   1-s-r-δ-1aq-H15-O26

3-s-w-a-γτ-1aq-O8   1-s-r-u-δ-1aq-H15   1-s-r-βδ-1aq-H27-O8   3-s-w-η-γ-1aq-O10-H27

Fig. 5 Representative adducts of arzanol with one water molecule attached to different donor or acceptor sites. The first two rows show adducts of the lowest energy conformer of arzanol (1-d-r-ξ-αδ), the other rows show other relevant types of arrangement of the water molecule, for different types of conformers

Supplementary Information. Nearly all the geometries are obtained with both HF and DFT, with similar shapes.

In many cases, the optimisation does not change the site to which the water molecule binds; it is the case of inputs in which it binds to O14 or to O23, but also to other favourable sites. There are, however, also several cases in which the

Table 2 Examples of changes occurring during optimization, highlighting the binding preferences of the water molecule

Reference acronym	Input geometry	Output geometry
ch-1	1-d-r-ξ-αδ-1aq-O19-O23	1-d-r-ξ-αδ-1aq-O23
ch-2	1-d-r-ξ-αδ-1aq-O8	1-d-r-ξ-αδ-1aq-O14
ch-3	3-s-r-b-γε-1aq-O12	3-s-r-b-γε-1aq-O14
ch-4	3-s-w-η-γτ-1aq-O8	3-s-w-η-γτ-1aq-O23
ch-5	1-d-r-b-αδ-1aq-O8	1-d-r-b-αδ-1aq-O26
ch-6	3-d-w-η-τ-1aq-O8	3-d-w-η-τ-1aq-O8-O23
ch-7	3-d-w-ξ-1aq-O23	3-d-w-ξ-1aq-O8-O23
ch-8	3-d-r-ξ-ε-1aq-O26	3-d-r-ξ-ε-1aq-H27
ch-9	3-d-w-η-τ-1aq-O12	3-d-w-η-τ-1aq-H17
ch-10	3-d-w-ξ-τ-1aq-O10	3-d-w-ξ-τ-1aq-H16
ch-11	1-s-r-βδ-1aq-O8	1-s-r-βδ-1aq-O8-H27
ch-12	1-s-r-βδ-1aq-O26	1-s-r-βδ-1aq-O8-H27
ch-13	3-d-w-η-τ-1aq-O26	3-d-w-η-τ-1aq-O26-H16
ch-14	3-d-r-u-b-ε-1aq-O26	3-d-r-u-b-ε-1aq-O10-H27
ch-15	1-s-r-u-δ-1aq-O8-O26	1-s-r-u-βδ-1aq-O8-H27
ch-16	3-s-r-γε-1aq-O10-π1	3-s-r-γε-1aq-O10-H27
ch-17	3-s-w-η-γ-1aq-O10-O26	3-s-w-η-γ-1aq-O10-H27
ch-18	1-d-w-ξ-α-1aq-O10-O23	1-d-ξ-α-1aq-H16-O23
ch-19	1-d-w-η-α-1aq-O10-O23	1-d-w-η-α-1aq-H16-O23
ch-20	1-s-w-u-η-α-1aq-H15	1-s-w-u-η-α-1aq-H15-O26
ch-21	3-s-w-u-η-τ-1aq-H15	3-s-w-u-η-τ-1aq-H15-O23
ch-22	1-d-w-α-1aq-H16	1-d-w-α-1aq-H16-O23
ch-23	1-s-r-βδ-1aq-H27	1-s-r-βδ-1aq-O8-H27
ch-24	1-d-w-ξ-1aq-H27	1-d-w-ξ-1aq-O14-H27
ch-25	3-s-w-η-γ-1aq-H27	3-s-w-η-γ-1aq-O10-H27
ch-26	1-d-w-ξ-α-1aq-O12	1-d-w-α-1aq-O12
ch-27	1-d-r-b-αδ-1aq-O10-π1	1-d-r-ξ-αδ-1aq-O10'
ch-28	1-d-w-η-α-1aq-O10-O23	1-d-w-ξ-α-1aq-O12
ch-29	1-d-r-b-αδ-1aq-H17	1-d-r-αδ-1aq-H17-π1
ch-30	3-s-w-a-γτ-1aq-H16	3-s-w-γτ-1aq-H16-π1
ch-31	3-d-w-τ-1aq-H16-π1	3-d-w-ξ-τ-1aq-H16
ch-32	1-d-w-1aq-H17-π1	3-d-w-η-τ-1aq-H17'
ch-33	3-s-w-a-γτ-1aq-O12-π1	3-s-w-η-γτ-1aq-O12
ch-34	3-d-r-b-ε-1aq-O10-π1	3-d-r-ξ-ε-1aq-O10-H27

optimisation changes the site to which the water molecule binds, with respect to the input. Since these changes highlight binding preferences, they are given detailed attention in the current analysis. Examples are reported in Table 2, considering types of changes which appeared more than once. The changes are denoted with acronyms ('ch', which stands for 'change', followed by a number) to enable easy referencing to them within this text.

The water molecule appears to prefer to bind to an sp^2 O rather than to an sp^3 O. Thus, inputs in which the water molecule is attached to O19, or to both O19 and O23, optimize to adducts with the water molecule attached to O23 (ch-1). Inputs in which the water molecule is attached to the donor of the first IHB (O8 or O12, for d and s conformers respectively), in conformers in which O14 is the only sp^2 O in the vicinity, often optimize to adducts in which water binds to O14 (ch-2, ch-3). If O23 is also available in the vicinity (e.g., in inputs involving 3-d conformers), the water molecule often shifts from O8 to O23 (ch-4) or, sometimes, to O26 (ch-5). These changes may be related to the hydrophobic character of IHB regions for hydroxybenzenes in general [33], and of the first IHB of ACPLs in particular [17].

A clear preference appears for the water molecule to bind to two sites simultaneously, when two sites are geometrically suitable; this may lead to adducts in which it binds to an sp^2 O and an sp^3 O simultaneously (e.g., O8 and O23; ch-6, ch-7). The water molecule also appears to prefer to be acceptor to an OH group (consistently also with the known tendency of phenol OHs to be donors in intermolecular H-bonds [16, 31]). This may involve rotation of H15, H16, H17 or H27, to enable the formation of the H-bond with the water molecule (it is interesting to recall that the phenol OHs in ACPLs do not usually rotate to form IHBs [16], whereas the adducts calculated here show that the OHs in ARZ may rotate to form an intermolecular H-bond with the solvent). The water molecule may change binding site completely in order to be acceptor to the H of an OH (ch-8, ch-9, ch-10). On the other hand, its tendency to bind to two geometrically suitable sites simultaneously may lead to adducts in which it binds simultaneously to the H of an OH and to a suitably close O atom (ch-11 to ch-19). The tendency to bind to two sites appears also for inputs in which the water molecule is initially placed as acceptor to an OH (ch-20 to ch-25).

The water molecule does not break O-H$\cdots$O IHBs. It appears, however, to be able to act on the O-H$\cdots\pi$ interaction, by breaking it (ch-26) or prompting it (ch-27), or changing its pattern (ch-28, in which it changes from η to ξ). The water molecule itself may interact with the C29=C30 π bond; the interaction usually appears during optimisation (ch-29, ch-30), whereas inputs having the interaction do not often optimise to adducts in which it is maintained (ch-31 to ch-34).

The changes just outlined appear both with HF and with DFT optimization. In most cases, the same change occurs with the same input, i.e., HF and DFT optimisations lead to the same changed output. In some cases, the changes are different with the two methods, leading to different outputs. Figure 6 shows a case in which HF and DFT lead to outputs that are different from the input and different from each other. In such cases, it is not easy to estimate a priori which of the two outcomes might be closer to reality; for the specific case shown in Fig. 6, the HF result seems more probable, as it shows preference for an sp^2 O, which can form a stronger H-bond than an sp^3 O like O8 or O26.

Table 3 reports the relative energy and the ARZ-water interaction energy for representative adducts, selected in such a way as to comprise nearly all the

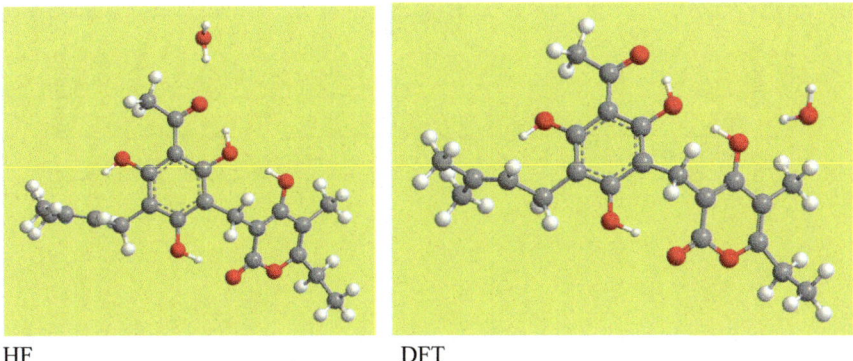

HF   DFT

Fig. 6 An example in which the HF and DFT optimisations of the same input with one water molecule lead to different results. In the input, the water molecule was attached to O8. The HF optimisation moves it to O14 and the DFT optimisation moves it to O26. The conformer is the lowest energy conformer of arzanol (1-d-r-ξ-αδ)

identified binding options of the water molecule. The relative energy depends to a considerable extent on the type of conformer and its relative energy in vacuo, whereas the interaction energy depends largely on the binding site (although the geometry of the conformer influences the approach of the water molecule to a given binding site). Table 4 reports the ranges of the interaction energy for the different binding sites. The interaction energy values confirm the binding preferences of the water molecule highlighted by the changes occurring during optimisation, such as the preference for simultaneous binding to two atoms of ARZ (and, among these, for one of the sites being a donor OH), and the preference for H atoms of OH groups, followed by sp^2 O atoms, when it binds only to one site.

Table 5 reports the ranges of the length of the ARZ-water H-bonds. The binding sites are listed in the same sequence as in Table 4 to facilitate comparison in terms of length of the H-bond and strength of the interaction energy (a few sites present in Table 4 are not reported in Table 5 for space reasons). The correspondence is meaningful because, for adducts with only one water molecule, the molecule-water interaction energy can be viewed as the energy of the molecule-water H-bond, and an H-bond length is an indication of its strength. The H-bond lengths are shorter when the water molecule is acceptor to an OH group of ARZ. The H-bond lengths for sp^2 O are shorter than those of sp^3 O, consistently with the ability of sp^2 O to form stronger H-bonds. Although the trends (comparison of lengths across adducts) are largely similar in the HF and DFT results, the HF values are longer than the DFT values for the same adducts. This is consistent with the known tendency of HF to underestimate the strength of H-bonds and of DFT to overestimate it; therefore, it appears reasonable to assume that the actual H-bond distance for a given adduct will be somehow intermediate between the HF and the DFT values.

Table 3 Relative energy and arzanol-water interaction energy of representative adducts of arzanol with one explicit water molecule, in the HF/6-31G(d,p) and DFT/B3LYP/6-31+G(d,p) results

HF/6-31G(d,p) results			DFT/B3LYP/6-31+G(d,p) results		
Adduct	Relative energy (kcal/mol)	Interaction energy (kcal/mol)	Adduct	Relative energy (kcal/mol)	Interaction energy (kcal/mol)
1-d-r-ξ-αδ-1aq-O23	0.000	−6.523	1-d-r-ξ-αδ-1aq-O23	0.000	−6.521
1-d-r-ξ-αδ-1aq-O14	0.111	−5.502	1-d-r-ξ-αδ-1aq-O14	0.814	−5.707
1-d-r-αδ-1aq-H17-π1	1.210	−11.926	1-d-r-ξ-αδ-1aq-O10	1.300	−5.221
1-d-r-b-αδ-1aq-O12	1.357	−5.259	1-d-r-αδ-1aq-H17-π1	1.511	−7.511
3-s-w-η-γτ-1aq-O23	1.453	−6.412	3-s-w-a-γτ-1aq-H16-π1	2.129	−9.689
1-d-r-ξ-αδ-1aq-O10	1.575	−5.029	1-d-r-ξ-αδ-1aq-O26	2.329	−4.192
3-s-w-γτ-1aq-H16-π1	1.788	−8.695	3-s-w-η-γτ-1aq-O12	3.066	−5.430
1-d-r-ξ-αδ-1aq-O14-O26	2.915	−3.689	1-d-w-η-α-1aq-O23-H16	4.911	−6.850
3-s-w-η-γτ-1aq-O26	3.075	−4.790	3-s-w-u-η-τ-1aq-O23-H15	5.449	−16.818
1-d-w-ξ-α-1aq-O23-H16	4.256	−15.101	1-d-r-u-b-αδ-1aq-H17	5.819	−6.722
3-s-r-γε-1aq-O10-H27	4.526	−12.600	3-s-r-γε-1aq-O10-H27	6.463	−12.089
3-s-w-u-η-τ-1aq-O23-H15	4.755	−17.118	3-s-w-a-γτ-1aq-O8	7.358	−4.461
3-s-r-γε-1aq-O10-π1	5.510	−11.616	3-s-r-u-ε-1aq-O10-H27	7.484	−11.111
3-s-r-b-γε-1aq-O10-O26	5.510	−11.601	3-d-w-η-α-1aq-O10-O26	9.467	−6.220
3-s-w-a-γτ-1aq-O8	6.232	−4.250	1-s-r-βδ-1aq-O8-H27	9.477	−9.879
3-s-w-η-γ-1aq-O10-H27	7.285	−9.8685	3-s-w-η-γ-1aq-H27	10.687	2.191
1-s-r-βδ-1aq-O8-H27	8.020	−10.144	1-r-ξ-βδ-1aq-O14-H27	12.480	−19.134
1-d-w-α-1aq-H17	11.500	−7.857	3-s-r-b-γε-1aq-O12-π1	13.431	−5.165
3-s-r-γε-1aq-O12-π1	12.038	−5.088	1-s-w-u-η-α-1aq-O26-H15	14.059	−8.849
1-s-w-u-η-α-1aq-O26-H15	12.622	−10.546	3-d-w-ξ-τ-1aq-H16	15.863	−9.941

(continued)

Table 3 (continued)

HF/6-31G(d,p) results			DFT/B3LYP/6-31+G(d,p) results		
Adduct	Relative energy (kcal/mol)	Interaction energy (kcal/mol)	Adduct	Relative energy (kcal/mol)	Interaction energy (kcal/mol)
3-w-ξ-γτ-1aq-H16	15.167	−8.338	1-s-r-u-αδ-1aq-H15	17.161	−8.936
3-d-w-η-τ-1aq-O8-O23	15.354	−7.793	3-d-w-u-η-τ-1aq-O8-O23	17.971	−7.504
3-d-w-η-τ-1aq-O26-H16	16.297	−6.850	1-s-r-u-δ-1aq-O8-O26	18.197	−6.378
1-d-w-η-1aq-O8-H27	17.040	−12.144	3-d-w-ξ-τ-1aq-O8-O23	18.246	−7.558
1-d-w-η-1aq-O14-H27	20.485	−8.699	3-w-ξ-γτ-1aq-O12-O14	19.408	−7.010
1-r-ξ-βδ-1aq-O12-O14	20.620	−7.202	1-d-w-η-1aq-O23-H16	19.972	−9.698

Table 4 Ranges of the magnitude (absolute values) of the arzanol-water interaction energy for the adducts of arzanol with one explicit water molecule, in the HF/6-31G(d,p) and DFT/B3LYP/6-31 +G(d,p) results in vacuo. When only two values are available, they are reported individually, separated by a comma. When only one value is available, it is reported individually

Binding site	HF range (kcal/mol)	DFT range (kcal/mol)	Binding site	HF range (kcal/mol)	DFT range (kcal/mol)
O23-H15	15.4, 17.1	15.2, 16.8	H16-π1	7.2–8.7	7.3–9.9
O10-H27	7.6–15.6	9.3–12.1	O12	3.2–8.1	3.0–8.5
O23-H16	10.2–15.3	9.2–16.3	O8-O23	6.5–7.9	7.2–7.8
O8-H27	10.1–13.5	9.9	O23	3.8–7.8	4.0–11.1
O8-O26	12.3	5.4, 5.4	O14-O23	7.8	7.0
H17-π1	7.3–11.9	7.0–8.8	O12-O14	7.2, 7.5	7.5–7.8
O10-π1	6.1, 11.6		O14	3.9–7.1	3.7–7.4
O10-O26	11.6	6.1, 6.2	O8-O14	7.1	7.0
O26-H15	9.5–10.5	6.1–8.9	O10	3.9–7.0	5.2–8.2
H15	8.4–10.2	7.2	O26-H16	6.9	
O10-O23	10.1	7.0–9.9	O12-π1	5.1–5.3	5.1–7.5
H27	5.7–10.0	6.6–11.2	O26	2.4–5.2	3.5–8.0
H16	8.3–9.5	7.6–9.9	O12-H17	4.6	
O14-H27	8.7, 9.2	19.1	O8	4.3–4.4	4.5–5.1
H17	6.7–9.1	6.7–7.5	O14-O26	3.7	

3.3   Results for the Adducts of Arzanol with More Than One Water Molecule

A total of 84 adducts of ARZ with 4–12 water molecules were calculated, with greater number of adducts with 7, 8 or 9 water molecules. What the calculation of such adducts may contribute in a study of this type is not an exhaustive inclusion of all possible options, but a sufficiently representative selection highlighting the variety of possibilities and some recurrent patterns. The geometries present a high variety of possible arrangements of the water molecules around the ARZ molecule. Figure 7 shows some geometries of adducts with different numbers of water molecules.

Figure 8 illustrates some relevant aspects in the arrangement of water molecules, including some recurrent patterns. The water molecules tend to keep away from the region of an IHB, as it is a hydrophobic region [17, 33]. Shapes already encountered for adducts of other ACPLs [17] appear with a certain frequency, such as a pentagon of O atoms formed by the atoms engaged in an IHB and the O atoms of the water molecules keeping away from it. The whole region of two cooperative IHBs appears to be hydrophobic, and the water molecules may form a larger ring (e.g., seven O atoms) while keeping away from it.

Differently from the adducts with one water molecule, a water molecule may break an O-H···O IMHB and insert itself between the donor and the acceptor; the

Table 5 Ranges of the length of the hydrogen bonds between the arzanol molecule and the water molecule for selected binding sites in the calculated adducts of arzanol with one water molecule. When only two values are available, they are reported individually, separated by a comma. When only one value is available, it is reported individually

Binding site/s	Length considered	Range of values (Å)	
		HF results	DFT results
O23, H15	$O_{aq} \cdots H15$	1.878, 1.892	1.713, 1.754
	$H_{aq} \cdots O23$	1.918, 1.953	1.747, 1.819
O10, H27	$O_{aq} \cdots H27$	1.913–2.229	1.784–2.174
	$H_{aq} \cdots O10$	1.983–2.298	1.911–2.282
O23, H16	$O_{aq} \cdots H16$	1.851–2.190	1.690–1.751
	$H_{aq} \cdots O23$	1.904–2.164	1.731–1.984
O8, H27	$H_{aq} \cdots O8$	1.966–2.125	–
	$O_{aq} \cdots H27$	1.877–1.932	–
O8, O26	$H_{aq} \cdots O8$	–	2.177–2.401
	$H_{aq} \cdots O26$	–	2.033–2.177
H17, π1	$H17 \cdots O_{aq}$	1.885–1.901	1.792–1.848
O26, H15	$O_{aq} \cdots H15$	1.943–1.991	1.871–1.930
	$H_{aq} \cdots O26$	2.365–2.429	2.035–2.199
H15	$O_{aq} \cdots H15$	1.909–2.030	1.888–1.895
H27	$O_{aq} \cdots H27$	1.890–2.009	1.822–1.933
H16	$O_{aq} \cdots H16$	1.926, 1.935	1.905, 1.914
O14, H27	$H_{aq} \cdots O14$	2.253, 2.332	–
	$O_{aq} \cdots H27$	2.011, 2.025	–
H17	$O_{aq} \cdots H17$	1.939–2.005	1.873–1.941
H16, π1	$H_{aq} \cdots O10$	1.879–1.901	1.752–1.870
O12	$H_{aq} \cdots O12$	2.132–2.240	1.931–2.060
O8, O23	$H'_{aq} \cdots O23$	2.120–2.158	1.965–2.037
	$H_{aq} \cdots O8$	2.311–2.378	2.130–2.351
O23	$H_{aq} \cdots O23$	2.038–2.178	1.911–2.024
O14, O23	$H'_{aq} \cdots O23$	2.129, 2.182	2.008
	$H_{aq} \cdots O14$	2.573, 2.604	2.662
O12, O14	$H'_{aq} \cdots O14$	2.217, 2.289	2.438–2.570
	$H_{aq} \cdots O12$	2.285, 2.309	2.000–2.056
O14	$H_{aq} \cdots O14$	2.040–2.089	1.870–1.929
O8, O14	$H'_{aq} \cdots O14$	2.209	2.097
	$H_{aq} \cdots O8$	2.366	2.220
O10	$H_{aq} \cdots O10$	2.106–2.158	1.952–2.023
O12, π1	$H_{aq} \cdots O12$	2.366–2.478	1.837–2.145
O26	$H_{aq} \cdots O26$	2.152–2.257	1.987–2.052
O8	$H_{aq} \cdots O8$	2.087–2.112	1.926–1.957

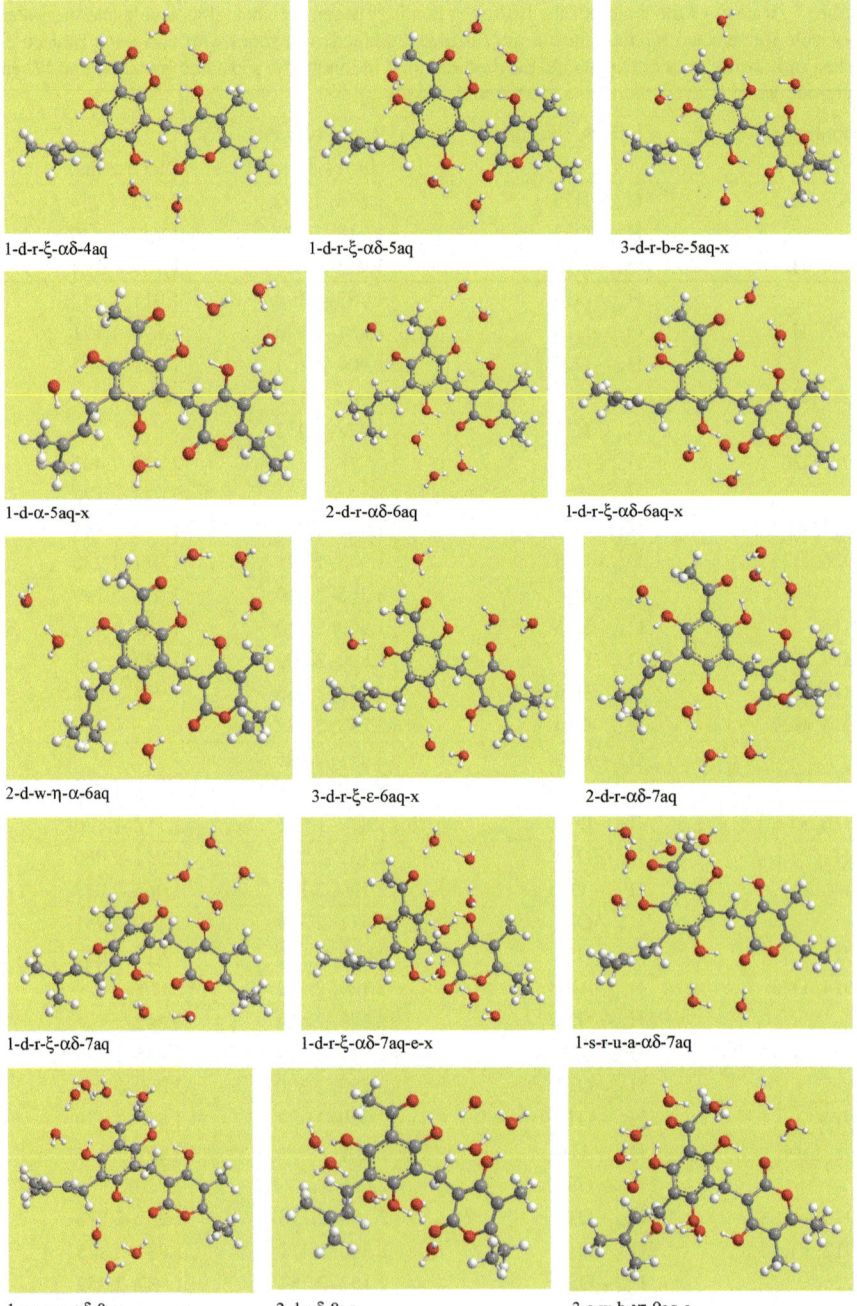

Fig. 7 Illustration of the variety of possible geometries for the arrangement of water molecules in the adducts

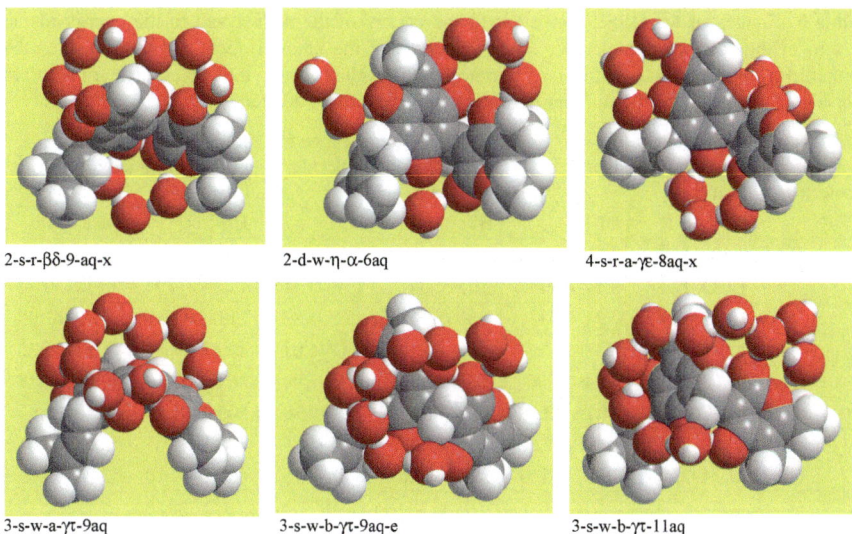

2-s-r-βδ-9-aq-x   2-d-w-η-α-6aq   4-s-r-a-γε-8aq-x

3-s-w-a-γτ-9aq   3-s-w-b-γτ-9aq-e   3-s-w-b-γτ-11aq

Fig. 8 Some relevant aspects in the arrangement of water molecules around the arzanol molecule. The water molecules tend to keep away from the IHB regions; a frequent result is a pentagon of O atoms (including those of the IHB), as in 2-s-r-βδ-9-aq-x, 4-s-r-a-γε-8aq-x and 3-s-w-a-γτ-9aq; in some cases, the ring may contain more O atoms, as the ring around two consecutive IHBs in 2-d-w-η-α-6aq. A water molecule may break an inter-monomer H-bond, as in 3-s-w-a-γτ-9aq, where it breaks the H27···O10 IHB. In a number of cases, the tendency of the water molecules to cluster together fosters arrangements in which some donor or acceptor sites of the arzanol molecule are not binding a water molecule, as in 3-s-w-b-γτ-9aq-e (where no water molecule attaches to O14, although it is a strong acceptor) and in 3-s-w-b-γτ-11aq. Although in adducts with one or few water molecules no water molecule binds to O19, it may happen that a water molecule binds to it if there are enough bridging water molecules to facilitate the arrangement, as in 3-s-w-b-γτ-11aq

cases observed concern IMHBs where H27 is the donor (H27···O8 or H27···O10). The length of the H-bond between H27 and the O of the water molecule attached to it is the shortest ARZ-water H-bond length in these adducts, suggesting that this H-bond might be among the strongest if not the strongest. No breaking has been observed for the first IHB (which is considerably stronger than H27···O8 or H27···O10), consistently with the results for the adducts of other ACPLs [16, 17]. For similar reasons (the acceptor being an sp^2 O), no breaking is observed for IMHBs in which O23 is the acceptor. The O-H···π interaction is broken in a number of cases, with a water molecule inserting itself between the H and the C29=C30 π-bond. The tendency of the water molecules to cluster together may prompt major changes during optimization, leading to chains of water molecules alternatingly binding to a site of ARZ and playing bridging roles.

The relative energies of adducts with the same number of water molecules depend on their geometrical arrangement and binding sites more than on the relative energy of the isolated conformer. The BSSE correction increases as the number of

Table 6 Ranges of the BSSE correction to the energy of the adduct and to the energy of the system of the sole water molecules, and percentage of the latter correction with respect to the former, for the calculate adducts of arzanol with explicit water molecules. The ranges are considered according to the number of water molecules in the adduct

Number of water molecules	BSSE correction range for the adduct (kcal/mol)		BSSE correction range for the sole water molecules (kcal/mol)		Range of percentage of the contribution of water	
	HF	DFT	HF	DFT	HF	DFT
5	8.50–11.05	5.15–6.46	0.95–3.99	0.99–1.84	9.63–47.25	16.56–55.52
6	10.32–12.84	6.14–8.11	2.25–6.64	1.61–3.51	17.84–53.74	25.17–50.87
7	12.52–15.45	7.34–9.71	3.00–7.70	2.65–5.83	21.00–52.65	31.93–62.99
8	12.45–18.28	8.43–11.17	4.85–8.91	4.17–6.01	30.57–57.59	45.51–71.32
9	14.62–21.96	8.94–13.06	5.53–11.12	4.48–7.99	36.58–57.60	50.11–64.28
10	19.87–20.42	11.16–14.52	11.86–12.29	7.25–9.39	50.84–61.86	60.18–64.97
11	20.00–25.85	14.68–15.70	9.23–13.97	9.37–10.52	45.41–55.07	53.68–67.01
12	24.71–25.77	15.81–16.51	11.46–12.62	8.42–9.75	46.31–49.18	53.30–60.65

water molecules in the adduct increases, although it can vary broadly for adducts with the same number of water molecules. The ranges of values are somewhat narrower in the DFT results and somewhat broader in the HF results (Table 6). The correction is often greater for adducts with lower relative energy.

The ARZ-water interaction energy depends on the conformer of ARZ, on the sites to which the water molecules bind and on the number of water molecules. The dominant factor appears to be the arrangement of the water molecules, which includes the binding sites and also the types of water-water interactions. Table 7 reports the ranges of the ARZ-water interaction energy as identified from the calculated adducts; the values show a tendency to higher upper-limit of the range as the number of water molecules increases, but without a straightforward relationship between the two. The dominance of qualitative aspects (characteristics of the ARZ conformer, types of binding sites, geometry of the water molecules arrangements) may hamper the possibility of identifying more definite types of relationship. The magnitude of the DFT values tend to be greater than that of the HF values for corresponding conformers, which may be ascribed to the tendency of HF and DFT to respectively underestimate and overestimate the strength of H-bonds.

The length of the ARZ-water H-bonds depends on the binding site and appears to be fairly consistent with the patterns highlighted by the adducts with one water molecule, although the other water molecules binding to a water molecule attached to ARZ may influence the length of its H-bond with the ARZ site. This is the case of H-bonds between H27 and the O of a water molecule, which are the shortest in adducts with several water molecules, but not in the adducts with one water molecule.

Table 7 Ranges of the magnitude of the arzanol-water interaction energy (corrected for BSSE) according to the number of water molecules in the adduct

Number of water molecules	Magnitude of the interaction energy (kcal/mol)		Number of water molecules	Magnitude of the interaction energy (kcal/mol)	
	HF	DFT		HF	DFT
5	13.64–30.52	15.82–36.70	9	16.14–38.63	21.10–45.31
6	13.40–35.77	17.93–39.47	10	16.79–24.24	18.92–33.62
7	6.32–35.95	8.64–48.69	11	22.16–37.55	24.23–47.61
8	11.97–41.16	15.16–41.93	12	26.13–39.39	31.12–57.73

3.4 Results in Water Solution for the Adducts of Arzanol with More Than One Water Molecule

When solute–solvent intermolecular H-bonds are possible, the combination of explicit consideration of those solvent molecules that are expected to be more closely linked to the solute molecule (as those H-bonded to it or bridging them), and a bulk effect for the rest of the solvent, can provide a better picture of the situation in solution than the sole PCM calculations on the central molecule, while remaining within QM calculations affordability because of the limited number of discrete solvent molecules. Calculations in water solution were performed on selected adducts considering the entire adduct as a solute and utilising the PCM model. They were performed as SP calculations, because of affordability reasons in view of the size of the supermolecular structures of the adducts.

The results show a decrease in the relative energy for most adducts, with greater decrease for adducts having higher relative energy in vacuo. This is consistent with common behaviours for isolated molecules. Few exceptions may appear for lower energy adducts, more frequently in the DFT results. The identification of the lowest energy adduct may differ from that in vacuo, but remains among the lower energy adducts in vacuo. No significant patterns for the energy decrease can be identified in terms of types of the conformers of ARZ.

The solvent effect (free energy of solvation, ΔG_{solv}) is mostly negative in the DFT results; in the HF results, it is mostly negative for adducts with smaller numbers of water molecules and positive for several adducts with higher numbers of water molecules. Given the expectation that a molecular unit incorporating water molecules in its outer region would have some solubility in water, the DFT results are likely more realistic. The electrostatic component of ΔG_{solv} (G_{el}) is always negative, with values always smaller than -20 kcal/mol; its magnitude varies rather randomly with adducts of different conformers, but shows a tendency to an average increase as the number of water molecules increases.

4 Discussion and Conclusions

A study of the type considered in this work involves a variety of challenges because of the nature of the adducts and of the character of the information that can be obtained.

The high number of conformers of the ARZ molecule and the high number of its H-bond donor or acceptor site implies an enormous number of possible adducts with a given number of water molecules, and this number increases rapidly as the number of water molecules in the adducts increases. Calculations show high sensitivity of the optimisation procedure to small differences in the inputs, which would recommend the consideration of several adducts with similar or very similar (but not identical) input geometries for each relevant geometry, thus further multiplying the number of potentially interesting adducts. On the other hand, the tendency of water molecules to cluster on optimisation may yield similar adducts from different inputs, which decreases the informative role of the result. It may also lead to adducts where some water molecules cluster beyond the boundaries of the adopted criterion for the definition of 'first solvation layer', resulting in high water-water interaction energy and poor solute-water interaction energy; since the latter phenomenon is more extensive as the number of water molecules in the adduct increases, it prevents the possibility of considering a higher number of water molecules than the one for which their clustering becomes extensive or dominant.

The huge number of adducts that would be needed to provide a comprehensive panoramic taking into account all the relevant energy-influencing features (all the geometrical features of all the conformers of ARZ, and all the possible arrangements of water molecules around each conformer) would imply enormous computational costs. Therefore, it was opted to select representative adducts for each relevant characteristic. This corresponds to a sampling approach more than to an unaffordable exhaustive approach. All the same, a sampling approach can be informative for a variety of aspects.

Within the reality of a water solution, the supermolecular structures of the adducts are not 'fixed' in time, because the water molecules H-bonded to a solute molecule do not remain the same in time (there is continuous fast interchange with the surrounding water molecules). Therefore, the calculated adducts represent time-averaged probable possibilities rather than permanent structures.

Despite all these challenges, the calculation of adducts with explicit water molecules provides information on the relative strength with which a water molecule can bind to each donor or acceptor site of the solute molecule, on the distance to which a water molecule preferably approaches each site, and on the preferred arrangements of water molecules around each donor or acceptor sites or around the region of two or more spatially close donors or acceptors. Since it results from optimisation and H-bonds are directional, this information can be considered as responding to the more common situations in the vicinity of the donors or acceptors of the solute molecule. The adducts also provide indications about whether a certain IHB tends to remain or to break in water solution—a type of information which may be relevant for other investigations, including the investigation of possible mechanisms for the biological activity of the molecule.

For the specific case of ARZ, the adducts confirm that the binding of a water molecule to the central molecule is stronger when it binds to two sites simultaneously, or when it is acceptor to an OH, or donor to an sp^2 O. They also confirm that the regions around IHBs are largely hydrophobic. They show the possibility that the clustering of water molecules (when these remain within the first solvation layer) may lead to a continuous frame of alternating water molecules H-bonded to ARZ and water molecules bridging them—an arrangement with enhanced stability. Finally, the results also contribute to the general ensemble of information on the interactions of ACPLs with explicit water molecules and with water as a solvent.

References

1. Appendino G, Ottino M, Marquez N, Bianchi F, Giana A, Ballero M, Sterner O, Fiebich BL, Munoz E (2007) J Nat Prod 70:608–612
2. Bauer J, Koeberle A, Dehm F, Pollastro F, Appendino G, Northoff H, Rossi A, Sautebin L, Werz O (2011) Biochem Pharmacol 81:259–268
3. Rosa A, Pollastro F, Atzeri A, Appendino G, Melis MP, Deiana M, Incani A, Loru D, Dessì MA (2011) Chem Phys Lipids 164:24–32
4. Singh IP, Bharate SB (2006) Nat Prod Rep 23:558–591
5. Mammino L, (2017) Molecules.22, 1294. https://doi.org/10.3390/molecules22081294
6. Mammino L, Kabanda MM (2009) J Mol Struct (Theochem) 901:210–219
7. Mammino L, Kabanda MM (2009) J Phys Chem A 113(52):15064–15077
8. Mammino L, Kabanda MM (2012) Int J Quant Chem 112:2650–2658
9. Kabanda MM, Mammino L (2012) Int J Quant Chem 112:3691–3702
10. Mammino L, Kabanda MM (2013) Molec Simul 39(1):1–13
11. Tomasi J, Persico M (1994) Chem Rev 94:2027–2094
12. Tomasi J, Mennucci B, Cammi R (2005) Chem Rev 105:2999–3093
13. Mennucci B (2010) J Phys Chem Lett 1:1666–1674
14. Alagona G, Ghio C (2002) Int J Quant Chem 90:641–656
15. Mammino L (2009) Chem Phys Letters 473:354–357
16. Mammino L, Kabanda MM (2007) J Mol Struct (Theochem) 805:39–52
17. Mammino L, Kabanda MM (2010) Int J Quant Chem 110(13):2378–2390
18. Mammino L (2017) J Mol Model. DOI. https://doi.org/10.1007/s00894-017-3443-4
19. Mammino L, Kabanda MM (2009) WSEAS Transact Biol Biomed 6(4):79–88
20. Mammino L, Kabanda MM (2008) Int J Quant Chem 108:1772–1791
21. Mammino L, Kabanda MM (2012) Int J Biol Biomed Engin 1(6):114–133
22. Mammino L (2013) Int J Biol Biomed Engin 2(7):15–25
23. Mammino L (2013) J Molec Model 19:2127–2142
24. Mammino L (2014) Curr Bioact Compd 10(3):163–180
25. Mammino L (2015) Current Phys Chem 5:274–293
26. Becke AD (1992) J Chem Phys 96:9489
27. Becke AD (1993) J Chem Phys 98:5648–5652
28. Lee C, Yang W, Parr RG (1998) Phys Rev B 37:785–789
29. Alagona G, Ghio C (2006) J Phys Chem A 110:647–659
30. Boys SF, Bernardi F (1970) Mol Phys 19:553
31. Alagona G, Ghio C (2009) J Phys Chem A 113:15206–15216

32. Frisch MJ, Trucks GW, Schlegel HB, Scuseria GE, Robb MA, Cheeseman JR, Montgomery JA, Vreven T, Kudin KN, Burant JC, Millam JM, Iyengar SS, Tomasi J, Barone V, Mennucci B, Cossi M, Scalmani G, Rega N, Petersson GA, Nakatsuji H, Hada M, Ehara M, Toyota K, Fukuda R, Hasegawa J, Ishida M, Nakajima T, Honda Y, Kitao O, Nakai H, Klene M, Li X, Knox JE, Hratchian HP, Cross JB, Adamo C, Jaramillo J, Gomperts R, Stratmann RE, Yazyev O, Austin AJ, Cammi R, Pomelli C, Ochterski JW, Ayala PY, Morokuma K, Voth GA, Salvador P, Dannenberg JJ, Zakrzewski VG, Dapprich S, Daniels AD, Strain MC, Farkas O, Malick DK, Rabuck AD, Raghavachari K, Foresman JB, Ortiz JV, Cui Q, Baboul AG, Clifford S, Cioslowski J, Stefanov BB, Liu G, Liashenko A, Piskorz P, Komaromi I, Martin RL, Fox DJ, Keith T, Al-Laham MA, Peng CY, Nanayakkara A, Challacombe M, Gill PMW, Johnson B, Chen W, Wong MW, Gonzalez C, Pople JA (2003) Gaussian 03. Gaussian Inc, Pittsburgh
33. Mammino L, Kabanda MM (2011) Int J Quant Chem 111:3701–3716

Computational Study of Jozimine A$_2$, a Naphthylisoquinoline Alkaloid with Antimalarial Activity

Mireille K. Bilonda and Liliana Mammino

Abstract Jozimine A$_2$ is a dioncophyllaceae-type naphthylisoquinoline alkaloid isolated from the root bark of an *Ancistrocladus* species from the Democratic Republic of Congo and exhibiting high antimalarial activity. It is the first naturally occurring dimeric naphthylisoquinoline of this type to be discovered. Its molecule consists of two identical 4′-O-demethyldioncophylline A units, with each unit containing an isoquinoline moiety and a naphthalene moiety. A thorough conformational study of this molecule was performed in vacuo and in three solvents with different polarities and different H-bonding abilities (chloroform, acetonitrile and water), using two levels of theory, HF/6-31G(d,p) and DFT/B3LYP/6-31+G(d,p). Intramolecular hydrogen bond (IHB) patterns were investigated considering all the possible options. Preferences for the mutual orientations of the moieties were identified through the potential energy profiles for the rotation of the single bonds between moieties. Harmonic vibrational frequencies were calculated to confirm the true-minima nature of stationary points, to obtain the zero point energies and to get indications about IHB strengths from red shifts. Intramolecular hydrogen bonds (O−H···O IHBs and O−H···π interaction) are the most stabilizing factors. The mutual orientations of the four moieties also have considerable influence and they prefer to be perpendicular to each other.

Keywords Alkaloids · Antimalarials · Intramolecular hydrogen bond
Jozimine A$_2$ · Naphthyl-isoquinoline alkaloids · Mutual orientation
Solute-solvent interactions

Electronic supplementary material The online version of this chapter (https://doi.org/10.1007/978-3-319-74582-4_17) contains supplementary material, which is available to authorized users.

M. K. Bilonda · L. Mammino (✉)
Department of Chemistry, University of Venda, P/bag X5050, Thohoyandou, South Africa
e-mail: sasdestria@yahoo.com

M. K. Bilonda
e-mail: mireillebilonda@yahoo.fr

© Springer International Publishing AG, part of Springer Nature 2018   305
Y. A. Wang et al. (eds.), *Concepts, Methods and Applications of Quantum Systems in Chemistry and Physics*, Progress in Theoretical Chemistry and Physics 31, https://doi.org/10.1007/978-3-319-74582-4_17

1 Introduction

Malaria is an infectious disease caused by *plasmodia*, among which *Plasmodium falciparum* is the most dangerous and responsible for most deaths. According to WHO 2014 reports, 97 countries and territories are affected by malaria transmission [1]. 98 million cases and 584 000 deaths were recorded in 2013 [1] and 212 million new cases and 429 000 deaths in 2015 [2]. 90% of the cases occurred in African countries and most of the deaths concern children under 5 years [1, 2].

The major challenge for malaria treatment is the fast development of resistance to new drugs, just within few years after their introduction into clinical use. Some drugs have already lost their efficacy and resistance has started appearing for artemisinin, which is the current drug of choice. The development of new drugs should at least keep the pace with the rate at which Plasmodium develops resistance to existing drugs. This implies urgent need for new compounds having original modes of action. Natural products constitute a potentially immense source of new compounds, with diverse molecular structures and pharmacophores.

Jozimine A_2 (Fig. 1, denoted by the acronym JZM in the rest of the text) is a dioncophyllaceae-type naphthylisoquinoline alkaloid isolated from a plant belonging to the *ancistrodaceae* family. It has confirmed antimalarial activity, with the lowest IC_{50} (0.0014 µm) among antimalarial naphthylisoquinoline alkaloids [3]. (The IC_{50} indicates how much of a particular drug is needed to inhibit 50% of a given biological process or component of a process such as an enzyme, cell, cell receptor or microorganism). The molecule has a dimeric structure: it consists of the two units, each consisting of a naphthalene moiety and an isoquinoline moiety. The two units are identical (have the same substituents in corresponding positions) making the molecule a C_2 symmetric dimer [3]. These characteristics make the computational study of this molecule particularly interesting. The analysis of the results will give specific attention to symmetry aspects.

The electronic structure of a molecule and the properties related to it—such as dipole moment, molecular electrostatic potential, molecular orbital energies—provide important information for a better understanding of molecular interactions when biological recognition processes are involves [4–8]. Some studies have also already shown relationships between the electronic structure and the antimalarial activity of alkoxylated and hydroxylated chalcones [9], tetrahydropyridines [10] and the cinchona alkaloids [8]. Some Quantitative Structure-Activity Relationship (QSRA) studies have shown correlation between electronic structure, antimalarial activity and phototoxicity of selected quinolinemethanol derivatives and their analogs [11].

The current study is the first study of the electronic structure of JZM, and it is part of an ongoing computational investigation of naphthylisoquinoline alkaloids [12, 13]. The conformational preferences of JZM were studied in vacuo and in three solvents with different polarities and different H-bonding abilities (chloroform, acetonitrile and water). Two levels of theory, Hartree-Fock (HF/6-31G(d,p)) and Density Functional Theory with the B3LYP functional (DFT/B3LYP/6-31+G(d,p)),

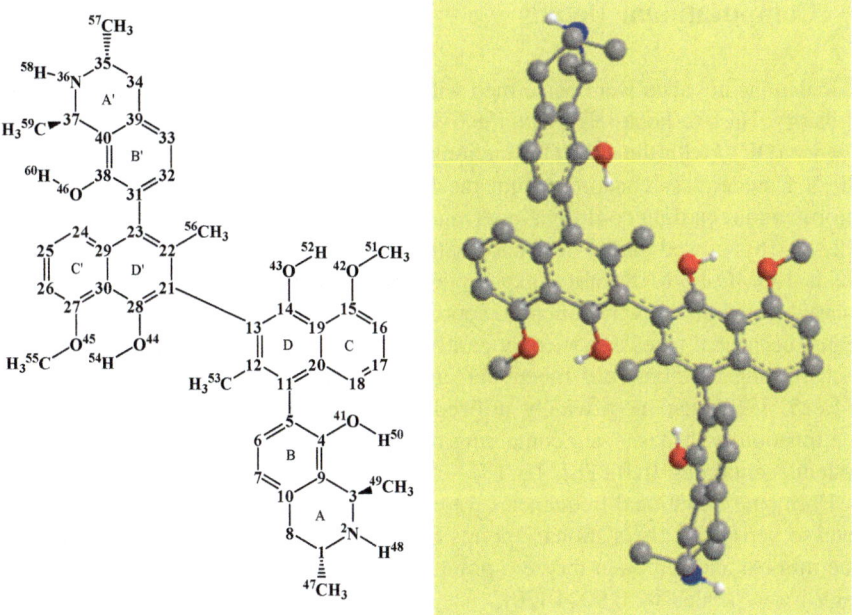

Fig. 1 Two representations of the jozimine A$_2$ molecule and atom numbering utilized in this work. In the structure on the left, the C atoms in the rings are represented only by the numbers denoting their positions. Only the H atoms attached to O or N are numbered separately, while the H atoms attached to C atoms are given the same number as the C atom and are not shown in the structure. The two units are identical. Identical rings are denoted by the same letters, primed for one of the units (A, B, C, D and C′, D′, A′ and B′)

were used to perform the calculations. Calculations in solution were performed using the PCM model.

The results highlight the dominant stabilizing effect of the O−H···O intramolecular hydrogen bonds (IHBs) and of other H-bond-type intramolecular interactions (for instance, O−H···π interaction) and the preference for all the moieties to be mutually perpendicular. Other computable molecular properties (dipole moments, HOMO-LUMO energy gaps, solvent effect, etc.) also show some dependence on IHB patterns.

Comparisons are also made with the results of a study of michellamine A [12]— a dimeric naphthylisoquinoline alkaloid with anti-HIV activity [13] and with the same ring systems as JZM. The two molecules differ by the substituents in the moieties and by the fact that the two naphthalene moieties in michellamine A are not identical. The stabilising effects show interesting similarities. The main difference is that no IHB between the two naphthalene moieties is possible in JZM, whereas it is possible in michellamine A.

2 Computational Details

Calculations in vacuo were performed with fully relaxed geometry using two levels of theory: Hartree-Fock (HF) with the 6-31G(d,p) basis set, and Density Functional Theory (DFT) with the B3LYP functional [14, 15] and the 6-31+G(d,p) basis set. HF is a moderately cheap quantum mechanical method which can yield accurate information regarding conformational analysis. Previous studies on other molecules [12, 16–18] showed that HF can successfully handle intramolecular H-bonding and yields HOMO-LUMO energy gaps approaching those of experiments. DFT—an alternate method to wavefunction approaches—is often used in conformational search because it takes into account part of the correlation effects at a relatively low cost. Among the numerous functionals available for the DFT framework, B3LYP [14, 15, 19] is the most widely utilized; it can provide better quality results in combination with basis sets containing diffuse functions, above all for molecular systems containing IHBs [12, 16–18].

Harmonic vibrational frequencies were calculated in vacuo at the HF/6-31G(d,p) level to verify that the stationary points from optimization results corresponded to true minima and to obtain the zero-point energy (ZPE) corrections. The frequency values were scaled by 0.9024 [20].

A preliminary identification of conformers of interest was carried out by considering the potential energy profiles for the rotation of the C5−C11 bond (showing the minima for the mutual orientation of the two moieties within one unit, Fig. 2) and for the rotation of the C13−C21 bond (showing the minima for the mutual orientation of the two units). The rotation of the two bonds was carried out simultaneously, yielding a 3D potential energy profile (Fig. 2). It was carried out

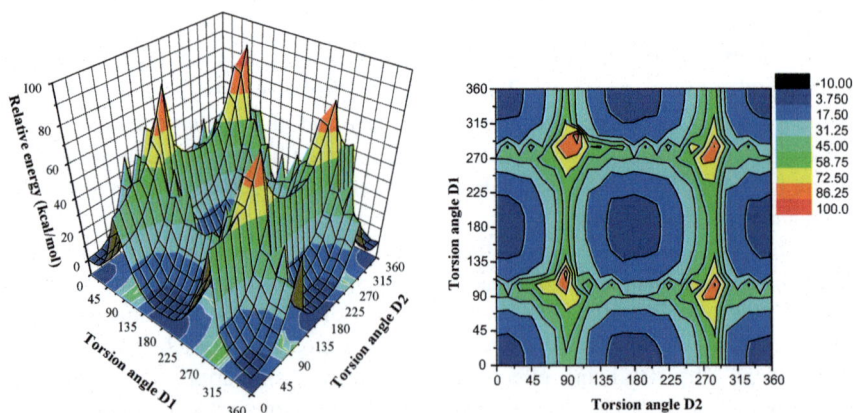

Fig. 2 3D potential energy profile for the scan of the rotation of the C5–C11 and C13–C21 bonds (energy versus D1 and D2, with D1 being the C4–C5–C11–C12 torsion angle and D2 being the C14–C13–C21–C28 torsion angle) and 2D potential energy profile for the same rotation scan. The complete 360° rotations were performed at 15° pace

for a complete rotation (360°) of the bond with a 15° pace. A conformer without O−H···π interaction between the moieties (a conformer with the O43−H52···O42 and O44−H54···O45 IHBs) was chosen as input, to avoid the effect of the removal of the interaction during the rotation. The scan highlighted five minima (Fig. 2, considering that angles differing by 360° are the same). Each of the corresponding geometries was optimized to find the actual geometry corresponding to a given minimum (as the optimization involves fully relaxed geometry). Since interactions between S and S′ were not present in the input, this provides information only on the preferences for the mutual orientations of the moieties and identifies five different combinations of orientations. More conformers were then identified by changing the IHB patterns in each of the five conformers obtained from the potential energy profile.

Calculations in solution utilized the Polarizable Continuum Model (PCM, [21–26]). In this model, the solvent is considered immeasurable and is modelled by a continuous isotropic dielectric into which the solute is inserted. Thus, the solute molecule is embedded in a cavity surrounded by the continuum solvent. The geometry of the cavity follows the geometry of the solute molecule, considering its solvent accessible surface. The calculations utilised the default settings of Gaussian03 [27] for PCM, namely, Integral Equation Formalism model (IEF, [23–26]) and Gepol model for building the cavity around the solute molecule [28–30], with simple United Atom Topological Model (UAO) for the atomic radii and 0.200 Å² for the average area of the *tesserae* into which the cavity surface is subdivided. The SCFVAC option was selected to obtain more thermodynamic data.

Calculations in solution were performed as single point (SP) calculations on the in-vacuo optimised geometries, with the same levels of theory utilised in vacuo. It was opted to use SP calculations because the size of the molecule makes re-optimisation in solution computationally expensive. Although SP calculations cannot provide information on the geometry changes caused by the solvent, they can provide reasonable information on the energetics, such as the conformers' relative energies in solution and the energy aspects of the solution process (the free energy of solvation, ΔG_{solv}, and its components).

The three solvents considered (chloroform, acetonitrile and water) cover the ranges of polarity and of hydrogen bonding abilities interesting for biologically active molecule. Chloroform is an apolar aprotic solvent with low relative permittivity ($\varepsilon_r = 4.90$) and low dipole moment ($\mu = 1.04$ D [31]). Acetonitrile is a dipolar aprotic solvent with large relative permittivity ($\varepsilon_r = 36.64$) and high dipole moment ($\mu = 3.92$ D [31]). Water is a protic solvent with high relative permittivity ($\varepsilon_r = 78.39$) and a sizeable dipole moment ($\mu = 1.83$ D [31]).

Calculations were performed using GAUSSIAN 03, Revision D 01 [27].

All the energy values reported are in kcal/mol and all the distances are in Å. Acronyms are utilized for the calculation methods and for the media, for conciseness sake on reporting values: HF for HF/6-31G(d,p), DFT for DFT/B3LYP/6-31+G(d,p), 'vac' for vacuum, 'chlrf' for chloroform, 'actn' for acetonitrile and 'aq' for water. Tables with all the numerical values of the properties of the calculated conformers (relative energy, dipole moments, free energy of solvation in the

solvents considered, HOMO-LUMO energy gaps, etc.), in the results of both calculation methods, and figures showing all the conformers, are included in the
Supplementary Information.

3 Results

3.1 Naming of Conformers

A system of symbols is introduced to be able to clearly and easily identify the
geometric characteristics of each conformer when analysing the results; thus, each
conformer is denoted by an acronym summarising its characterising features. Since
it is useful to be able to mention each ring individually and concisely, the rings are
denoted by uppercase letters [12]. Since the two units are identical, identical rings
are denoted with the same letter, primed for one of the units. Thus, the rings are
denoted as A, B, C, D, A′, B′, C′ and D′ as shown in Fig. 1. In order to be able to
mention each of the two units of the dimeric molecular structure individually, the
whole unit comprising the A, B, C and D rings is called S and the whole unit
comprising the A′, B′, C′ and D′ rings is called S′.

The O–H···O IHBs and other IHB-type interactions such as O–H···π and C
–H···O are the most important factors stabilizing the conformers. Different conformers have different combinations of these intramolecular interactions and are
characterized by them. A specific lowercase letter is used to denote each IHB or
IHB-type interaction in the acronyms denoting the conformers. The letters (a–h)
and their meanings are listed in Table 1. Their use is illustrated by the names of the
representative conformers of JZM shown in Fig. 3.

Table 1 Letters utilized in the acronyms denoting the conformers of Jozimine A_2 in this work,
and their meanings

Letter	Meaning	Letter	Meaning
a	Presence of the O43 –H52···O42 IHB	h	Presence of C–H59···O46
b	Presence of the O44 –H54···O45 IHB	t	C14−C13−C21−C28 torsion angle close to +90°
c	Presence of the O43−H52···π interaction	v	C14−C13−C21−C28 torsion angle close to −90°
d	Presence of the O44−H54···π interaction	p	C4−C5−C1−C12 torsion angle close to +90°
e	Presence of the O41−H50···π interaction	q	C4−C5−C1−C12 torsion angle close to −90°
f	Presence of the O46−H60···π interaction	x	C22−C23−C31−C38 torsion angle close to +90°
g	Presence of C−H49···O41	y	C22−C25−C31−C38 torsion angle close to −90°

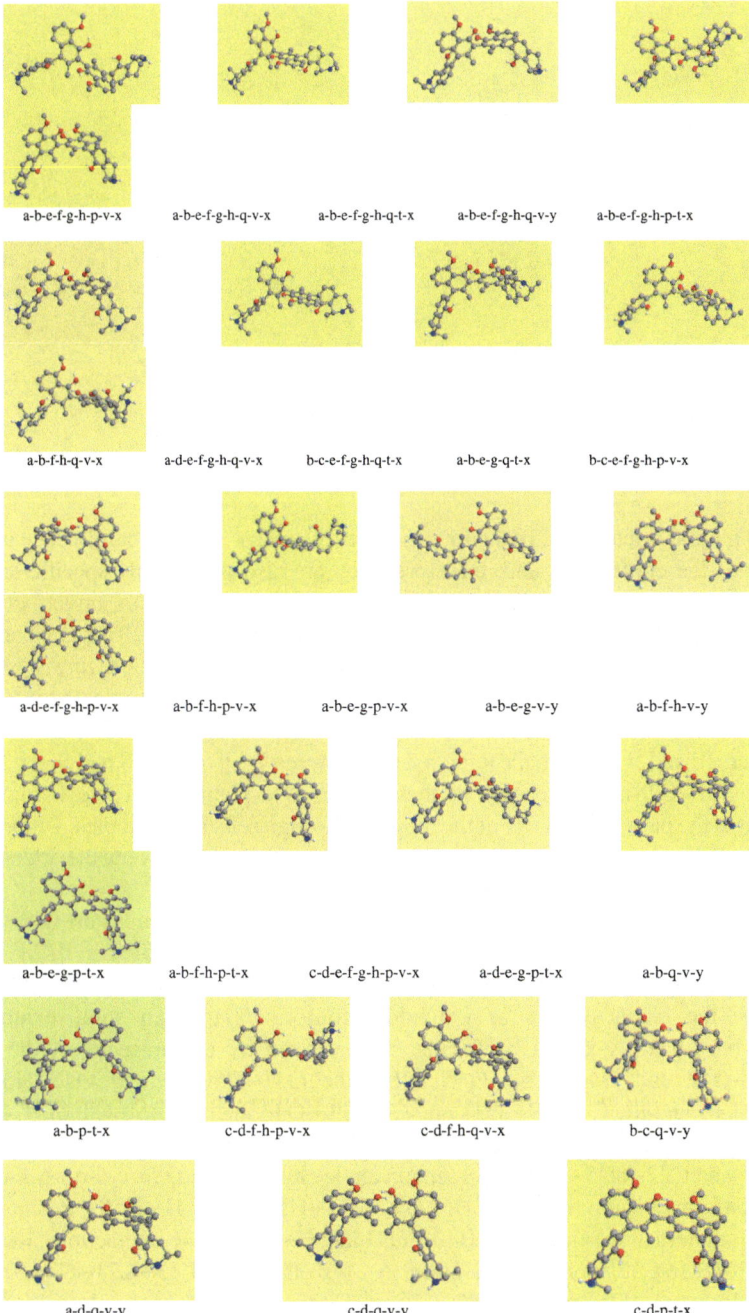

Fig. 3 Optimized geometries of representative conformers of jozimine A$_2$ and letter-combinations identifying their characteristics in the acronyms denoting them. Geometries from HF/6-31G(d,p) results. The images show the main combinations of interaction types and moieties' orientations which may be present in the conformers of jozimine A$_2$. The initial part of the acronyms is not reported under the images, because of space reasons; it is the same for all the conformers (JZM)

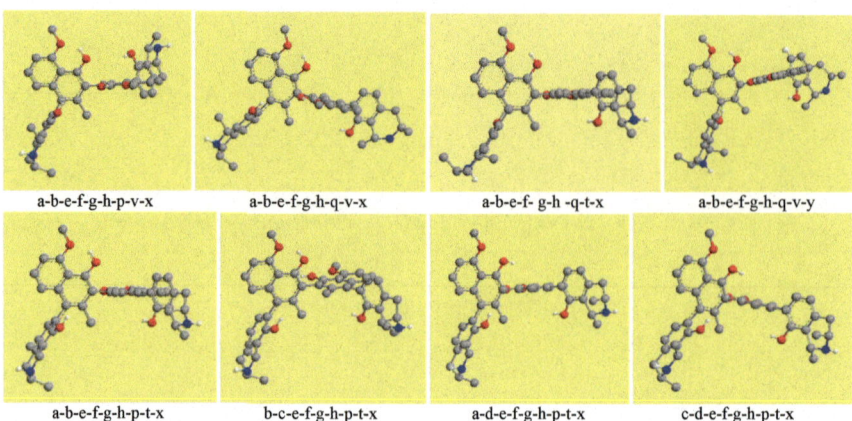

a-b-e-f-g-h-p-v-x   a-b-e-f-g-h-q-v-x   a-b-e-f- g-h -q-t-x   a-b-e-f-g-h-q-v-y

a-b-e-f-g-h-p-t-x   b-c-e-f-g-h-p-t-x   a-d-e-f-g-h-p-t-x   c-d-e-f-g-h-p-t-x

Fig. 4 Optimized geometries of selected conformers of jozimine A_2 highlighting the perpendicularity of the moieties

The mutual orientations of the four moieties also have significant influence on the energy of the conformers and, therefore, they are also denoted by specific letters. Lowercase letters are utilised to provide information about the torsion angles between moieties, i. e, information about their mutual orientation. The letters 'p' and 'q' refer to the torsion angle within the S unit (angle between the A, B and the D, C rings systems, i.e., C4−C5−C11−C12); the letters 'x' and 'y' refer to the torsion angle within S' unit (angle between the C', D' and the A', B' rings systems, i.e., C22−C25−C31−C38); the letters 't' and 'v' refer to the torsion angle between the two units S and S' (i.e., C14−C13−C21−C28); the meaning of these letters is explained in Table 1. Figure 4 highlights the perpendicularity of the moieties through selected examples. Differently from michellamine A [12], there is no specific correspondence between the orientation of the two moieties (within each unit) and the IHB patterns.

When the conformers are mentioned in the text, their acronyms start with JZM, followed by the letters denoting their characteristics. In tables and figures, 'JZM' is not reported for space-saving reasons. It may be expedient to illustrate how the acronyms describe the characteristics of individual conformers through some examples. JZM-a-b-e-f-g-h-p-v-x is a conformer characterized by the presence of the O43 −H52···O42 and O44−H54···O45 IHBs, the O41−H50···π and O46−H60···π interactions and the C−H49···O41 and C−H59···O46 interactions, with the C4−C5 −C11−C12 torsion angle close to 90°, the C14−C13−C21−C28 torsion angle close to −90° and C22−C25−C31−C38 torsion angle close to 90°; JZM-c-d-e-f-g-h-q-v-y is a conformer with the O43−H52···π, O44−H54···π, O41−H50···π and O46 −H60···π interactions and the C−H49···O41 and C−H59···O46 interactions, with the C4−C5−C11−C12 torsion angle close to −90°, the C14−C13−C21−C28 torsion angle close to −90° and the C22−C25−C31−C38 torsion angle close to −90°; JZM-b-c-q-t-x is a conformer characterized by the presence of the O44−H54···O45 IHB and the O43−H52···π interaction, and with the C4−C5−C11−C12 torsion angle close to −90°, the C14−C13−C21−C28 torsion angle close to +90° and the C22 −C25−C31−C38 torsion angle close to +90°.

3.2   Results in Vacuo

80 conformers were identified, and the frequency calculations confirm that they correspond to true minima. The relative energies of selected calculated conformers of JZM in vacuo and in the three solvents considered are reported in Table 2.

Table 2 Relative energies of the selected conformers of jozimine A_2 shown in Fig. 3. HF/6-31G (d,p) and DFT/B3LYP/6-31+G(d,p) results in vacuo and in the solvents considered, respectively denoted as HF and DFT in the columns' headings. The results in vacuo are from full optimization calculations; the results in solution are from single point PCM calculations on the in-vacuo-optimized geometries. The conformers are listed in order of increasing relative energies in the HF results in vacuo

Conformer	Relative energy (kcal/mol)							
	HF				DFT			
	vac	chlrf	actn	aq	vac	chlrf	actn	aq
a-b-e-f-g-h-p-v-x	0.000	0.000	0.000	0.112	0.000	0.000	0.000	0.095
a-b-e-f-g-h-q-v-x	0.743	0.467	0.278	0.020	0.608	0.404	0.320	0.000
a-b-e-f-g-h-q-t-x	0.793	0.591	0.303	0.243	0.606	0.411	0.290	0.058
a-b-e-f-g-h-q-v-y	1.463	0.637	0.618	0.000	1.202	0.777	0.567	0.005
a-b-e-f-g-h-p-t-x	1.466	0.920	0.707	0.075	1.180	0.825	0.578	0.187
a-b-f-h-q-v-x	5.783	5.411	3.856	1.772	5.634	4.645	3.970	1.860
a-d-e-f-g-h-q-v-x	5.857	5.469	5.429	3.921	5.944	5.799	$-^a$	$-^a$
b-c-e-f-g-h-q-t-x	5.934	5.759	5.547	3.902	5.990	5.840	5.679	4.132
a-b-e-g-q-t-x	5.962	4.497	4.003	1.987	5.770	4.757	4.121	1.900
b-c-e-f-g-h-p-v-x	6.209	5.756	5.725	4.095	6.267	6.053	5.902	4.341
a-d-e-f-g-h-p-v-x	6.209	5.756	5.725	4.095	6.267	6.053	5.897	4.375
a-b-e-g-p-v-x	6.269	5.549	4.132	2.037	6.028	4.978	4.266	2.160
a-b-f-h-p-v-x	6.269	4.914	4.192	2.096	6.028	5.004	$-^a$	2.134
a-b-e-g-v-y	6.476	5.168	4.123	1.625	6.202	$-^a$	$-^a$	1.819
a-b-f-h-v-y	6.476	5.157	4.037	1.706	6.202	4.917	4.289	1.813
a-b-e-g-p-t-x	6.552	5.316	4.308	1.969	6.204	5.008	4.295	1.999
a-b-f-h-p-t-x	6.552	5.194	4.294	1.945	6.204	5.021	4.278	1.959
c-d-e-f-g-h-p-v-x	11.627	11.590	11.467	8.363	11.508	11.728	11.623	8.798
a-d-e-g-p-t-x	11.682	10.404	9.397	5.723	11.568	10.460	9.692	6.632
a-b-q-v-y	11.899	9.483	7.954	3.038	11.602	$-^a$	$-^a$	3.878
a-b-p-t-x	12.096	10.142	8.111	3.817	11.684	9.638	8.295	4.437
c-d-f-h-p-v-x	15.532	14.858	14.216	10.362	15.508	15.181	14.688	10.984
c-d-f-h-q-v-x	15.687	14.974	14.302	3.432	15.648	15.128	14.575	$-^a$
b-c-q-v-y	16.262	13.764	12.531	6.784	16.266	14.080	12.760	7.637
a-d-q-v-y	16.262	13.982	12.425	6.961	16.266	14.099	12.753	7.714
c-d-q-v-y	20.097	18.547	17.013	11.069	20.169	18.675	17.585	11.609
c-d-p-t-x	20.432	18.889	17.427	11.707	20.216	18.933	18.027	12.417

^{a}The calculation for this conformer did not converge in the given solvent

The reported conformers comprise the first fourteen lower-energy conformers and representative conformers of other types (other combinations of IHBs) not appearing among the first fourteen. The IHB patterns (numbers and types of IHBs present in a conformer) are the dominant stabilizing factors. Others factors influencing conformational preferences are the mutual orientation of the two naphthalene moieties (C, D and C′, D′ ring systems), which determines the mutual orientation of the two units (S and S′) and the orientation of the two moieties (naphthalene and isoquinoline) within each unit. The isoquinoline moiety prefers to be perpendicular to the naphthalene moiety and the two naphthalene moieties (linked by the inter-units biaryl axis) also prefer to be perpendicular to each another; this results in the two units (S and S′) preferring to be mutually perpendicular, which excludes the possibility of an IHB between them.

As mentioned previously, three types of IHBs interactions may be present: $O-H \cdots O$ IHBs (O43−H52$\cdots$O42 and O44−H54$\cdots$O45), $O-H \cdots \pi$ interactions between the O−H in an isoquinoline moiety and the closest π system in the naphthalene moiety of the same unit (O41−H50$\cdots\pi$, O46−H60$\cdots\pi$) or between an O−H in the naphthalene moiety of one unit and the closest π system in the naphthalene moiety of the other unit (O43−H52$\cdots\pi$, O44−H54$\cdots\pi$); and $C-H \cdots O$ interactions within an isoquinoline moiety (C−H49$\cdots$O41 and C−H59$\cdots$O46). The $O-H \cdots O$ IHBs have the strongest stabilizing effect. The two $O-H \cdots O$ IHBs (O43 −H52$\cdots$O42 and O44−H54$\cdots$O45) have practically the same geometric parameters in a conformer having both of them simultaneously.

The lowest energy conformers are the conformers having all the IHB-types interactions simultaneously. The first five lowest energy conformers have relative energy below 1.5 kcal/mol in both the HF and the DFT results. They have the same types of IHBs interactions and differ only by the orientations of the moieties. Altogether, they account for 99.95% of the population in vacuo (Table S12).

The reported X-ray structure of JZM [3] corresponds to the a-b-f-h-p-v-x conformer, which has 6.269 kcal/mol relative energy and 0.0014 population (Table 2). This conformer has the two $O-H \cdots O$ IHBs (O43−H52$\cdots$O42 and O44 −H54$\cdots$O45), one $O-H \cdots \pi$ interaction and one $C-H \cdots O$ interaction. Compared to the five lowest energy ones (which have all the interactions), this conformer has one $O-H \cdots \pi$ and $C-H \cdots O$ interaction less, which leaves one OH group free. This phenomenon is observed frequently (e.g., in the case of caespitate [32]): the crystalline structure differs from the geometry in vacuo by not having one or more weaker IHBs or other IHB-type interactions that are present in the gas phase. A possible reason is the need, for molecules in the crystal structure, to have some donors or acceptors not engaged in intramolecular interactions, so that they are available for intermolecular interactions with the surrounding molecules.

The symmetry of the molecule results in a number of pairs of conformers with the same characteristics (IHBs) in the S moiety of one conformer and in the S′ moiety of the other conformer. Pairs of this type will be termed S/S′ symmetric pairs in the rest of the text.

The removal of the O43−H52···O42 IHB brings about the O43−H52···π interaction and the removal of the O44−H54···O45 IHB brings about the O44 −H54···π interaction. Each of these removals causes an increase in the energy of the conformer. Comparison of the energies of conformers JZM-a-b-e-f-g-h-p-v-x and JZM-b-c-e-f-g-h-p-v-x (removal of O43−H52···O42 and formation of O43 −H52···π) shows an energy increase of ≈6 kcal/mol. Since the two units are identical, the presence or the absence of a given IHB in only one of the units has the same effects whether it occurs in unit S (O43−H52···O42) or in unit S′ (O44 −H54···O45). Therefore, the conformers JZM-a-d-e-f-g-h-p-v-x and JZM-b-c-e-f-g-h-p-v-x (or the conformers JZM-a-d-e-f-g-h-q-v-y and JZM-b-c-e-f-g-h-q-v-y, or JZM-a-d-e-f-g-h-p-t-x and JZM-b-c-e-f-g-h-p-t-x, or JZM-a-d-q-t-x and JZM-b-c-q-t-x, or JZM-b-c-q-v-y and JZM-a-d-q-v-y, or JZM-b-c-p-t-x and JZM-a-d-p-t-x, or JZM-b-c-p-v-x and JZM-a-d-p-v-x, or JZM-b-c-q-t-x and JZM-a-d-q-v-x (Table S13) have the same relative energy and similar molecular properties such as dipole moment and HOMO-LUMO energy gap, and also rather close values of ΔG$_{solv}$.

Since the removal of an O−H···O IHB brings about an O−H···π IHB and vice versa, it is not possible to evaluate the energy of an IHB by comparison with a conformer in which it is removed by 180° rotation of the donor [33–42]. It is only possible to compare the strength of the O−H···O and O−H···π IHBs utilizing conformers in which only one or the other is present and the other geometry characteristics are identical, as in the previously considered case of JZM-a-b-e-f-g-h-p-v-x and JZM-b-c-e-f-g-h-p-v-x. Other analogous cases can be considered. For instance, JZM-a-b-e-f-g-h-p-v-x and JZM-c-d-e-f-g-h-p-v-x differ only because the former has two O−H···O IHBs and two O−H···π IHBs and the latter has four O−H···π IHBs. The relative energy of the latter is 11.627/HF and 11.508/DFT kcal/mol higher than that of the former, suggesting that in this case the stabilisation brought by an O−H···π IHB is ~5.8 kcal/mol smaller than that of an O−H···O IHB. Conformers JZM-a-b-e-f-g-h-q-v-x and JZM-c-d-e-f-g-h-q-v-x, JZM-a-b-e-f-g-h-q-t-x and JZM-c-d-e-f-g-h-q-t-x, JZM-a-b-e-f-g-h-q-v-y and JZM-c-d-e-f-g-h-q-v-y, and JZM-a-b-e-f-g-h-p-t-x and JZM-c-d-e-f-g-h-p-t-x are other pairs of this type with the relative energy of the latter in each pair ~10 kcal/ mol higher than that of the corresponding former, with both HF and DFT.

Table 3 reports the ranges of the lengths of all the IHB types present in the conformers. Complete data of the parameters of all the conformers are reported in the Supplementary Information. The parameters of the O43−H52···O42 and O44 −H54···O45 IHBs in the same conformer are very close. For S/S′ symmetric pairs, the parameters are identical whether these IHBs are in S or S′. The bond length (Å) is close to 1.73/DFT and 1.77/HF, the O···O distance (Å) is close to 2.58 with both HF and DFT and the OĤO bond angle is ~143°/HF and ~145°/DFT for conformers having also other types of IHBs. On the basis of their bond lengths, O···O distances and bond angles, the O−H···O IHBs in JZM can be classified in the lower region of moderate H-bonds [43].

Table 3 Ranges of the length of the IHBs in the conformers of Jozimine A$_2$. For the O–H···π interaction, the length is taken as distance between the H atom and the closest C atom in the π system, which is indicated in parentheses after the π symbol

IHB	H···O (Å) or H···C (Å)	
	HF	DFT
H52···O42	1.764–1.770[a]	1.722–1.731[a]
	1.771–1.773[b]	1.731–1.735[b]
H54···O45	1.764–1.770[a]	1.722–1.732[a]
	1.768–1.773[b]	1.731–1.735[b]
H49···O41	2.367–2.372	2.866–3.008
H59···O46	2.369–2.372	3.000–3.008
H50···π (C11)	2.323–2.343	2.284–2.312
H52···π (C21)	2.222–2.233	2.189–2.203
H54···π (C13)	2.221–2.325	2.186–2.288
H60···π (C23)	2.324–2.404	2.288–2.315

[a]In conformers having other IHBs
[b]In conformers having only O–H···O IHBs

For the O–H···π interactions, a bond length is not defined, as the acceptor is a whole π system and not an individual atom. It may be convenient to consider the distance between the H atom of the donor OH and the C atom in the aromatic system closest to it. The ranges of the distances thus defined are reported in Table 3. It can be inferred from these ranges that the H···C distance is slightly shorter when the interaction involves an O–H in the naphthalene moiety of one unit and the closest π system in the naphthalene moiety of the other unit (O43–H52···π and O44 –H54···π) than when it involves an O–H in the isoquinoline moiety and the closest π system in the naphthalene moiety of the same unit (O41–H50···π and O46 –H60···π).

The ranges of the H···O distance for the C–H···O interactions within the iso-quinoline moiety (C–H49···O41 and C–H59···O46) are also reported in Table 3. This distance is considerably longer than the H···O distance for O–H···O IHBs, consistently with the fact that the C–H···O interaction is considerably weaker. The donor–acceptor C···O distance for C–H49···O41and C–H59···O46 is ≈3.0 Å in both the HF and DFT results. The CĤO bond angle is ≈115.0°/HF and ≈116.0°/ DFT for both C–H49···O41 and C–H59···O46. Their parameters show that the C –H···O IHBs are weak H-bonds.

Table 4 lists the ranges of the calculated harmonic vibrational frequencies of the O–H bonds. Only HF frequencies are available in this study because frequency calculations with the DFT method did not complete (which is probably due to the high number of atoms in this molecule). When IHBs are present, it is interesting to consider also the red-shift (lowering of the vibrational frequency of the donor OH) caused by the IHBs. The red shift is evaluated with respect to the frequency of a free OH of the same type. Since the OHs in JZM are never free, two model structures with free OHs and with the other features similar to those of the moieties in JZM (including a CH$_3$ to mimic the presence of other moieties attached to the one considered) were used to calculate a reference frequency. These structures are

Table 4 Ranges of the harmonic vibrational frequencies of the OH groups in jozimine A₂ and of the red shifts caused by IHBs

OH	Frequency (cm^{-1})	Red shift when engaged in O–H···O IHB	Red shift when engaged in O–H···π interactions
O41–H50	3735.02–3804.55	–	39.08–64.31
O43–H52	3707.53–3758.99	27.37–60.59	9.13–46.53
O44–H54	3707.73–3759.84	30.03–60.39	8.23–51.28
O46–H60	3736.13–3838.26	–	40.34–65.17

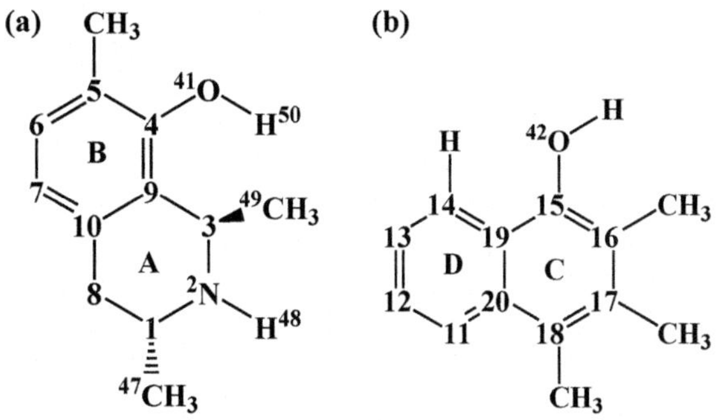

Fig. 5 Model structures used to calculated a reference frequency for the vibration of the free O41–H50 and O46–H50 (**a**) or O43–H52 and O44–H54 (**b**)

shown in Fig. 5. The ranges of the red shifts of O43–H52 and O44–H54 when engaged in the different types of IHBs are reported in Table 4. The values of the red shifts for O–H···O IHBs and the O–H···π interactions are comparable, which is consistent with the stabilizing effect of both types of IHBs (although the strength of the O–H···O IHB is expected to be greater than that of the O–H···π IHB).

Table 5 reports the values of the ZPE and of the relative energies corrected for ZPE for the conformers listed in Table 2. The ZPE corrections are very close for all the conformers. Their values (kcal/mol) are in the 571.572–572.602 kcal/mol range, with the greater values corresponding to lower energy conformers. The relative energies corrected for ZPE have the same trends as the uncorrected ones.

Table 6 reports the values of the dipole moment for the conformers listed in Table 2. The ranges (debye) of the values of the dipole moments in vacuo are 0.50 –77.53/HF and 0.59–77.33/DFT. Both the magnitude and the direction of the dipole moment vector change according to the conformer and are largely influenced by the orientation of the OH groups (although the orientation of the aromatic rings also plays a role).

Conformers with O–H···O IHBs have higher dipole moment than conformers with O–H···π interactions, with few exceptions. Conformers with only O–H···π interactions (no O–H···O IHBs) are among the conformers with smaller dipole

Table 5 Relative energy ($\Delta E_{correct}$, kcal/mol) corrected for ZPE and ZPE corrections (kcal/mol) for conformers of jozimine A_2 selected among those reported in Table 2. Results from HF/6-31G (d,p) frequency calculations

Conformer	$\Delta E_{correct}$	ZPE correction	Conformer	$\Delta E_{correct}$	ZPE correction
a-b-e-f-g-h-p-v-x	0.000	572.602	a-b-e-g-p-t-x	6.100	572.150
a-b-e-f-g-h-q-v-x	0.644	572.503	c-d-e-f-g-h-p-v-x	11.036	572.011
a-b-e-f-g-h-q-t-x	0.688	572.497	a-d-e-g-p-t-x	11.053	571.973
a-b-e-f-g-h-q-v-y	1.268	572.407	a-b-q-v-y	11.089	571.791
a-b-e-f-g-h-p-t-x	1.294	572.430	a-b-p-t-x	11.313	571.819
a-d-e-f-g-h-q-v-x	5.545	572.289	c-d-f-h-p-v-x	14.776	571.846
b-c-e-f-g-h-q-t-x	5.616	572.284	c-d-f-h-q-v-x	14.866	571.781
a-b-e-g-q-t-x	5.575	572.215	b-c-q-v-y	15.280	571.621
b-c-e-f-g-h-p-v-x	5.871	572.264	c-d-q-v-y	19.007	571.572
a-d-e-f-g-h-p-v-x	5.872	572.264	c-d-p-t-x	19.433	571.603

moments. The same phenomenon was observed in the previous study of naphthylisoquinoline alkaloids with antimalarial activity [16] and in the study of michellamine A [12]. The mutual orientation of the moieties also has considerable influence on the dipole moment. Different orientations of the S and S′ units may cause 1−4 D difference in the dipole moments, and different orientations of the isoquinoline and naphthalene moieties within each unit may cause 1−2 D difference in the dipole moments. This trend is reversed with respect to what was observed for the conformers of michellamine A [12], where the orientation of the units caused a 1−2 D difference and the different orientation of the isoquinoline and naphthalene moieties within each unit caused a 1−4 D difference. When two conformers of JZM differ both by the orientation of the S and S′ units and by the orientations of the isoquinoline and naphthalene moieties, the dipole moment difference is ≈2 D. Conformers of S/S′ symmetric pairs have the same dipole moment.

Table 7 reports the HOMO-LUMO energy gap for the conformers listed in Table 2. The gap is influenced by the IHB patterns. Conformers with only one O−H···O IHB and other types of IHBs interactions (O−H···π and C−H···O), such as conformers JZM-b-c-e-g-p-v-x, JZM-c-f-h-q-t-x, JZM-a-d-e-g-p-t-x, JZM-a-d-e-g-q-v-y, have the smallest HOMO-LUMO energy gap. The gap is slightly greater for conformers with the two O−H···O IHBs and other IHB-type interactions and highest for conformers with only O−H···π interactions (which are accompanied by C−H···O interactions if they involve the two moieties within the same unit). The presence of the C−H···O interactions in a conformer slightly decreases the HOMO-LUMO energy gap with respect to a corresponding conformer where it is absent; for instance, the gap in JZM-c-d-f-h-p-v-x is ≈1 kcal/mol less than in JZM-c-d-p-v-x. Conformers of S/S′ have the same HOMO-LUMO energy.

The estimation of the HOMO-LUMO energy gap shows marked difference between HF and DFT values. This is a known phenomenon, as DFT substantially

Table 6 Dipole moment of the conformers of jozimine A₂ listed in Table 2, in vacuo and in the three solvents considered. HF/6-31G(d,p) and (DFT/B3LYP/6-31+G(d,p) results, respectively denoted as HF and DFT in the column headings. The results in vacuo are from full optimization calculations, the results in solution are from single point PCM calculations on the in-vacuo-optimized geometries

Conformer	Dipole moment (Debye)							
	HF				DFT			
	vac	chlrf	actn	aq	vac	chlrf	actn	aq
a-b-e-f-g-h-p-v-x	2.65	2.67	2.55	2.44	2.81	2.87	2.74	2.58
a-b-e-f-g-h-q-v-x	5.46	5.96	6.06	5.97	5.32	5.93	6.09	5.99
a-b-e-f-g-h-q-t-x	5.03	5.41	5.43	5.34	5.01	5.50	5.57	5.45
a-b-e-f-g-h-q-v-y	6.83	7.46	7.55	7.44	6.59	7.36	7.52	7.39
a-b-e-f-g-h-p-t-x	7.53	8.29	8.50	8.43	7.33	8.28	8.57	8.49
a-b-f-h-q-v-x	4.25	4.95	4.51	4.31	4.04	4.36	4.41	4.22
a-d-e-f-g-h-q-v-x	4.56	4.85	4.87	4.73	4.70	5.11	−ᵃ	−ᵃ
b-c-e-f-g-h-q-t-x	5.03	6.12	6.29	6.28	5.52	6.22	6.46	6.41
a-b-e-g-q-t-x	4.20	4.45	4.47	4.27	4.10	4.39	4.42	4.21
b-c-e-f-g-h-p-v-x	4.11	4.53	4.63	4.69	4.15	4.65	4.81	4.83
a-d-e-f-g-h-p-v-x	4.11	4.53	4.64	4.69	4.15	4.66	4.80	4.83
a-b-e-g-p-v-x	3.93	4.24	4.29	4.37	4.38	4.79	4.85	4.91
a-b-f-h-p-v-x	3.93	4.23	4.28	4.36	4.38	4.78	−ᵃ	4.90
a-b-e-g-v-y	4.84	5.08	5.04	4.89	4.36	−ᵃ	−ᵃ	4.27
a-b-f-h-v-y	4.84	5.07	5.04	4.89	4.36	4.59	4.55	4.27
a-b-e-g-p-t-x	4.96	5.21	5.18	4.88	4.78	5.12	5.10	4.78
a-b-f-g-p-t-x	4.95	5.22	5.17	4.88	4.77	5.11	5.09	4.78
c-d-e-f-g-h-p-v-x	0.50	0.20	0.02	0.17	0.70	0.40	0.20	0.21
a-d-e-g-p-t-x	4.62	5.10	5.24	5.20	4.03	4.54	4.70	4.62
a-b-q-v-y	1.38	1.13	−ᵃ	0.65	0.59	−ᵃ	−ᵃ	0.41
a-b-p-t-x	1.42	1.19	0.96	0.50	1.08	0.85	0.60	0.15
c-d-f-h-p-v-x	3.83	4.28	4.49	4.65	4.04	4.60	4.82	4.98
c-d-f-h-q-v-x	3.79	4.22	4.39	4.39	3.71	4.56	4.29	−ᵃ
b-c-q-v-y	3.91	4.40	4.56	4.67	3.97	4.69	5.00	5.25
a-d-q-v-y	3.91	4.40	4.56	4.67	3.97	4.70	5.00	5.24
c-d-q-v-y	1.28	1.81	2.03	2.40	2.07	2.83	3.19	3.70
c-d-p-t-x	1.33	1.86	2.09	2.64	1.84	2.52	2.87	3.48

ᵃThe calculation for this conformer did not converge in the given solvent

underestimates the values of the gaps [44, 45]. However, the two methods show similar trends.

Figure 6 shows the shapes of the HOMO and LUMO orbitals for the five lowest energy conformers, some representative higher energy ones and the conformers of some S/S′ symmetric pairs. The shapes indicate greater electron density and similar distribution in the two naphthalene moieties than in the isoquinoline moieties for

Table 7 HOMO-LUMO energy gap of the conformers of jozimine A$_2$ reported in Table 2, in vacuo and in the three solvents considered. HF/6-31G(d,p) and (DFT/B3LYP/6-31+G(d,p) results, respectively denoted as HF and DFT in the column headings. The results in vacuo are from full optimization calculations, the results in solution are from single point PCM calculations on the in-vacuo-optimized geometries

Conformer	HOMO-LUMO energy gap (kcal/mol)							
	HF				DFT			
	vac	chlrf	actn	aq	vac	chlrf	actn	aq
a-b-e-f-g-h-p-v-x	232.762	232.850	232.881	232.919	98.343	98.506	98.563	98.613
a-b-e-f-g-h-q-v-x	253.376	233.929	233.960	234.048	98.908	99.065	99.115	99.203
a-b-e-f-g-h-q-t-x	233.810	233.923	233.967	234.029	98.757	98.946	99.027	99.102
a-b-e-f-g-h-q-v-y	234.921	235.046	235.096	235.159	99.391	99.567	99.642	99.730
a-b-e-f-g-h-p-t-x	234.745	234.883	234.933	234.977	99.579	99.774	99.862	99.931
a-b-f-h-q-v-x	232.341	232.800	232.517	233.007	97.007	97.094	97.389	98.155
a-d-e-f-g-h-q-v-x	224.843	228.733	230.126	231.927	89.031	92.922	–a	–a
b-c-e-f-g-h-q-t-x	224.661	228.608	230.158	231.896	88.868	92.752	94.327	96.266
a-b-e-g-q-t-x	232.191	232.272	232.505	233.051	96.963	96.982	97.226	97.923
b-c-e-f-g-h-p-v-x	224.912	228.840	230.359	232.166	89.050	92.991	94.515	96.511
a-d-e-f-g-h-p-v-x	225.157	228.859	230.365	232.134	89.044	92.984	94.515	96.505
a-b-e-g-p-v-x	231.846	231.978	232.235	232.749	96.655	96.718	97.025	97.596
a-b-f-h-p-v-x	231.839	231.971	232.241	232.718	96.655	96.718	–a	97.615
a-b-e-g-v-y	232.938	232.856	232.969	233.709	96.837	–a	–a	98.036
a-b-f-h-v-y	232.938	232.850	232.969	233.697	96.843	96.944	97.214	98.036
a-b-e-g-p-t-x	232.498	232.618	232.925	233.640	96.756	96.850	97.163	97.954
a-b-f-h-p-t-x	232.498	232.611	232.900	233.659	96.756	96.850	97.151	97.973
c-d-e-f-g-h-p-v-x	236.558	236.508	236.483	236.433	100.552	100.502	100.489	100.502
a-d-e-g-p-t-x	221.166	225.226	227.064	229.618	85.435	89.257	91.020	93.536
a-b-q-v-y	234.017	234.306	234.506	234.707	98.776	–a	–a	99.460

(continued)

Table 7 (continued)

Conformer	HOMO-LUMO energy gap (kcal/mol)							
	HF				DFT			
	vac	chlrf	actn	aq	vac	chlrf	actn	aq
a-b-p-t-x	234.237	234.462	234.569	234.776	98.889	99.184	99.335	99.567
c-d-f-h-p-v-x	233.051	233.032	233.239	233.747	97.251	97.088	97.295	97.904
c-d-f-h-q-v-x	234.205	233.873	233.967	234.557	97.760	97.835	97.835	0.000
b-c-q-v-y	225.000	229.028	230.591	232.442	88.999	93.016	94.547	96.442
a-d-q-v-y	225.000	229.041	230.559	232.398	89.006	92.978	94.509	96.448
c-d-q-v-y	236.320	236.307	236.289	236.245	100.4077	100.364	100.326	100.282
c-d-p-t-x	236.452	236.401	236.370	236.301	100.3073	100.276	100.257	100.213

aThe calculation for this conformer did not converge in the given solvent

HOMO

LUMO

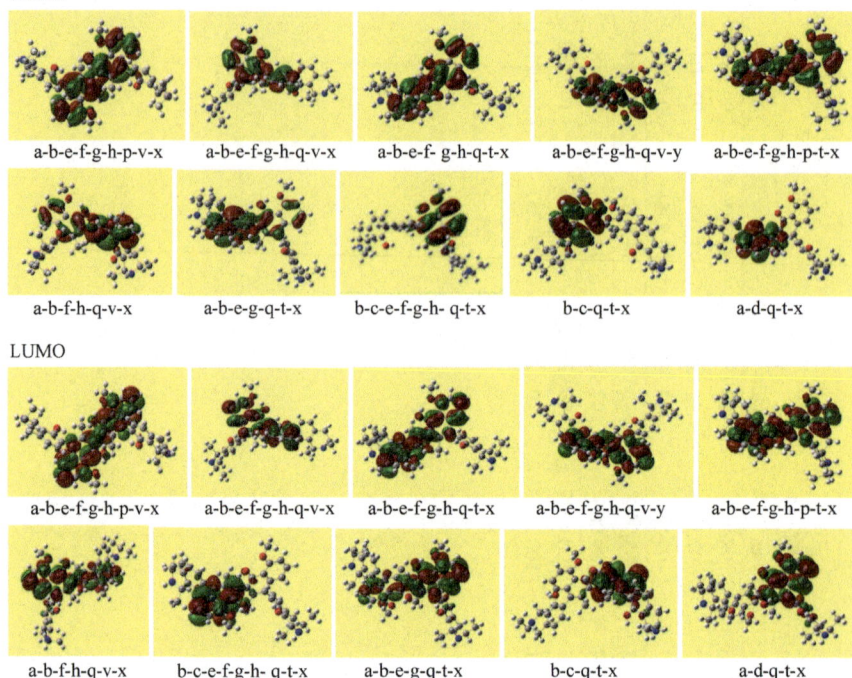

Fig. 6 Shapes of the HOMO and LUMO frontier orbitals of the lowest energy conformers of jozimine A_2. HF/6-31G(d,p) results. The initial part of the acronyms is not reported under the images, because of space reasons; it is the same for all the conformers (JZM)

the five lowest energy conformers—a phenomenon that was also observed for monomeric naphthylisoquinoline alkaloids with antimalarial activity [16] and for michellamine A [12]. For higher energy conformers, the electron distribution becomes considerably different for the two naphthalene moieties. For conformers of S/S′ symmetric pairs, electrons concentrate on one or the other of the naphthalene moieties.

3.3 Results in Solution

Table 2 shows the relative energy of the conformers in the three solvents considered (chloroform, acetonitrile and water). The relative energy of the conformer decreases with increasing solvent polarity—which is consistent with common behaviours. (Acetonitrile has a greater dipole moment than water, but its effect on solutes is often intermediate between that of chloroform and that of water; therefore, "increasing solvent polarities" refers to the chloroform-acetonitrile-water sequence in this discussion). The first five lowest energy conformers have relative energy below 1 kcal/mol in chloroform and acetonitrile, and below 0.24 kcal/mol in water. The identification of the lowest energy conformer does not change in chloroform and acetonitrile with respect to in vacuo, but may be different in water; however, the relative energies of the five lower-energy conformers are so small in water that a different identification of the lowest-energy one among them does not affect the interpretation of the results. Major changes in water occur for some conformers having relative energy in vacuo greater than 5 kcal/mol. These cases concern conformers in which O41−H50 or O46−H60 are not engaged in the O−H···π interaction and, therefore, are available to form H-bonds with water molecules. Although PCM does not take into explicit account solute-solvent H-bonds, it appears sometimes to take it into account implicitly through the energetics (possibly as the effect of the point-charges distribution in the areas of the cavity surface corresponding to an H-bond donor or acceptor) [46]. A confirmation about the interactions of these conformers with water molecules can be obtained through the consideration of adducts with explicit water molecules, which might be the object of a separate study.

The five lowest energy conformers account for 99.89% of the population in chloroform, 99.66% in acetonitrile and 90.71% in water. The population distribution of the other conformers (besides the five lowest energy ones) is also different. In *vacuo*, no other conformer has a population greater than 0.003%; in chloroform, there are two conformers with population of 0.02 and 0.01% respectively, and all the other populations are <0.007%; in acetonitrile, ten other conformers have a population between 0.02 and 0.05%; in water, three other conformers have popu-lation between 1.0 and 1.3% and seven conformers between 0.1 and 0.9%.

Conformers of S/S′ symmetric may have slightly different relative energy in solution, suggesting that the effect of their symmetric situation may somewhat decrease in solution.

If 3.5 kcal/mol is taken as a cautious threshold value for conformers which might be responsible for the biological activity, the results in water solution suggest that most conformers of JZM (all those with relative energy ≤ 3.5 kcal/mol in water solution) might be considered as potential responsibles for the antimalarial activity of JZM.

Table 8 reports the solvent effect (free energy of solvation, ΔG_{solv}) for the conformers listed in Table 2. ΔG_{solv} is positive in chloroform and acetonitrile and negative in water for all the conformers. The values in acetonitrile are considerably greater than those in chloroform. The electrostatic component of ΔG_{solv} (G_{el}) has negative values in all the three solvents, but considerably more negative in water. A quick estimation of the octanol/water partition coefficient of JZM (6.3743, [47]) suggests that JZM could be more soluble in non-polar solvents than in water. This could be due to presence of many aromatic rings and to the high molecular mass. On the other hand, the negative values of ΔG_{solv} in water suggest the possibility of some (although limited) solubility in water or, at least, the possibility that the molecule may be present also in the water phases of living organisms cannot be

Table 8 Solvent effect (free energy of solvation, ΔG_{solv}) of selected conformers of jozimine A_2 in the three solvents considered Results from HF/6-31G(d,p) and DFT/6-31+G(d,p) single point PCM calculations on the in-vacuo-optimized geometries

Conformer	ΔG_{solv} (kcal/mol)					
	HF			DFT		
	chlrf	actn	aq	chlrf	actn	aq
a-b-e-f-g-h-p-v-x	7.19	17.60	−4.12	7.37	18.30	−4.12
a-b-e-f-g-h-q-v-x	6.95	17.66	−4.39	7.29	18.16	−4.64
a-b-e-f-g-h-q-t-x	7.15	17.77	−4.04	7.32	18.14	−4.57
a-b-e-f-g-h-q-v-y	6.81	17.67	−4.70	7.19	17.93	−5.09
a-b-e-f-g-h-p-t-x	7.17	17.89	−4.50	7.22	17.96	−4.87
a-b-f-h-q-v-x	5.76	16.21	−7.62	6.65	16.98	−7.54
a-d-e-f-g-h-q-v-x	6.97	17.86	−5.37	7.61	−^{a}	−^{a}
b-c-e-f-g-h-q-t-x	7.19	17.91	−5.46	7.61	18.44	−5.50
a-b-e-g-q-t-x	5.78	16.21	−7.54	6.57	16.93	−7.72
b-c-e-f-g-h-p-v-x	6.84	18.05	−5.26	7.50	18.36	−5.60
a-d-e-f-g-h-p-v-x	7.17	17.95	−5.55	7.50	18.36	−5.57
a-b-e-g-p-v-x	6.38	15.89	−7.96	6.44	16.72	−7.83
a-b-f-h-p-v-x	5.74	15.95	−7.90	6.47	−^{a}	−7.86
a-b-e-g-p-t-x	6.01	15.96	−8.09	6.61	16.94	−7.73
a-b-f-h-p-t-x	5.88	15.94	−8.13	6.62	16.93	−7.76
c-d-e-f-g-h-p-v-x	7.57	18.43	−6.35	7.95	18.87	−6.33
a-d-e-g-p-t-x	6.13	16.10	−9.25	6.68	16.93	−8.49
a-b-q-v-y	4.89	−^{a}	−12.31	−^{a}	−^{a}	−11.36
a-b-p-t-x	5.65	14.61	−11.38	5.61	15.32	−10.92
c-d-f-h-p-v-x	6.74	17.09	−8.42	7.30	17.83	−8.25
c-d-f-h-q-v-x	6.80	17.12	−8.53	7.48	18.00	−^{a}
b-c-q-v-y	5.36	15.08	−12.30	5.85	15.53	−11.91
a-d-q-v-y	5.76	15.07	−12.03	5.88	15.54	−11.81
c-d-q-v-y	6.10	15.53	−12.02	6.59	16.50	−11.77
c-d-p-t-x	5.91	15.47	−11.86	6.56	16.70	−11.23

^{a}The calculation for this conformer did not converge in the given solvent

excluded. It is probable that the ΔG_{solv} takes into account the presence of OHs, which favours the solubility in water.

The presence and number of O−H···O IHBs, and the O−H···π and C−H···O interaction seem to influence ΔG_{solv} and this effect is different for different solvents. The first five lowest energy conformers (with a-b-e-f-g-h in common) have the smallest absolute values of ΔG_{solv} in water. Conformers in which either O41−H50 or O46−H60 is not engaged in the O−H···π IHB have the greatest absolute values of ΔG_{solv} in water. Conformers in which only O41−H50 or O46−H60 is engaged in O$-$H$\cdots\pi$ interaction have greater values of ΔG_{solv} than the conformers in which O41$-$H50 and O46$-$H60 are simultaneously engaged in O$-$H$\cdots\pi$ interaction.

In acetonitrile, the conformers with all the OHs engaged in IHBs have the greatest values of ΔG_{solv}. The conformers in which neither O41$-$H50 nor O46$-$H60 are engaged in O$-$H$\cdots\pi$ IHBs have the smallest absolute values of ΔG_{solv} and conformers in which only either O41$-$H50 or O46$-$H60 is engaged in the O$-$H$\cdots\pi$ IHB have values of ΔG_{solv} greater than the conformers in which they are not engaged in them. The behaviour in acetonitrile is different with respect to the behaviour in water. The importance of solute-solvent H-bonds for acetonitrile may not be as relevant as in water, both because acetonitrile can only be H-bond acceptor and because N is a weaker acceptor than O. The trend in chloroform is almost similar to the trend in acetonitrile, but the conformers having all the possible O$-$H$\cdots\pi$ interactions have slightly greater ΔG_{solv} than the other conformers. Conformers of S/S$'$ symmetric pairs may have slightly different ΔG_{solv} value$_s$.

The dipole moments (Table 6) increases in chloroform with respect to in vacuo and in acetonitrile with respect to chloroform, but decreases slightly in water with respect to acetonitrile. This might be related to the fact that acetonitrile has greater dipole moment than water (although several of its effects on solute molecules appear to be intermediate between those of chloroform and those of water). Conformers of S/S$'$ symmetric pairs have the same value of dipole moment in the solvents considered.

The HOMO-LUMO energy gap increases with the solvent polarity. Conformers with all the O$-$H$\cdots\pi$ and C$-$H$\cdots$O interactions (c-d-e-f-g-h) have the highest values of the HOMO-LUMO energy gap in all the media, and conformers with only one O$-$H$\cdots$O IHB (O43$-$H52$\cdots$O42 or O44$-$H54$\cdots$O45) have the lowest energy gap. The range of the HOMO-LUMO energy gap in vacuo is mostly 221.17$-$236.56/HF and 85.43$-$100.79/DFT. The gap mostly increases in solution, but it decreases in some cases; the change with respect to in vacuo is $-0.41$$-$4.34/HF and $-0.25$$-$4.15/DFT in chloroform; $-0.43$$-$6.04/HF and $-3.39$$-$1.84/DFT in acetonitrile and $-0.15$$-$8.56/HF and $-0.13$$-$8.34/DFT in water. Conformers of S/S$'$ symmetric pairs may have slightly different HOMO-LUMO gaps.

4 Discussion and Conclusions

A computational study of jozimine A₂—a naphthylisoquinoline alkaloid with antimalarial activity—has been carried out at the HF/6-31G(d,p) and the DFT/B3LP/6-31+G(d,p) levels in vacuo and in chloroform, acetonitrile and water. The results highlight the main properties of this molecule. Conformational preferences and other properties are influenced by the presence of O$-$H$\cdots$O IHBs (in the naphthalene moieties) and other IHB-type interactions, such as O$-$H$\cdots\pi$ (between some OH and a π ring in the same or in another moiety) and C$-$H$\cdots$O (between a CH and an OH of the same isoquinoline moiety), and by the mutual orientation of the moieties. The O—H···O IHBs are the dominant factor determining conformational preferences and energetics.

The low relative energy of many conformers in water solution suggests that a comparatively high number of conformers may be potential responsibles for the molecule's biological activity. The values of the free energy of solvation suggest greater solubility in water than in the other two solvents considered in this study, which would be consistent with the high polarity of the JZM molecule, but not with the octanol/water partition coefficient of JZM or with its high molecular mass.

It is interesting to note that (differently from a number of other molecules), JZM appears to respond to the three solvents considered according to their increasing dipole moment, so that the responses to water are slightly smaller than those to acetonitrile.

A typical feature of this molecule is the presence of axial chirality (stereoisomerism resulting from the non-planar arrangement of four groups in pairs about a chirality axis). The presence of axial chirality suggests that rotation about the C13 —C21 bond is not free because of steric hindrance. A space-filling visualisation with Chem3D [47] appears to confirm the hindrance, at least for lower energy conformers. On the other hand, the rotation scans about this bond (and about the other two biaryl bonds) were not disrupted (i.e., the steric hindrance did not appear in the bond-rotation model). A possible reason is the fact that the rotation scan input was a conformer with 12.811 kcal/mol relative energy (2.32×10^{-8} population); although expedient to identify minima in the orientation of the moieties, this conformer is poorly populated and, therefore, it does not contribute to the molecule's actual (or experimentally determined) behaviour.

An analysis of the differences with the behaviours and properties of michellamine A may be useful for a better understanding of the difference in the biological activities of these largely similar molecules.

Acknowledgements M. K. Bilonda expresses her gratitude to the National Research Foundation (NRF) of South Africa for a bursary to support her Ph.D. studies.

References

1. World Health Organization (2014) World malaria report 2014 summary. http://www.who.int/malaria/publications/world_malaria_report_2014/wmr-2014-no-profiles.pdf
2. World Health Organization (2016) World malaria report 2016. http://www.who.int/malaria/media/world-malaria-report-2016/en/
3. Bringmann G, Zhang G, Büttner T, Bauckmann G, Kupfer T, Braunschweig H, Brun R, Mudogo V (2013) Chem Eur J 19:916–923
4. Bierer DE, Dener JM, Dubenko LG, Gerber RE, Litvak J, Peterli S, Peterli-Roth P, Truong TV, Mao G, Bauer BE (1995) J Med Chem 38:2628
5. Ganellin CR, Mitchell RC, Young RC (1988) In: Melchiorre C, Giannella M (eds) Recent advances in receptor chemistry. Elsevier Science Publishers B.V., Amsterdam, pp 289–306
6. Guha S, Majumdar D, Bhattacharjee AK (1992) J Mol Struct (Theochem) 256:61
7. Sjoberg P, Murray JS, Brinck T, Evans P, Politzer P (1990) J Mol Graph 8:81

8. Karle JM, Bhattacharjee AK (1999) Bioorg Med Chem 7:1769–1774
9. Batagin-Neto A, Lavarda FC (2014) Med Chem Res 23:580–586
10. Naranjo-Montoya OA, Martins LM, da Silva-Filho LC, Batagin-Neto A, Lavarda FC (2015) J Braz Chem Soc 26:832–836
11. Purcell WP, Sundaram K (1969) J Med Chem 12:18–21. http://dx.doi.org/10.5935/0103-5053.20140263
12. Mammino L, Bilonda MK (2017) In: Tadjern A, Pavlov R, Maruani J, Brändas EJ, Delgado-Barrio G (eds) Quantum systems in physics, chemistry, and biology—advances in concepts and applications. Book series Progress in theoretical chemistry and physics. Springer, pp 303–316
13. Boyd MR (1994) J Med Chem 37:1740–1745
14. Becke AD (1993) J Chem Phys 98:5648–5662
15. Lee C, Yang W, Parr RG (1988) Phys Rev B 37:785–789
16. Mammino L, Bilonda MK (2016) Theor Chem Acc 135:101. https://doi.org/10.1007/s00214-016-1843-7
17. Mammino L, Kabanda MM (2009) J Mol Struct (Theochem) 901:210–219
18. Mammino L, Kabanda MM (2012) Int J Quantum Chem 112:2650–2658
19. Harvey JN (2004) In: Nkaltsoyanis N, McGrady JE (eds) Principles and applications of density functional theory in inorganic chemistry II. Springer, p 170
20. Irikura K, Johnson III RD, Kacker RN (2005) J Phys Chem A 109:8430–8437
21. Barone V, Cossi M (1997) J Chem Phys 107:3210–3221
22. Tomasi J, Mennucci B, Cammi R (2005) Chem Rev 105:2999–3093
23. Barone V, Cossi M, Tomasi J (1998) J Comput Chem 19:404–417
24. Cossi M, Scalmani G, Rega N, Barone V (2002) J Chem Phys 117:43–54
25. Cancès E, Mennucci B, Tomasi J (1997) J Chem Phys 107:3032–3041
26. Tomasi J, Mennucci B, Cancès E (1999) (Theochem) 464:211–226
27. Frisch MJ, Trucks GW, Schlegel HB, Scuseria GE, Robb MA, Cheeseman JR, Montgomery JA, Vreven T, Kudin KN, Burant JC, Millam JM, Iyengar SS, Tomasi J, Barone V, Mennucci B, Cossi M, Scalmani G, Rega N, Petersson GA, Nakatsuji H, Hada M, Ehara M, Toyota K, Fukuda R, Hasegawa J, Ishida M, Nakajima T, Honda Y, Kitao O, Nakai H, Klene M, Li X, Knox JE, Hratchian HP, Cross JB, Adamo C, Jaramillo J, Gomperts R, Stratmann RE, Yazyev O, Austin AJ, Cammi R, Pomelli C, Ochterski JW, Ayala PY, Morokuma K, Voth GA, Salvador P, Dannenberg JJ, Zakrzewski VG, Dapprich S, Daniels AD, Strain MC, Farkas O, Malick DK, Rabuck AD, Raghavachari K, Foresman JB, Ortiz JV, Cui Q, Baboul AG, Clifford S, Cioslowski J, Stefanov BB, Liu G, Liashenko A, Piskorz P, Komaromi I, Martin RL, Fox DJ, Keith T, Al-Laham MA, Peng CY, Nanayakkara A, Challacombe M, Gill PMW, Johnson B, Chen W, Wong MW, Gonzalez C, Pople JA (2003) Gaussian 03. Gaussian Inc, Pittsburgh
28. Pascual-Ahuir JL, Silla E (1990) J Comput Chem 11:1047–1047
29. Silla E, Villar F, Nilsson O, Pascual-Ahuir JL, Tapia O (1990) J Mol Graph 8:168–172
30. Silla E, Tunon I, Pascual-Ahuir JL (1991) J Comput Chem 12:1077–1088
31. Reichardt C (2003) Solvents effects in organic chemistry, 3rd edn. Wlley-VCH Verlag GmbH, Boschstr. 12, D-69469 Weinheim, Germany, p 82
32. Mammino L, Kabanda MM (2007) J Mol Struct (Theochem) 805:39–52
33. Buemi G, Zuccarello F (2002) J Mol Struct (Theochem) 581:71–85
34. Simperler A, Lampert H, Mikenda W (1998) J Mol Struct 448:191–199
35. Gilli G, Bellucci F, Ferretti V, Bertolasi V (1989) J Am Chem Soc 111:1023–1028
36. Bertolasi V, Gilli P, Ferretti V, Gilli G (1991) J Am Chem Soc 113:4017–4925
37. Gilli P, Bertolasi V, Ferretti V, Gilli G (1994) J Am Chem Soc 116:909–915
38. Nolasco MM, Ribeiro-Claro PJA (2005) Chem Phys Chem 6:496–502
39. Buemi G (2002) Chem Phys 282:181–195
40. Posokhov Y, Gorski A, Spanget-Larsen J, Duus F, Hansen PE, Waluk (2004) J Chem Phys Chem 5:495–502
41. Sobczyk L, Grabowski SJ, Krygowski TM (2005) Chem Rev 105:3513–3560

42. Jablonski M, Kaczmarek A, Sadlej AJ (2006) J Phys Chem A 110:10890–10898
43. Schalley CA, Springer A (2009) Mass spectrometry and gas-phase chemistry of non-covalent complexes. Wiley, Hoboken, NJ, p 17
44. https://www.wiki.ed.ac.uk/display/EaStCHEMresearchwiki/How+to+analyse+the+orbitals+from+a+Gaussian+calculation=25, cited in Synthetic Metals (2010) 160(7):643. https://doi.org/10.1016/j.synthmet.2009.12.026
45. Zhang G, Musgrave CB (2007) J Phys Chem A 111:1554–1561
46. Mammino L (2009) Chem Phys Lett 473:354–357
47. Chem3D Ultra Version 8.0.3., ChemOffice, Cambridge Software, 2003

Part IV
Fundamental Theory

Functional Derivatives and Differentiability in Density-Functional Theory

Ping Xiang and Yan Alexander Wang

Abstract Based on Lindgren and Salomonson's analysis on Fréchet differentiability [Phys Rev A 67:056501 (2003)], we showed a specific variational path along which the Fréchet derivative of the Levy-Lieb functional does not exist in the unnormalized density domain. This conclusion still holds even when the density is restricted within a normalized space. Furthermore, we extended our analysis to the Lieb functional and demonstrated that the Lieb functional is not Fréchet differentiable. Along our proposed variational path, the Gâteaux derivative of the Levy-Lieb functional or the Lieb functional takes a different form from the corresponding one along other more conventional variational paths. This fact prompted us to define a new class of *unconventional* density variations and inspired us to present a modified density variation domain to eliminate the problems associated with such unconventional density variations.

Keywords Density functional · Density variation · Functional differentiability
Functional derivative

1 Introduction

Functional differentiability plays crucial roles in density-functional theory (DFT) [1]. Within DFT, the total electronic energy is expressed as a functional of the electron density $\rho(\mathbf{r})$ [1, 2]. To find the ground-state (GS) density and its corresponding energy, we need to perform variational calculations based on the functional derivative [1–4]. Naturally, researchers have been interested in studying the differentiability of density functionals since the beginning of DFT.

More than twenty years ago, based on Lieb's early work [5], Englisch and Englisch proved the Gâteaux differentiability of a large class of density functionals and settled

P. Xiang · Y. A. Wang (✉)
Department of Chemistry, University of British Columbia, 2036 Main Mall,
Vancouver, BC V6T 1Z1, Canada
e-mail: yawang@chem.ubc.ca

© Springer International Publishing AG, part of Springer Nature 2018   331
Y. A. Wang et al. (eds.), *Concepts, Methods and Applications of Quantum Systems
in Chemistry and Physics*, Progress in Theoretical Chemistry and Physics 31,
https://doi.org/10.1007/978-3-319-74582-4_18

the issue regarding the Gâteaux differentiability of density functionals [6, 7]. However, the Fréchet differentiability of density functionals remained unresolved until Lindgren and Salomonson claimed its plausibility recently [8–10]. In this paper, we reexamine Lindgren and Salomonson's analysis to gain a better understanding about the Fréchet differentiability of density functionals in DFT.

Mathematically speaking, a functional is a mapping from a function to a number. $G[f]$, a functional of function $f(x)$, can be expressed as $f(x) \mapsto G[f]$. The differential of a functional, $dG[f, \delta f]$, is the part of the difference,

$$\delta G[f, \delta f] = G[f + \delta f] - G[f] \,, \tag{1}$$

that depends on δf linearly:

$$dG[f, \delta f] = \left\langle \frac{\delta G}{\delta f(x)} \delta f(x) \right\rangle \,, \tag{2}$$

where $\delta G / \delta f(x)$ is the functional derivative of $G[f]$ with respect to f at point x. For the sake of brevity, $\langle \cdot \rangle$ is adopted as a shorthand notation for integration throughout the text.

In DFT, there are two kinds of functional derivative: the Gâteaux derivative and the Fréchet derivative [6–10]. Following Lindgren and Salomonson [8–10], all the density functionals are defined on a convex space of densities,

$$\mathcal{Y} = \{\rho \,|\, \rho \geq 0, \sqrt{\rho} \in \mathcal{H}^1(\mathcal{R}^3)\} \,, \tag{3}$$

where $\mathcal{H}^1(\mathcal{R}^3)$ is a Sobolev space [5]:

$$\mathcal{H}^1(\mathcal{R}^3) = \{q \,|\, q \in \mathcal{L}^2(\mathcal{R}^3), \nabla q \in \mathcal{L}^2(\mathcal{R}^3)\} \,, \tag{4}$$

and $\mathcal{L}^2$ and $\mathcal{R}^3$ denote the spaces of square-integrable functions and three-dimensional real coordinates, respectively. Rather than the general definitions given in Appendix 1, we have slightly different definitions for functional differentiability and functional derivatives [11].

Let G be a functional from $\mathcal{Y}$ to the real numbers. If the limit

$$dG(\rho_0; \delta\rho) = \lim_{\beta \to 0+} \frac{G(\rho_0 + \beta\delta\rho) - G(\rho_0)}{\beta} \tag{5}$$

exists, it is called the Gâteaux differential of G at ρ_0 in the direction $\delta\rho$. If the limit exists for any $\delta\rho$ such that $\rho_0 + \beta\delta\rho \in \mathcal{Y}$, we say G is Gâteaux differentiable at ρ_0. Stated in another way, provided the functional G is Gâteaux differentiable at ρ_0, the functional difference upon a density variation, $\rho_0(\mathbf{r}) \to \rho_0(\mathbf{r}) + \beta\delta\rho(\mathbf{r})$, has two terms,

$$\delta G[\rho_0, \beta\delta\rho] = G[\rho_0 + \beta\delta\rho] - G[\rho_0] = \beta dG[\rho_0, \delta\rho] + R[\rho_0, \beta\delta\rho] \,, \tag{6}$$

where the last term satisfies the following limiting condition:

$$\lim_{\beta \to 0+} \frac{R[\rho_0, \beta \delta \rho]}{\beta} = 0 , \tag{7}$$

for a scaling parameter $0 \leq \beta \leq 1$. Usually we choose another density $\rho' \in \mathcal{Y}$ to be $\delta \rho$ and require $\rho = \rho_0 + \beta \rho'$ to be always in $\mathcal{Y}$ during the variational process; it follows immediately from the definition of a convex subspace (see Appendix 1) that the range of β must be $[0, 1]$. In a rigorous sense, the differential dG needs to be neither linear nor continuous in $\delta \rho$. However, most applied literature adopt the convention that dG is linear and continuous in $\delta \rho$. Consequently, dG can be written as

$$dG[\rho_0, \delta \rho] = \left\langle \frac{\delta G}{\delta \rho_0(\mathbf{r})} \delta \rho(\mathbf{r}) \right\rangle , \tag{8}$$

in which $\delta G / \delta \rho_0(\mathbf{r})$ is called the Gâteaux derivative of G at $\rho_0(\mathbf{r})$. Similar to a directional derivative, the Gâteaux derivative is a functional of $\rho_0(\mathbf{r})$ only, although along various directions it might take different expressions (e.g., different functions of the variable $\mathbf{r}$).

If the last term of the functional difference,

$$\delta G[\rho_0, \delta \rho] = G[\rho_0 + \delta \rho] - G[\rho_0] = \left\langle \frac{\delta G}{\delta \rho_0(\mathbf{r})} \delta \rho(\mathbf{r}) \right\rangle + R[\rho_0, \delta \rho] , \tag{9}$$

instead satisfies

$$\lim_{||\delta \rho|| \to 0} \frac{R[\rho_0, \delta \rho]}{||\delta \rho||} = 0 , \tag{10}$$

for the norm of $\delta \rho(\mathbf{r})$, $||\delta \rho|| = \langle |\delta \rho(\mathbf{r})| \rangle$, $\delta G / \delta \rho_0(\mathbf{r})$ is called the Fréchet derivative. The Fréchet derivative is a global derivative: all directions approaching $\rho_0(\mathbf{r})$ yield the same derivative (with the same expression). By default, Fréchet differentiability is *stronger* than Gâteaux differentiability.

2   Controversy over Fréchet Differentiability

Consider an N-electron quantum system under the influence of a local electron-nuclear potential $v(\mathbf{r})$. In the adiabatic connection formulation [12–17], the total Hamiltonian operator $\hat{H}$ is a sum of three terms, the kinetic-energy operator $\hat{T}$, the potential-energy operator $\hat{V}$, and the inter-electron coulombic repulsion operator $\hat{W}$ with an adiabatic connection parameter ω:

$$\hat{H} = \hat{T} + \omega \hat{W} + \hat{V} = -\frac{1}{2} \sum_{i=1}^{N} \nabla_i^2 + \omega \sum_{i<j}^{N} \frac{1}{|\mathbf{r}_i - \mathbf{r}_j|} + \sum_{i=1}^{N} v(\mathbf{r}_i) , \tag{11}$$

where i and j are dummy electron indices. For convenience, the first two operators in Eq. (11) can be grouped into a single Hohenberg-Kohn (HK) universal operator [2],

$$\hat{F}^{\omega} = \hat{T} + \omega\hat{W} . \tag{12}$$

In Eqs. (11) and (12), when $\omega = 0$, we have the noninteracting system; when $\omega = 1$, we have the fully interacting system instead. It is then straightforward to show that the Schrödinger equation governing this system,

$$\hat{H}\Psi = E_v\Psi , \tag{13}$$

can be reduced to a set of single-electron Kohn-Sham equations [3, 4] with a local effective external potential $v_{\mathit{eff}}(\mathbf{r})$,

$$\left[-\frac{1}{2}\nabla^2 + v_{\mathit{eff}}(\mathbf{r}) \right]\phi_i(\mathbf{r}) = \varepsilon_i\phi_i(\mathbf{r}) , \tag{14}$$

where electron i is described by spin orbital ϕ_i with orbital energy ε_i. It is understood that any wavefunction, Ψ, considered here is constructed from spin orbitals. However, for simplicity, we do not indicate any spin dependence explicitly throughout the entire text unless otherwise noted.

Consider first the nondegenerate case. The corresponding total energy can be expressed as a density functional,

$$E_v[\rho] = F^{\omega}[\rho] + V[\rho] , \tag{15}$$

with the electron density defined as

$$\rho(\mathbf{r}) = N\langle\Psi|\Psi\rangle_{N-1} , \tag{16}$$

where the subscript "$N - 1$" indicates the integration to be carried out over all spatial and spin coordinates except for one spatial coordinate of a single electron.

In Eq. (16), the normalization constraint of the electron density can be relaxed to allow $\langle\Psi|\Psi\rangle \neq 1$. As a result, the potential-energy density functional $V[\rho]$ and the HK universal density functional $F^{\omega}[\rho]$ are defined in the domain of unnormalized densities:

$$V[\rho] = \langle\Psi|\hat{V}|\Psi\rangle = \langle v(\mathbf{r})\,\rho(\mathbf{r})\rangle , \tag{17}$$

and

$$F^{\omega}[\rho] = F_{LL}^{\omega}[\rho] = \inf_{\Psi\to\rho} \langle\Psi|\hat{T} + \omega\hat{W}|\Psi\rangle , \tag{18}$$

where "inf" is the infimum or the greatest lower bound and the subscript "LL" denotes the Levy-Lieb functional [5, 18]. It should be noted that the density of concern, ρ, belongs to $\mathcal{Y}$, not the convex set of N-representable densities, $\mathcal{J}_N$:

$$J_N \equiv \{\rho \mid \rho \geq 0, \langle \rho \rangle = N, \sqrt{\rho} \in \mathcal{H}^1(\mathcal{R}^3)\}. \tag{19}$$

If $\omega = 0$, Eq. (18) reduces to the definition of the noninteracting kinetic-energy density functional,

$$T_s[\rho] = \inf_{\Psi \to \rho} \langle \Psi | \hat{T} | \Psi \rangle. \tag{20}$$

Built upon the following paradoxical statement [4]:

$$\frac{\delta T_s[\rho]}{\delta \rho(\mathbf{r})} = \varepsilon_i - v_{\mathit{eff}}(\mathbf{r}), \tag{21}$$

Nesbet asserted that the functional derivative of the noninteracting kinetic-energy functional in unnormalized density domain is a Gâteaux derivative rather than a Fréchet derivative [19, 20]. Shortly after, Lindgren and Salomonson [8–10] refuted Nesbet by pointing out that the noninteracting kinetic-energy functional used by Nesbet was not a proper density functional and further reasoned the Fréchet differentiability of the noninteracting kinetic-energy functional.

Based on Eqs. (16) and (17), Lindgren and Salomonson [8–10] obtained the expression of the HK universal density functional for the GS density $\rho_0(\mathbf{r})$ and wavefunction Ψ_0,

$$F^\omega[\rho_0] = \langle \Psi_0 | \hat{F}^\omega | \Psi_0 \rangle = \langle \Psi_0 | \hat{H} - \hat{V} | \Psi_0 \rangle = \left\langle \left[\frac{E_0}{N} - v(\mathbf{r}) \right] \rho_0(\mathbf{r}) \right\rangle, \tag{22}$$

where E_0 is the GS energy. Because of the identity for arbitrary $\Psi = \Psi_0 + \delta\Psi$,

$$\begin{aligned}
\langle \Psi | \hat{F}^\omega | \Psi \rangle &= \langle \Psi | \hat{H} | \Psi \rangle - \langle v(\mathbf{r}) \rho(\mathbf{r}) \rangle \\
&= E_0 \langle \Psi | \Psi \rangle + \langle \delta\Psi | \hat{H} - E_0 | \delta\Psi \rangle - \langle v(\mathbf{r}) \rho(\mathbf{r}) \rangle \\
&= \left\langle \left[\frac{E_0}{N} - v(\mathbf{r}) \right] \rho(\mathbf{r}) \right\rangle + \langle \delta\Psi | \hat{H} - E_0 | \delta\Psi \rangle,
\end{aligned} \tag{23}$$

the corresponding Levy-Lieb density functional takes the following form:

$$\begin{aligned}
F_{LL}^\omega[\rho_0 + \delta\rho] &= \inf_{\Psi \to \rho_0 + \delta\rho} \langle \Psi | \hat{F}^\omega | \Psi \rangle \\
&= \left\langle \left[\frac{E_0}{N} - v(\mathbf{r}) \right] [\rho_0(\mathbf{r}) + \delta\rho(\mathbf{r})] \right\rangle + \inf_{\Psi_0 + \delta\Psi \to \rho_0 + \delta\rho} \langle \delta\Psi | \hat{H} - E_0 | \delta\Psi \rangle. \tag{24}
\end{aligned}$$

Here, the GS density, ρ_0, is a pure-state v-representable (PS-v-representable) density, which corresponds to a single GS wavefunction of a Hamiltonian with a physically reasonable potential v in the space of $\mathcal{L}^\infty + \mathcal{L}^{3/2}$ [5, 9, 21]. Subtracting Eq. (22) from Eq. (24) leads to

$$\delta F_{LL}^{\omega} = F_{LL}^{\omega}[\rho_0 + \delta\rho] - F_{LL}^{\omega}[\rho_0]$$
$$= \left\langle \left[\frac{E_0}{N} - v(\mathbf{r})\right] \delta\rho(\mathbf{r}) \right\rangle + \inf_{\Psi_0 + \delta\Psi \to \rho_0 + \delta\rho} \langle \delta\Psi | \hat{H} - E_0 | \delta\Psi \rangle . \qquad (25)$$

Then, the condition for Fréchet differentiability requires the last term in Eq. (25) to satisfy

$$\inf_{\Psi_0 + \delta\Psi \to \rho_0 + \delta\rho} \frac{\langle \delta\Psi | \hat{H} - E_0 | \delta\Psi \rangle}{||\delta\rho||} \to 0 , \quad \text{as } ||\delta\rho|| \to 0 , \qquad (26)$$

for all density variations in the neighborhood of $\rho_0(\mathbf{r})$. Lindgren and Salomonson argued that Eq. (26) was plausible because the numerator is quadratic in $\delta\Psi$ whereas $\delta\rho$ is only linear in $\delta\Psi$ [8], and they analyzed this issue further based on their proof of the Gâteaux differentiability of the Levy-Lieb functional [9].

When the ground state is degenerate, it can no longer be represented by a PS-v-representable density. Instead, we should use an ensemble v-representable (E-v-representable) density [6, 21],

$$\rho_0 = \sum_k s_k \rho_0^k , \quad s_k \geq 0 , \quad \sum_k s_k = 1 , \qquad (27)$$

where ρ_0^k is a PS-v-representable density of the Hamiltonian in Eq. (11). For conciseness, we use the same notation ρ_0 to represent either a PS-v-representable density or an E-v-representable density [6, 21], and whenever needed, we will specify the type of the density. For this degenerate case, by using a similar method as above, Lindgren and Salomonson [9] has also shown the plausibleness of the Fréchet differentiability of the Lieb functional [5],

$$F_L^{\omega}[\rho] = \inf_{s_k, \Psi^k \to \rho} \sum_k s_k \left\langle \Psi^k \left| \hat{T} + \omega\hat{W} \right| \Psi^k \right\rangle , \qquad (28)$$

where $\{\Psi^k\}$ is any set of orthonormal eigenfunctions of the Hamiltonian in Eq. (11).

3 Analysis of Lindgren-Salomonson's Assessment

Unfortunately, Lindgren and Salomonson's assessment is incomplete. Before exposing the limitation of their assessment, we would like to introduce the concept of strongly orthogonal function (SOF) [22]. Two functions $f(x_1, x_2, \ldots, x_m, x_{m+1}, \ldots, x_N)$ and $g(x_1, x_2, \ldots, x_m, x_{m+1}, \ldots, x_N)$ are mutually order-m strongly orthogonal, if they satisfy the following strongly orthogonal condition for any arbitrary $(N - m)$ variables:

$$\langle f | g \rangle_{N-m} = 0 , \qquad (29)$$

where N is the total number of variables and m is the number of variables excluded in the integration. If f is chosen to be the seed function, we can define the order-m SOF space that consists all functions of order-m strongly orthogonal to the same seed function f.

Because spin operator commutes with the Hamiltonian, they share a common complete set of eigenfunctions. In the forthcoming discussion, we only consider wavefunctions in Hilbert space spanned by this complete set. Let the seed function be the N-electron GS wavefunction Ψ_0 and the integration in Eq. (29) be over all spatial and spin coordinates except m spatial coordinates, and label $\mathcal{D}$ as its associated order-1 SOF space and $\mathcal{S}$ as its order-0 SOF space with $\mathcal{D}$ excluded. For a general Hamiltonian ($0 \le \omega \le 1$), there is no doubt that $\mathcal{S}$ exists. Since we integrate over all spin coordinates, if any non-GS eigenfunction Ψ has a different spin multiplicity from that of Ψ_0, we have

$$\langle \Psi_0 | \Psi \rangle_{N-1} = 0 \,, \tag{30}$$

then Ψ is in $\mathcal{D}$. Therefore, $\mathcal{D}$ does exist in general. For the noninteracting system ($\omega = 0$) in particular, Ψ_0 is a single Slater determinant, $\mathcal{S}$ includes all singly excited determinants, and $\mathcal{D}$ has all doubly and higher excited determinants. Thus, the complete set of eigenfunctions $\{\Psi_i\}$ of the Hamiltonian can be partitioned into three parts: Ψ_0, $\mathcal{S}$, and $\mathcal{D}$.

Let a normalized wavefunction Ψ_D be a linear combination of eigenfunctions in $\mathcal{D}$. Hereafter, we are going to show that along one particular variational path defined by

$$\Psi_p = \Psi_0 + \beta \Psi_D \,, \tag{31}$$

the Fréchet derivative does not exist. For the wavefunction Ψ_p, its corresponding electron density is

$$\begin{aligned}
\rho_p(\mathbf{r}) &= N \left\langle \Psi_p \middle| \Psi_p \right\rangle_{N-1} \\
&= N \left\langle \Psi_0 | \Psi_0 \right\rangle_{N-1} + \beta^2 N \left\langle \Psi_D | \Psi_D \right\rangle_{N-1} + 2\beta N Re\left(\left\langle \Psi_D | \Psi_0 \right\rangle_{N-1} \right) \\
&= \rho_0(\mathbf{r}) + \beta^2 \rho_D(\mathbf{r}) \,,
\end{aligned} \tag{32}$$

where $Re\left(\langle \Psi_D | \Psi_0 \rangle_{N-1} \right)$, the real part of $\langle \Psi_D | \Psi_0 \rangle_{N-1}$, is zero because of the nature of Ψ_D. When β approaches 0, $\rho_p(\mathbf{r})$ also approaches $\rho_0(\mathbf{r})$. Clearly, $\rho_p(\mathbf{r})$ lies in the neighborhood of $\rho_0(\mathbf{r})$ within $\mathcal{Y}$. Equation (32) specifies the density variation path for the forthcoming discussion. For later convenience, we label $\mathcal{B}$ as the set of all legitimate $\rho_p(\mathbf{r})$ for a given Ψ_0 or $\rho_0(\mathbf{r})$. Throughout the text, $Re(\cdot)$ will be used to denote the real part of the quantity involved.

Along this particular variational path, let $\widetilde{\Psi}$ be a trial wavefunction yielding the same density $\rho_p(\mathbf{r})$:

$$\widetilde{\Psi} = \Psi_0 + \lambda \Psi_t \, \longmapsto \, \rho_p(\mathbf{r}) \,, \tag{33}$$

where Ψ_t is normalized to 1 and λ is a real scaling parameter (potentially different from β). We can expand Ψ_t in terms of the complete set of normalized eigenfunctions $\{\Psi_i\}$ of $\hat{H}$,

$$\Psi_t = \sum_{i=0}^{\infty} c_i \Psi_i , \tag{34}$$

where c_i is the expansion coefficient of Ψ_i. Without loss of generality, we can choose all expansion coefficients $\{c_i\}$ to be real because any complex phases in $\{c_i\}$ can be attributed to the corresponding $\{\Psi_i\}$ instead. The electron density for $\widetilde{\Psi}$ then takes the following form:

$$
\begin{aligned}
\widetilde{\rho}(\mathbf{r}) &= N\langle\widetilde{\Psi}|\widetilde{\Psi}\rangle_{N-1} \\
&= N\langle\Psi_0|\Psi_0\rangle_{N-1} + \lambda^2 N\langle\Psi_t|\Psi_t\rangle_{N-1} + 2\lambda N Re\big(\langle\Psi_0|\Psi_t\rangle_{N-1}\big) \\
&= \rho_0(\mathbf{r}) + \lambda^2 \rho_t(\mathbf{r}) + 2\lambda N Re\big(\langle\Psi_0|\Psi_t\rangle_{N-1}\big) .
\end{aligned}
\tag{35}
$$

Clearly, as $\lambda \to 0$, $\widetilde{\rho}(\mathbf{r})$ also approaches $\rho_0(\mathbf{r})$.

Because the density variation is along the path with the same density (but with different wavefunctions),

$$\widetilde{\rho}(\mathbf{r}) = \rho_p(\mathbf{r}) \to \rho_0(\mathbf{r}) , \tag{36}$$

the normalization of the two densities must be identical,

$$\langle\widetilde{\rho}(\mathbf{r})\rangle = \langle\rho_p(\mathbf{r})\rangle . \tag{37}$$

Substituting Eqs. (32) and (35) into Eq. (37), one has

$$\lambda^2 \langle\rho_t(\mathbf{r})\rangle + 2\lambda N Re\big(\langle\Psi_0|\Psi_t\rangle\big) = \beta^2 \langle\rho_D(\mathbf{r})\rangle . \tag{38}$$

Since $\langle\rho_D(\mathbf{r})\rangle$ and $\langle\rho_t(\mathbf{r})\rangle$ are all equal to N, Eq. (38) can be readily simplified to

$$\beta^2 = \lambda^2 + 2\lambda Re\big(\langle\Psi_0|\Psi_t\rangle\big) = \lambda^2 + 2\lambda \sum_{i=0}^{\infty} Re\big(c_i\langle\Psi_0|\Psi_i\rangle\big) = \lambda^2 + 2\lambda c_0 . \tag{39}$$

At any specific point along the variational path, the value of β is fixed, and we can solve λ in terms of β based on Eq. (39):

$$\lambda = -c_0 \pm \sqrt{c_0^2 + \beta^2} . \tag{40}$$

If $c_0 \neq 0$ when β approaches 0, λ takes a Taylor-series expansion,

$$\lambda = -c_0 \pm \left[c_0 + \frac{1}{2c_0}\beta^2 - \frac{1}{8c_0^3}\beta^4 + \cdots \right] . \tag{41}$$

Because both λ and β approach 0 concurrently near the end of the variational path, only the positive sign (see Appendix 3) is allowed in Eq. (41). Thus, as $\beta \to 0$, we get

$$\lambda = \frac{1}{2c_0}\beta^2 - \frac{1}{8c_0^3}\beta^4 + \cdots .\tag{42}$$

Immediately, we can conclude that towards the end of the variational path, λ is of the same magnitude as β^2/c_0.

If $c_0 = 0$, Eq. (39) immediately reduces to $\lambda^2 = \beta^2$. In this case, the wavefunction variation $\delta\Psi$ in Eq. (26) can be regarded as linear in β and the density variation $\delta\rho = \beta^2\rho_D$ is quadratic in β as shown by Eq. (32). This immediately invalidates Fréchet differentiability by destroying Eq. (26). Nonetheless, we are able to gain much deeper understanding about the structure of Ψ_t through the following analysis.

Consider the case when $c_0 \neq 0$. Because of Eqs. (32), (35), and (36), we obtain

$$\lambda^2 \rho_t + 2\lambda N Re\left(\left\langle\Psi_0|\Psi_t\right\rangle_{N-1}\right) = \beta^2\rho_D .\tag{43}$$

Substituting Eq. (39) into Eq. (43) and grouping terms according to the powers of λ, we have

$$2\lambda\left[c_0\rho_D - N\sum_i^{S_0} c_i Re\left(\left\langle\Psi_0|\Psi_i\right\rangle_{N-1}\right)\right] = \lambda^2(\rho_t - \rho_D) ,\tag{44}$$

where the summation on the left-hand side (LHS) is only within the combined space of S and Ψ_0,

$$S_0 \equiv S \cup \Psi_0 .\tag{45}$$

Separating the Ψ_0 contribution from the summation, we can simplify Eq. (44) to

$$2c_0(\rho_0 - \rho_D) + 2N\sum_i^{S} c_i Re\left(\left\langle\Psi_0|\Psi_i\right\rangle_{N-1}\right) = \lambda\left(\rho_D - \rho_t\right) .\tag{46}$$

Because ρ_0, ρ_D, and all Ψ_i in Eq. (46) have no dependence on λ, we must admit that c_i corresponding to any wavefunction in S_0 must be a function of λ.

Based on perturbation theory, $c_i(\lambda)$ can be expanded as

$$c_i(\lambda) = c_i^{(0)} + c_i^{(1)}\lambda + h.o. ,\tag{47}$$

where "$h.o.$" represent higher-order terms in λ. It can be readily shown that all the $c_i^{(0)}$ must be zero in order to satisfy Eq. (46) for any $\lambda \to 0$, whereas $c_i^{(1)}$ can be zero or non-zero. Consequently, all $\{c_i\}$ for Ψ_i in S_0 vary at least linearly in λ. Some c_i

may have a higher-than-linear dependence on λ, but at least one c_i is linear in λ for Eq. (46) to be valid. At this point, we do not know the exact behavior of those $\{c_i\}$ for Ψ_i in D. Actually, in the above derivation, we only considered one constraint along the variational path, namely, Eq. (36), but ignored the other constraint, $\langle \Psi_t | \Psi_t \rangle = 1$. What will happen if both constraints are enforced concurrently?

By the minimization process, we can find the best set of coefficients $\{c_i\}$ that delivers the lowest energy out of the following energy functional for the given $\rho_p(\mathbf{r})$ with a fixed β,

$$
\begin{aligned}
\langle \widetilde{\Psi} | \hat{H} | \widetilde{\Psi} \rangle &= \langle \Psi_0 + \lambda \Psi_t | \hat{H} | \Psi_0 + \lambda \Psi_t \rangle \\
&= \langle \Psi_0 + \lambda \Psi_t | \hat{H} | \Psi_0 \rangle + \langle \Psi_0 + \lambda \Psi_t | \hat{H} | \lambda \Psi_t \rangle \\
&= E_0 \langle \widetilde{\Psi} | \Psi_0 \rangle + \langle \Psi_0 | \hat{H} | \lambda \Psi_t \rangle + \lambda^2 \langle \Psi_t | \hat{H} | \Psi_t \rangle \\
&= E_0 \langle \widetilde{\Psi} | \widetilde{\Psi} \rangle - E_0 \langle \widetilde{\Psi} | \lambda \Psi_t \rangle + E_0 \langle \Psi_0 | \lambda \Psi_t \rangle + \lambda^2 \langle \Psi_t | \hat{H} | \Psi_t \rangle \\
&= E_0 \langle \widetilde{\Psi} | \widetilde{\Psi} \rangle - \lambda^2 E_0 \langle \Psi_t | \Psi_t \rangle + \lambda^2 \langle \Psi_t | \hat{H} | \Psi_t \rangle \\
&= (1 + \beta^2) E_0 + \lambda^2 \left(\langle \Psi_t | \hat{H} | \Psi_t \rangle - E_0 \right).
\end{aligned}
\tag{48}
$$

To achieve this task, one only needs to minimize $\lambda^2 \left(\langle \Psi_t | \hat{H} | \Psi_t \rangle - E_0 \right)$ in Eq. (48) under the following two constraints:

$$
\sum_{i=0}^{\infty} c_i^2 = 1 ,
\tag{49}
$$

and

$$
\widetilde{\rho}(\mathbf{r}) = \rho_p(\mathbf{r}) .
\tag{50}
$$

The second constraint, Eq. (50), is equivalent to the following equation based on our previous analysis:

$$
2\lambda \left[\frac{c_0(\rho_0 - \rho_D)}{N} + \sum_{i}^{S} c_i Re\left(\langle \Psi_0 | \Psi_i \rangle_{N-1} \right) \right]
$$
$$
= \lambda^2 \left(\frac{\rho_D}{N} - \sum_{i,j}^{\infty} c_i c_j \langle \Psi_i | \Psi_j \rangle_{N-1} \right) .
\tag{51}
$$

We will use the Euler-Lagrange multiplier method to find the set of coefficients $\{c_i\}$ that minimizes the value of $\lambda^2 \left(\langle \Psi_t | \hat{H} | \Psi_t \rangle - E_0 \right)$.

Define

$$
\mathbf{A} = \sum_{i=0}^{\infty} c_i^2 - 1 ,
\tag{52}
$$

$$\mathbf{B} = 2\lambda \left[\frac{c_0(\rho_0 - \rho_D)}{N} + \sum_{i}^{S} c_i Re(\langle \Psi_0 | \Psi_i \rangle_{N-1}) \right]$$
$$- \lambda^2 \left(\frac{\rho_D}{N} - \sum_{i,j}^{\infty} c_i c_j \langle \Psi_i | \Psi_j \rangle_{N-1} \right) , \tag{53}$$

and

$$\begin{aligned}
\mathbf{\Omega} &= \lambda^2 \left(\langle \Psi_t | \hat{H} | \Psi_t \rangle - E_0 \right) - h\mathbf{A} - \langle g(\mathbf{r}) \mathbf{B} \rangle \\
&= \lambda^2 \left(\sum_{i,j}^{\infty} c_i c_j \langle \Psi_i | \hat{H} | \Psi_j \rangle - E_0 \right) - h\mathbf{A} - \langle g(\mathbf{r}) \mathbf{B} \rangle \\
&= \lambda^2 \left(\sum_{i,j}^{\infty} c_i c_j E_j \delta_{ij} - E_0 \right) - h\mathbf{A} - \langle g(\mathbf{r}) \mathbf{B} \rangle \\
&= \lambda^2 \left[\sum_{i=1}^{\infty} c_i^2 (E_i - E_0) \right] - h\mathbf{A} - \langle g(\mathbf{r}) \mathbf{B} \rangle ,
\end{aligned} \tag{54}$$

where h and $g(\mathbf{r})$ are the Lagrange multipliers corresponding to the two constraints in Eqs. (52) and (53), and Eq. (49) has been used to derive the last expression. Minimizing Eq. (54) with respect to $\{c_i\}$, one obtains

$$\lambda \left\langle \left[\frac{\rho_0 - \rho_D}{N} + \lambda \sum_{j=0}^{\infty} c_j Re(\langle \Psi_0 | \Psi_j \rangle_{N-1}) \right] g(\mathbf{r}) \right\rangle = -hc_0 , \tag{55}$$

and

$$\lambda \left\langle \left[Re(\langle \Psi_i | \Psi_0 \rangle_{N-1}) + \lambda \sum_{j=0}^{\infty} c_j Re(\langle \Psi_i | \Psi_j \rangle_{N-1}) \right] g(\mathbf{r}) \right\rangle$$
$$= \left[\lambda^2 (E_i - E_0) - h \right] c_i \qquad (\text{for } i \neq 0) . \tag{56}$$

Because c_0 is at least linear in λ as we showed above, we can readily infer from Eq. (55) that $g(\mathbf{r})$ must take the following form,

$$g_\lambda(\mathbf{r}) = g^{(0)}(\mathbf{r}) + \sum_{k=1}^{\infty} \frac{g^{(k)}(\mathbf{r})}{k!} \lambda^k , \tag{57}$$

where $g^{(0)}$ can be zero depending on whether c_0 is higher-than-linear in λ or not. Substituting Eq. (57) back into Eq. (56) and ignoring the higher-order terms when λ approaches 0, we obtain a much simplified expression of c_i for $\Psi_i \in S$,

$$-hc_i = \lambda \langle Re(\langle \Psi_i | \Psi_0 \rangle_{N-1}) g^{(0)}(\mathbf{r}) \rangle + h.o. , \tag{58}$$

where "*h.o.*" denotes higher-order terms in λ. Obviously, we reach the same conclusion as we derived before: $\{c_i\}$ for $\Psi_i \in S$ is at least linear in λ towards the end of the variational path. For those $\{c_i\}$ for $\Psi_i \in D$, where Ψ_i is order-1 strongly orthogonal to Ψ_0, the first term in the square brackets on the LHS of Eq. (56) disappears, and Eq. (56) reduces to

$$\lambda^2 \left\langle \sum_j^{S_0} c_j Re\left(\langle \Psi_i | \Psi_j \rangle_{N-1}\right) g(\mathbf{r}) \right\rangle + \lambda^2 \left\langle \sum_j^{D} c_j Re\left(\langle \Psi_i | \Psi_j \rangle_{N-1}\right) g(\mathbf{r}) \right\rangle$$
$$= \lambda^2 (E_i - E_0)c_i - hc_i . \tag{59}$$

For this equation to be valid at $\lambda \to 0$, the LHS and the right-hand side (RHS) should have the same dependence on λ. On the RHS, the first term decays faster than the second term, and the second term will dominate when λ approaches 0. Therefore, we must match the magnitude of the second term on the RHS to the LHS. Of course, we cannot match it with the second term on the LHS because doing so will lead to self inconsistency. Then, the second term on the RHS must decay the same way as the first term on the LHS. So, $\{c_i\}$ for $\Psi_i \in D$ are proportional to λ^3 or higher-than-cubic terms in λ. Unfortunately, such a behavior is contradictory to the normalization constraint in Eq. (49). Otherwise, $\sum_i c_i^2$ will become 0 as $\lambda \to 0$. Hence, we conclude that this contradiction must come from the initial assumption: $\Psi_t = \sum_i c_i \Psi_i$, where the expansion is over the complete set of eigenfunctions of $\hat{H}$.

To resolve this contradiction, we have to modify our assumption about Ψ_t. We notice that if the summation $\sum_i c_i \Psi_i$ includes any wavefunction from S_0, the same problem will persist. Thus, Ψ_t can only be expanded in D,

$$\Psi_t = \sum_i^{D} c_i \Psi_i . \tag{60}$$

In this case, Eq. (50) is equivalent to

$$\lambda^2 \rho_t(\mathbf{r}) = \beta^2 \rho_D(\mathbf{r}) . \tag{61}$$

Integrating both sides of Eq. (61) over the entire space of $\mathbf{r}$, one finds

$$\lambda^2 = \beta^2 , \tag{62}$$

which further ensures that

$$\rho_t(\mathbf{r}) = \rho_D(\mathbf{r}) . \tag{63}$$

Now, the original minimization process is reduced to minimizing the following term,

$$\Xi = \left(\sum_i^D |c_i|^2 E_i\right) - h\left(\sum_i^D |c_i|^2 - 1\right) - \left\langle \left(\frac{\rho_D}{N} - \sum_{i,j}^D c_i^* c_j \langle \Psi_i | \Psi_j \rangle_{N-1}\right) g(\mathbf{r}) \right\rangle ,$$

(64)

where all traces of λ (or β) are completely gone. The minimization will yield the optimal set of expansion coefficients $\{\bar{c}_i\}$, which has no dependence on λ or β from the appearance of Eq. (64). At this stage, we are ready to test the condition for Fréchet differentiability shown in Eq. (26):

$$\inf_{\Psi_0 + \delta\Psi \to \rho_0 + \delta\rho} \frac{\langle \delta\Psi | \hat{H} - E_0 | \delta\Psi \rangle}{||\delta\rho||} = \inf_{\Psi_0 + \delta\Psi \to \rho_0 + \delta\rho} \frac{\langle \delta\Psi | \hat{H} - E_0 | \delta\Psi \rangle}{||\rho_p - \rho_0||}$$

$$= \inf_{\Psi_t \to \rho_D} \frac{\lambda^2 \langle \Psi_t | \hat{H} - E_0 | \Psi_t \rangle}{||\beta^2 \rho_D||} = \frac{\inf_{\Psi_t \to \rho_D} \langle \Psi_t | \hat{H} - E_0 | \Psi_t \rangle}{||\rho_D||}$$

$$= \frac{1}{||\rho_D||} \left\langle \sum_i^D \bar{c}_i \Psi_i \middle| \hat{H} - E_0 \middle| \sum_j^D \bar{c}_j \Psi_j \right\rangle = \frac{1}{||\rho_D||} \left(\sum_i^D |\bar{c}_i|^2 E_i - E_0\right)$$

$$> \frac{1}{||\rho_D||} \left(\sum_i^D |\bar{c}_i|^2 E_0 - E_0\right) = 0 ,$$

(65)

where Eq. (62) is used to simplify the expression after the second equal sign. Evidently, Eq. (65) suggests that the condition for Fréchet differentiability is *not* fulfilled.

Since the above infimum approaches a nonzero constant towards the end of the variational path, it is possible to combine the linear-order term of $\delta\rho$ from the second term with the first term in Eq. (25), and the new resulting residual term might satisfy the condition for Fréchet differentiability. However, the corresponding *path-dependent* functional derivative will be different from the one obtained by Lindgren and Salomonson [8–10]. This is contradictory to the fact that the Fréchet derivative is a global derivative, *independent* of variational path. Therefore, we have no choice but to conclude that the Levy-Lieb density functionals F_{LL}^ω are *not* Fréchet differentiable at PS-v-representable densities. The places where Fréchet differentiability breaks down (e.g., along our specially designed variational path in $\mathcal{B}$) are exactly the same locations where the Gâteaux derivative espouses different forms.

In the above analysis, we worked within unnormalized density domain. The situation in normalized density domain is almost identical. This can be straightforwardly proven by normalizing the wavefunction after finishing the minimization process in the unnormalized wavefunction space, simply because all the wavefunctions of our concern are normalizable in Hilbert space and the normalization factor does not affect the expectation values of observables. In Appendix 2, we have offered a much more detailed but rather lengthy proof to further confirm our assessment here.

After showing the non-Fréchet differentiability of the Levy-Lieb functionals, we are ready to examine the differentiability of the Lieb functionals, shown in Eq. (28).

Using similar derivations as shown in Sect. 2, we get the variation of the Lieb functional:

$$
\delta F_L[\rho_0] = \sum_k s_k \left\langle \left[\frac{E_0}{N} - v(\mathbf{r}) \right] \delta \rho^k(\mathbf{r}) \right\rangle + \sum_k s_k \inf_{\Psi_0^k + \delta \Psi^k \to \rho_0^k + \delta \rho^k} \langle \delta \Psi^k | \hat{H} - E_0 | \delta \Psi^k \rangle
$$

$$
= \left\langle \left[\frac{E_0}{N} - v(\mathbf{r}) \right] \delta \rho_0(\mathbf{r}) \right\rangle + \sum_k s_k \inf_{\Psi_0^k + \delta \Psi^k \to \rho_0^k + \delta \rho^k} \langle \delta \Psi^k | \hat{H} - E_0 | \delta \Psi^k \rangle , \quad (66)
$$

based on Eq. (27), where the total density variation $\delta \rho$ is a linear combination of individual variation of ρ_0^k,

$$
\delta \rho = \sum_k s_k \delta \rho^k , \qquad (67)
$$

where s_k is the same as that in Eq. (27). Then, the condition for the Lieb functional to be Fréchet differentiable is

$$
\sum_k s_k \inf_{\Psi_0^k + \delta \Psi^k \to \rho_0^k + \delta \rho^k} \frac{\langle \delta \Psi^k | \hat{H} - E_0 | \delta \Psi^k \rangle}{||\delta \rho||} \to 0 , \text{ as } ||\delta \rho|| \to 0 . \quad (68)
$$

For each PS-v-representable density ρ_0^k in the expansion of the E-v-representable density ρ_0, we use the same kind of variational path as shown before. Let the density $\rho_p^k = \rho_0^k + \delta \rho^k$ along the variational path be generated by the wavefunction:

$$
\Psi_p^k = \Psi_0^k + \beta \Psi_D^k , \qquad (69)
$$

as β approaches zero. The corresponding total density variation is

$$
\delta \rho = \beta^2 \sum_k s_k \rho_D^k . \qquad (70)
$$

Along each variational path, let the trial wavefunction to be

$$
\widetilde{\Psi}^k = \Psi_0^k + \lambda_k \Psi_t^k . \qquad (71)
$$

By the same analysis, we can show that each scaling parameter λ_k must satisfy

$$
\lambda_k^2 = \beta^2 , \qquad (72)
$$

and each trial wavefunction Ψ_t^k can only be expanded in the corresponding D space,

$$
\Psi_t^k = \sum_i^D c_i^k \Psi_i^k . \qquad (73)
$$

Consequently, we have

$$
\sum_k s_k \inf_{\Psi_0^k + \delta\Psi^k \to \rho_0^k + \delta\rho^k} \frac{\langle \delta\Psi^k | \hat{H} - E_0 | \delta\Psi^k \rangle}{||\delta\rho||} = \sum_k s_k \inf_{\rho_t^k \to \rho_D^k} \frac{\lambda_k^2 \langle \Psi_t^k | \hat{H} - E_0 | \Psi^k \rangle}{\beta^2 || \sum_k s_k \delta\rho_D^k ||}
$$
$$
= \sum_k s_k \inf_{\rho_t^k \to \rho_D^k} \frac{\langle \Psi_t^k | \hat{H} - E_0 | \Psi_t^k \rangle}{|| \sum_k s_k \delta\rho_D^k ||} , \tag{74}
$$

which is positive and cannot be zero for any positive semi-definite set of $\{s_k\}$ because

$$
\inf_{\rho_t^k \to \rho_D^k} \frac{\langle \Psi_t^k | \hat{H} - E_0 | \Psi_t^k \rangle}{|| \sum_k s_k \delta\rho_D^k ||} > 0 . \tag{75}
$$

Therefore, the condition for Fréchet differentiability of the Lieb functional is not fulfilled, and the Lieb functional is not Fréchet differentiable at any E-v-representable densities.

4 Unconventional Density Variations

In this section, we will examine the implications of our above results.

Density variation in Hilbert space is defined as the difference of a trial density $\rho(\mathbf{r})$ from a GS density $\rho_0(\mathbf{r})$,

$$
\delta\rho(\mathbf{r}) = \rho(\mathbf{r}) - \rho_0(\mathbf{r}) = \beta\eta(\mathbf{r}) , \tag{76}
$$

where the scaling parameter β can take any real value as long as the resultant density $\rho(\mathbf{r})$ is not negative,

$$
\rho(\mathbf{r}) = \rho_0(\mathbf{r}) + \beta\eta(\mathbf{r}) \geq 0 , \tag{77}
$$

everywhere in the entire space, for a given $\eta(\mathbf{r})$. As said before, if we let $\eta(\mathbf{r})$ to be a density in $\mathcal{Y}$, then β must be in the range $[0, 1]$ because of the convexity of $\mathcal{Y}$.

Perdew and Levy proposed a kind of *unconventional* density variations (UDVs) which satisfy [23]:

$$
\frac{\eta(\mathbf{r})}{\sqrt{\rho_0(\mathbf{r})}} \notin \mathcal{L}^2 . \tag{78}
$$

This class of UDVs gives rise to an unconventional energy variation, δE_v, of sub-quadratic order of β [25],

$$
\delta E_v[\rho_0, \delta\rho] = E_v[\rho_0 + \delta\rho] - E_v[\rho_0] \propto O(\beta^b), \; 1 < b < 2 , \tag{79}
$$

as opposed to the $O(\beta^2)$ behavior of conventional density variations (CDVs).

In general, the energy variation can be written as

$$\delta E_v[\rho_0, \delta\rho] = \left\langle \left. \frac{\delta E_v}{\delta\rho(\mathbf{r})} \right|_{\rho_0} \delta\rho(\mathbf{r}) \right\rangle + h.o. , \tag{80}$$

where "*h.o.*" encompasses all higher-order terms in $\delta\rho(\mathbf{r})$. Zhang and Wang [25] recently discovered that Eq. (80) cannot be expanded to second order in $\delta\rho(\mathbf{r})$ for the UDVs of the first kind [23]. In other words, the second (or any higher-order) functional derivative of the total energy functional may not exist for all allowed density variations within the current DFT framework. This imposes a serious problem in DFT since the effort to interpret the second functional derivative as the chemical hardness [1] will fail for the UDVs of the first kind [23]. Some modification of the density variation domain must be in place to rescue the situation.

From a different perspective, we introduce a new definition of UDV based on our discussion in Sect. 3. If a density variation, $\delta\rho(\mathbf{r}) = \rho(\mathbf{r}) - \rho_0(\mathbf{r})$, comes from a density $\rho(\mathbf{r}) \in B$, then the density variation is an UDV. For these UDVs of the second kind, their corresponding Gâteaux derivatives are different from the conventional one [8–10], or simply, not Fréchet derivatives.

Let us write the wavefunction in Eq. (31) differently as $\Psi'_p = \Psi_0 + \sqrt{\beta}\Psi_D$, the associated density variation will become linear in β,

$$\delta\rho'(\mathbf{r}) = \beta\rho_D(\mathbf{r}) . \tag{81}$$

Because of Eqs. (31), (33), and (62), $\delta\Psi' = \sqrt{\beta}\Psi_t$, and Eq. (65) is invariant,

$$\inf_{\Psi_0+\delta\Psi'\to\rho_0+\delta\rho'} \frac{\langle \delta\Psi'|\hat{H} - E_0|\delta\Psi'\rangle}{||\delta\rho'||} = \inf_{\Psi_t\to\rho_D} \frac{\langle \sqrt{\beta}\Psi_t|\hat{H} - E_0|\sqrt{\beta}\Psi_t\rangle}{||\beta\rho_D(\mathbf{r})||}$$

$$= \frac{\inf_{\Psi_t\to\rho_D} \langle \Psi_t|\hat{H} - E_0|\Psi_t\rangle}{||\rho_D||} = \text{constant} \neq 0 . \tag{82}$$

Equation (25) then becomes

$$\delta F_{LL}^\omega = \left\langle \left[\frac{E_0}{N} - v(\mathbf{r}) \right] \delta\rho'(\mathbf{r}) \right\rangle + \inf_{\Psi_0+\delta\Psi'\to\rho_0+\delta\rho'} \langle \delta\Psi'|\hat{H} - E_0|\delta\Psi'\rangle . \tag{83}$$

From Eq. (82), we know that the second term on the RHS of Eq. (83) is of linear order in $\delta\rho'(\mathbf{r})$ when the density is approaching $\rho_0(\mathbf{r})$. We can then split the second term in Eq. (83) into a first-order term of $\delta\rho'(\mathbf{r})$ and a higher-order residual:

$$\inf_{\Psi_0+\delta\Psi'\to\rho_0+\delta\rho'} \langle \delta\Psi'|\hat{H} - E_0|\delta\Psi'\rangle = \langle K(\mathbf{r})\delta\rho'(\mathbf{r})\rangle + R[\rho_0, \delta\rho'] , \tag{84}$$

where $K(\mathbf{r})$ has no dependence on $\delta\rho'(\mathbf{r})$ and $R[\rho_0, \delta\rho']$ satisfies:

$$\lim_{\beta \to 0} \frac{R[\rho_0, \delta\rho']}{\beta} = 0 . \tag{85}$$

Substituting Eq. (84) into Eq. (83), we obtain

$$\delta F_{LL}^\omega = \left\langle \left. \frac{\delta F_{LL}^\omega}{\delta\rho(\mathbf{r})} \right|_{\rho_0} \delta\rho'(\mathbf{r}) \right\rangle + R[\rho_0, \delta\rho'] , \tag{86}$$

where the new Gâteaux derivative is

$$\left. \frac{\delta F_{LL}^\omega}{\delta\rho(\mathbf{r})} \right|_{\rho_0} = \frac{E_0}{N} - v(\mathbf{r}) + K(\mathbf{r}) . \tag{87}$$

If a CDV path is chosen instead, we will get the Gâteaux derivative previously obtained by Lindgren and Salomonson [8–10]:

$$\left. \frac{\delta F_{LL}^\omega}{\delta\rho(\mathbf{r})} \right|_{\rho_0} = \frac{E_0}{N} - v(\mathbf{r}) . \tag{88}$$

Equations (87) and (88) clearly indicate that the nonuniqueness of the Gâteaux derivative along different paths. For the Lieb functionals F_L^ω, it can be easily shown that the same conclusion is still valid.

Because of the existence of UDVs of the first kind [23] and the second kind, density variation domain has to be cleansed so that consistent results can be obtained. However, there remains one problem: What is the relationship between these two kinds of UDVs? In other words, does the set of Perdew and Levy's UDVs contains our UDVs or is the opposite true? Consider a noninteracting two-electron atom with nuclear charge $Z = 2$, its ground state is $1s^2$ and the GS wavefunction is denoted by Φ_{1s^2}. If both electrons are excited to the $2s$ orbital, we would get a wavefunction Φ_{2s^2} from the D space of Φ_{1s^2}. The density variation,

$$\rho(\mathbf{r}) = \rho_0(\mathbf{r}) + \beta\delta\rho(\mathbf{r}) = \rho_{1s^2}(\mathbf{r}) \mid \beta\rho_{2s^2}(\mathbf{r}) , \tag{89}$$

with $0 \leq \beta \leq 1$, would belong to the UDVs of the second kind. Due to the asymptotic behavior of wavefunctions [1], we have

$$\lim_{r \to \omega} \rho_{1s^2}(\mathbf{r}) \sim e^{-2\mathbf{r}\sqrt{2I_{1s^2}}} \tag{90}$$

and

$$\lim_{r \to \infty} \rho_{2s^2}(\mathbf{r}) \sim e^{-2\mathbf{r}\sqrt{2I_{2s^2}}} , \tag{91}$$

where I_{1s^2} and I_{2s^2} are first ionization potentials for $1s^2$ and $2s^2$ configurations, respectively. Because the system concerned here is a noninteracting one, it can be readily shown that $I_{1s^2} = 4I_{2s^2}$. In this case, we have

$$\lim_{r \to \infty} \frac{\eta(\mathbf{r})}{\sqrt{\rho_0(\mathbf{r})}} = \lim_{r \to \infty} \frac{\rho_{2s^2}(\mathbf{r})}{\sqrt{\rho_{1s^2}(\mathbf{r})}} = \lim_{r \to \infty} e^{-\mathbf{r}(\sqrt{8I_{2s^2}} - \sqrt{2I_{1s^2}})} = 1 . \tag{92}$$

Consequently, Perdew and Levy's condition for the UDVs of the first kind, Eq. (78), is satisfied and the density variation, Eq. (89), also belongs to the UDVs of the first kind.

However, not all UDVs of the second kind are UDVs of the first kind. Here is an example. Replacing the atom in the last example with a noninteracting 4-electron hydrogen-like atom and letting $\delta\rho$ to be from the wavefunction $\Phi_{1s^2 3s^2}$, which is in the D space of GS wavefunction $\Phi_{1s^2 2s^2}$, then we have a new density variational path:

$$\rho'(\mathbf{r}) = \rho_0(\mathbf{r}) + \beta \delta\rho(\mathbf{r}) = \rho_{1s^2 2s^2}(\mathbf{r}) + \beta \rho_{1s^2 3s^2}(\mathbf{r}) , \tag{93}$$

with $0 \leq \beta \leq 1$. Then, we instead have

$$\lim_{r \to \infty} \frac{\eta(\mathbf{r})}{\sqrt{\rho_0(\mathbf{r})}} = \lim_{r \to \infty} \frac{\rho_{1s^2 3s^2}(\mathbf{r})}{\sqrt{\rho_{1s^2 2s^2}(\mathbf{r})}} = \lim_{r \to \infty} e^{-\mathbf{r}(\sqrt{8I_{1s^2 3s^2}} - \sqrt{2I_{1s^2 2s^2}})} = 0 , \tag{94}$$

because the first ionization potentials of the two configurations, $1s^2 2s^2$ and $1s^2 3s^2$, satisfy a different equation:

$$4I_{1s^2 2s^2} = 9I_{1s^2 3s^2} = I_{1s^2} . \tag{95}$$

Clearly, the condition for Perdew and Levy's UDVs cannot be fulfilled in this case.

From the discussion above, we can see that the set of the UDVs of the second kind is not enclosed in the set of the UDVs of the first kind. However, the question of whether the UDVs of the second kind fully contain the UDVs of the first kind is still left open. Most likely, these two sets of UDVs share some common elements, but not mutually inclusive.

The current definition of density variation domain [25] is based on the pioneer work of Lieb [5] and of Englisch and Englisch [6, 7, 24]. For wavefunctions in Hilbert space, the density of concern belongs to the convex set of N-representable densities, $\mathcal{J}_N$. For wavefunctions in Fock space, the density domain is the direct sum of $\mathcal{J}_N$:

$$\mathcal{J} \equiv \bigoplus_{N \in \mathcal{R}^+} \mathcal{J}_N , \tag{96}$$

where $\mathcal{R}^+$ is the space of positive real numbers.

To ensure density functionals to be analytic through second order, we should require the density to stay in the following modified variational domain (without the UDVs of the first kind [23]):

$$\mathcal{M}_N \equiv \left\{ \rho \,\middle|\, \rho_0 \in \mathcal{A}_N, \rho \in \mathcal{J}_N, \frac{\rho - \rho_0}{\sqrt{\rho_0}} \in \mathcal{L}^2 \right\}, \tag{97}$$

where $\mathcal{A}_N$ is the set of GS densities for an N-particle quantum system [5]. In addition, to guarantee Fréchet differentiability of density functionals, the density should be further restricted within

$$\mathcal{M}'_N \equiv \left\{ \rho \,\middle|\, \rho_0 \in \mathcal{A}_N, \rho \in \mathcal{J}_N, \rho \notin \mathcal{B}_N \right\}, \tag{98}$$

without the UDVs of the second kind, where $\mathcal{B}_N$ is defined in Appendix 2. In other words, the density must be in the nexus of the above two restricted density domains:

$$\rho(\mathbf{r}) \in \mathcal{J}'_N \equiv \mathcal{M}_N \cap \mathcal{M}'_N \equiv \left\{ \rho \,\middle|\, \rho_0 \in \mathcal{A}_N, \rho \in \mathcal{J}_N, \frac{\rho - \rho_0}{\sqrt{\rho_0}} \in \mathcal{L}^2, \rho \notin \mathcal{B}_N \right\}, \tag{99}$$

for wavefunctions in Hilbert space. Accordingly, we have to restrict the density for wavefunctions in Fock space:

$$\rho(\mathbf{r}) \in \mathcal{J}' \equiv \bigoplus_{N \in \mathcal{R}^+} \mathcal{J}'_N. \tag{100}$$

5 Conclusions

Within the current framework of DFT, the Levy-Lieb functionals are not Fréchet differentiable at PS-v-representable densities and the Lieb functionals are not Fréchet differentiable at E-v-representable densities. For the Levy-Lieb functionals, when the density variation comes from a wavefunction in $\mathcal{D}$, the Gâteaux derivatives will become path-dependent, taking a different form from the conventional ones. For the Lieb functionals, when the variation of each individual PS-v-representable density that comprises the total density comes from a wavefunction in its corresponding $\mathcal{D}$ space, the Gâteaux derivatives will take a different form from the conventional ones. Based on the analysis on UDVs, we have proposed necessary modifications on the density variational domain on which density functionals are Fréchet differentiable and possess the conventional analytic density expansion through second order in density variation.

Acknowledgements Financial support for this project was provided by a grant from the Natural Sciences and Engineering Research Council (NSERC) of Canada.

Appendix 1

Here, we will briefly introduce some mathematical concepts relevant to our discussion in the main text. All the following content are adopted from an introductory book on functional analysis [26].

Definition 1 A **vector space** V is a set of elements called vectors with two operations called addition and scalar multiplication, which satisfy the following axioms.

- Addition axioms: To every pair of vectors $x, y \in V$, there corresponds a unique vector $x + y \in V$, the sum of x and y, such that

  1. $x + y = y + x$;
  2. $(x + y) + z = x + (y + z)$;
  3. there exists a unique zero vector $\theta \in V$ such that $x + \theta = \theta + x = x, \forall x \in V$;
  4. for every vector x there exists a unique vector $(-x) \in V$ such that $x + (-x) = \theta$.

- Scalar multiplication axioms: To every scalar α and every vector $x \in V$ there corresponds a unique vector $\alpha x \in V$ such that

  1. $\alpha(\beta x) = (\alpha\beta)x$ for every scalar β;
  2. $1x = x, 0x = 0, \forall x \in V$;
  3. $\alpha(x + y) = \alpha x + \alpha y$ and $(\alpha + \beta)x = \alpha x + \beta x$.

Definition 2 If x and y are two points of a vector space, then the **line segment** joining them is the set of elements $\{\beta x + (1 - \beta)y \mid 0 \leq \beta \leq 1\}$. A subset S of a vector space is **convex** if the line segment of joining any two points in S is contained in S.

Definition 3 Let U and V be two vector spaces with the same system of scalars. Then a function (or mapping) that maps uniquely the elements of V onto elements of U,

$$T : V \to U \tag{101}$$

is called a **linear transformation** of V into U if

1. $T(x + y) = Tx + Ty, \forall x, y \in V$;
2. $T(\alpha x) = \alpha Tx, \forall x \in V$ and for all scalars α.

Definition 4 A **metric** (or distance function) on a set S is a real-valued function $d(x, y)$ defined for all pairs of elements x and y in S and which satisfies the following axioms:

1. $d(x, y) > 0; d(x, y) = 0$, if and only if $x = y$;
2. $d(x, y) = d(y, x), \forall x, y \in S$;
3. $d(x, z) \leq d(x, y) + d(y, z), \forall x, y, z \in S$.

A **metric space** denoted by (S, d) consists of a set S and a metric d on S.

Definition 5 Let T be an operator (mapping, transformation) whose domain $Dom(T)$ and range $Ran(T)$ belong to metric spaces (X, d_X) and (Y, d_Y), respectively. The operator T is **continuous** at point $x_0 \in Dom(T)$ if, for every $\epsilon > 0$, there exists $\delta > 0$ such that

$$d_Y(Tx, Tx_0) < \epsilon \tag{102}$$

whenever

$$d_X(x, x_0) < \delta . \tag{103}$$

Definition 6 A sequence $\{x^{(k)}\}$ in a metric space (S, d) is said to be a **Cauchy sequence** if $d(x^{(k)}, x^{(l)}) \to 0$ as $k, l \to \infty$. This means that for every $\delta > 0$ there exists N_δ such that $d(x^{(k)}, x^{(l)}) \leq \delta$ for any $k, l \geq N_\delta$.

Definition 7 A metric space (S, d) is said to be **complete** if every Cauchy sequence in (S, d) has a limit in (S, d).

Definition 8 A **norm** (or length function) on a vector space $\mathcal{V}$ is a real-valued function, $||x||$, defined for all vectors $x \in \mathcal{V}$ and which satisfies the following axioms:

1. $||x|| > 0$; $||x|| = 0$ if and only if $x = \theta$;
2. $||x + y|| \leq ||x|| + ||y||$, $\forall x, y \in \mathcal{V}$;
3. $||\lambda x|| = |\lambda| \cdot ||x||$, for an arbitrary scalar λ.

A **normed vector space**, denoted by $(\mathcal{V}, || \cdot ||)$ consists of a vector space $\mathcal{V}$ and a norm $|| \cdot ||$ on $\mathcal{V}$.

Definition 9 A complete (with respect to the norm) normed vector space is called a **Banach space**.

Definition 10 Let $T : \mathcal{V} \to \mathcal{U}$ be a **bounded linear transformation**, that is,

$$||Tx|| \leq K||x||. \tag{104}$$

The smallest value of K which satisfies this inequality is denoted by $||T||$ and called the norm of T. It can be verified that this norm for operators satisfies the axioms for a norm function and that we may therefore talk of the **vector space of bounded linear transformations** $T : \mathcal{V} \to \mathcal{U}$. This normed vector space is denoted by $\mathcal{L}(\mathcal{V}, \mathcal{U})$.

Definition 11 Consider an operator $T : \mathcal{V} \to \mathcal{U}$ where $\mathcal{V}$ is a vector space and $\mathcal{U}$ is a normed vector space. Let the domain of the operator T, $Dom(T) \subset \mathcal{V}$, and $s \in \mathcal{V}$: if the limit

$$dT(x; s) = \lim_{\lambda \to 0} \frac{T(x + \lambda s) - T(x)}{\lambda} \tag{105}$$

exists, it is called the **Gâteaux differential** of T at x in the direction s. The limit is to be understood in the sense of convergence with respect to the norm in $\mathcal{U}$. The differential may exist for some s and fail to exist for others: if the differential exists at x for all s we say that T is **Gâteaux differentiable** at x.

The Gâteaux differential is homogeneous in s in the sense that

$$dT(x; \alpha s) = \alpha dT(x; s) \tag{106}$$

but is in general *neither linear nor continuous* in s. Nor does the existence of the Gâteaux differential at x ensure continuity of T at x. For example,

$$f(\xi_1, \xi_2) = \begin{cases} \dfrac{\xi_1^3}{\xi_2} & (\xi_1, \xi_2 \neq 0) \\ 0 & (\xi_1 = \xi_2 = 0) \end{cases} \ . \tag{107}$$

At point $(0,0)$, it can be easily shown that the Gâteaux differential exists and it is zero. Clearly, the Gâteaux differential is a continuous linear operator. However, f is not continuous at $(0,0)$. Therefore, we cannot relate the Gâteaux differentiability of T to the continuity of T.

Let us go forward on the basis that $\mathcal{V}$ is also a normed vector space. Suppose $dT(x; s)$ is linear and continuous in s for some $x \in \mathcal{V}$, then we may write

$$dT(x; s) = \lim_{\lambda \to 0} \frac{T(x + \lambda s) - T(x)}{\lambda} = T_G'(x)s \ . \tag{108}$$

The operator T_G' is by definition, a mapping $\mathcal{V} \to \mathcal{U}$ and is linear and continuous: we may conclude that

$$T_G'(x) \in \mathcal{L}(\mathcal{V}, \mathcal{U}) \ . \tag{109}$$

This operator is called the ***Gâteaux*** or *weak derivative* of T at x. It is very important to note that when speaking of the linearity and continuity of $T_G'(x)$, we means those properties in the operator sense with respect to a fixed s. T_G' itself may be a function of x, but its continuity and linearity with respect to the variable x are complete different things from the continuity and linearity we discussed here.

When $T_G'(x)$ exists, it is certainly true that

$$T(x + \lambda s) - T(x) = T_G'(x)\lambda s + \epsilon(x, s, \lambda) \ , \tag{110}$$

where $\epsilon/\lambda \to 0$ as $\lambda \to 0$ with x and s fixed. However, the convergence may not be uniform with respect to s and in that case T cannot be approximated by a linear operator with uniform accuracy in the neighborhood of x. If we further demand uniform convergence then we arrive at the *strong* derivative.

Definition 12 Let $\mathcal{V}$ and $\mathcal{U}$ be normed vector spaces. An operator $T : \mathcal{V} \to \mathcal{U}$ is **Fréchet differentiable** at $x \in Dom(T) \subset \mathcal{V}$ if there exists a continuous linear operator $T_F'(x) \in \mathcal{L}(\mathcal{V}, \mathcal{U})$ such that, for all $s \in \mathcal{V}$,

$$T(x + s) - T(x) = T_F'(x)s + \epsilon(x; s) \tag{111}$$

with

$$\lim_{||s||_{\mathcal{V}} \to 0} \frac{||\epsilon(x;s)||_{\mathcal{U}}}{||s||_{\mathcal{V}}} = 0 . \tag{112}$$

The operator $T'_F(x)$ is called the **Fréchet** or strong derivative of T at x. The Fréchet derivative at x is unique. It can be shown that the existence of the Fréchet derivative of T at x implies continuity of T at x.

Theorem 1 *If the Gâteaux derivative $T'_G(x)$ exists in the neighborhood of x and is continuous with respect to the norm in $\mathcal{L}(\mathcal{V}, \mathcal{U})$ at x, then the Fréchet derivative $T'_F(x)$ exists and is equal to $T'_G(x)$.*

Appendix 2

In this appendix, we show that the Fréchet derivative does not exist in the normalized density domain, $\mathcal{J}_N$.

Define a normalized path wavefunction,

$$\Psi_p = \sqrt{1 - \beta^2}\Psi_0 + \beta\Psi_D , \tag{113}$$

where Ψ_0 is the GS wavefunction for an N-electron quantum system, Ψ_D is a linear combination of eigenfunctions in D of Ψ_0, and $0 \leq \beta \leq 1$. Both Ψ_0 and Ψ_D are normalized to 1. The corresponding path density is

$$\begin{aligned}
\rho_p(\mathbf{r}) &= N\langle\Psi_p|\Psi_p\rangle_{N-1} \\
&= (1 - \beta^2)N\langle\Psi_0|\Psi_0\rangle_{N-1} + \beta^2 N\langle\Psi_D|\Psi_D\rangle_{N-1} \\
&= (1 - \beta^2)\rho_0(\mathbf{r}) + \beta^2\rho_D(\mathbf{r}) .
\end{aligned} \tag{114}$$

When β approaches 0, $\rho_p(\mathbf{r})$ also approaches $\rho_0(\mathbf{r})$. Letting β changes continuously from 1 to 0, we obtain the desired density variational path. Equation (114) shows that the path density is automatically normalized to N, therefore the density variation stays within the normalized space. Clearly, $\rho_p(\mathbf{r})$ lies in the neighborhood of $\rho_0(\mathbf{r})$ within $\mathcal{J}_N$. For convenience, we label $\mathcal{B}_N$ as the set of all legitimate N-representable $\rho_p(\mathbf{r})$ defined for a given Ψ_0 or $\rho_0(\mathbf{r})$ in Eq. (114).

A trial wavefunction is then assumed to yield the same path density:

$$\widetilde{\Psi} = \sqrt{1 - \beta^2}\Psi_0 + \lambda\Psi_t = \sqrt{1 - \beta^2}\Psi_0 + \lambda\sum_{i=0}^{\infty} c_i\Psi_i \longmapsto \rho_p(\mathbf{r}) , \tag{115}$$

where Ψ_i is the ith normalized eigenfunction of $\hat{H}$, $\langle\Psi_t|\Psi_t\rangle = 1$, and the expansion coefficients $\{c_i\}$ are chosen to be real. The complete set of $\{\Psi_i\}$ can be divided into three parts: Ψ_0, S, and D. The electron density (the trial density) for $\widetilde{\Psi}$ takes the following form:

$$
\begin{aligned}
\widetilde{\rho}(\mathbf{r}) &= N\langle\widetilde{\Psi}|\widetilde{\Psi}\rangle_{N-1} \\
&= (1-\beta^2)N\langle\Psi_0|\Psi_0\rangle_{N-1} + \lambda^2 N\langle\Psi_t|\Psi_t\rangle_{N-1} + 2\lambda\sqrt{1-\beta^2}NRe\big(\langle\Psi_0|\Psi_t\rangle_{N-1}\big) \\
&= (1-\beta^2)\rho_0(\mathbf{r}) + \lambda^2\rho_t(\mathbf{r}) + 2\lambda\sqrt{1-\beta^2}NRe\big(\langle\Psi_0|\Psi_t\rangle_{N-1}\big) .
\end{aligned}
\tag{116}
$$

At any point, the trial density is identical to the path density to ensure that the density variation is actually along the path we designed:

$$
\widetilde{\rho}(\mathbf{r}) = \rho_p(\mathbf{r}) \to \rho_0(\mathbf{r}) .
\tag{117}
$$

Therefore, we have

$$
\langle\widetilde{\rho}(\mathbf{r})\rangle = \langle\rho_p(\mathbf{r})\rangle .
\tag{118}
$$

Substituting Eqs. (114) and (116) into Eq. (118) and simplifying the result, one derives

$$
\beta^2 = \lambda^2 + 2\lambda c_0\sqrt{1-\beta^2} .
\tag{119}
$$

At one specific point on the variational path, the value of β is fixed, we can solve λ in terms of β based on Eq. (119):

$$
\lambda = -c_0\sqrt{1-\beta^2} \pm \sqrt{c_0^2(1-\beta^2)+\beta^2} .
\tag{120}
$$

Near the end of the variational path, when $\beta \to 0$ and $c_0 \neq 0$,

$$
\lambda \to -c_0\sqrt{1-\beta^2} \pm \left[c_0\sqrt{1-\beta^2} + \frac{1}{2c_0}\beta^2 + \cdots \right] .
\tag{121}
$$

Again (see Appendix 3), the positive sign is chosen in Eq. (121), and we have

$$
\lambda \to \frac{1}{2c_0}\beta^2 + \cdots , \quad \text{as } \beta \to 0 .
\tag{122}
$$

Immediately, we can conclude that towards the end of variational path, λ is of the same magnitude of β^2/c_0. In other words, λ also approaches zero at nearly the same rate as β^2/c_0 approaches zero.

Because of Eqs. (114), (116), and (117), we obtain

$$
\lambda^2\rho_t + 2N\lambda\sqrt{1-\beta^2}Re\big(\langle\Psi_0|\Psi_t\rangle_{N-1}\big) = \beta^2\rho_D .
\tag{123}
$$

Substituting Eq. (119) into Eq. (123) yields

$$2\sqrt{1-\beta^2}\left[c_0\rho_D - N\sum_i^{S_0} c_i Re(\langle\Psi_0|\Psi_i\rangle_{N-1})\right] = \lambda(\rho_t - \rho_D), \quad (124)$$

where the summation on the LHS is only within S_0. At $\beta \to 0$, we find that the coefficients $\{c_i\}$ for $\Psi_i \in S_0$ are linear in λ.

After knowing the property of $\{c_i\}$ for wavefunctions in S_0, we then investigate other remaining $\{c_i\}$ for wavefunctions in D. At one particular point on the variational path (β fixed), we optimize trial wavefunction to find out the set of coefficients $\{c_i\}$ that yields the lowest energy for

$$\begin{aligned}
\langle\widetilde{\Psi}|\hat{H}|\widetilde{\Psi}\rangle &= \left\langle\sqrt{1-\beta^2}\Psi_0 + \lambda\Psi_t \left|\hat{H}\right| \sqrt{1-\beta^2}\Psi_0 + \lambda\Psi_t\right\rangle \\
&= E_0 - \left[\beta^2 - 2\lambda c_0\sqrt{1-\beta^2}\right]E_0 + \lambda^2\langle\Psi_t|\hat{H}|\Psi_t\rangle \\
&= E_0 - \lambda^2 E_0 + \lambda^2\langle\Psi_t|\hat{H}|\Psi_t\rangle = E_0 + \lambda^2\left(\langle\Psi_t|\hat{H}|\Psi_t\rangle - E_0\right), \quad (125)
\end{aligned}$$

where Eq. (119) has been used to simplify the expression after the second equal sign. Obviously, we only need to minimize the last term in Eq. (125) under the following two constraints:

$$\sum_{i=0}^{\infty} c_i^2 = 1, \quad (126)$$

and

$$\widetilde{\rho}(\mathbf{r}) = \rho_p(\mathbf{r}). \quad (127)$$

The density constraint, Eq. (127), is equivalent to the following identity based on our previous analysis:

$$\begin{aligned}
2\lambda\sqrt{1-\beta^2}&\left[\frac{c_0(\rho_0 - \rho_D)}{N} + \sum_i^{S} c_i Re(\langle\Psi_0|\Psi_i\rangle_{N-1})\right] \\
&- \lambda^2\left(\frac{\rho_D}{N} - \sum_{i,j}^{\infty} c_j c_i\langle\Psi_j|\Psi_i\rangle_{N-1}\right). \quad (128)
\end{aligned}$$

We will use the Euler-Lagrange multiplier method to find the set of coefficients $\{c_i\}$ that minimizes the value of $\lambda^2\left(\langle\Psi_t|\hat{H}|\Psi_t\rangle - E_0\right)$. Let

$$A = \sum_{i=0}^{\infty} c_i^2 - 1, \quad (129)$$

$$\mathbf{B} = 2\lambda\sqrt{1-\beta^2}\left[\frac{c_0(\rho_0 - \rho_D)}{N} + \sum_i^S c_i Re\left(\left\langle \Psi_0 | \Psi_i \right\rangle_{N-1}\right)\right]$$

$$-\lambda^2\left(\frac{\rho_D}{N} - \sum_{i,j}^\infty c_i c_j \left\langle \Psi_i | \Psi_j \right\rangle_{N-1}\right), \tag{130}$$

and

$$\begin{aligned}
\Omega &= \lambda^2\left(\left\langle \Psi_t | \hat{H} | \Psi_t \right\rangle - E_0\right) - h\mathbf{A} - \langle g(\mathbf{r})\,\mathbf{B}\rangle \\
&= \lambda^2\left(\sum_{i,j}^\infty c_i c_j \langle \Psi_i | \hat{H} | \Psi_j \rangle - E_0\right) - h\mathbf{A} - \langle g(\mathbf{r})\,\mathbf{B}\rangle \\
&= \lambda^2\left(\sum_{i,j}^\infty c_i c_j E_j \delta_{ij} - E_0\right) - h\mathbf{A} - \langle g(\mathbf{r})\,\mathbf{B}\rangle \\
&= \lambda^2\left[\sum_{i=1}^\infty c_i^2(E_i - E_0)\right] - h\mathbf{A} - \langle g(\mathbf{r})\,\mathbf{B}\rangle,
\end{aligned} \tag{131}$$

where h and $g(\mathbf{r})$ are the Lagrange multipliers corresponding to the two constraints in Eqs. (129) and (130). Minimizing Eq. (131) with respect to $\{c_i\}$, one obtains

$$\lambda\left\langle\left[\frac{\sqrt{1-\beta^2}(\rho_0 - \rho_D)}{N} + \lambda\sum_{j=0}^\infty c_j Re\left(\left\langle \Psi_0 | \Psi_j \right\rangle_{N-1}\right)\right] g(\mathbf{r})\right\rangle = -hc_0, \tag{132}$$

and

$$\lambda\left\langle\left[\sqrt{1-\beta^2}Re\left(\left\langle \Psi_i | \Psi_0 \right\rangle_{N-1}\right) + \lambda\sum_{j=0}^\infty c_j Re\left(\left\langle \Psi_i | \Psi_j \right\rangle_{N-1}\right)\right] g(\mathbf{r})\right\rangle$$

$$= \left[\lambda^2(E_i - E_0) - h\right] c_i \qquad (\text{for } i \neq 0). \tag{133}$$

Because c_0 is linear in λ as we previously showed, we can readily infer from Eq. (132) that $g(\mathbf{r})$ must take the following form:

$$g_\lambda(\mathbf{r}) = g^{(0)}(\mathbf{r}) + \sum_{k=1}^\infty \frac{g^{(k)}(\mathbf{r})}{k!}\lambda^k. \tag{134}$$

Substituting Eq. (134) into Eq. (133) and ignoring the higher-order terms as $\lambda \to 0$, we obtain an equation for $\Psi_i \in S$,

$$-hc_i = \lambda\sqrt{1-\beta^2}\left\langle Re\left(\left\langle \Psi_i | \Psi_0 \right\rangle_{N-1}\right) g^{(0)}(\mathbf{r})\right\rangle + h.o., \tag{135}$$

where "*h.o.*" denotes higher-order terms in λ. Therefore, we reach the same conclusion as before: $\{c_i\}$ for $\Psi_i \in S$ is linear in λ towards the end of variational path. For those $\{c_i\}$ for $\Psi_i \in D$, utilizing the additional fact that Ψ_i is order-1 strongly orthogonal to Ψ_0, we can further simplify Eq. (133) to

$$\lambda^2 \left\langle \sum_j^{S_0} c_j Re\left(\langle \Psi_i|\Psi_j \rangle_{N-1}\right) g(\mathbf{r}) \right\rangle + \lambda^2 \left\langle \sum_j^{D} c_j Re\left(\langle \Psi_i|\Psi_j \rangle_{N-1}\right) g(\mathbf{r}) \right\rangle$$
$$= \lambda^2 (E_i - E_0)c_i - hc_i \, . \tag{136}$$

For this equation to be valid at $\lambda \to 0$, the LHS and the RHS must have the same dependence on λ. On the RHS, the first term decays faster than the second term, and the second term will dominate when λ approaches 0. Therefore, we must match the magnitude of the second term on the RHS to the LHS. Of course, we cannot match it with the second term on the LHS because doing so will lead to self inconsistency. Then, the second term on the RHS must decay in the same way as the first term on the LHS. Thus, $\{c_i\}$ for $\Psi_i \in D$ are proportional to λ^3. Unfortunately, such a λ^3-behavior is contradictory to the normalization constraint in Eq. (126), because $\sum_i c_i^2$ will become 0 as $\lambda \to 0$. Hence, we conclude that this contradiction must come from the assumption: $\Psi_t = \sum_i c_i \Psi_i$, where the expansion is over the complete set of eigenfunctions of $\hat{H}$.

To resolve the contradiction, we have to modify our assumption about the expansion of Ψ_t. We notice that if the summation $\sum_i c_i \Psi_i$ includes any wavefunction from S_0, the same problem will persist. Therefore, Ψ_t can only be expanded in D,

$$\Psi_t = \sum_i^{D} c_i \Psi_i \, . \tag{137}$$

In this case, Eq. (127) is equivalent to

$$\lambda^2 \rho_t(\mathbf{r}) = \beta^2 \rho_D(\mathbf{r}) \, . \tag{138}$$

Integrating both sides of Eq. (138) over the entire space, one obtains

$$\lambda^2 = \beta^2 \, , \tag{139}$$

which further ensures that

$$\rho_t(\mathbf{r}) = \rho_D(\mathbf{r}) \, . \tag{140}$$

Now, the original minimization process is reduced to minimizing the following term,

$$\Xi = \left(\sum_i^D |c_i|^2 E_i \right) - h \left(\sum_i^D |c_i|^2 - 1 \right) - \left\langle \left(\frac{\rho_D}{N} - \sum_{i,j}^D c_i^* c_j \langle \Psi_i | \Psi_j \rangle_{N-1} \right) g(\mathbf{r}) \right\rangle .$$

$$(141)$$

Suppose this minimization will yield the optimal set of expansion coefficients, $\{\bar{c}_i\}$, which have no dependence on λ and β from the appearance of Eq. (141). Then, we have

$$\inf_{\Psi_0 + \delta\Psi_0 \to \rho_0 + \delta\rho} \frac{\langle \delta\Psi | \hat{H} - E_0 | \delta\Psi \rangle}{||\delta\rho||} = \inf_{\Psi_0 + \delta\Psi \to \rho_0 + \delta\rho} \frac{\langle \widetilde{\Psi} - \Psi_0 | \hat{H} - E_0 | \widetilde{\Psi} - \Psi_0 \rangle}{||\rho_p - \rho_0||}$$

$$= \inf_{\Psi_t \to \rho_D} \frac{\left\langle (\sqrt{1 - \beta^2} - 1)\Psi_0 + \lambda\Psi_t \left| \hat{H} - E_0 \right| (\sqrt{1 - \beta^2} - 1)\Psi_0 + \lambda\Psi_t \right\rangle}{||\beta^2(\rho_D - \rho_0)||}$$

$$= \inf_{\Psi_t \to \rho_D} \frac{\lambda^2 \langle \Psi_t | \hat{H} - E_0 | \Psi_t \rangle}{\beta^2 ||\rho_D - \rho_0||} = \frac{\inf_{\Psi_t \to \rho_D} \langle \Psi_t | \hat{H} - E_0 | \Psi_t \rangle}{||\rho_D - \rho_0||}$$

$$= \frac{1}{||\rho_D - \rho_0||} \left\langle \sum_i^D \bar{c}_i \Psi_i \left| \hat{H} - E_0 \right| \sum_j^D \bar{c}_j \Psi_j \right\rangle = \frac{1}{||\rho_D - \rho_0||} \left[\sum_i^D |\bar{c}_i|^2 E_i - E_0 \right]$$

$$> \frac{1}{||\rho_D - \rho_0||} \left[\sum_i^D |\bar{c}_i|^2 E_0 - E_0 \right] = 0 ,$$

$$(142)$$

where Eq. (139) is used to simplify the expression after the third equal sign. Evidently, Eq. (142) suggests that the condition for Fréchet differentiability proposed by Lindgren and Salomonson [8–10] is not fulfilled. In other words, the Fréchet derivative does not exist in the normalized density domain $\mathcal{J}_N$ either.

Appendix 3

In this appendix, we analyze the consequence of choosing the negative sign in Eqs. (41) and (121). In the end, we will conclude that this particular choice is fully equivalent to the more natural decision made in the main text and Appendix 2.

Let us start from a unified version of Eqs. (40) and (120):

$$\lambda = -ac_0 \pm \sqrt{a^2 c_0^2 + \beta^2} ,$$

$$(143)$$

where constant $a = 1$ and $\sqrt{1 - \beta^2}$ in the main text and Appendix 2, respectively. Obviously, if $c_0 = 0$ or $c_0 \to 0$ as $\beta \to 0$, both λ and β approach 0 concurrently near the end of the variational path.

We only need to further examine the situation when $c_0 \neq 0$ as $\beta \to 0$ with the choice of the negative sign in Eq. (143):

$$\lambda = -2ac_0 - \lambda' , \tag{144}$$

where the residual term λ' approaches 0 as $\beta \to 0$:

$$\lambda' = \left[\frac{1}{2ac_0}\beta^2 - \frac{1}{8a^3c_0^3}\beta^4 + \cdots \right] \to 0 . \tag{145}$$

Consequently, Eqs. (46) and (124) can be rewritten as

$$2a\left[c_0(\rho_0 - \rho_t) + N \sum_i^S c_i Re\big(\langle \Psi_0 | \Psi_i \rangle_{N-1}\big) \right] = \lambda' \left(\rho_t - \rho_D \right) , \tag{146}$$

which immediately suggests that as $\beta \to 0$, $(\rho_0 - \rho_t)$ and the coefficients $\{c_i\}$ for $\Psi_i \in S$ are linear in λ'. Because $\rho_t \to \rho_0$, $\Psi_t \to c_0\Psi_0$ with $|c_0| \to 1$, as $\beta \to 0$. Therefore, at the end of the variational path ($\beta = 0$ and $\lambda' = 0$), $\lambda = -2ac_0$, $\Psi_t = c_0\Psi_0$, $|c_0| = 1$, and $\widetilde{\Psi} = -a\Psi_0$.

Evidently, the choice of the negative sign in Eqs. (41) and (121) yields a fully equivalent, alternative trial wavefunction,

$$\widetilde{\Psi}' = -a\Psi_0 - \lambda' \Psi_t , \tag{147}$$

where $\lambda' \to 0$ as $\beta \to 0$. Then, we can carry out the discussion on the basis of $\lambda' \to 0$ instead.

References

1. Parr RG, Yang W (1989) Density-functional theory of atoms and molecules. Oxford University Press, New York
2. Hohenberg P, Kohn W (1964) Phys Rev 136:B864
3. Kohn W, Sham LJ (1965) Phys Rev 140:A1133
4. Wang YA, Xiang P (2013) In: Wesolowski TA, Wang YA (eds) Recent advances in orbital-free density functional theory, Chap. 1. World Scientific, Singapore, pp 3–12
5. Lieb EH (1983) Int J Quantum Chem 24:243
6. Englisch H, Englisch R (1983) Phys Stat Sol 123:711
7. Englisch H, Englisch R (1984) Phys Stat Sol 124:373
8. Lindgren I, Salomonson S (2003) Phys Rev A 67:056501
9. Lindgren I, Salomonson S (2003) Adv Quantum Chem 43:95
10. Lindgren I, Salomonson S (2004) Phys Rev A 70:032509
11. Ekeland I, Temam R (1976) Convex analysis and variational problems. North-Holland, Amsterdam
12. Harris J, Jones RO (1974) J Phys F 4:1170

13. Harris J (1984) Phys Rev A 29:1648
14. Gunnarsson O, Lundqvist BI (1976) Phys Rev B 13:4274
15. Langreth DC, Perdew JP (1980) Phys Rev B 21:5469
16. Wang YA (1997) Phys Rev A 55:4589
17. Wang YA (1997) Phys Rev A 56:1646
18. Levy M (1979) Proc Natl Acad Sci USA 76:6062
19. Nesbet RK (2001) Phys Rev A 65:010502
20. Nesbet RK (2003) Adv Quantum Chem 43:1
21. Dreizler RM, Gross EKU (1990) Density functional theory. Springer, Berlin
22. Davidson ER (1976) Reduced density matrices in quantum chemistry. Academic, New York
23. Perdew JP, Levy M (1985) Phys Rev B 31:6264
24. Englisch H, Englisch R (1983) Physica A 121:253
25. Zhang YA, Wang YA (2009) Int J Quantum Chem 109:3199
26. Milne RD (1980) Applied functional analysis: an introductory treatment. Pitman Publishing, UK

The Dirac Electron and Elementary Interactions: The Gyromagnetic Factor, Fine-Structure Constant, and Gravitational Invariant: Deviations from Whole Numbers

Jean Maruani

Abstract In previous papers, we revisited the Dirac equation and conjectured that the electron can be viewed as a massless charge spinning at light speed, this internal motion being responsible for the rest mass involved in external motions and interactions. Implications of this concept on basic properties such as time, space, electric charge, and magnetic moment were considered. The present paper investigates the deviations of the resulting gyromagnetic factor, fine-structure constant, and gravitational invariant from their integer approximates, and their implication in a better understanding of the electromagnetic, gravitational, and other interactions.

Keywords Dirac equation · Spin momentum · Magnetic moment
Matter antimatter · Wave beat · *Zitterbewegung* · Light velocity
Compton diameter · Planck units · Catalan numbers · Casimir force
Nuclear forces · Gyromagnetic factor · Fine-structure constant
Gravitational invariant · Quantum electrodynamics · General relativity

1 The Heuristic Road to the Dirac Equation

In previous papers [1–4], we revisited the Dirac equation and developed a model for the electron, making conjectures about its rest mass, spin motion, effective size, and electric charge. This led us to investigate universal constants related to this model. In the present paper, after recalling specificities related to the Dirac equation, we elaborate on the light shed by peculiarities of these constants on the relations between the electromagnetic and gravitational interactions.

J. Maruani (✉)
Laboratoire de Chimie Physique-Matière et Rayonnement,
CNRS & UPMC, 4 Place Jussieu, #32-42, 75005 Paris, France
e-mail: jean.maruani@upmc.fr

© Springer International Publishing AG, part of Springer Nature 2018
Y. A. Wang et al. (eds.), *Concepts, Methods and Applications of Quantum Systems in Chemistry and Physics*, Progress in Theoretical Chemistry and Physics 31,
https://doi.org/10.1007/978-3-319-74582-4_19

361

Today, the Dirac equation can be derived from more general theoretical frameworks [5–8]. But the inductive derivation originally given by Dirac [9, 10] has shown great heuristic value: it has explained the spin kinetic momentum and magnetic moment, predicted antimatter, and set the ground for quantum electrodynamics.

In the 1920s, *light* appeared alternatively as geometric *rays* following Fermat's principle of least optical path (stemming from Snell-Descartes' laws for reflection and refraction), or as electromagnetic *waves* obeying Maxwell's differential equations, or as massless *particles* following Planck-Einstein's quantum relations:

$$E = h\nu = hc/\lambda, \quad p = E/c; \tag{1}$$

whereas *matter* was considered as made of *particles* following Maupertuis' principle of least action integral (or Hamilton's differential equations for position and momentum), as well as Einstein-Poincaré's relativistic relation:

$$E = mc^2. \tag{2}$$

Through a detailed analysis of the similarities between Fermat's and Maupertuis' principles and making use of the above relations, de Broglie came to the conjecture that matter also is associated with waves, according to the formula [11]:

$$\lambda_B = h/p, \quad p = mv. \tag{3}$$

It is to be noted that this de Broglie wavelength λ_B differs from the Compton wavelength λ_C, introduced a year earlier in the theory of x-ray inelastic scattering [12], in that the latter involves the *rest mass* and the *speed of light*:

$$\lambda_C = h/m_0 c; \quad \lambdabar_C \equiv \lambda_C/2\pi = \hbar/m_0 c. \tag{4}$$

De Broglie's *matter-wave formula* succeeded to explain Bohr's *quantization rule* as due to *stationary waves* and it led to predicting *electron diffraction*. That brought Schrödinger to formulate his equation for *Wave Mechanics*. In the meanwhile, Heisenberg's phenomenological approach had led to *Matrix Mechanics*, and Dirac's bra/ket approach to *Operator Mechanics*. Schrödinger eventually showed that these three approaches are equivalent to a common pattern, nowadays known as *Quantum Mechanics*.

However, although *special relativity was originally involved* in de Broglie's matter-wave derivation, further formulations of Quantum Mechanics used the *non-relativistic* kinetic and potential energies:

$$T = p^2/2m, \quad V = k_e e^2/r, \tag{5}$$

and were not Lorentz-invariant. This hybrid character was partly corrected by adding *ad hoc* spin-symmetry conditions for the resulting eigenfunctions Ψ.

A further step was taken by Klein and Gordon, who used as Hamiltonian H the full energy: $mc^2 = (m_0^2 c^2 + p^2)^{1/2} c$, where $p^2 = p_1^2 + p_2^2 + p_3^2$ with $p_i = mv_i$ along x_i, $m = m_0\gamma$, $\gamma = (1 - v^2/c^2)^{1/2}$. This led to an equation that was relativistic (it had time and space operators on the same footing), but not quantic (it was quadratic in the time operator and did not allow to apply the probability principle). Dirac's feat was to design a road towards an equation that was both symmetric and linear in time and space operators, by introducing 4-D matrices multiplying the momentum operators.

Dimensionwise, an *invariant* 'momentum' $p_0 \equiv m_0c$ can be defined for a particle *at rest*, and an *overall* 'momentum' $p_4 \equiv mc$ related to the *time* coordinate $x_4 \equiv ct$. With these notations, it can be written:

$$p_4{}^2 = p_0{}^2 + p_1{}^2 + p_2{}^2 + p_3{}^2. \tag{6}$$

This expression for the *invariant* (rest mass) 'momentum' p_0 is similar to that for the *invariant* (proper interval) 'coordinate' x_0:

$$x_4{}^2 = x_0{}^2 + x_1{}^2 + x_2{}^2 + x_3{}^2. \tag{7}$$

Using the notations recalled in Eq. (8) (note that there is no coordinate derivative associated with the invariant momentum p_0), the Schrödinger (9), Klein-Gordon (10), and Dirac (11) equations can be written as:

$$p_1 \sim -i\hbar\, \partial/\partial x,\ p_2 \sim -i\hbar\, \partial/\partial y,\ p_3 \sim -i\hbar\, \partial/\partial z,\ p_4 \sim i\hbar\, \partial/\partial(ct),\ p_0 \equiv m_0c, \tag{8}$$

$$\left[p_4 - \left(p_1{}^2 + p_2{}^2 + p_3{}^2\right)/2mc\right]\Psi = 0. \tag{9}$$

$$\left[p_4{}^2 - \left(p_0{}^2 + p_1{}^2 + p_2{}^2 + p_3{}^2\right)\right]\Psi = 0. \tag{10}$$

$$\left[p_4 - \left(\alpha_0 p_0 + \alpha_1 p_1 + \alpha_2 p_2 + \alpha_3 p_3\right)\right]\Psi = 0. \tag{11}$$

It can be seen that the Schrödinger equation, being linear in p_4 but quadratic in the p_i's, is rather a *diffusion* equation, while the Klein-Gordon equation reduces to a *wave* equation for $p_0 = 0$. In the Dirac equation, the α_μ matrices are independent of the p's and x's as well as Hermitian and normalized. For Eq. (11) to be equivalent to Eq. (10), these matrices must also be *four-dimensional* and *anticommutative*.

There results that any vector representative of an eigenfunction Ψ must have *four components* or, alternatively, that Ψ contains a variable that may take on *four values*. Dirac explained that these are the well-known two components of the spin ($\pm\frac{1}{2}$) and in addition positive and negative values for the mass energy ($\pm mc^2$). The existence of a spin kinetic momentum thus appears unseparable from negative-energy states: both stem from the matrix linearization of a wave equation involving a quadratic form for the energy.

2   The Internal Motion of the Dirac Electron

Before the Dirac equation elucidated its origin, the electron spin had entered
quantum mechanics in two different ways [13]. (1.) A non-energetic, symmetry
requirement for systems of identical particles (Pauli 1925), antisymmetry of the
wave function, this implying an internal dynamical variable with two possible
values. (2.) The deflection of the trajectories of silver atoms by an inhomogeneous
magnetic field (Stern and Gerlach 1922) and the splitting of the spectral lines of
atoms by a magnetic field (Goodsmit and Uhlenbeck 1925), which implied an
intrinsic magnetic moment interacting with the field. The electron spin magnetic
moment is responsible for most of the macroscopic magnetism, from the oxygen
that we breathe to the hard disks of our computers.

To have the spin *magnetic moment* show up, Dirac made it interact with a
magnetic field. And to have its *spin* kinetic momentum appear, he had it combined
with an *orbital* kinetic momentum [10]. Equation (10) was thus extended to include
interactions with an electromagnetic potential $(A_4, \underline{A})$:

$$\left[(p_4 + eA_4/c) - \alpha_0 p_0 - \underline{\boldsymbol{\alpha}} \cdot (\underline{p} + e\underline{A}) \right] \Psi = 0. \tag{12}$$

Note that the invariant momentum p_0 is not affected by the external potential $(A_4, \underline{A})$.
Writing $H = m_0 c^2 + H'$, Dirac showed that, to first order:

$$H' = p_4 c - p_0 c = -eA_4 + (\underline{p} + e\underline{A})^2/2m_0 + (e\hbar/2m_0)\underline{\boldsymbol{\sigma}} \cdot \underline{B}. \tag{13}$$

In addition to the classical *potential* and *kinetic* energies, there appears an extra term,
which he interpreted as due to the interaction of the magnetic field $\underline{B}$ with an *intrinsic*
magnetic moment: $\underline{\boldsymbol{\mu}}_s = -(e\hbar/2m_0)\underline{\boldsymbol{\sigma}} = -\mu_B \underline{\boldsymbol{\sigma}}$, μ_B being the Bohr magneton.

The spin *kinetic* momentum does not give rise to any *potential* energy. To show
its existence, Dirac computed the angular momentum integrals for an electron
moving in a central electric field (e.g., that of a nucleus):

$$H = p_4 c = -eA_4(r) + c\alpha_0 p_0 + c\underline{\boldsymbol{\alpha}} \cdot \underline{p}. \tag{14}$$

For any component l_k of the *orbital* momentum: $\underline{l} = -i\hbar \underline{r} \times \underline{\nabla}$, Dirac obtained a
non-zero expression for $i\hbar \, \partial l_k/\partial t$, and similarly for the component σ_k of the Pauli
matrix vector used to build the Dirac matrices α_μ; thus, neither $\underline{l}$ nor $\underline{\sigma}$ was a
constant of the motion. But the sum was:

$$\partial l_k/\partial t + (\hbar/2)\partial\sigma_k/\partial t = 0. \tag{15}$$

Dirac interpreted this as meaning that the electron has a *spin* kinetic momentum: $\underline{s} = (\hbar/2)\,\underline{\sigma}$, which is to be added to the *orbital* kinetic momentum $\underline{l}$ to get a constant of the motion. The directions of $\underline{s}$ and $\underline{\mu}_s$ being defined by the *same* matrix vector $\underline{\sigma}$, one has:

$$\underline{\mu}_s = -(e/m_0)\underline{s}. \tag{16}$$

The *spin* magnetic moment $\underline{\mu}_s$ differs from the *orbital* magnetic moment $\underline{\mu}_l$ by a factor 2:

$$\underline{\mu}_l = -(e/2m_0)\underline{l}, \tag{17}$$

as if the 'loop' described by the electron in its spin motion had half the length of its orbital 'loop'. The total magnetic moment can then be written:

$$\underline{\mu}_t = -(\mu_B/\hbar)\,(\underline{l} + g_e\underline{s}), \tag{18}$$

where we introduce the *dimensionless factor* g_e, whose deviations from 2 (the value given in Dirac's theory) will be discussed in this paper. This factor distinguishes the spin magnetic moment from the orbital magnetic moment derived classically. Being the ratio of $\underline{\mu}_s$ (in units of μ_B) to $\underline{s}$ (in units of $\hbar$), it is called *gyromagnetic factor*.

In another computation, Dirac used a field-free Hamiltonian to determine *at which velocity* the electron 'spins' to acquire *kinetic* and *magnetic* momenta [10]:

$$H = c(\alpha_0 p_0 + \alpha_1 p_1 + \alpha_2 p_2 + \alpha_3 p_3). \tag{19}$$

Making use of the properties of the α_k's he obtained, for any component v_k of the electron velocity:

$$i\hbar\,\partial x_k/\partial t = [x_k, H] = i\hbar\,c\,\alpha_k \rightarrow v_k = \partial x_k/\partial t = \pm c. \tag{20}$$

The paradox of an electron moving *at light velocity* was elucidated by Schrödinger [14] while investigating the Dirac velocity operators $v_k = c\alpha_k$. He showed that:

$$i\hbar\,\partial^2 \alpha_k/\partial t^2 = 2\,(\partial\alpha_k/\partial t)\,H. \tag{21}$$

This differential equation can be integrated twice, yielding the explicit time dependence of the velocity and then of the position. One first obtains:

$$v_k = c\,\alpha_k = c^2 p_k H^{-1} + (i\hbar c/2)\gamma_k^0 e^{-i\omega t} H^{-1}, \tag{22}$$

where $\omega = 2H/\hbar$ and $\gamma_k^0 = \partial\alpha_k/\partial t$ at $t = 0$. As $H = mc^2$, the first term is a constant of the order of p_k/m, the classical relation between momentum and velocity. But here also there is an extra term, oscillating at the 'Zitterbewegung' frequency [14]:

$$\nu_e = 2\,mc^2/h. \tag{23}$$

The *constant part* gives the *average velocity* through a time interval larger than ν_e^{-1}, which is observed in practical measurements, while the *oscillatory part* explains why the *instantaneous velocity* has eigenvalues $\pm c$. Further integration yields the time dependence of the electron coordinate x_k, and it appears that the *Zitterbewegung* amplitude is of the order of the Compton radius: $r_C = \lambdabar_C/2 = \lambda_C/4\pi$.

Thus, in addition to its *external motion* (e.g., an orbital motion around a nucleus), governed by de Broglie's wavelength λ_B, the electron is endowed with an *internal motion* (*Zitterbewegung*), governed by Compton's wavelength λ_C. Schrödinger showed that *Zitterbewegung* vanishes when one takes expectation values over wave packets made up entirely of positive or negative energy states. This was understood by de Broglie [15] as it resulting from a *wave beat* between the *two coupled* matter and antimatter energy states, the beat frequency ν_e being the difference of the two frequencies. This oscillation may be pictured as a Lissajous curve $L_{\pi/2,\,1:2}$.

Then, the *average* mass of the vibrating entity can be considered as *null*, departures from this value being allowed by Heisenberg's uncertainty principle, i.e.:

$$2\,m_0 c \cdot c\,\tau_0 \approx \hbar \rightarrow \tau_0 \approx \hbar/2m_0 c^2 = 1/2\pi\nu_e \approx 0.64 \times 10^{-21}\text{s}.$$

To the *rest mass* 'momentum' $m_0\,c \equiv p_0$ is associated an *internal time* 'coordinate' $c\,\tau_0 \equiv x_0$. One may then see Eq. (7) as involving *three* space dimensions and *two* time dimensions, space 'emerging', so to say, from a Pythagorean substraction of an 'internal time' from the 'external time'.

Among the various authors who speculated on the electron internal motion [16], Barut and coworkers [17] described the *spin* as the *orbital* momentum associated with *Zitterbewegung* and the *rest mass* as the *internal energy* in the *rest frame*. In previous papers [1–4], we conjectured that the *rest mass* observed in external motions (inertia) and interactions (gravitation) essentially results from the *kinetic energy* of a *massless charge* spinning at *light speed*. Thus, if mass is linked to spin, spin to *Zitterbewegung*, and this latter to a wave beat between the particle and its antiparticle, then *there is no matter without antimatter*, which together result in both spin and mass.

Antimatter is thus around us and in us, like the two faces of a same coin. What we call matter (an electron) would be the visible face of the coin, and what we call antimatter (a positron) would result from a *dephased oscillation* between the two energy states. In modern quantum field theories [8], the *entanglement of matter and antimatter* appears in the necessity to include antiparticles to cope with infinities.

3 The *Gyromagnetic Factor* g_e

The *value ½ of the spin* of the electron (Eq. 15) and the *related value* 2 of its gyromagnetic factor (Eq. 13) stem from the *beat frequency* ν_e of the positive and negative energy waves being twice the electron mass-energy frequency (Eq. 23).

However, in Dirac's theory, the electron interacts with an electromagnetic field that fulfills relativistic but not quantum requirements. Further consistency was reached by quantizing the electromagnetic field, which led to quantum electrodynamics (QED) [7]. Resulting zero-point field (ZPF) oscillations entail 'radiative corrections' that are responsible for the Lamb shift between the $2s_{1/2}$ and $2p_{3/2}$ levels of hydrogenoid atoms [13] and for the departure of g_e from the Dirac integer value 2. Several authors have used Feynman diagrams to compute increasingly accurate corrections to g_e, yielding the following expansion [18]:

$$\varepsilon_g \equiv (g_e - 2)/2 \approx 0.001\,159\,652\,181 \approx 1/2a\pi + P/2(a\pi)^2 + L/2(a\pi)^3 + \cdots, \quad (24)$$

which is accurate within 1 ppb. Here a is the fine-structure constant inverse: $a \equiv a^{-1} \approx 137.036$, and P and L are coefficients involving hyperlogarithms. It took half a century to arrive at this expansion: the first term was obtained by Schwinger in 1948, the second one by Peterman in 1957, and the third one by Laporta in 1996.

But the deviation of the measured value of g_e from 2 does not come only from QED effects. Very accurate measurements [19] have shown that, after subtracting these effects, the free electron g_e value is still slightly larger than 2:

$$g_e(\text{measured}) \approx 2.002\,319\,304\,362; \quad g_e(\text{corrected}) \approx 2.000\,000\,000\,110\,(60).$$

This was interpreted by endowing the Dirac 'point charge' with a tiny but finite size: $\rho_e \sim 10^{-22}$ m, much smaller than the electron classical radius: $r_0 \approx 2.82 \times 10^{-15}$ m, but larger than the Planck length: $l_P \approx 1.62 \times 10^{-35}$ m. It would be that tiny charge which undergoes *Zitterbewegung* in a range set by the Compton radius: $r_C \approx 1.93 \times 10^{-13}$ m.

This discussion deals solely with the free electron interacting with an applied field. For electrons bound in paramagnetic systems, the measured values of g_e are effective values including, to second order, the orbital momentum and spin-orbit coupling [20]. The effective factor measured may then be a *tensor* $\mathbf{g}_e$, whose principal values and axes will depend on the anisotropy of the molecule or of the crystal site bearing the unpaired electron. Then the extreme accuracy of the value measured for g_e on the free electron is a sign of the extreme isotropy of vacuum fluctuations.

4 The *Fine-Structure Constant*

The *fine-structure constant a* was surmised by several quantum physicists (Planck, Haas, Bohr, …) before it was explicitly introduced by Sommerfeld in 1916 [21] to express the relativistic line splittings in hydrogenoid atomic spectra:

$$W(n,k) \approx -R\,hc\left(Z^2/n^2\right)\left[1 + \alpha^2\left(Z^2/n^2\right)(n/k - 3/4)\right], \tag{25}$$

where R is Rydberg's constant, n is the main quantum number, $k = 1, 2, …, n$, and:

$$\alpha \equiv k_e e^2/\hbar c \equiv 1/a \approx 1/137.0359991. \tag{26}$$

The *dimensionless quantity a* was identified with the ratio of the electron velocity in the first Bohr orbit to that of light. Its inverse a was given a more general significance by Eddington [22], who proposed the integer value 137 on speculative grounds. In atomic units, a measures the *velocity of light c*; and in Planck units, the *electron charge* $e = q_P\, a^2$. Then a can be seen as expressing the strength of the electromagnetic interaction between charged particles.

The fine-structure constant α (or its inverse, the electric parameter a) shows up in various domains in physics. For instance, it occurs (together with π) in the expansion of ε_g given in Eq. (24). In earlier papers [1–4], we recalled that the electron Compton diameter (or reduced wavelength) $2r_C$ (λ_C) is the geometric average of the classical electrostatic radius: $r_0 = k_e e^2/m_0 c^2$, and the hydrogen Bohr radius: $a_0 = \hbar^2/k_e m_e e^2$, the ratio of this harmonic relation being α:

$$2r_C/a_0 = r_0/2r_C = \alpha. \tag{27}$$

It is commonly believed that $Z = 137$ sets a limit for the periodic table. This is due to the fact that, when using the *point-nucleus model*, the energy expression includes a factor: $[1 - (aZ)^2]^{1/2}$, which becomes imaginary for $Z > 137$. However, this model is a poor approximation for superheavy elements. If one estimates the $1s$ orbital radius in these elements and compares it with actual nuclear sizes, it displays a strong overlap with the nucleus, and the factor above appears as an artifact [23]. There are several models of a finite nuclear charge distribution, none of which gives this factor [24]. Therefore, there may in principle be elements with $Z > 137$, although they might be too unstable to be observed. Nevertheless, the fact that $Z = 137$ *is a limit for point nuclei* remains *a peculiarity of a*.

Another reservation about the number 137 is that a is close to this value only when it is measured in our *low-energy world*. At a W boson energy (≈ 81 Gev), a decreases to about 128; and at grand-unification energy, it merges with similar constants defining the strong and weak nuclear forces: $a_e \approx a_w \approx a_s$ [25]. But, here again, the fact that $a \approx 137$ *at the limit of low energy* remains *a peculiarity of a* [26].

For the gyromagnetic factor g_e, both the integer value 2 resulting from the Dirac equation and the shift of measured values from 2 have theoretical explanations: the integer value is ultimately the expression of the duality matter/antimatter, and the corrections listed in Eq. (24) express the interactions with the quantum vacuum. For the electric parameter a, no such understanding is available, neither for the integer value 137 nor for the measured shift of ~ 0.26 ppt (Eq. 26). Feynman referred to the constant a in these terms:

> It has been a mystery ever since it was discovered more than fifty years ago [...]. You would like to know where this number for a coupling comes from: is it related to π or perhaps to the base of natural logarithms? Nobody knows; it is one of the greatest damn mysteries of physics: a magic number that comes to us with no understanding by man. You might say the 'hand of God' wrote that number [...]. We know what kind of a dance to do experimentally to measure this number very accurately; but we don't know what kind of dance to do on the computer to make this number come out without putting it in secretly!
>
> Richard Feynman, *The Strange Theory of Light and Matter* (Princeton UP, 1985).

Over the years, a kind of mystic has developed about the peculiar value $a \approx 137$, as illustrated in the quotation below:

> The fine-structure constant holds a special place among cult numbers: [it] seduces otherwise sedate people into seeking mystical truths and developing uncollaborated theories .... The cult of 137 began with scientists who already had quite a reputation, including Pauli, Jung, Heisenberg, and most notably Eddington [...]. In more recent years, the tradition has spread to the larger community of science theory hobbyists ....
>
> Robert Munato, http://www.mrob.com/pub/num/n-b137_035.html.

However, although Eddington's speculations about the integer 137 are disputable, this number has, even more than 2, a number of mathematical properties that make it rather unlikely that its physical occurrences be merely due to chance. In addition to being the 33rd prime number, it is a superadditive prime in base 10 ($1 + 3 + 7 = 11$ and $1 + 1 = 2$). It is also a Chen prime (twinned with 139), a Stern prime (the 4th among 8), an Euler prime, an Eisenstein prime, a Pythagorean prime, a binary quadratic prime, and an optimal primeval prime (the 10th of the 15 primes generated with its 3 digits).

It has been shown [27] that a and 137 are related to the conspicuous numbers π and e (related themselves through the magic relation: $e^{i\pi} = -1$), and to the harmonic sum 137/60 and golden ratio Φ (defined from the equation: $\Phi = \Phi^{-1} + 1$), through the relations (precise within ~ 0.24 ppm, 0.15 ppm, 0.33 ppt and 0.15 ppt, respectively):

$$a^2 \approx 137^2 + \pi^2; \; Log\, a / Log\, 137 \approx a^2 / (a^2 - 1); \; 3\, a \approx 60\, \Phi^4; \quad \Phi^{137/60} \approx 3.$$

Sanchez [27] has also disclosed puzzling occurrences of a (or 137) in the mass distributions of nucleotide bases and amino-acids in terms of the hydrogen mass m_H. For instance, every amino-acid being coded by three couples of nucleotide bases (in pairs A-T and G-C), the average mass $\underline{M}_C$ of all codons appears to be (within $<0.2\%$):

$$\underline{M}_C/m_H \approx 3^3 \, a! \tag{28}$$

The number 137 was known to the Egyptians, and 60 to the Chaldeans [4]. Both occur in the 5th sum of the harmonic series: $\Sigma_n(1/n)$, i.e.: $\Sigma_5 = 137/60$. 137 is also a central polygonal number within the series $C_n = n(n + 1)/2 + 1$. More precisely, one has: $C_2 = 4$, $C_4 = 11$, $C_{16} = 137$, $C_{60} = 1831$, then $4^2 + 11^2 = 137$ and [27]:

$$137^2 + 1831^2 \approx 36 \, \pi^{10} \approx 1836.12^2 \approx \left(m_p/m_e\right)^2 (1836.15^2). \tag{29}$$

The number 137 also relates to the Catalan series built on the Mersenne numbers: $M_n = 2^n - 1$. The sequence of its terms is: $M_2 = 3$, $M_3 = 7$, $M_7 = 127$, and (the ending term) $M_{127} \approx 1.701411835 \times 10^{38}$ [28]. Now M'$_7 \equiv M_2 + M_3 + M_7 = 137 \approx a$ (Eq. 26), and M'_{127} yields (within <0.6%) the Hubble radius R_U of the Universe in terms of the Compton radius r_C of the electron [27]:

$$M'_{127}(4r_C) \approx 1.31403 \times 10^{26} \text{m} \approx 13.89 \, \text{Gly} \approx R_U(13.81 \, \text{Gly}). \tag{30}$$

Thus, the special integer 137 (and hence the electric constant a) is related to quantities that play a major role in microphysics, biophysics, or astrophysics. This has two main consequences: (1) There might be a deep connection between the various levels of complexity, as had been surmised by Dirac, Schrödinger, Eddington, and others. (2) Following Pythagoras' conjecture that 'everything proceeds from numbers', constants occurring in these fields might be determined not by some cosmic 'natural selection' or 'anthropic principle' [29], but by numerical properties of specific numbers.

We have looked empirically for an expansion of the relative shift of a from 137 similar to that of g_e from 2 given by Eq. (24). The following expansion:

$$\varepsilon_a \equiv (a - 137)/137 \approx 0.0002627671533 \approx (1/2)(\pi/137)^2 - (9/16)(\pi/137)^4 + \cdots, \tag{31}$$

which holds within 0.4 ppb, is surprisingly simpler and more accurate than Eq. (24). A theoretical explanation is in progress.

5   The *Gravitational Invariant*

One of the major problems in modern physics is the relation between electromagnetism, governed by quantum electrodynamics [7], and gravitation / inertia, governed by general relativity [30]. In this paper, we shall recall two unconventional attempts to shed some light on this relation.

The Dirac equation for the electron [10], which was derived in the frame of special relativity theory, introduced four-component spinors expressing quantum properties. On the other hand, Einstein's general relativity theory [30] expressed space-time curvature, entailed by local equivalence of acceleration and gravity, in the language of Riemann tensors. Among others, Chapman and Leiter [31] made the Dirac equation conform to general relativity by expanding the principle of covariance to spinor transformations.

For Mendel Sachs [32], the formal structure of quantum theory, including the Pauli exclusion principle, can be derived from a low-energy, linear approximation of a generally covariant, nonlinear field theory of inertia based on the basic ideas of general relativity. On the other hand, John Macken [33] starts with the high-energy, nonlinear vacuum fluctuations of quantized space-time and then derives the basic ideas of general relativity, including the equivalence of acceleration and gravity, as a low-energy, smooth limit.

Expressing the distance r between two identical particles (e, m_0) as a multiple N of their Compton diameter (reduced Compton wavelength): $2r_C = \lambdabar \equiv \hbar/m_0 c$, and scaling the electrostatic force: $F_e = k_e \cdot e^2/r^2$, and the gravitational force: $F_g = G \cdot m_0^2/r^2$, to Planck units [34]: $F_P = c^4/G, E_P = (\hbar c^5/G)^{1/2}$, Macken [33] has managed to express the two widely different forces as simply two different powers of the rest mass energy of the two particles, $E_0 = m_0 c^2$:

$$\underline{F}_e = \alpha \underline{E}_0^2/N^2, \quad \underline{F}_g = \underline{E}_0^4/N^2, \tag{32}$$

where $\underline{F}_e \equiv F_e/F_P, \underline{F}_g \equiv F_g/F_P$, and $\underline{E}_0 \equiv E_0/E_P$ are dimensionless quantities. The fact that in Eq. (32) the larger $\underline{F}_e$ appears as the square of the reduced rest mass $\underline{E}_0$ while the smaller $\underline{F}_g$ appears as its fourth power results from the fact that in $\underline{E}_0$ the rest mass energy E_0 is much smallet than the Planck energy E_P.

For $N = 1$, the two particles being contiguous: $r = 2r_C$, Eq. (32) yields a harmonic relation similar to that between the Bohr radius and the classical radius, Eq. (27):

$$a F_e/F_P = F_g/a F_e = \delta. \tag{33}$$

Here, the Compton diameter $2r_C$ is replaced by the electric (quantum) force $a F_e$, the larger Bohr radius a_0 by the Planck force F_P, and the smaller classical radius r_0 by the gravific (relativistic) force F_g. According to Eq. (33), *the gravific force is to the electric force as the electric force is to the Planck force*, the ratio of this relation being a gravitational invariant: $\delta \equiv d^{-1}$, similar to the fine-structure constant: $\alpha \equiv a^{-1}$, Eq. (26):

$$\delta \equiv G m_0^2/\hbar c = \underline{E}_0^2 = (m_0/m_P)^2 \approx 1/5.7087 \times 10^{44}, \tag{34}$$

where m_0 is the electron rest mass and m_P is the Planck limit mass: $m_P = (\hbar c/G)^{1/2}$. Equation (33) is an indication that *the two forces are deeply related through the Compton diameter* and then, *to the particle spin*, this diameter defining the amplitude of *Zitterbewegung*, responsible for the spin properties (§ 2).

In previous papers [1, 2], we showed that $2r_C$ is also the geometric average of the space-time curvatures, defined from general relativity [30], 'inside' the electron: $2r_G = 2(G/c^2)\, m_0$, and 'outside' a volume of radius r_Q: $2R_G = 2r_Q^2/r_G$. For $r_Q = r_C$, one obtains a harmonic relation similar to Eqs. (27) and (33):

$$2r_C/2R_G = 2r_G/2r_C = 2\delta, \tag{35}$$

δ being defined in Eq. (34). Auxiliary relations resulting from Eqs. (27) and (35) can be written:

$$r_G/r_0 = \delta/\alpha, \quad r_G/2r_C = \delta, \quad r_G/a_0 = \delta\,\alpha. \tag{36}$$

As the fine-structure constant α (or its inverse a) is considered as defining the electric force, the gravitational invariant δ (or its inverse d) can be viewed as defining the gravific force. However, the constant introduced in Eq. (34) involves the rest mass of the electron, a lepton, while for dealing with gravity it may be more appropriate to use that of a baryon. Most authors defined d by using the proton mass [29] or sometimes a cross product of the proton and the neutron (or proton and hydrogen) masses [27].

In earlier papers [3, 4], we showed that the δ_n defined by using the cross product of the proton and neutron masses and the δ_X defined from that of the electron and Universe masses obey (within ~ 0.4 ppt) the simple relation: $\delta_n \times \delta_X \approx 1$. It may be interesting to compare the proton and neutron gravitational invariants (accurate to the 4th decimal only due to the poor accuracy of the measurements of G). For the proton, we obtain:

$$\delta_p \equiv G\,m_p^2/\hbar\,c \equiv 1/d_p \approx 1/1.69328 \times 10^{38}. \tag{37}$$

From Eqs. (26, 37), canonical forms of the electric and gravific forces can be drawn:

$$F_e = \hbar c/ar^2, \quad F_g = \hbar\,c/d_p r^2. \tag{38}$$

The value of d_p in Eq. (37) is very close to the Catalan sum M'_{127} occurring in Eq. (30) [28]. We have looked empirically for a series expansion of the relative shift of d_p from M'_{127} similar to that of a from $M'_7 = 137$ or of g_e from 2, given in Eqs. (31) and (24), respectively. The following expansions:

$$
\begin{aligned}
\varepsilon_p &\equiv (M'_{127} - d_p)/M'_{127} \approx 0.00478021 \\
&\approx (1/3)(2/137) - (2/5)(2/137)^2 + \cdots \approx (g_e/a\pi) + 6(g_e/a\pi)^2 - \cdots,
\end{aligned}
\tag{39a}
$$

which hold within ~ 0.7 ppm and ~ 0.6 ppm respectively, are surprisingly simpler and more accurate than one would expect from the poor accuracy of the measured value of G used in Eq. (37). It can be noticed that $2/137 \approx g_e/a$ within $<0.1\%$ and that $g_e/a \approx 6^{-3}$ (6 being the first 'perfect' number) within $<0.5\%$.

If now we use the neutron (instead of the proton) mass m_n in Eq. (37), we obtain: $d_n \approx 1.68862 \times 10^{38}$, and a resulting $\varepsilon_n \approx 0.00751719$ which could be expanded as:

$$\varepsilon_n \approx (1/2)(2/137) + (2/2)(2/137)^2 + (3/2)(2/137)^3 + \cdots, \qquad (39b)$$

accurate to ~ 5 ppm with the first 2 terms (~ 2 ppm if $2/137$ is replaced by g_e/a) and to ~ 0.1 ppm with the 3 terms. This expansion looks more consistent than that of Eq. (39a).

6   Other Interactions and Force Constants

Nothing can be thought that doesn't exist or coudn't exist.

Parmenides, *Peri Physeos* (5th Century BC).

The gravitational and electromagnetic forces, which are the oldest known (though rationalized only in modern times), thus share three features in common: (1) they are long-distance and decrease as $1/r^2$ (Hooke-Newton and Coulomb laws); (2) in Planck units, their magnitudes can be expressed as even powers of the rest-mass energy (Eq. 32); (3) the parameters defining their strengths can be expressed as Catalan integers of the Mersenne series: $2^n - 1$, with small shifts that can be expanded, on the model of g_e, as simple series involving a (or 137) and π (Eqs. 24, 31, 39). However, these forces are governed by very different theories [7, 30], although various conciliation schemes have been attempted [e.g., 31–33].

6.1   The Casimir Force

Another force, of pure QED origin, is the Casimir force, which induces attraction between two conducting plates separated by vacuum. It was proposed by Casimir in 1948 as a relativistically retarded van der Waals force. It has then been the subject of various theoretical and experimental studies [35, 36], the first accurate measurements being made not earlier than 1997. In the ideal case of perfect plates in perfect vacuum at 0 °K, the force per unit area can be written as:

$$F_C/A = \kappa\, \hbar\, c/r^4, \kappa = \pi^2/240. \qquad (40)$$

Contrary to those above, this force is very short ranged (<1 μm). However, three interesting analogies can be drawn. (1) Similarly as empty space curvature (induced by massive objects) is responsible for the gravity force [30], vacuum radiation confinement (induced by conducting objects) is responsible for the Casimir force.

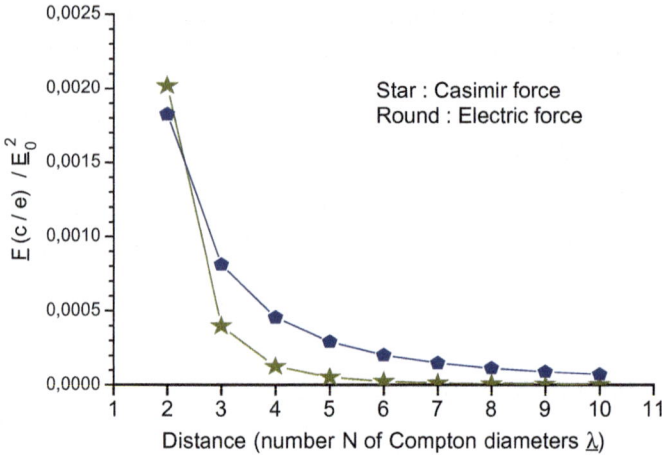

Fig. 1 Reduced Casimir force: $\underline{F}_C/E_0^2$ (Eq. 41 with $A \sim \pi\lambda^2/4$), versus reduced electric force: $\underline{F}_e/E_0^2$ (Eq. 32), as functions of the number N of Compton diameters between two electrons. The curves cross after $N = 2$. The value $N = 7$ corresponds to $\sim 5\%$ of a Bohr radius (Eq. 27)

(2) Similarly as an effective inertia is endowed to light by confinement in an optical resonator [33] or in a photonic crystal [37], an effective force is exerted over two close objects by ZPF confinement between them. (3) If one uses the same scalings as those that led to Eq. (32), the distance between the plates being expressed as an integer number of Compton diameters: $r = N\lambda$, one obtains:

$$\underline{F}_C/A = \kappa\,\underline{E}_0^2/N^4\lambda^2, \kappa \cong 0.0411234. \tag{41}$$

This means that the reduced Casimir force $\underline{F}_C$ exerted on a Compton-size square (the size of an electron): $A = \lambda^2$, has the same expression as $\underline{F}_e$ but with the factor α replaced by κ/N^2. This factor can then be seen as a *dimensionless parameter characterizing the Casimir force*. At a distance equal to a Compton diameter: $N = 1$, the Casimir force is ~ 6 times larger than the electric force. But it decreases much faster with increasing N (Fig. 1). It may be worth investigating the role that a *Casimir attraction 'inside' the electron* may play in its stability, as the strong force in that of nuclei.

6.2   The Intranuclear Forces

These forces, which are responsible for nuclear stability, are also very short ranged. The strong nuclear force overcomes the electrostatic repulsion between protons, while the weak nuclear force allows neutrinos to bind to nucleons [38].

A number of model potentials have been proposed for the strong force (Yukawa 1934, Woods-Saxon 1954, Reid 1968, ...), most of which are central scalar potentials involving some exponential decrease. Nowadays, in the frame of the standard model (where protons and neutrons are seen as made up of tight-bonded quarks), the strong force appears as a residual, dispersion-like force. Its mean magnitude, which depends on various factors, is known with poor accuracy [38], and its dimensionless coupling constant: $\alpha_s \approx 0.1185$ [39], is ~ 16 times larger than the fine-structure constant α_e. Although this force does not follow a $1/r^2$ decrease, its magnitude with respect to the electric force is consistent with the first sum M'_3 of the combinatorial hierarchy [28]: $M'_7/M'_3 = 137/10 \sim 14$ [40].

The weak nuclear force was introduced in 1933 by Fermi to explain β decay [38]. It has a number of specific features: (1) contrary to the other interactions, it does not create bound states, but it allows transformation of a neutron to a proton or vice versa by changing a quark flavor, this inducing β decay or electron capture; (2) it is the only interaction that can violate parity symmetry; (3) contrary to the other interactions, mediated by massless bosons, it is mediated by three very heavy, short-lived bosons, two charged and one neutral. Besides, it is expected to decrease with distance even faster than the strong nuclear force. However, in spite of these crucial differences with the electromagnetic force, it has been unified with it in an *electroweak theory*.

The weak force coupling constant is usually expressed in terms of a Fermi mass: $m_F \approx 5.730073 \times 10^5 \, m_e \approx 5.219743 \times 10^{-25}$ kg, i.e.: $G_F \equiv \hbar^3/c \, m_F^2$, which is in J.m^3 while $\hbar$ is in J.s [27]. One often uses the reduced constant: $G_F/(\hbar c)^3 = 1/m_F^2 c^4 \approx 0.454380 \times 10^{15} \, \text{J}^{-2}$ [39]. But this is not dimensionless, as were α and δ. One could also use the more familiar form: $G_F/\hbar c = (\hbar/m_F c)^2 \approx 0.454164 \times 10^{-36} \, \text{m}^2$, which is a Fermi-mass Compton-like area.

In order to get a dimensionless constant, some authors [29] have used the ratio α_w of this area to that for the electron mass (or its inverse a_w). However, since the weak force acts at the nucleon level, a more relevant choice would be to use the ratio β_w (or its inverse b_w) involving the neutron mass:

$$\alpha_w \equiv (m_e/m_F)^2 \approx 3.045648 \times 10^{-12}; \beta_w \equiv (m_n/m_F)^2 \approx 1.029660 \times 10^{-5}. \quad (42)$$

The second ratio is ~ 700 times smaller than the electric force constant α and 11,000 times smaller than the strong force constant α_s, which is conform to expected values.

Sanchez [27] has proposed the following relation (precise within 0.6 ppt) between the gravitational, nucleoweak and electromagnetic force constants:

$$d_e^5(\approx 6.063 \times 10^{223}) \approx a_w^7 u_e^{67}(\approx 6.056 \times 10^{223}), \quad (43)$$

where the inverses of d_e, a_w and a_e are defined in Eqs. (34), (42) and (26), respectively. It can be noted that the exponents in Eq. (43) are related to the consecutive Catalan numbers 3, 7 and 127 by: $5 = (3 + 7)/2$; $67 = (7 + 127)/2$.

6.3 Hypothetical Infraelectric and Supragravific Forces

The question has often been raised as to whether the Newton and Coulomb forces may not behave differently at very large or very short distances, or whether there may not be other, undisclosed forces. Cosmological observations have led to the conclusions that there is a dark energy and a dark matter constituting, respectively, 75% and 21% of the observable Universe. Could there also be forces still larger than the electric and nuclear forces or smaller than the gravific force?

Looking back at the relation between F_e and F_g expressed by Eqs. (32, 33), one may wonder whether the electric and gravific forces may not be just two of a series of r^{-2}-dependent forces acting at various levels, but with powers of $\underline{E}_0$ other than 2 or 4. By extrapolation, one would then write:

$$\underline{F}_{total} = \underline{F}_a + \underline{\boldsymbol{F}_e} + \underline{\boldsymbol{F}_g} + \underline{F}_j + \cdots = \alpha^2 + \alpha \underline{E}_0^2 + \underline{E}_0^4 + \alpha^{-1}\underline{E}_0^6 + \cdots$$
$$= \alpha^2 + \alpha \delta_p + \delta_p^2 + a^{-1}\delta_p^3 + \cdots, \tag{44}$$

where the electric and gravific force terms are boldface. The divider N^2, which is the same for all r^{-2}-dependent forces, has been omitted. Note that neither the Casimir nor the nuclear forces, which are residual forces and decrease much faster than r^{-2}, fall in this frame. Each term in this series is derived from the previous one by multiplication by $\alpha^{-1}\underline{E}_0^2$. These extra forces could then be expressed in terms of the electric and gravific forces and, as a result of Eqs. (27) and (35), as ratios of particle radii:

$$\underline{F}_a = \underline{F}_e^2/F_g, \, F_j = F_g^2/\underline{F}_e. \tag{45}$$

Here it should be recalled that, following a stochastic electrodynamics approach of *Zitterbewegung*, Haisch et al. [16] interpreted inertia as a resistance of ZPF to spectral distorsion in an accelerated frame, and identified the *Newtonian force* with the *van der Waals force* generated by this motion. The gravific force then appeared as a kind of residue from the electromagnetic interaction, and the inertial and gravific masses thus derived were equivalent. Similarly, according to Macken [33], the highest-order term in Eq. (44) would amount to some relativistic residue from the gravific force.

If extra forces do exist, they should have realistic strengths. Below are listed the respective strengths (in Newtons) derived from Eq. (44) and from the definitions of the reduced forces in terms of the Planck force: $F_P \approx 1.21035 \times 10^{45}$ N.

$$F_a^{-1} \approx 1.8779 \times 10^4 \rightarrow F_a \approx 6.4450 \times 10^{39} \text{ N};$$
$$\overline{F_e}^{-1} \approx 2.3204 \times 10^{40} \rightarrow F_e \approx 5.2159 \times 10^3 \text{ N};$$
$$\overline{F_g}^{-1} \approx 2.8672 \times 10^{76} \rightarrow F_g \approx 4.2212 \times 10^{-33} \text{ N};$$
$$F_j^{-1} \approx 3.5428 \times 10^{112} \rightarrow F_j \approx 3.4162 \times 10^{-69} \text{ N}.$$

The first force is 30 orders of magnitude larger than the strong nuclear force but, as it is still 6 orders of magnitude smaller than the Planck force, it is not excluded that it exists at a subnuclear level. The last force is 36 orders of magnitude smaller than the gravity force and it appears much too small to show up in observable effects, except maybe in the vicinity of neutron stars or black holes.

One may now turn back to the combinatorial hierarchy connection between a and d_p involved in Eqs. (31) and (39). Writing $\underline{F}_x^{-1} \approx 2^n - 1$ (with 1 negligible for n large) yields: $n_a \approx 14.18 \approx 2 \times 7^+$ (the increment to 7 being due to $137 > 127$); $n_e \approx 134.09 \approx 127 + 7^+$; $n_g \approx 253.99 \approx 127 \times 2$, and $n_j \approx 373.88 \approx 127 \times 3$ (from d_p^3) $- 7^+$ (from a^{-1}). However, none of the compound exponents 14 and 374 yields peculiar numbers comparable to those associated with a and d_p.

7 Conclusions

In this paper, we first recalled the origins, features and main outcomes of the Dirac equation: an *underlying antimatter* coupled with ordinary matter [9, 10], this yielding a *wave beat* between the positive and negative energy states [15], resulting in an *internal motion at light speed within a Compton diameter* [14]. Hence, the spin *kinetic momentum* with quantum number $s = \frac{1}{2}$ and *magnetic moment* with gyromagnetic factor $g_0 = 2$ [9, 10]. Various investigators had featured that the rest mass would be related to the spin motion [16, 17], and we conjectured that the electron can actually be seen as a *massless charge spinning at light speed*, the observed rest mass stemming mainly from this very internal motion [1–4].

In this framework the *Compton diameter* $\lambdabar$ [12] plays a special role: as the range of the internal motion, but also as the geometric average of the electron classical radius r_0 and the hydrogen Bohr radius a_0, the ratio of this harmonic relation being the *fine-structure constant* $\alpha - k_e m_0^2 / \hbar c$: $\lambdabar / a_0 = r_0 / \lambdabar = \alpha$. It is also the geometric average of the gravitational curvature diameters 'inside' and 'outside' the electron, $2r_G$ and $2R_G$, the ratio of this harmonic relation being the *gravitational invariant* $2\delta_e = 2 Gm_0^2 / \hbar c$: $\lambdabar / 2R_G = 2r_G / \lambdabar = 2\delta_e$.

Expressing the electric and gravific forces in Planck units, and the distance between interacting particles as an integer number of Compton diameters, Macken has shown [33] that they take the form of even powers of the rest mass energy. This entails [2] a new harmonic relation, involving both α and δ, between the Planck, electric and gravific forces: $\alpha^{-1} F_e / F_P = F_g / \alpha^{-1} F_e = \delta_e$. In this paper, we show that a similar relation holds with the Casimir force but with α^{-1} replaced by a constant rapidly decreasing with distance. Other forces that would be expressed as similar products of α and δ_p appear too large or too small to be observable.

Another link between the electric and gravific forces [27] results from the integer parts of a and d_p being the sums of Catalan numbers in the Mersenne series: $2^2 - 1 = 3$, $2^3 - 1 = 7$, $2^7 - 1 = 127$, then $M'_7 = 3 + 7 + 127 = 137 \approx a$, and the sum M'_{127} up to $2^{127} - 1 \approx d_p$ where $d_p \equiv \delta_p^{-1}$ and δ_p is defined as δ_e but with the electron mass replaced by that of the proton (or the neutron). The magnitude of the strong nuclear force, although it is a short distance one, seems consistent with the sum $M'_3 = 10$.

In this paper, we have discussed the deviations, from their integer approximates, of the measured values of the free electron gyromagnetic factor g_e, the fine-structure constant inverse a, and the proton gravitational invariant inverse d_p. After recalling that the relative deviation of the measured g_e from 2 ($\sim 0.1\%$) was explained by quantum field and size effects, we have tried to express that of a from 137 and that of d_p from M'_{127} as similar expansions.

Surprisingly, the relative deviation of measured a from 137 ($< 0.03\%$) could even better be expressed (with 0.4 ppb accuracy) as a simple 2-term series involving even powers of $\pi/137$. Similarly, the relative deviation of measured d_p from M'_{127} ($< 0.50\%$) could be expressed (with 0.6 ppm accuracy) as 2-term series involving powers of either $2/137$ ($\approx g_e/a$) or $g_e/a\pi$ ($\approx 1/6^3$). For reasons yet unexplained, the numbers g_e, a, d_p and π seem to be deeply related.

Acknowledgements I wish to thank the colleagues who helped me clarify these ideas at QSCP meetings and elsewhere. Erkki Brändas, Uzi Kaldor, John Macken, Francis Sanchez, and Ivan Todorov made especially useful comments. Thanks are due to my wife, Marja Rantanen, for stimulating my speculations with inspiring piano playing.

References

1. Maruani J (2012) The Dirac electron: spin, Zitterbewegung, the Compton wavelength, and the kinetic foundation of rest mass. Prog Theor Chem Phys B 26:23–46
2. Maruani J (2013) The Dirac electron as a massless charge spinning at light speed: implications on some basic physical concepts. Prog Theor Chem Phys B 27:53–74
3. Maruani J (2015) The Dirac electron as a privileged road to the understanding of quantum matter. Quantum Matter 4:3–11
4. Maruani J (2016) The Dirac electron: from quantum chemistry to holistic cosmology. J Chin Chem Soc 63:33–48 and references therein
5. Thaller B (1992) The Dirac equation. Springer, Berlin
6. Sakurai J (1967) Advanced quantum mechanics. Addison-Wesley, Reading, MA, ch. 3
7. Feynman RP (1998) Quantum electrodynamics. Addison-Wesley, Reading, MA
8. Weinberg S (1995) The quantum theory of fields. Cambridge U P
9. Dirac PAM (1928) Quantum theory of the electron. Proc Roy Soc (London) A 117:610–624; Quantised singularities in the electromagnetic field: ibid (1931) A 133:60–72; Theory of electrons and positrons. Nobel lectures (1933) pp 320–325
10. Dirac PAM The principles of quantum mechanics. Clarendon Press, Oxford, 1st edn 1930, 4th edn 1958, chs 11–12

11. de Broglie L (1924) Recherches sur la Théorie des Quanta. Thesis, Sorbonne, Paris; Ann Phys 10(III):22–128
12. Compton AH (1923) A quantum theory of the scattering of x-rays by light elements. Phys Rev 21:483–502
13. Brandsden BH, Joachain CJ (1983) Physics of atoms and molecules. Longman, Harlow, England, 1st edn, chs 1 and 5
14. Schrödinger E (1930) Über die kräftefreie Bewegung in der relativistischen Quantenmechanik: Sitzungsber. Preuss Akad Wiss Berlin, Phys-Math Kl 24:418–428; Zur Quantendynamik des Elecktrons: ibid (1931) 25:63–72
15. de Broglie L (1934) l'Electron Magnétique: Théorie de Dirac, Hermann, Paris, chs 9–22
16. Haisch B, Rueda A, Puthoff HE (1994) Inertia as a zero-point field Lorentz force. Phys Rev A 49:678–694 and references therein
17. Barut AO, Bracken AJ (1981) Zitterbewegung and the internal geometry of the electron. Phys Rev D 23:2454–2463; D 24:3333–3342; Barut AO, Zanghi N (1984) Classical model of the Dirac electron. Phys Rev Lett 52:2009–2012; Barut AO, Pavšič M (1987) Quantization of the Zitterbewegung in the Schrödinger picture. Class Quant Grav 4:L131–L136
18. Todorov I (2015) Hyperlogarithms and periods in Feynman amplitudes. Opening lecture at QSCP XX, Varna; published as CERN-TH-2016–042; and private communication
19. Dehmelt H (1993) Is the electron a composite particle? an experiment. Hyperfine Interact 81:1–3; updated and condensed version of the Nobel lecture in Les Prix Nobel 1989, Stockholm, 1990, p 95
20. Maruani J (1980) Magnetic resonance and related techniques. In: Becker P (ed) Electron and magnetization densities in molecules and crystals, NATO ASI series. Plenum, and references therein
21. Sommerfeld A (1919) Atombau und Spektrallinien. Friedrich Vieweg, Braunschweig; translated as Atomic structure and spectral lines. Methuen, London, 1923; and references therein
22. Eddington AS (1935) New pathways in science, Cambridge U P, ch. 11
23. Kaldor U, Eliav E, Landau A (2004) Relativistic electronic structure theory. Elsevier, ch. 2, and private communication
24. Visscher L, Dyall KG (1997) At Data Nucl Data Tables 67:207
25. Kragh H (2003) Magic number—a partial history of the fine-structure constant. Arch Hist Exact Sci 57:395–431
26. Gilson JG (1996) Calculating the fine-structure constant. Spec Sci Tech, 23 pp
27. Sanchez FM, Kotov VA, Bizouard C (2009) Evidence for a steady-state, holographic, tachyonic, and supersymmetric cosmology. GED 20(3):43–53; Sanchez FM (2017) A coherent resonant cosmology approach and its implications in microphysics and biophysics. Prog Theor Chem Phys B 30:375–407, references therein, and private communication
28. Bastin T, Kilmister CW (1995) Combinatorial physics. World Scientific, Singapore
29. Carr BJ, Rees MJ (1979) The anthropic principle and the structure of the physical world. Nature 278:605–612. See also: Barrow JD, Tipler FK (1986) The anthropic cosmological principle. Oxford U P, and references therein
30. Einstein A (1916) Die Grundlage der allgemeinen Relativitätstheorie. Ann Phys (Leipzig) 49:769–823. See also: Eddington AS (1920) Space, time, and gravitation. Cambridge U P; French edition completed with an updated mathematical presentation of general relativity. Hermann, Paris, 1921. See also: Special issue commemorating Einstein A (2005). Ann Phys (Leipzig) 14:1–204
31. Chapman TC, Leiter DJ (1976) On the generally covariant Dirac equation. Am J Phys 44:858–862
32. Sachs M (1986) Quantum mechanics from general relativity. Reidel, Dordrecht
33. Macken JA (2013) The universe is only spacetime. http://www.onlyspacetime/home; see also: Spacetime-based foundation of quantum mechanics and general relativity. Prog Theor Chem Phys A (2015) 29:219–245; and private communication
34. Pavšič M (2001) The landscape of theoretical physics: a global view. Kluwer, pp 347–352. See also: http://en.wikipedia.org/wiki/planck_units

35. Milton KA (2001) The Casimir effect: physical manifestations of zero-point energy. World Scientific, Singapore
36. Genet C, Intravaia F, Lambrecht A, Reynaud S (2004) Electromagnetic vacuum fluctuations, Casimir and van der Waals forces. Annales Fondation Louis de Broglie 29:311–328
37. André J-M, Jonnard Ph (2011) Effective mass of photons in a one-dimensional photonic crystal. Phys Scr 84:035708 and references therein
38. Brown LM, Rechenberg H (1996) The origin of the concept of nuclear forces. Institute of Physics Publishing, Bristol and Philadelphia
39. Beringer J et al (2012) Particle data group. Phys Rev D 86, 010001; for updates, see: http://pdg.lbl.gov
40. It seems that Catalan numbers were known in the Antiquity. Not only the conspicuous 3 and 7 and their sum 10 (and various products and multiples), which occur in many traditions, but also 127 (the number of provinces in Ahasuerus' empire, cf *Esther* **1**, 1) and the sum 137 (the number of years Abraham's elder son lived, cf *Genesis* **25**, 17). The hypostyle room of Ammon's temple in Karnak, Egypt, has a total of 136 columns, the number that Eddington initially proposed for the electric constant a [22]

A Simple Communication Hypothesis: The Process of Evolution Reconsidered

Erkki J. Brändas

Consciousness is a transparent brain representation of the world from a privileged egocentric perspective.

Arnold Trehub

Abstract The scientific basis of Darwinian evolution is reconsidered from the recent progress in chemistry and physics. The idea, promoting a stochastic communication hypothesis, reflects Kant's famed insight that 'space and time are the two essential forms of human sensibility', translated to modern practices of quantum science. The formulation is commensurate with pioneering quantum mechanics, yet extended to take account of dissipative dynamics of open systems incorporating some fundamental features of special and general relativity. In particular we apply the idea to a class of Correlated Dissipative Structures, CDS, in biology, construed to sanction fundamental processes in biological systems at finite temperatures, ordering precise space-time scales of free energy configurations subject to the Correlated Dissipative Ensemble, CDE. The modern scientific approach is appraised and extended incorporating both the material- as well as the immaterial parts of the Universe with significant inferences regarding processes governed by an evolved program. The latter suggests a new understanding of the controversy of molecular versus evolutionary biology. It is demonstrated by numerous examples that such an all-inclusive description of Nature, including the law of self-reference, widens the notion of evolution from the micro to the cosmic rank of our Universe.

Keywords Density matrix · Space-time · Stochastic communication
Evolution · Correlated dissipative structure · CDS · Correlated dissipative
ensemble · CDE

Dedicated to Arnold Trehub for his great visions and intuitions displayed in his Retinoid Model of vision.

E. J. Brändas (✉)
Department of Chemistry, Uppsala University, Uppsala, Sweden
e-mail: erkki.brandas@gmail.com

381

1 Introductory Remarks

A simple Communication Hypothesis is developed, explained and illustrated. It is based on a quantum mechanical formulation of ensemble representable density matrices qualifying the emergence of quantum- and thermal correlations, engrained in long-range off-diagonal order, with precise thermal conditions satisfied at specific temperatures. In analogy with condensed matter phenomena, like e.g. superconductivity, connectors occur between mechanical- and electromagnetic fluctuations. The subsequent ensemble, the Correlated Dissipative Ensemble, CDE, subject to Poissonian statistics, provides stochastic communication channels for cellular recognition including a proliferated knowledge of neuron correlates. Since this document is entrenched in a *modus operandi* built on the scientific evaluation and the conclusion of the process of Darwinian evolution, some general comments are enclosed below as an introduction.

It is quite remarkable that exacting correlations between modern science and present philosophical enquiry are still largely wanting due, not only to the presumed lack of interaction between the disciplines, but also owing to the disordered state of commensurate agreements between the various scientific subjects. A satisfactory elucidation of the situation should indeed be accomplished by clearing up these issues from the appropriate knowledge of the state-of-the-art of physics, chemistry and biology. While important suggestions to this challenge have been propounded by e.g. Tegmark [1], Deutsch [2], Primas [3] and others, the conundrum still remains. The general goal concerns the ultimate nature of reality; say our mathematical universe [1], the consequences of the emergence of knowledge [2] or the concrete need for a conceptual recasting of the logical and philosophical foundations of physics as a result of the high degree of incommensurate specializations that persist in isolation ignoring most other areas of pursuits [3].

In particular Primas' critique, well-formulated and certainly thought provoking, gives a personal characteristic that is rooted in his long career as chemistry professor at ETH, starting with problem solving in physical- and theoretical chemistry and ending up with issues in the cognitive sciences and the philosophy of mind [4]. He came to view the dual epistemic aspects of the mental and the material world to be conceived as one underlying ontic[1] reality, named dual aspect monism. In reference [3] there is an interesting foreword by the philosopher Paul Feyerabend, where he places Hans Primas as a representative of the structural approach with 'a comprehensive picture of the world and man's place in it'.

The notorious anarchist is widely known for his declaration 'anything goes' and his book 'Against Method' [6], where he proposed the thesis: 'the events, procedures and results that constitute the sciences have no common structure'. Although Feyerabend

---

[1]The ontic conceptualization is part of Heidegger's neologisms [5]. His Being (Dasein) has a pre-ontological-ontic signification, however, not to be understood as a biological human being. This distinction is probably the reason why Heidegger takes a poetic turn as he investigates the *Question Concerning Technology* and finds it to be but 'a means to an end'.

contributed his visions with a tongue in cheek, it should not give him absolute immunity and exemption from liability. His attack on the claims of a philosophical legislation for the exercise of science is stimulating and inspirational, yet it ignores the foundation of abstracting science as an art of knowledge. To avoid the most bizarre forms of philosophical relativism the scientific practice should focus on scientific disciplines and their evolution rather than narrowing the perspective to a comparison and review of various individual methodological rules and standards.

In the development of a scientific field, the historian of science, Mary Jo Nye [7] identifies essentially six characteristics of a scientific discipline, i.e. genealogy, core literature, practices, physical location, recognition and shared values. Within this perspective it is clear that disciplines, as attractors for accumulated knowledge, communicated between its members, focused on a particular sphere of Mother Nature, do evolve with respect to styles and traditions compatible with existing scientific paradigms and their inevitable changes. Hence scientific disciplines, as examples of evolutionary lineages, imbedded in the human genetic code, do arise, develop, evolve, die and diversify. For a detailed example of this process, see e.g. the historical account of the emergence and evolution of quantum chemistry as a sub-discipline and its importance in the province of science [8].

Recently the author advocated an evolutionary approach to represent and interpret the origin and emergence of life, established by modern advances in chemical physics, from unstable chemical states to biological evolution and order [9]. The Zero Energy Universe Scenario (ZEUS) was elaborated commensurable with Darwin's theory of evolution, from the microscopic ranks to the cosmological domain.

In this contribution we will present an authentic stochastic communication hypothesis, building on the results portrayed in [9], i.e. combining the novel formulation of quantum mechanics for open systems, relativity theories, Gödel's theorems, Off-Diagonal Long-Range Order (ODLRO), the Correlated Dissipative Ensemble, CDE, the Poisson distribution and associate encoding- decoding protocols for communication between complex enough systems defining generic life forms. Some examples of the various levels of organization, that abides by our Communication

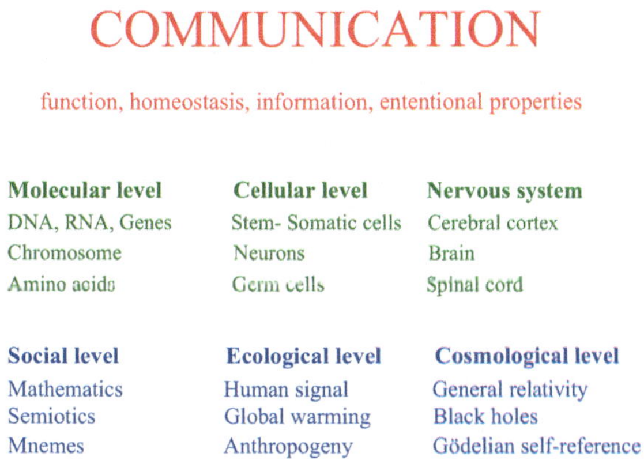

Fig. 1 Communication levels from the micro-to the cosmic rank

COMMUNICATION

function, homeostasis, information, ententional properties

Molecular level	**Cellular level**	**Nervous system**
DNA, RNA, Genes	Stem- Somatic cells	Cerebral cortex
Chromosome	Neurons	Brain
Amino acids	Germ cells	Spinal cord

Social level	**Ecological level**	**Cosmological level**
Mathematics	Human signal	General relativity
Semiotics	Global warming	Black holes
Mnemes	Anthropogeny	Gödelian self-reference

Hypothesis, will be discussed in the final section, see also Fig. 1. We will also address the longstanding disagreements between molecular- and evolutionary biologists [10] on how natural selection acts, while supporting a trans-level semiotics that is appropriate for geno-phenotypic translations including the role of somatic programs in a Darwinian perspective. We will also compare and discuss our findings with other results and strategies made recently in connection with the self-referential paradox associated with the concept of consciousness. To make the presentation self-contained, yet without bequeathing too many excruciating details, some brief reviews are incorporated for the convenience of the prospective reader.

2   The Ontological Framework

Today Darwin's theory of evolution, in spite of initially being a badly split field, has reached a certain consensus and acquired a synthetic unification known as the evo-lutionary synthesis [10]. Nevertheless it is still classed as a sort of unfinished business as the materialistic Neo-Darwinism[2] has recently been seriously critiqued and appraised [11]. Our answer to the issues and questions rendered above, has been considered conceptually as the Paradigm of Evolution, see e.g. Refs. [9, 11] and references therein. The formulation derives from a general extension of standard mathematical representations of stationary states in separation from the environment to so-called open (dissipative) systems.[3] The logical framework should be general enough to include the axioms of quantum mechanics. Below, we will, as already mentioned, give a short account, but for more details we refer to [9, 11, 13].

Before activating a reading of the specifics needed for a more basic portrait, we must recognize already at the start that there are many controversial undertakings instigated by the various practitioners of science and their views of what should be considered as fundamental concepts in the scientific description of Mother Nature. For instance Barbour [14] rhetorically asks whether time is 'real', while at the same time advocating a timeless theory of the universe. Furthermore Primas, in his Mind and Matter article [15] projects a holistic reality, where time is not taken to be an a priori concept. These examples of deviation from standard practice are well enunciated, yet they run the risk of drawing unwanted consequences that incur more problems than it solves, e.g. the unavoidable emergence of internal times, the imperative dependence on boundary conditions, or paraphrasing Primas: 'the death of natural laws' by being trapped in a Gödelian self-reference.[4] We will return to this question further below.

---

[2]For discussions on the somewhat controversial concept of Neo-Darwinism, see e.g. [10, 11].

[3]The concept of dissipative systems and their irreversible processes has been the focus of Pri-gogine and the Brussels–Austin School, [12].

[4]Note that the Gödel (self-referential) paradox can be translated to a consistently formulated mathematical singularity of a suitably extended logic [16].

Since the dawn of civilization dichotomies like atom—void, realism—idealism, reality—logos, body—soul, nature—spirit, res extensa—res cogitans, sein—zeit, material—immaterial, objective—subjective etc., have been at the centre of attention for philosophers and naturalists alike. However, Kant's scientific worldview [17] engrained in the Enlightenment did set the terms for a metaphysics and philosophy for the coming years with fundamental motifs and patterns surviving until today. He told us, already more than two hundred years ago, that 'space and time are the two essential forms of human sensibility', i.e. everything we, as evolved life forms perceive and experience, concerns phenomena set in this perspective.

Of course it is impossible for us to comprehend what Kant and his contemporaries actually had in mind and meant with such a statement, nevertheless an orthodox starting point for doing science would rightfully study material representations in terms of energy and momentum as imagined and sensed by us in Nature. Obviously energy-momentum constitutes the material part, while space and time comprise the immaterial part of the world. Moreover, as any student of quantum mechanics must learn today, it is imperative to realize that the two aspects referred to are entwined through the non-commutative conjugate relationships of their representations as linear operators.

This distinction is therefore mandatory in modern physics, however, not only in connection with quantum physics, but also in preserving a fundamental interpretation within classical physics, where key correspondences are discerned through the celebrated Fourier Transform. It is perhaps somewhat surprising that this feature extended to incorporate special and general relativity admits generalizations to open system dynamics, see e.g. Ref. [18] and in the following section.

3 A Consistent Formulation of Energy-Momentum and Space-Time

Based on the argument just given above it follows that once space and time are specified, momentum and energy are inevitable consequences of the declaration—and vice versa. For instance, if some object, defined by its energy and momentum, permits a portrayal within the basic deductive axioms of science as practised at present in physics and formulated by the most succinct language we know today, mathematics, we cannot avoid *deductio in domum* of the immaterial degrees of freedom as given by time and space. Likewise, if we are able to differentiate some non-stationary entities in space, we are in principle capable to discuss their energy-mass-momentum relationships. The obvious conclusion is thus: inducing the question whether time (or space) is not fundamental in some general setting invites situations where the conjugate relationship viz. time–energy (or space–momentum) would be relaxed. Although such situations might occur, e.g. in a black hole, it might still be preferable to keep the deductive structure of a physical theory intact, in order to investigate and analyse the emergence of any violations as they might appear. In what follows we will see by a simple realization how such situations arise under the most trivial conditions.

In particular we will analyse what happens when a non-zero rest-mass particle, satisfying the Klein–Gordon (wave) equation, via some physical process (pair annihilation etc.) produces photons. The interest in this question is motivated by what happens to the degrees of freedom, i.e. the loss of the longitudinal degree exhibited by massless particles (photons). The starting point will be a consideration of the abstract Dirac kets in terms of the coordinate vector $\vec{x}$ and the linear momentum $\vec{p}$

$$|\vec{x}, ict\rangle, |\vec{p}, iE/c\rangle$$

with the scalar product for a free particle given by

$$\psi(\vec{x}, t|\vec{p}, E) = (2\pi\hbar)^{-2} e^{\frac{i}{\hbar}(\vec{p}\cdot\vec{x} - Et)}$$

Here the energy is given by the mass relation $E = mc^2$, m the mass, c the velocity of light and t, τ the time variable (operator).

The Klein–Gordon equation for a non-zero rest-mass particle is given in obvious notation by

$$-\frac{E_0^2}{c^2} = \vec{p}^2 - \frac{E_{op}^2}{c^2} \tag{3.1}$$

where $E_0 = m_0 c^2$ and $\vec{p} = m\vec{v}$. Since we are interested in the immaterial degrees of freedom we also write down the analogous relation for the conjugate variables (operators) introducing the familiar eigentime expression τ_0 given by

$$-c^2\tau_0^2 = \vec{x}^2 - c^2\tau^2 \tag{3.2}$$

In order to analyse the intrinsic character of these two equations when the non-zero rest-mass m_0 goes to zero, we use Dirac's trick to "take the square root of the equation" by rewriting the observables in matrix form with Eqs. (3.1) and (3.2) being their associated secular equation.

Hence our concern is the conjugate pair of observables usually represented in operator form as (note that the time operator is trivially defined when the energy interval is $(-\infty, +\infty)$ as is also the time interval)

$$E_{op} = i\hbar\frac{\partial}{\partial t}; \quad \vec{p}_{op} = -i\hbar\nabla_{\vec{x}} \tag{3.3}$$

and

$$\tau = t_{\text{op}} = -i\hbar \frac{\partial}{\partial E}; \quad \vec{x}_{\text{op}} = i\hbar \nabla_{\vec{p}} \tag{3.4}$$

In writing down the ansatz, we recognize two[5] fundamental characteristics of the relativistic formulation. The first is due to Alexander S. Davydov, who asserted in his celebrated text on quantum mechanics: *"we show the inapplicability of the concept of an essentially relativistic motion of a single particle"* [19]. The second does result from PerOlov Löwdin's treatment of general binary products in Chap. 5 of his book Linear Algebra for Quantum Theory [20], where he discusses *the indefinite metric associated with the Minkowski space.* It is easy to see that our open system dynamics based on a complex symmetric matrix representation is commensurate with an indefinite metric [18]. Hence we set forth the ansatz as (note that the insertion of $-i$ in the off-diagonal elements is just a convention in the construction of a complex symmetric matrix)

$$\begin{pmatrix} E_{\text{op}} & -i\vec{p}_{\text{op}}c \\ -i\vec{p}_{\text{op}}c & -E_{\text{op}} \end{pmatrix} \tag{3.5}$$

Hence we recognize Eq. (3.1) as the secular equation of the matrix defined in Eq. (3.5) with the two eigenvalues $\pm\lambda$, given by $\lambda^2 = m_0^2 c^2$, with $m_0 \neq 0$. Since the formulation works for both variables and for operators, we will not explicitly single out which reading is made unless the situation calls for a preference.

Similarly one obtains for the conjugate operators in (3.4) that Eq. (3.2) becomes the secular equation of the (operator) matrix

$$\begin{pmatrix} c\tau & -i\vec{x} \\ -i\vec{x} & -c\tau \end{pmatrix} \tag{3.6}$$

with $\pm c\tau_0$ being the associated eigenvalues exhibiting the two time directions and the left- and right-handed coordinate systems connected by space inversion symmetry. From (3.5) follows directly the usual approximations and estimates that result in the conventional time-dependent Schrödinger equation as restricted to the non-relativistic domain. With this inherent conjugate structure in mind, we remind the readers of the attempts, see e.g. Refs. [14, 15, 21] to derive pioneering quantum mechanics from general algebraic structures asserting that time (or space-time) is not a primary concept. Yet the present description, as consistently built above, is fundamental and commensurate with the deductive nature of the Lorentz

---

[5]In fact a third comment is due, viz. the use of operators in the matrix calling for logical extensions, see more in the next section. Note also the juxtaposition of the "arrow" of time and the parity of space.

transformation, see e.g. Refs. [18, 22], since from Eqs. (3.1) and (3.2) one obtains straightforwardly the relativistic space-time scales of special relativity

$$m = \frac{m_0}{\sqrt{1 - \beta^2}} \tag{3.7}$$

with $\beta = v/c = p/mc$, and

$$\tau = \frac{\tau_0}{\sqrt{1 - \beta^2}}; \quad x = \frac{x_0}{\sqrt{1 - \beta^2}} \tag{3.8}$$

In this setting Eqs. (3.7) and (3.8) are valid irrespective of whether we are representing classical wave propagation, quantum matter waves or classical particles.

It is now possible to investigate the limit $m_0 \to 0$, observing the connection with Maxwell's equation for scalar and vector fields, by placing $m_0 = 0$ in Eq. (3.1). Something extraordinary happens with the character of the matrices of Eqs. (3.5) and (3.6), when inserting the relation $E = pc$, with the momentum $\vec{p}$ assumed to be directed along the x-axis. The result is a complex symmetric degenerate matrix (reduced to the dimension of mass)

$$\begin{pmatrix} p/c & -ip/c \\ -ip/c & -p/c \end{pmatrix} = p/c \begin{pmatrix} 1 & -i \\ -i & -1 \end{pmatrix} \tag{3.5'}$$

which is similar to the classical canonical form given by[6]

$$2p/c \begin{pmatrix} 0 & 1 \\ 0 & 0 \end{pmatrix} \tag{3.5''}$$

In other words (3.5') cannot be diagonalized since the two column vectors in the matrix are linear dependent.

Analogously we obtain for (3.6) (since for the photon $c^2 = \vec{x}^2/\tau^2$)

$$\begin{pmatrix} c\tau & -\vec{x} \\ -\vec{x} & -c\tau \end{pmatrix} = c\tau \begin{pmatrix} 1 & -i \\ -i & -1 \end{pmatrix} \to 2c\tau \begin{pmatrix} 0 & 1 \\ 0 & 0 \end{pmatrix} \tag{3.6'}$$

We find to our surprise that that the complex symmetric matrices in Eqs. (3.5') and (3.6') are nothing but a Jordan block of order 2, (a non-zero matrix, whose square is zero) or, in more technical language, of Segrè characteristic 2. The linear dependency leaves space-time with one less spatial dimension, i.e. along the x-axis. Hence there is no longitudinal degree of freedom for a zero rest-mass particle with

---

[6]The actual transformation is unitary, for details see e.g. Ref. [13].

speed c, like the photon! As will be clear below, this implies that zero- and non-zero rest-mass particles behave fundamentally different, which will become of crucial importance in the case of the theory of general relativity to be reviewed below.

It is straightforward to extend our conjugate operator arrays above, to the general case by the following modifications, see below, where μ is the gravitational radius, G the gravitational constant, $v = p/c$, $r = |\vec{x}|$, M a (usually large) spherically symmetric (non-rotating) mass,[7] independent of m, i.e.

$$\begin{pmatrix} m(1 - \kappa(r)) & -iv \\ -iv & -m(1 - \kappa(r)) \end{pmatrix} \tag{3.9}$$

with[8]

$$m\kappa(r) = \frac{m\mu}{r}; \quad \mu = G \cdot \frac{M}{c^2} \tag{3.10}$$

Note that as before, space-time and energy-momentum spaces associate the two partitions into the material- and the immaterial sections of the Universe. Furthermore, as the area velocity multiplied by m is a constant of motion, one obtains for local circular motion the boundary condition:

$$v = \kappa(r)c \tag{3.11}$$

to be incorporated in Eq. (3.9).

To be precise one must proceed by solving the corresponding secular equations for the constituent partners of the representation of the Universe,[9] as the space-time background must be simultaneously incorporated with the energy-mass dynamics so as to conform to background independence. Considering that Einstein's law of general relativity equates the matter-energy content with the metric of a curved space-time, one encounters an interpretative difference in contrast to the present formulation. While the equations of general relativity asserts that space-time are regarded as the forms of existence of a real world, with Matter as its substance, our material-immaterial partitioning imparts a conjugate structure that does not presuppose one conjugate entity to be consigned to the other. This is an important 'quality' as we will see later.

Consequently one must address the problem of conjugate consistency and the proper division of zero- and non-zero rest-mass particles. Though being not too difficult, we skip the details, see e.g. [9, 11, 16, 18], and quote the result below.

---

[7]Representing M by a 'black hole' shows the consistency between particle—antiparticle symmetry.

[8]The matrix (3.9) has a direct link to Gödel's self-referential paradox see e.g. [9].

[9]Concepts like Nature, World, Universe, Cosmos, etc. are here used interchangeably.

Leaving out the term $r^2 d\Omega^2$ the outcome becomes, not unexpectedly, the so-called Schwarzschild[10] line element $(m_0 \neq 0)$:

$$-c^2 ds^2 = -c^2 d\tau^2 (1 - 2\kappa(r)) + dr^2 (1 - 2\kappa(r))^{-1} \tag{3.12}$$

Rewriting Eq. (3.9) with the condition (3.11) one obtains

$$m \begin{pmatrix} (1 - \kappa(r)) & -i\kappa(r) \\ -i\kappa(r) & -(1 - \kappa(r)) \end{pmatrix} \to m \begin{pmatrix} \sqrt{1 - 2\kappa(r)} & 0 \\ 0 & -\sqrt{1 - 2\kappa(r)} \end{pmatrix} \tag{3.13}$$

trivially diagonalized, provided $\kappa(r) \neq 1/2$. However at $\kappa(r) = 1/2$, which occurs at the Schwartzschild radius $r = 2\mu$, one encounters an old 'friend', i.e. a Jordan block of order two (independent of m)

$$\frac{1}{2} m \begin{pmatrix} 1 & -i \\ -i & -1 \end{pmatrix} \to m \begin{pmatrix} 0 & 1 \\ 0 & 0 \end{pmatrix} \tag{3.14}$$

The singular behaviour occurs at the Schwarzschild radius signifying a boundary, inside which the material particle, with $m_0 \neq 0$, becomes immaterial, defining a black-hole-like object with at most rotational degrees of freedom.[11]

In analogy with the treatment of the special theory, we write for the line element, see [18] for details

$$\begin{pmatrix} cds & 0 \\ 0 & -cds \end{pmatrix} = \begin{pmatrix} cAd\tau & -iBd\vec{x} \\ -iBd\vec{x} & -cAd\tau \end{pmatrix} \tag{3.15}$$

with

$$A = B^{-1} = \sqrt{1 - 2\kappa(r)} \tag{3.16}$$

and observing that the conjugate operators are modified as follows

$$i\hbar \frac{\partial}{\partial t} = E_{op}(t); \quad t_{op} = -i\hbar \frac{\partial}{\partial E_t} \tag{3.16}$$

$$i\hbar \frac{\partial}{\partial s} = E_{op}(s); \quad s_{op} = -i\hbar \frac{\partial}{\partial E_s} \tag{3.17}$$

with

$$E_t = E_s \sqrt{1 - 2\kappa(r)}; \quad \frac{ds}{dt} = \sqrt{1 - 2\kappa(r)} \tag{3.18}$$

---

[10]As it is usually projected today.

[11]In retrospect the singularity shares the same self-referential conundrum as we associate with Gödel's incompleteness theorem(s) [9, 16].

where E_s and E_t represents the energy at the space time-points s and t respectively. With the notation $E_s = m_s c^2$, one might identify the energy at a space-time point s with the mass m_s. One can see that the matrix Eq. (3.15) is commensurate with the metric given by Eq. (3.12). For the case of photons (zero-rest-mass) one obtains for $\kappa(r) \neq 1/2$ that $ds = 0$ and

$$\begin{pmatrix} cAd\tau & -iBd\vec{x} \\ -iBd\vec{x} & -cAd\tau \end{pmatrix} = cAd\tau \begin{pmatrix} 1 & -i \\ -i & -1 \end{pmatrix} \rightarrow 2cAd\tau \begin{pmatrix} 0 & 1 \\ 0 & 0 \end{pmatrix} \qquad (3.19)$$

which is consistent with Eq. (3.6′). One notes that the energy formula for photons becomes

$$m(1 - 2\kappa(r)) = p/c$$

in correspondence with the law of gravitational light bending. Furthermore at $\kappa(r) = 1/2$ the metric is conditionally singular for both mass- and massless particles.

In summary we have demonstrated that the a priori choice of conjugate relationships, like energy-time, impart simple relations, commensurate with the theory of relativity, including the emergence of mathematical singularities at the precise physical conditions commensurate with constraints like the limit velocity of light, the associated loss of longitudinal spatial dimension and the emergence of a black hole entity trapping energetic particles. Although not emphasized here, see also footnotes 8 and 11, the matrix to the left of Eq. (3.13) displays a direct analogy with the self-referential enigma exhibited by the Gödel's inconsistency theorem(s),[12] see e.g. Refs. [9, 11, 13, 16]. In the next section we will show how to extend the degenerate formulation to any order including its interpretation in relation to the Paradigm of Evolution.

4　The Correlated Dissipative Structure

As we have maintained above, time-energy and space-momentum provide a priori observables of primary descriptions of the scientific discourse. In this portrayal of processes in the natural world, one preferred option of a distinct constituent involves by definition its conjugate companion. Within the theory of evolution, time is the key variable with a fundamental spatial significance. To appreciate this idea one needs to understand the emergence of spatio-temporal scales from the

---

[12]The common inconsistency in the vernacular is also known as the liar's paradox and it has recently been discussed under the name of The Pinocchio Paradox [23].

dynamics of their conjugate partners. The strategy will be to develop the appropriate thermodynamics, invoking temperature and entropy to derive apt scales in concert with appropriate ensembles and to find out how complex enough systems interact or rather communicate with each other. Since we have presented the various derivations in earlier contributions [9, 11, 16], we will for the most part only state the mathematical results, as they are required as well as their physical interpretation.

Though our formulation has a rigorous origin in the axioms of quantum theory as a trace algebra, see Ref. [20], in terms of general system operators, various ensembles etc., there are important generalizations to quantum logic as engrained in the illustrious theorem due to Gleason [24]. Hence the present formulation does not only refer to pioneering quantum mechanical interpretations, but it also covers interpretations that go beyond classical Boolean structures.[13]

A convenient starting point is the second order, reduced for N fermions, characterized by the space-spin coordinates[14] x_k normalized to the number of pairings (for details see [9, 11, 16])

$$\Gamma^{(2)}\left(x_1, x_2 | x_1', x_2'\right) =$$
$$\binom{N}{2} \int \Psi(x_1, x_2, x_3, \ldots, x_N) \Psi^*\left(x_1', x_2', x_3, \ldots, x_N\right) dx_3, \ldots, dx_N \tag{4.1}$$

with

$$E = \mathrm{Tr}\left\{H_2 \Gamma^{(2)}\right\} \tag{4.2}$$

for a suitable reduced Hamiltonian—so far all in a standard setting of quantum mechanics. An essentially wave-function representable two-matrix can be written

$$\Gamma^{(2)} = \sum_{k,l=1}^{n} |h_k\rangle \gamma_{kl} \langle h_l| = |h\rangle \gamma \langle h| \tag{4.3}$$

where the n-dimensional matrix γ (in principle specified below) is represented in the space of the preferred basis $|h\rangle$, the latter referring to appropriate pairs of light carriers, like electrons described by paired orbitals denoted as geminals localized at various nuclear centers. In fact the density matrix should describe, together with the nuclear motion, the full dynamics of the system,[15] although we have here only

---

[13]For instance the demonstrated connection with Gödel's incompleteness theorem contains a non-Boolean probability that is extended to describing the interactions and communications between Complex Enough Systems, CES. This prompts the system to be density matrixdenoted as a Gödelian network [25].

[14]We will denote the spatial coordinates with a vector notation, i.e. $\vec{x}_k$.

[15]Usually one invokes the Born-Oppenheimer approximation, i.e. separating the nuclear motion from the many electron quantum problem. However, this will not be adequate here as will be seen further below.

indicated the electronic variables. Examples of such representable density matrices are those related to Yang's celebrated notion of Off-Diagonal Long-Range Order, ODLRO [26], see also an alternant derivation by Sasaki [27] and Coleman's concept of extreme states [28], of direct relevance for the understanding of superconductivity and superfluidity. The extreme state corresponds to a degenerate state with one large eigenvalue, λ_L, approaching (for large n) the number of pairs, $N/2$, and other the $(n - 1)$-degenerate[16] one, $\lambda_S \to 0$, where the unitary matrix B will be defined further below, i.e.

$$B\gamma B^{-1} = d = \begin{pmatrix} \lambda_L & 0 & & 0 & 0 \\ 0 & \lambda_S & \cdots & 0 & 0 \\ \vdots & & \ddots & & \vdots \\ 0 & 0 & & \lambda_S & 0 \\ 0 & 0 & \cdots & 0 & \lambda_S \end{pmatrix} \tag{4.4}$$

Next we resume the exploration by confronting two major problems, i.e. dealing with temperature dependences or accounting for the presence of quantum-thermal correlations, and at the same time removing the notorious Born-Oppenheimer approximation, in principle treating the nuclear degrees of freedom on an equal footing with the electrons.

Let us treat the last problem first by exercising the so-called mirror theorem employed by Carlson and Keller, [29], in connection with the reduced degrees of freedom in connection with density matrix theory.[17] In particular an application to the entangled activities between the electronic motion and the movements of the nuclei, implies that they are coupled through mirroring dynamics [31]. For instance an electron orbital or geminal, h_k projected around nucleus l, i.e. described locally by the spatial coordinate $\vec{x}_l$ writes in Dirac notation

$$h_k(\vec{x}_l) = \langle \vec{x}_l \mid h_k \rangle$$

to be viewed as the kth electron orbital (geminal) as anticipated around the lth nucleus, can also be interpreted as the scalar product between the lth nucleus described by the Dirac ket,[18] given by $|\vec{x}_l\rangle$ and the electronic motion characterized by $|h_k\rangle$. Such scalar products suggest the key ingredients for the mapping between the electronic and nuclear degrees of freedom. Hence system operators of the kind Eq. (4.1) should in principle contain both electrons and nuclei, one of which, electrons or nuclei, could be traced away in order to study the remaining dynamics by suitable master equations, see e.g. Ref. [12]. The conclusion from the mirror

---

[16]For fermionic systems there is also a $2n (n - 1)$ dimensional tail of unphysical pairings, which is omitted.

[17]The actual theorem goes back to Erhard Schmidt [30], the mathematician who was behind the Gram-Schmidt orthogonalization.

[18]The Dirac ket denotes a quantum state and its significance originates from the bra-c-ket form as a scalar product.

theorem is that the mappings between the electron- and the nuclear degrees of freedom, and back, exhibits the same classical canonical form. This imparts the possibility of a dual interpretation of the actual matrix representations of either the electronic motion or the associated nuclear motion in principle bypassing the Born-Oppenheimer approximation.

One may also view the entwined dynamics as a general scattering problem with electronic particles scattered on a number of nuclear targets yielding consistent scattering data, see e.g. Ref. [9]. This leads to the second problem of merging quantum and thermal correlations at precise temperatures invoking associated time scales and rigorous dissipative (open system) dynamics, which will be done in two steps. First we need to match the relevant time scales of the system, via a proper thermalization procedure obtaining the system operator for the open system at the relevant temperature exhibiting authentic time scales for a realistic description of non-equilibrium situations of relevance for, what we will denote, Complex Enough Systems, CES, and their interpretation.

Consider an open system involving n bosonic or paired fermionic degrees of freedom as an "incoming beam" impinging on a set of nuclear sites. The whole scattering arrangement is characterized by a relaxation process[19] with the time scale τ_{rel}, assumed to be distinctly larger than the smaller thermal timescale $\tau_{corr} = \hbar/k_B T$. The protocol describes a process that one will, on the average, detect one quasi particle degree of freedom in the differential solid-angle element $d\Omega$ during the thermal timescale, e.g. with $\tau_{corr} \approx 2.46 \times 10^{-14}$ s at 310 K. Straightforwardly one obtains, with the incident flux, N_{inc}, being the number of particles/(degrees of freedom) per unit area and time, $N_s d\Omega$, the number of particles scattered into $d\Omega$ per unit time being the standard relations between the differential- and the total cross sections σ_Ω and σ_{tot}, the following formulas

$$N_{inc} = \frac{n}{\sigma_{tot}\tau_{rel}}; \quad \sigma_\Omega d\Omega = N_s d\Omega = \frac{d\Omega}{\tau_{corr}}; \quad \sigma_{tot} = \int \sigma_\Omega d\Omega = \int \frac{N_s}{N_{inc}} d\Omega \qquad (4.5)$$

yielding the simple relationship between the two characteristic times, i.e.

$$\frac{n}{4\pi} = \frac{k_B T}{\hbar}\tau_{rel} = \frac{\tau_{rel}}{\tau_{corr}} \qquad (4.6)$$

Assuming that the correlated cluster of nuclei perform harmonic oscillations distributed over the various energies $\hbar\tau_l^{-1}$, from the zero point energy, with equidistant harmonic levels displaying a spectrum from the zero-point energy to $\hbar\tau_{corr}^{-1}$. Straightforward examination of the situation reveals

$$\tau_{rel} = (l-1)\tau_l = \tau_2 = \frac{n\tau_{corr}}{4\pi}; \quad l = 2, 3, \ldots, n \qquad (4.6')$$

---

[19]Generally speaking a specific molecular process is considered as a local perturbation out of the quantum state which then relaxes back to thermodynamic equilibrium after a certain time τ_{rel}.

Utilizing the standard Heisenberg relation between life times and energy widths of the state, i.e. $\varepsilon_k = \hbar/2\tau_k$ one may express Eq. (4.6) as (with $\beta = 1/k_B T$, T the absolute temperature and k_B the Boltzmann constant)

$$\beta\varepsilon_l = \frac{2\pi(l-1)}{n} \tag{4.7}$$

Relations (4.5)–(4.7) display the relation between the time scales, the temperature T and the relevant dimension n of the non-equilibrium dissipative system.

The thermalization procedure, extended to the density matrix subject to the Bloch equation with a Hamiltonian, H, producing thermal fluctuations commensurate with a given temperature, gives (note that the usual condition of the eigenstate thermalization hypothesis is not fulfilled due to ODLRO)

$$-\frac{\partial\varrho}{\partial\beta} = \mathcal{L}_{B}\varrho; \mathcal{L}_{B}\varrho = \frac{1}{2}(H\varrho + \varrho H) \tag{4.8}$$

which together with (4.7) yields the surprising result, with $\gamma \to \gamma_{\text{term}}$

$$\varrho = e^{-\beta\mathcal{L}_B}\Gamma^{(2)} = |h\rangle\gamma_{\text{term}}\langle h| = |f\rangle B^{-1}\gamma_{\text{term}}B\langle f| = |f\rangle\omega_d\langle f| \tag{4.9}$$

with

$$\omega_d = \begin{pmatrix} 0 & \lambda_S & & 0 & \lambda_L \\ 0 & 0 & \cdots & 0 & 0 \\ \vdots & & \ddots & & \vdots \\ 0 & 0 & & 0 & \lambda_S \\ 0 & 0 & \cdots & 0 & 0 \end{pmatrix} = \lambda_L J^{n-1} + \lambda_S J \tag{4.10}$$

with J denoting the standard nilpotent matrix with zeros everywhere except with ones above the diagonal. Note that $n \to N/2$ imparts, in contrast to $n \to \infty$ that $\lambda_L \to \lambda_S \to 1$. In the basis f the density matrix ϱ in Eq. (4.10) writes, while here leaving out the second term[20] above, since it will here only play a minor role in the time evolution, $n \approx N/2$, $\rho \propto \varrho$ normalized to $\text{Tr}\{\rho\rho^\dagger\} = 1$

$$\rho = \frac{1}{\sqrt{n-1}}|h\rangle Q\langle h| = \frac{1}{\sqrt{n-1}}|f\rangle J\langle f| = \frac{1}{\sqrt{n-1}}\sum_{k=1}^{n-1}|f_k\rangle\langle f_{k+1}| \tag{4.11}$$

with

---

[20]When $n \to \infty$, $\lambda_L \to N/2$ will dominate $\rho \propto \frac{N}{2}|f_1\rangle\langle f_n|$ activating e.g. large-scale coherent axonal firing.

$|h\rangle = |f\rangle B$ and the complex symmetric Jordan block given by

$$Q_{kl} = \left(\delta_{kl} - \frac{1}{n}\right) e^{\frac{i\pi}{n}(k+l-2)} \tag{4.12}$$

The unitary transformation B is finally given by[21] with $\omega = e^{i\pi/n}$

$$B = \frac{1}{\sqrt{n}} \begin{pmatrix} 1 & \omega & \omega^2 & \cdots & \omega^{n-1} \\ 1 & \omega^3 & \omega^6 & \cdots & \omega^{3(n-1)} \\ \vdots & \vdots & \vdots & \vdots & \vdots \\ 1 & \omega^{2n-1} & \omega^{2(2n-1)} & \cdots & \omega^{(n-1)(2n-1)} \end{pmatrix} \tag{4.13}$$

In summary we have derived an irreducible representation Eq. (4.11), which will be denoted as a Correlated Dissipative Structure, CDS, that is commensurate with the physical time scales τ_{corr} and τ_{rel} related through Eqs. (4.6) and (4.7). The CDS configuration depends on T releasing a thermal oscillation into the complex enough system, CES. The transformation from the "local" preferred basis $|h\rangle$ to the canonical one $|f\rangle$ is given by B^{-1}, or since it is unitary $B^{\dagger}$. Note that B also transforms $\Gamma^{(2)}$ to diagonal form, Eq. (4.4). Finally we observe an interesting factor property of B, which will be of vital importance in the next section.

5   The Correlated Dissipative Ensemble

To illustrate the Gödelian Network as a Correlated Dissipative Structure we will employ the CDS as base units for a "higher level" Liouville formulation based on the Liouville generator

$$\mathcal{L}\rho = [H\rho - \rho H]$$

For instance applying $\mathcal{L}$ above to Eq. (4.11) gives to first order the sum of the energy differences $E_i - E_j$. Since the thermalization, leading up to the successive transitions in ρ, is brought about by the exchange of a thermal oscillation, due to the energy super-operator in (4.8), the final outcome becomes the overall change $E_1 - E_n$, which is nothing but the thermal frequency associated with τ_{corr}. Hence the present Liouville picture describes the CDS by this frequency and with the characteristic lifetime[22] τ_{rel}.

As a result one obtains an entity, which we will call the Correlated Dissipative Ensemble, CDE. The latter, by its construction, integrates a principal basis set of

---

[21]This form was obtained in collaboration with C. E. Reid [32] see also [33].

[22]A rigorous analytic continuation of $\mathcal{L}$ is given in [34]. While the real part of the eigenvalue appear as energy differences, the imaginary parts add up here to be consistent with Eq. (4.6).

CDS's defined by Eq. (4.11), and denoted by $|H\rangle$, where each CDS is commensurate with the time scales τ_{corr} and τ_{rel} related through Eqs. (4.6) and (4.7). In terms of the frequency[23] $\omega_0 = \hbar/k_B T$ and $\tau = \tau_{rel}$ each CDS is characterized by

$$\frac{n}{4\pi} = \frac{kT}{\hbar}\tau_{rel} = \omega_0\tau \tag{5.1}$$

with $Q = \int \omega_0 \tau d\Omega = n$ serving as a quality factor for the CDE.[24] Deriving the CDE from CDS units, i.e. molecular aggregates, like the DNA and/or its protein overcoats, in a biological system, will permit the definition of a particular cellular *quality value*, Q. This is to some extent analogous to signal processing in communication systems, where the quality aspects of cavity resonators, like musical instruments etc., play a vital role in transmitting resonating qualities. For that reason the *quality value* $Q = n$, for e.g. somatic cells in a multicellular organism, transferring vital details regarding their traits, provides cell recognition through the process of "molecular communication".

The CDE exhibits an analogical irreducible unit Q in the basis H of CDS base units. A simplified analysis of the associated system operator shows that its diagonal elements can be determined by a general probability measure, let us say $p(A) = \mathrm{Tr}(\rho P(A))$, ($\rho$ a given density operator and $P(A)$ a suitable self-adjoint projection on Hilbert Space for the event A), and $(1 - p(A))p(A)$ for any off-diagonal one displaying ODLRO. An analogous analysis, cf. the CDS, leads through diagonalization and thermalization to a similar irreducible unit Q as in Eq. (4.11) but generally with a different dimension m, in terms of H and the transformed F, cf. the relations between h and f. The key difference is that $\tau = \tau_{rel}$ play the role of the short (relatively speaking) time scale with a longer scale emerging through Q see more below. We can now write down the propagator corresponding as (with $I = \sum_{k=1}^{m} |F_k\rangle\langle F_k|$)

$$\mathcal{P} = (\omega_0\tau - i)I + iJ \tag{5.2}$$

$$J = \sum_{k=1}^{m-1} |F_k\rangle\langle F_{k+1}| \tag{5.3}$$

$$|H\rangle = |F\rangle B \tag{5.4}$$

In analogy with classical dynamics we have the equivalent of a causal propagator $\mathcal{G}(t)$ and a resolvent $\mathcal{G}_R(z)$ defined by

$$\mathcal{G}(t) = e^{-i\mathcal{P}\frac{t}{\tau}}; \quad \mathcal{G}_R(\omega) = (\omega\tau I - \mathcal{P})^{-1} \tag{5.5}$$

---

[23]The notation $\omega_0 = \hbar/k_B T$ should not be confused with $\omega = e^{i\pi/n}$ defined in Eq. (4.13).

[24]The Q-value should not be confused with the complex symmetric matrix Q.

which in the classical case are related through the standard Fourier transform. In the present case the extension requires the separation of positive and negative times focusing our interest on the retarded propagator.[25] Inserting the Liouvillian, Eq. (5.2) in (5.5), one obtains

$$e^{-i\mathcal{P}\frac{t}{\tau}} = e^{-i\omega_0 t} e^{-\frac{t}{\tau}} \sum_{k=0}^{m-1} \left(\frac{t}{\tau}\right)^k \frac{1}{k!} J^k \tag{5.6}$$

$$(\omega\tau I - \mathcal{P})^{-1} = \sum_{k=1}^{m} [(\omega - \omega_0)\tau + i]^{-k} (iJ)^{(k-1)} \tag{5.7}$$

where one notes that the expansions are finite, limited by the dimension m. The occurrence of higher order poles in Eq. (5.7) is reflected by the build up of a polynomial in front of the decay factor in (5.6). As a result the usual microscopic law of evolution $dN(t) = -(1/\tau)N(t)dt$ modifies according to the highest power $m - 1$ of J, see [13] for details

$$dN(t) = t^{m-2}\left(m - 1 - \frac{t}{\tau}\right)N(t)dt; \quad dN(t) > 0; t < (m-1)\tau \tag{5.8}$$

The consequence of the irreducible perturbation in Eq. (5.2) is the emergence of a new basic "communication" time scale $\tau_{\text{com}} = (m-1)\tau_{\text{rel}}$ or $m\tau_{\text{rel}}$ adding the decay time.

It might seem a misnomer to refer to Eqs. (5.5)–(5.7) as a CDE, i.e. a statistical ensemble associated with a perturbation out of equilibrium. However it represents assembled stochastic features construed for Complex Enough Systems with programmed timescales at precise temperatures. Yet the mystery of "molecular communication" remains, i.e. how does transmission of crucial molecular traits extend from microscopic levels generating significant cellular information? An answer is given by comparing the two transformations

$$|h\rangle = |f\rangle B \tag{5.9}$$

or the CDS, relating n light carriers in a nuclear skeleton of n sites, treating the nuclear and the electronic system on par, and

$$|H\rangle = |F\rangle B \tag{5.10}$$

for the CDE, relating e.g. m cells in a certain organ localised at key positions in a particular organ or organism. If m and n reveal no relation whatsoever, one might not anticipate any affinity between the cells with non-commensurate Q-values.

---

[25]A detailed exposition of the connection between the correlation function and the corresponding spectrum, including analyticity requirement and appropriate integration contours have been discussed in the appendix of Ref. [35].

However if they are the same—or as we will see containing many least common multiples—they share the same time scale $\tau = \tau_{\text{rel}}$ as well as displaying common time factorizations as shown by the transformation B. Actually any perception, depending on the rows of the matrix B, should amount to "translating" the cyclic vectors of B via the transformation back to the "physical" basis h or H localized in the cell and in the organism respectively. Hence one might envisage the translation of compatible molecular information via the communication bearing transformation B as authorized "messages" between the molecular-DNA-RNA-gene level and the higher-level hierarchical organization leading to cellular function and biological order.

On the condition that the *Q-values* in (5.9) and (5.10) are commensurate, it suggests the possibility of a communication protocol given by the cyclic properties and their nestings of the columns of B. For instance analysing the transformation in Eq. (4.13), one observes the way the elements repeat themselves due to the cyclicity of complex numbers. For instance choosing $n = 12$, computing the columns of B, one may display the result as the simplified diagram below, in which only the dimensions of the recurring vectors are indicated. Removing the first column of one-dimensional units "1", the resulting graph, containing 11 columns, will look like

$$
12\,{\begin{smallmatrix}&&&2&&&\\&&3&2&3&\\&4&3&2&3&4\\6&4&3&&3&4&6\\6&4&3&12&2&12&3&4&6\\6&4&3&&2&&3&4&6\\&4&3&2&3&4\\&&3&2&3&\\&&&2&&&\end{smallmatrix}}\,12 \tag{5.11}
$$

where we do observe the obvious column symmetry of the table. From Eq. (5.6) one finally realizes the typical behaviour of Poisson statistics that suggests a communication concept that will share some analogy with the dynamics of a telephone Call or Contact Center.[26]

6 Examples and Conclusions

In summary we have derived and suggested an explication of a possible programme for the transmission of proper gens between two distinct levels of scientific formulation, i.e. the molecular microscopic- and the cellular mesoscopic ones. Many biologists flirt with quantum mechanics without coming to terms with its consequences. This, in many cases, turn out to be quite confusing, since it is not easy to

---

[26]This analogy has been carried out in some detail in relation to Trehub's Retinoid Model of the cognitive brain [36, 37]. The representation of neurons with a chemical synapse onto itself is of particular significance for the Gödelian network [25].

ascertain the Grand Master's inner thoughts about the cornerstones of physics. While Einstein did say that 'time is an illusion', Prigogine on the other hand asserted that 'we are the children of time'. As the prospective reader of this contribution might have realized, energy-momentum and space-time are inevitably coupled to each other as conjugate entities and that the present formulation, as a consequence, invokes the characteristics of the Correlated Dissipative Ensemble, CDE.

In previous studies we have emphasized the opportunity of cell recognition suggested by the possibilities offered by the cellular *Q-value* prearranged by the CDS as building blocks in every cell. This problem involves both inter- and intra-cell communication. The lower- as well as the higher-level structures are subject to Poisson statistics defining physical communication channels between and within them. In particular we have focused on the CNS, the central nervous system, in the presence of a special type of cells, i.e. neurons, with the dynamics characterized by short time scale oscillations, building up pulses of light carriers (electrons), i.e. spikes that correlate the basic dissipative systems e.g. cells/neurons, see also footnote 21, and providing an irreducible coupling that arrange for the communication between them. For more details regarding the usage of stochastic backgrounds as objective physical communication channels, the former commensurate with the Poisson statistics, authorizing a direct call centre analogy, i.e. the distribution of the rate of incoming phone calls received at each neuron, the switchboard, see Refs. [9, 11, 25]. Note that the code for exchanging "communication" is given by the transformation B, Eq. (4.13), which applies in both limits:

$$\lambda_L \to \frac{N}{2}; \quad n \to \infty$$

and

$$\lambda_L \to 1; \quad \lambda_S \to 1; \quad n \to \frac{N}{2}$$

while not being contained in the actual "phone call".

Special attention has been given to Arnold Trehub's celebrated Retinoid Model for vision [36, 37] constituted by a special type of neurons called the autapse, i.e. with a chemical synapse linked onto itself. In particular we have utilized the symmetry[27] of the diagram (5.11), where the columns to the left of the centre are minus one times the complex conjugate of the ones to the right.[28] By projecting the neural dynamics onto Trehub's "3D egocentric space" one realizes a certain

---

[27]It is important to realize that this is not a "biological" symmetry, but induced by the perception of mirror neuron images produced by the autapse.

[28]This is a crucial feature, since according to the Charge, Parity, Time reversal, CPT, theorem, believed to be the fundamental symmetry of physical laws, see also Ref. [38] the brain perceive time reversal symmetry as a space parity inversion, which explains the Necker cube illusion [25].

mirroring symmetry that is instigated by the feedback loops of the autapse. Hence every trajectory in egocentric space exhibits a time-reversed copy in agreement with the above-mentioned symmetry. This yields a simple understanding of the Necker Cube Illusion [25] and the fact that our eyes see everything upside-down. The examples given above relate to communication between cells connected to the central nervous system, CNS. In what follows, we will concentrate on the other question, the trans level communication.

As our conclusion concerns open questions in the philosophy of biology, see Ayala and Arp [39] for recent debates, there appear some highly interesting work, where the communication between the cells plays a preferential role [40]. In his formulation of the First Principles of Physiology, integrating the self-organizing status of the unicellular state, see also Ref. [41], John Torday, asserts that biological organisms exist far from equilibrium, 'circumventing' the Second Law of Thermodynamics generating negative entropy within them, sustained by chemiosmosis. Moreover he finds, going through the life cycles of biological phenomena, that evolution on this level is deterministic, while at the same time arguing that the unicell "finds itself" in an ambiguous situation with respect to its environment.[29]

Along these lines of reasoning, while acquiring epigenetic marks, one discerns Lamarckian traits emphasizing environmental input over genetic evolution advancing the so-called Epigenetic Inheritance Systems, EIS, [42], suggesting alternative possibilities for extending the boundaries in evolutionary biology. For instance, recent studies on genomic imprinting in mammals, see the recent review by Li and Sasaki [43], is redolent of mechanisms relating to the reprogramming of pluripotent stem cells. One should emphasize, however, that these models have no direct relation to quantum chemical processes inside a cell. Nevertheless the general question remains, i.e. how does selection act, what sort of entities are selected? To answer this question, one needs to go to the intracellular level translating the intercellular signals on the level of the DNA-protein coding to the higher-order level of cell-interactions. For instance, Torday demonstrated how the parathyroid hormone-related protein PTHrP and leptin signalling mechanisms, do facilitate a somatic program of lung evolution, offering ontogenic/phylogenic links. In consideration of the present idea, it would be intriguing to find a way to transcend the intracellular level, translating the molecular signals from the microcirculation of the alveolus to alveolar metabolism.

One may enquire what time scale τ would be appropriate for the CDE in general. While the CDS engender a relaxation time commensurate with its process forming dynamics, it might be interesting to investigate e.g. the important role of the primary cilium in modulating neurogenesis. They consist of micro-tubule-based organelles, the latter also being part of the neuronal structure, which has been particularly emphasized by Hameroff and Penrose [44] in connection with their "Orch OR" theory for the appearance of consciousness in the universe.[30] In eukaryotes the structure of

---

[29]This is an intriguing declaration, since this is precisely "communication" formulated in this article and programmed by Gödel-like sentences.

[30]See also the discussions following their article in the Physics of Life Reviews.

microtubules are long cylinders of α- and β-tubulin dimers,[31] each with a molecular weight of about 50 kDa formed in parallel association of thirteen protofilaments. In total we have a weight of around 1000 kDa, corresponding to about 10^6 protons. With a nuclear average of approximately 10 Da, one might expect roughly 10^5 nuclear degrees of freedom to be at work in each of the ca 10^9 microtubules. Using Eq. (4.6) one finds cilia beating frequencies of the order $\tau_{rel} \approx 25$ ms at human body temperatures, which appears to be in the right range for motile cilia mediated pathways. It has also been noted that defects in the structure of the cellular primary cilium leads to various disorders and impairments [45]. Hence looking at recurrent frequencies, around 10–40 Hz, one will conclude that obstructing cell organelle movements might severely disturb the channel for stochastic communication. Despite the view that perhaps 10–20,000 neurons are involved in a perception instigated by specific spike trains, the present CDS concept, nevertheless, implies a more complex understanding in that a Gödelian network must have all neurons 'on alert' in case of need.

Another critical challenge, while retaining quantum chemistry applications to biological systems, is the temperature dependence, the thermoregulation, and the associated problem of the necessary prohibition of short-time[32] decoherences. The development and appearances of Correlated Dissipative Structures, CDS, not only suggests answers to this dilemma, it also incorporates realistic time scales and the linked nested 'code-encode bearing' transformation B. In this connection it is also interesting to observe that so-called poikilotherms, i.e. organisms whose internal temperatures varies considerably, have several different enzyme systems operating at different temperatures, hence representable by different CDE's.

Finally, a reminder regarding our biophysical organisation, termed a complex enough system, CES, subject to the CDE, viz. the interactions between the cells commensurate with the dynamics inside the cell, derived from a CDS, is neither deterministic nor probabilistic.[33] This incurs no contradiction, even if the biological process may appear deterministic at the mesoscopic level of understanding. The CDE is an irreducible ensemble defined in terms of the CDS, the latter exhibiting an analogous irreducible structure. Within this framework one might finally be reminded of the words of Ernst Mayr [10, 46] regarding the characteristics of biological processes: *a teleonomic process or behaviour is one that owes its goal-directedness to the influence of an evolved program.* Our interpretation suggests that the stochastic hypothesis depicted here, i.e. a "Communication

---

[31]Since the microtubules have a distinct polarity one obtains a direct coupling between the mechanical- and the electromagnetic oscillations generating the train of neuronal spikes.

[32]The notion 'short' is here in relation to any other time scales discussed in this article.

[33]It has been stated that the mathematical structure of cosmos, with the origin of the physical laws being in pure numbers, may discard the theory of biological evolution as well as being incommensurate with the non-deterministic interpretation of quantum mechanics [47]. It is clear, however, from the present work that numerical structures, like the formula $1 + 1 = 2$, reveals important and necessary facts about our universe that are essential for the communication between life forms and contingent on evolution, see e.g. a recent scientific analysis of 'Science and Music' [48] focusing on musical transcriptions in the living world.

Simpliciter", derived from first principles, can be understood and generated as such an authentic process.

Acknowledgements I am grateful to the Chair of QSCP-XXI, Prof. Yan Alexander Wang, and the Cochair Prof. Jean Maruani, for generously allowing me to present this work in the present proceedings from the Vancouver meeting. This work has, over time, been supported by the Swedish Natural Science Research Council, Swedish Foundation for Strategic Research, the European Commission, and the Nobel Foundation.

References

1. Tegmark M (2003) Parallel universes. Sci Am
2. Deutsch D (2011) The beginning of infinity. Viking, Penguin, New York
3. Primas H (1983) Chemistry, quantum mechanics and reductionism. Perspectives in theoretical chemistry. Springer, Berlin
4. Atmanspacher H, Müller-Herold U (eds) (2016) From chemistry to consciousness. The legacy of Hans Primas. Springer, Berlin
5. Heidegger M (1927) Sein und Zeit. Max Niemeyer Verlag, Tübingen
6. Feyerabend P (1975) Against method. New Left Books, London
7. Nye MJ (1993) From chemical philosophy to theoretical chemistry. University of California Press, Berkeley
8. Brändas EJ (2017) Löwdin—Father of quantum chemistry. Mol Phys 115:1
9. Brändas EJ (2015) A zero energy universe scenario: from unstable chemical states to biological evolution and cosmological order. In: Nascimento MAC, Maruani J, Brändas EJ, Delgado-Barrio G (eds) Frontiers in quantum methods and applications in chemistry and physics, vol 29. Springer, Dordrecht, p 247
10. Mayr E (1988) Towards a new philosophy of biology. Observations of an evolutionist. Harvard University Press, Cambridge
11. Brändas EJ (2017) The origin and evolution of complex enough systems. In: Tadjer A, Pavlov R, Maruani J, Brändas EJ, Delgado-Barrio G (eds) Quantum systems in physics, chemistry, and biology, vol 30. Springer, Berlin, p 413
12. Prigogine I (1996) The end of certainty: time, chaos, and the new laws of nature. The Free Press, New York
13. Brändas EJ (2012) Examining the limits of physical theory: analytical principles and logical implications. In: Nicolaides CA, Brändas EJ (eds) Unstable states in the continuous spectra Part II: Interpretation, theory, and applications. Advances in quantum chemistry, vol 63. Elsevier, Amsterdam, p 33
14. Barbour JB (1999) The end of time the next revolution in physics. Oxford University Press, Oxford
15. Primas H (2003) Time-entanglement between mind and matter. Mind and Matter 1(1):81
16. Brändas EJ (2011) Gödelian structures and self-organization in biological systems. Int J Quant Chem 111:1321
17. Kant I (1781) Kritik der reinen Vernunft. Johann Friedrich Hartknoch, Riga
18. Brändas EJ (2016) A comment on background independence in quantum theory. J Chin Chem Soc 63:11
19. Davydov AS (1965) Quantum mechanics. Pergamon Press, Oxford
20. Löwdin P-O (1998) Linear algebra for quantum theory. Wiley, New York
21. Baumgarten C (2015) Minkowski spacetime and QED from ontology of time. arXiv:1409.5338v5[physics.hist-ph] 30 Nov

22. Löwdin P-O (1998) Some comments on the foundations of physics. World Scientific, Singapore
23. Eldridge-Smith P, Eldridge-Smith V (2010) The Pinocchio paradox. Analysis 70(2):212
24. Gleason AM (1957) Measures on the closed subspaces of a Hilbert space. J Math Mech 6:885
25. Brändas EJ (2015)) Proposed explanation of the Phi phenomenon from a basic neural viewpoint. Quant Biosyst 6(1):160
26. Yang CN (1962) Concept of off-diagonal long-range order and the quantum phases of liquid helium and of superconductors. Rev Mod Phys 34:694
27. Sasaki F (1965) Eigenvalues of fermion density matrices. Phys Rev 138B:1338
28. Coleman AJ (1963) Structure of fermion density matrices. Rev Mod Phys 35:668
29. Carlson BC, Keller JM (1961) Eigenvalues of density matrices. Phys Rev 121:659
30. Schmidt E (1907) Math Ann 63:433
31. Brändas EJ, Hessmo B (1998) Indirect measurements and the mirror theorem. Lecture notes in physics, vol 504, p 359
32. Reid CE, Brändas EJ (1989) On a theorem for complex symmetric matrices and its relevance in the study of decay phenomena. Lecture notes in chemistry. Springer, Berlin, pp 325–475
33. Brändas EJ (2009) A theorem for complex symmetric matrices revisited. Int J Quant Chem 109:28960
34. Obcemea CH, Brändas EJ (1983) Analysis of Prigogine's theory of subdynamics. Ann Phys 151:383
35. Brändas EJ (1997) Resonances and dilation analyticity in Liouville space. Adv Chem Phys 99:211
36. Trehub A (1991) The cognitive brain. MIT Press
37. Trehub A (2007) Space, self, and the theatre of consciousness. Conscious Cogn 16:310
38. Quack M (2011) Fundamental symmetries and symmetry violations from high resolution spectroscopy. Handbook of high-resolution spectroscopy, vol 1, p 659
39. Ayala JF, Arp R (eds) (2010) Contemporary debates in philosophy of biology. Wiley-Blackwell, Chichester, West Sussex
40. Torday JS, Miller WB Jr (2016) The unicellular state as a point source in a quantum biological system. Biology 5(2):25
41. Torday JS (2016) Life is simple—biologic complexity is an epiphenomenon. Biology 5(2):17
42. Jablonka E, Lamb M (2005) Evolution in four dimension—genetic, epigenetic, behavioral, and symbolic variation in the history of life. The MIT Press, Cambridge
43. Li Y, Sasaki H (2011) Genomic imprinting in mammals: its life cycle, molecular mechanisms and reprogramming. Cell Res 21:466
44. Hameroff S, Penrose R (2014) Consciousness in the universe a review of the 'Orch OR' theory. Phys Life Rev Mar 11(1):39
45. Lee JH, Gleeson JG (2010) The role of primary cilia in neuronal function. Neurobiol Dis 38 (2):167
46. Mayr E (2004) What makes biology unique? Cambridge University Press, New York
47. Sanchez FM (2017) A coherent resonant cosmology approach and its implications in microphysics and biophysics. In: Tadjer A, Pavlov R, Maruani J, Brändas EJ, Delgado-Barrio (eds), Quantum systems in physics, chemistry, and biology, vol 30. Springer, Dordrecht, p 379
48. Maruani J, Lefebvre R, Rantanen M (2003) Science and music: from the music of the depths to the music of the spheres. In: Maruani J, Lefebvre R, Brändas EJ (eds), Advanced topics in theoretical chemical physics and physics, vol 12. Kluwer, Dordrecht, p 479

Index

© Springer International Publishing AG, part of Springer Nature 2018
Y. A. Wang et al. (eds.), *Concepts, Methods and Applications of Quantum Systems in Chemistry and Physics*, Progress in Theoretical Chemistry and Physics 31, https://doi.org/10.1007/978-3-319-74582-4

Printed by Printforce, the Netherlands